AF183730

Lecture Notes in Physics

Editorial Board

R. Beig, Vienna, Austria
J. Ehlers, Potsdam, Germany
U. Frisch, Nice, France
K. Hepp, Zürich, Switzerland
R. L. Jaffe, Cambridge, MA, USA
R. Kippenhahn, Göttingen, Germany
I. Ojima, Kyoto, Japan
H. A. Weidenmüller, Heidelberg, Germany
J. Wess, München, Germany
J. Zittartz, Köln, Germany

Managing Editor

W. Beiglböck
Assisted by Ms. Monika Eisenächer
c/o Springer-Verlag, Physics Editorial Department II
Tiergartenstrasse 17, D-69121 Heidelberg, Germany

Springer-Verlag Berlin Heidelberg GmbH

The Editorial Policy for Proceedings

The series Lecture Notes in Physics reports new developments in physical research and teaching – quickly, informally, and at a high level. The proceedings to be considered for publication in this series should be limited to only a few areas of research, and these should be closely related to each other. The contributions should be of a high standard and should avoid lengthy redraftings of papers already published or about to be published elsewhere. As a whole, the proceedings should aim for a balanced presentation of the theme of the conference including a description of the techniques used and enough motivation for a broad readership. It should not be assumed that the published proceedings must reflect the conference in its entirety. (A listing or abstracts of papers presented at the meeting but not included in the proceedings could be added as an appendix.)
When applying for publication in the series Lecture Notes in Physics the volume's editor(s) should submit sufficient material to enable the series editors and their referees to make a fairly accurate evaluation (e.g. a complete list of speakers and titles of papers to be presented and abstracts). If, based on this information, the proceedings are (tentatively) accepted, the volume's editor(s), whose name(s) will appear on the title pages, should select the papers suitable for publication and have them refereed (as for a journal) when appropriate. As a rule discussions will not be accepted. The series editors and Springer-Verlag will normally not interfere with the detailed editing except in fairly obvious cases or on technical matters.
Final acceptance is expressed by the series editor in charge, in consultation with Springer-Verlag only after receiving the complete manuscript. It might help to send a copy of the authors' manuscripts in advance to the editor in charge to discuss possible revisions with him. As a general rule, the series editor will confirm his tentative acceptance if the final manuscript corresponds to the original concept discussed, if the quality of the contribution meets the requirements of the series, and if the final size of the manuscript does not greatly exceed the number of pages originally agreed upon. The manuscript should be forwarded to Springer-Verlag shortly after the meeting. In cases of extreme delay (more than six months after the conference) the series editors will check once more the timeliness of the papers. Therefore, the volume's editor(s) should establish strict deadlines, or collect the articles during the conference and have them revised on the spot. If a delay is unavoidable, one should encourage the authors to update their contributions if appropriate. The editors of proceedings are strongly advised to inform contributors about these points at an early stage.
The final manuscript should contain a table of contents and an informative introduction accessible also to readers not particularly familiar with the topic of the conference. The contributions should be in English. The volume's editor(s) should check the contributions for the correct use of language. At Springer-Verlag only the prefaces will be checked by a copy-editor for language and style. Grave linguistic or technical shortcomings may lead to the rejection of contributions by the series editors. A conference report should not exceed a total of 500 pages. Keeping the size within this bound should be achieved by a stricter selection of articles and not by imposing an upper limit to the length of the individual papers. Editors receive jointly 30 complimentary copies of their book. They are entitled to purchase further copies of their book at a reduced rate. As a rule no reprints of individual contributions can be supplied. No royalty is paid on Lecture Notes in Physics volumes. Commitment to publish is made by letter of interest rather than by signing a formal contract. Springer-Verlag secures the copyright for each volume.

The Production Process

The books are hardbound, and the publisher will select quality paper appropriate to the needs of the author(s). Publication time is about ten weeks. More than twenty years of experience guarantee authors the best possible service. To reach the goal of rapid publication at a low price the technique of photographic reproduction from a camera-ready manuscript was chosen. This process shifts the main responsibility for the technical quality considerably from the publisher to the authors. We therefore urge all authors and editors of proceedings to observe very carefully the essentials for the preparation of camera-ready manuscripts, which we will supply on request. This applies especially to the quality of figures and halftones submitted for publication. In addition it might be useful to look at some of the volumes already published. As a special service, we offer free of charge LaTeX and TeX macro packages to format the text according to Springer-Verlag's quality requirements. We strongly recommend that you make use of this offer, since the result will be a book of considerably improved technical quality. To avoid mistakes and time-consuming correspondence during the production period the conference editors should request special instructions from the publisher well before the beginning of the conference. Manuscripts not meeting the technical standard of the series will have to be returned for improvement.

For further information please contact Springer-Verlag, Physics Editorial Department II, Tiergartenstrasse 17, D-69121 Heidelberg, Germany

Jan van Paradijs Johan A. M. Bleeker (Eds.)

X-Ray Spectroscopy in Astrophysics

Lectures Held at the
Astrophysics School X
Organized by the European Astrophysics Doctoral Network (EADN) in Amsterdam, The Netherlands,
September 22 - October 3, 1997

Springer

Editors

Jan van Paradijs
Astronomical Institute "Anton Pannekoek"
University of Amsterdam
Kruislaan 403
NL-1098SJ Amsterdam, The Netherlands
and
Physics Department
University of Alabama in Huntsville
Huntsville, AL 35899, USA

Johan A. M. Bleeker
SRON Space Research Laboratory
Sorbonnelaan 2
NL-3584CA Utrecht, The Netherlands

Library of Congress Cataloging-in-Publication Data.

Die Deutsche Bibliothek - CIP-Einheitsaufnahme

X-ray spectroscopy in astrophysics : lectures held at the Astrophysics School X in Amsterdam, The Netherlands, September 22 - October 3, 1997 / Jan van Paradijs ; Johan A. M. Bleeker (ed.). Organized by the European Astrophysics Doctoral Network (EADN).

(Lecture notes in physics ; Vol. 520)
ISBN 978-3-642-08486-7 ISBN 978-3-540-49199-6 (eBook)
DOI 10.1007/978-3-540-49199-6

ISSN 0075-8450
ISBN 978-3-642-08486-7

This work is subject to copyright. All rights are reserved, whether the whole or part of the material is concerned, specifically the rights of translation, reprinting, reuse of illustrations, recitation, broadcasting, reproduction on microfilm or in any other way, and storage in data banks. Duplication of this publication or parts thereof is permitted only under the provisions of the German Copyright Law of September 9, 1965, in its current version, and permission for use must always be obtained from
Springer-Verlag Berlin Heidelberg GmbH.
Violations are liable for prosecution under the German Copyright Law.

© Springer-Verlag Berlin Heidelberg 1999
Originally published by Springer-Verlag Berlin Heidelberg New York in 1999
Softcover reprint of the hardcover 1st edition 1999

The use of general descriptive names, registered names, trademarks, etc. in this publication does not imply, even in the absence of a specific statement, that such names are exempt from the relevant protective laws and regulations and therefore free for general use.

Typesetting: Camera-ready by the authors/editors
Cover design: *design & production*, Heidelberg

SPIN: 10644296 55/3144 - 5 4 3 2 1 0 – Printed on acid-free paper

Preface

The tenth summer school of the European Astrophysics Doctoral Network (EADN) took place from September 22 till October 3, 1997, at the Astronomical Institute 'Anton Pannekoek' of the University of Amsterdam, with the participation of seven teachers and thirty-two students from eleven European countries.

The subject of the school, 'X-ray Spectroscopy in Astrophysics', was selected in view of the high level of activity in this field of research, with ASCA and BeppoSAX currently in orbit, and the large observatories AXAF, XMM and ASTRO-E due for launch in the near future.

Following the tradition of the EADN summer schools, the subject matter of the lectures was both theoretical and observational/instrumental. This volume contains all lectures presented at the school. We wish to thank the teachers for the excellent lectures which have found their expression in this book.

All students gave a short presentation of their recently started research project. On Wednesday October 1st they visited ESTEC to receive a first-hand account of the XMM project.

We acknowledge with grateful appreciation financial support from:
- the Training and Mobility of Researchers and SOCRATES programmes of the European Union;
- the Space Research Organization of the Netherlands (SRON);
- the Astronomical Institute 'Anton Pannekoek';
- the Granholm Foundation of Sweden.

Special thanks are due to Ms Jane Ayal, whose organizational support made the summer school not only possible, but also a pleasant occasion for the teachers, the students and the organizers.

We also thank Tom Ray, secretary of the EADN for his help in getting the school started; Vincent Icke and Jacco Vink for assisting with the design of the poster; and Frank van der Hooft for his help with TeXing part of the manuscript.

Amsterdam/Utrecht
October 1998

Jan van Paradijs
Johan Bleeker

Contents

Continuum Processes in X-Ray and γ-Ray Astronomy

M.S. Longair

The X-Ray Spectral Properties of Photoionized Plasmas and Transient Plasmas

D.A. Liedahl

X-Ray Spectroscopic Observations with ASCA and BeppoSAX

J.S. Kaastra

Future X-Ray Spectroscopy Missions

F. Paerels

New Developments in X-Ray Optics

R. Willingale

Continuum Processes in X-ray and γ-ray Astronomy

Malcolm S. Longair

Cavendish Laboratory, Madingley Road, Cambridge CB3 0HE
&
Space Telescope Science Institute, 3700 San Martin Drive, Baltimore 21818.

1 Introduction

My assignment is to discuss the physics and astrophysics of continuum radiation processes in X-ray and γ-ray astronomy. This is an enormous subject and is central to the interpretation of observations in these wavebands. The astrophysics of X-ray and γ-ray sources is developing very rapidly and so I will concentrate upon basic physical processes as well as some examples of their application to current issues in X-ray and γ-ray astrophysics. I have no illusions about the transient nature of some of these topics, but I hope the underlying concepts and ideas may prove fruitful for understanding the literature. I should also emphasise that I am not a specialist in X-ray and γ-ray astronomy and so my impressions are from the outside, rather than from someone working at the coal-face every day. The topics I will cover are as follows:

- Overview of continuum processes from hot plasmas;
- Basic concepts in the radiation of charged particles;
- Bremsstrahlung;
- Synchrotron radiation;
- Inverse Compton radiation;
- Synchro-Compton radiation;
- γ-ray processes;
- Relativistic beaming;
- Acceleration of charged particles.

Aspects of some of these topics will be dealt with by other contributors to this volume, and you will find it illuminating to compare and contrast their treatments with mine. I make no apology for adopting a somewhat pedagogical approach to the topics listed above – my excuse is that there are numerous points where newcomers sometimes have problems and these are usually the apparently simple pieces of the story rather than the difficult bits. Finally, I make no pretense that this is a complete exposition of what you need to know. The bibliography includes texts which provide much further detail on all the above topics.

2 Continuum Radiation Processes from Hot and Relativistic Plasmas

By definition, X-ray and γ-ray astronomy involve the radiation of very hot and relativistic plasmas. The energies of the photons correspond to $\varepsilon \geq 0.1$ keV, so that the radiating particles must have at least this energy, or equivalently, the thermal plasmas responsible for the radiation must have $k_B T_e \geq 0.1$ keV, in other words, $T_e \geq 10^6$ K. Thus, the radiating plasmas must be very hot, relativistic or ultra-relativatistic. Perhaps the most important realisation of the pioneering X-ray and γ-ray observations of the 1970s was that hot and relativistic plasmas are found essentially everywhere in the Universe. Here is a list of some of the more important classes of objects from which X-ray and γ-ray emission have been discovered.

- *The Sun and the Solar Corona.* The Sun was the first X-ray object to be observed in the earliest rocket flights immediately after the War. It has now become the subject of very detailed studies, thanks to the magnificent images, and movies, produced by the Japanese *YOHKOH* satellite. The solar corona provides a key example of the heating of the environments of normal stars to very high temperatures by activity on their surfaces.
- *Normal Stars.* One of the most important results of the *Einstein X-ray Observatory* was the discovery that essentially all classes of normal star are sources of X rays. The processes we observe on the surface of the Sun appear to be very common phenomena. X-ray emission has been observed from stars in the process of formation, as well as from main-sequence stars.
- *X-ray binaries.* The *Uhuru* satellite provided the first glimpses of the enormous wealth of new phenomena to be discovered through X-ray observations and none has been more important for many branches of astrophysics than the X-ray binaries. Not only are these objects of the greatest interest in their own right, but they have also provided the most successful means of discovering black holes with masses $M \sim 10\,M_\odot$. Their existence indicated immediately the importance of accretion as the energy source for compact objects, and these concepts have had a profound impact upon the understanding of the astrophysics of active galactic nuclei.
- *Supernova Remnants.* Supernova remnants were among the first objects to be detected as X-ray sources. These catastrophic explosions cause strong shock waves which heat the surrounding interstellar gas to temperatures $T_e \sim 10^7$ K. These sources have been beautifully imaged by the *Einstein* and *ROSAT Observatories.* The thermal nature of their emission has been convincingly demonstrated by X-ray spectral observations of highly ionised species such as 25-times ionised iron Fe XXVI. In addition to thermal sources, there are also cases, of which the Crab Nebula is perhaps the most striking example, in which the emission is *non-thermal,*

in the sense that the emission is associated with the synchrotron emission of relativistic electrons accelerated in the nebula. A convincing case can be made that the ultimate source of energy for these particles is the rotating neutron star, which is the remnant of the progenitor star of the supernova which was observed to explode in 1054 AD.

- *The Interstellar Medium of our own Galaxy* is a source of continuum radiation, the most common assumption being that this hot phase of the gas is associated with heating by the expanding shells of supernova remnants. The temperature of the diffuse hot gas is about 10^6 K and so it is only observable at the soft X-ray energies, $\varepsilon < 1$ keV.
- *Intergalactic Gas in Clusters of Galaxies.* Another key discovery of the *Uhuru* Observatory was the diffuse X-ray emisssion of clusters of galaxies. This emission has been convincingly associated with the bremsstrahlung of diffuse intergalactic gas gravitationally bound within the potential well of the cluster of galaxies. As we will show, this emission is now providing one of the best methods of measuring the total gravitating mass in clusters, as well as an important means of identifying clusters of galaxies at large redshifts. As such, these observations are of the greatest importance for astrophysical cosmology.
- *Active Galaxies.* Galaxies are expected to be sources of X-ray emission because of the presence of stellar X-ray sources and the emission of the interstellar medium. The most spectacular extragalactic X-ray sources are, however, the active galactic nuclei. All types of active galactic nuclei are intense X-ray emitters and provide a unique means of probing physical conditions close to the last stable orbit about supermassive black holes. Indeed, it is confidentally expected that X-ray observations of active galaxies may be the most effective way of probing the structures of the accretion discs about supermassive black holes and so providing new tests of relativistic gravity.
- *Cosmological Applications.* Finally, there are cosmological applications of observations in X-ray and γ-ray astronomy. One continuing problem is the origin of the X-ray background and its spectrum. The X-ray background radiation is second only to the Cosmic Microwave Background Radiation in terms of its radiation energy density and its origin is not yet fully understood. At soft X-ray energies $\varepsilon \sim 1$ keV, the background radiation can be convincingly associated with the emission of discrete sources, but at energies, $\varepsilon \geq 1$ keV, there remain problems in accounting for both its intensity and spectral energy distribution. In addition, extragalactic X-ray sources display strong evolutionary changes with cosmic epoch, similar to those observed in the radio and optical wavebands for active galaxies and quasars.

This list indicates how X-ray and γ-ray astronomy are central to many key issues in Galactic and extragalactic high-energy astrophysics and cosmology.

Before beginning the exposition of the role of continuum processes in X-ray and γ-ray astrophysics, it is useful to give a list of basic references which I have used in preparing these lecture notes. These are:

- *Radiative Processes in Astrophysics* (1979) by G.B. Rybicki & A.P. Lightman. This book can be strongly recommended for a wide range of basic topics in radiation processes.
- *High Energy Astrophysics, Vols. 1, 2 and 3* (1992, 1994, 1999) by the present author (Cambridge: Cambridge University Press). These volumes describe my own approach to understanding radiation processes and a very wide range of other topics relevant to High-Energy Astrophysics. I will quote very extensively from the material in these volumes in these lecture notes. These volumes contain much more complete discussions and references than are possible here on essentially all topics. Note that the 1997 reprints of Vols. 1 and 2 have been revised and brought up-to-date as of Spring 1997 and so I will make reference to the 1997 reprints of these volumes (*HEA1* and *HEA2* hereafter). Vol. 3 is in the course of preparation and will concern extragalactic high-energy astrophysics.
- *Active Galactic Nuclei* (1990) by R.D. Blandford, H. Netzer & L. Woltjer (Berlin: Springer Verlag). This is an excellent volume which, regrettably, is currently out of print. The chapter of particular interest for these notes is that by Blandford.
- *Classical Electrodynamics* (1975) by J.D. Jackson. I use Jackson as the basic text for all fundamental issues concerning the radiation of charged particles.
- *Galaxy Formation* (1998) by the present author (Berlin: Springer Verlag). This volume contains useful material for those interested in the cosmological aspects of the studies described in these notes.

In addition to these books, there are innumerable conference proceedings and excellent reviews in *Annual Reviews of Astronomy and Astrophysics*, to which I will make reference in the text.

3 Basic Radiation Concepts

Much of what we need to understand radiation processes in X-ray and γ-ray astronomy can be derived using classical electrodynamics and central to that development is the physics of the radiation of accelerated charged particles. The central relation is the radiation loss rate of an accelerated charged particle in the non-relativistic limit

$$-\left(\frac{\mathrm{d}E}{\mathrm{d}t}\right)_{\mathrm{rad}} = \frac{|\ddot{\mathbf{p}}|^2}{6\pi\varepsilon_0 c^3} = \frac{q^2|\ddot{\mathbf{r}}|^2}{6\pi\varepsilon_0 c^3}. \tag{1}$$

$\mathbf{p} = q\mathbf{r}$ is the *dipole moment* of the accelerated electron with respect to some origin. This formula is very closely related to the radiation rate of a dipole radio antenna and so is often referred to as the radiation loss rate for *dipole radiation*. Note that I will use *SI units* in all the derivations, although it will be necessary to convert the results into the conventional units used in X-ray and γ-ray astronomy when they are confronted with observations. Thus, I will normally use metres, kilograms, Teslas and so on.

Expression (1) is the key result needed for understanding all classical continuum radiation processes. Notice that, in the non-relativistic limit, the radiation loss rate depends only upon the acceleration of the charged particle. This is such an important formula that I will give a simple derivation and then indicate how it can be derived using the full panoply of classical electrodynamics.

3.1 The radiation of an accelerated charged particle - J.J. Thomson's treatment

The normal derivation of (1) proceeds from Maxwell's equations and involves writing down the retarded potentials for the electric and magnetic fields at some distant point $\mathbf{r}$ from the accelerated charge. It is, however, instructive to begin with the remarkable argument due to J.J. Thomson, which indicates very clearly the origin of the radiation from an accelerated charged particle.

We consider a charge q stationary at the origin O of some inertial frame of reference S at time $t = 0$. The charge then suffers a small acceleration to velocity Δv in the short time interval Δt. Thomson visualised the resulting field distribution in terms of the electric field lines attached to the accelerated charge. After a time t, we can distinguish between the field configuration inside and outside a sphere of radius $r = ct$ centred on the origin of S. Outside this sphere, the field lines do not yet know that the charge has moved away from the origin because information cannot travel faster than the speed of light and therefore the field lines are radial, centred on O. Inside this sphere, the field lines are radial about the origin of the frame of reference centred on the moving charge. Between these two regions, there is a thin shell of thickness $c\,\Delta t$ in which we have to join up corresponding electric field lines. This field configuration is indicated schematically in Fig. 1a. Geometrically, it is clear that there must be a component of the electric field in the circumferential direction, that is, in the $\mathbf{i}_\theta$ direction. This pulse of electromagnetic field is propagated away from the charge at the speed of light and represents an energy loss from the accelerated charged particle.

Let us work out the strength of the electric field in the pulse. We assume that the increment in velocity Δv is very small, that is, $\Delta v \ll c$, and therefore it is safe to assume that the field lines are radial at $t = 0$ and also at time t in the frame of reference S. There will, in fact, be small aberration effects associated with the velocity Δv, but these are second-order compared with the gross effects we are discussing here. We may therefore consider a small cone of electric field lines at angle θ with respect to the acceleration vector of the charge at $t = 0$ and at some later time t when the charge is moving

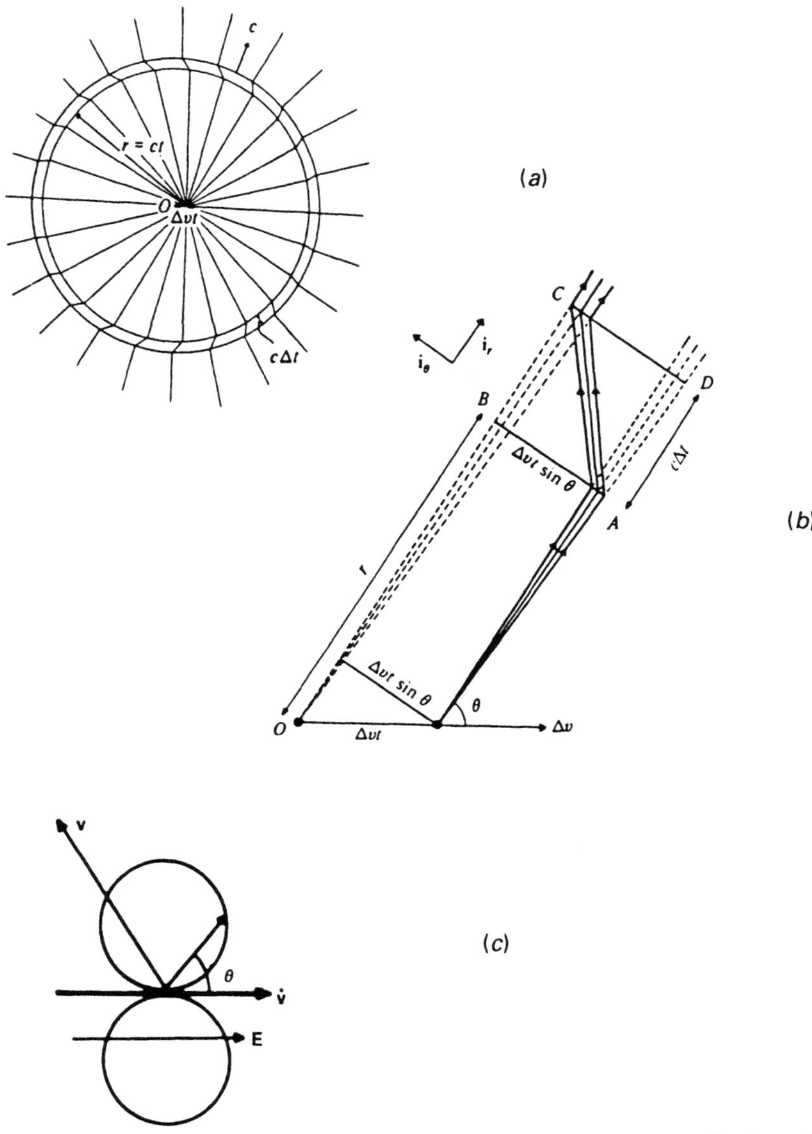

Fig. 1. (a) Illustrating J.J. Thomson's method of evaluating the radiation of an accelerated charged particle. The diagram shows schematically the configuration of the electric field lines at time t due to a charge accelerated to a velocity Δv in time Δt at $t = 0$. **(b)** An expanded version of (a) used to evaluate the strength of the E_θ component of the electric field due to the acceleration of the electron. **(c)** The polar diagram of the radiation emitted by an accelerated electron. The polar diagram shows the magnitude of the electric field strength as a function of polar angle θ with respect to the instantaneous acceleration vector **a**. The polar diagram $E \propto \sin\theta$ corresponds to circular lobes with respect to the acceleration vector. (Longair 1984, 1997a.)

at a constant velocity Δv (Fig. 1b). We now have to join up electric field lines between the two cones through the thin shell of thickness $c\,dt$ as shown in the diagram. The strength of the E_θ component of the field is given by number of field lines per unit area in the $\mathbf{i}_\theta$ direction. From the geometry of Fig. 1b, which exaggerates the discontinuities in the field lines, the E_θ field component is given by the relative sizes of the sides of the rectangle $ABCD$, that is

$$E_\theta / E_r = \Delta v\, t \sin\theta / c\Delta t. \tag{2}$$

But, E_r is given by Coulomb's law,

$$E_r = q/4\pi\varepsilon_0 r^2 \quad \text{where} \quad r = ct, \tag{3}$$

and so

$$E_\theta = \frac{q(\Delta v/\Delta t)\sin\theta}{4\pi\varepsilon_0 c^2 r}. \tag{4}$$

$(\Delta v/\Delta t)$ is the acceleration $\ddot{r}$ of the charge and hence

$$E_\theta = \frac{q\ddot{r}\sin\theta}{4\pi\varepsilon_0 c^2 r}. \tag{5}$$

Notice that the radial component of the field decreases as r^{-2}, according to Coulomb's law, but the field in the pulse decreases only as r^{-1} because the field lines become more and more stretched in the E_θ-direction, as can be seen from Eq. (2). Alternatively we can write $p = qr$, where p is the dipole moment of the charge with respect to some origin, and hence

$$E_\theta = \frac{\ddot{p}\sin\theta}{4\pi\varepsilon_0 c^2 r}. \tag{6}$$

This is a pulse of electromagnetic radiation and hence the energy flow per unit area per second at distance r is given by the Poynting vector $\mathbf{E}\times\mathbf{H} = E^2/Z_0$, where $Z_0 = (\mu_0/\varepsilon_0)^{1/2}$ is the impedance of free space. The rate of loss of energy through the solid angle $\mathrm{d}\Omega$ at distance r from the charge is therefore

$$-\left(\frac{\mathrm{d}E}{\mathrm{d}t}\right)_{\mathrm{rad}} \mathrm{d}\Omega = \frac{|\ddot{\mathbf{p}}|^2 \sin^2\theta}{16\pi^2 Z_0 \varepsilon_0^2 c^4 r^2}\, r^2 \mathrm{d}\Omega = \frac{|\ddot{\mathbf{p}}|^2 \sin^2\theta}{16\pi^2 \varepsilon_0 c^3}\,\mathrm{d}\Omega. \tag{7}$$

To find the total radiation rate, we integrate over all solid angles, that is, we integrate over θ with respect to the direction of the acceleration. Integrating over solid angle means integrating over $\mathrm{d}\Omega = 2\pi\sin\theta\,\mathrm{d}\theta$ and so

$$-\left(\frac{\mathrm{d}E}{\mathrm{d}t}\right)_{\mathrm{rad}} = \int_0^\pi \frac{|\ddot{\mathbf{p}}|^2 \sin^2\theta}{16\pi^2 \varepsilon_0 c^3}\, 2\pi\sin\theta\,\mathrm{d}\theta. \tag{8}$$

We find the key result

$$-\left(\frac{\mathrm{d}E}{\mathrm{d}t}\right)_{\mathrm{rad}} = \frac{|\ddot{\mathbf{p}}|^2}{6\pi\varepsilon_0 c^3} = \frac{q^2|\ddot{\mathbf{r}}|^2}{6\pi\varepsilon_0 c^3}. \tag{9}$$

This result is sometimes called *Larmor's formula* – precisely the same result comes out of the full theory. These formulae embody the three essential properties of the radiation of an accelerated charged particle.

1. The total radiation rate is given by Larmor's formula (9). Notice that, in this formula, the acceleration is the *proper acceleration* of the charged particle and that the radiation loss rate is that measured in the instantaneous rest frame of the particle.
2. The *polar diagram* of the radiation is of *dipolar* form, that is, the electric field strength varies as $\sin\theta$ and the power radiated per unit solid angle varies as $\sin^2\theta$, where θ is the angle with respect to the acceleration vector of the particle (Fig. 1c). Notice that there is no radiation along the acceleration vector and the field strength is greatest at right angles to the acceleration vector.
3. The radiation is *polarised* with the electric field vector lying in the direction of the acceleration vector of the particle, as projected onto a sphere at distance r from the charged particle (see Fig. 1b).

These are very useful rules which enable us to understand the radiation properties of particles in different astrophysical situations. It is important to remember that these rules are applicable in the *instantaneous rest frame* of the particle and therefore we will have to look carefully at what an external observer sees when the particle is moving at a relativistic velocity.

3.2 Thomson scattering

As an example of the power of these results, let us apply them to the case of *Thomson scattering*, the scattering of electromagnetic waves by free electrons in the classical limit. Thomson first published the formula for the *Thomson cross section* in 1906. He used it in a major paper of that year to show that the number of electrons in each atom is roughly the same as the element's atomic number by interpreting X-ray scattering experiments carried out by Barkla. He attributed the scattering to the re-radiation of the X rays by all the electrons in the sample which he assumed could be considered free particles.

The formula we seek describes the scattering of a beam of radiation incident upon a stationary electron. The problem is to find the intensity of radiation scattered through an angle α by the electron. We assume that the beam of incident radiation propagates in the positive z-direction (Fig. 2). Without loss of generality, we arrange the geometry of the scattering so that the scattering angle α lies in the $x-z$ plane. In the case of unpolarised radiation, we resolve the electric field strength into components of equal intensity in the $\mathbf{i}_x$ and $\mathbf{i}_y$ directions (see Fig. 2).

The electric fields experienced by the electron in the x and y directions, $E_x = E_{x0}\exp(\mathrm{i}\omega t)$ and $E_y = E_{y0}\exp(\mathrm{i}\omega t)$, respectively, cause the electron to oscillate and the accelerations in these directions are:

$$\ddot{r}_x = eE_x/m_\mathrm{e} \quad \ddot{r}_y = eE_y/m_\mathrm{e}. \tag{10}$$

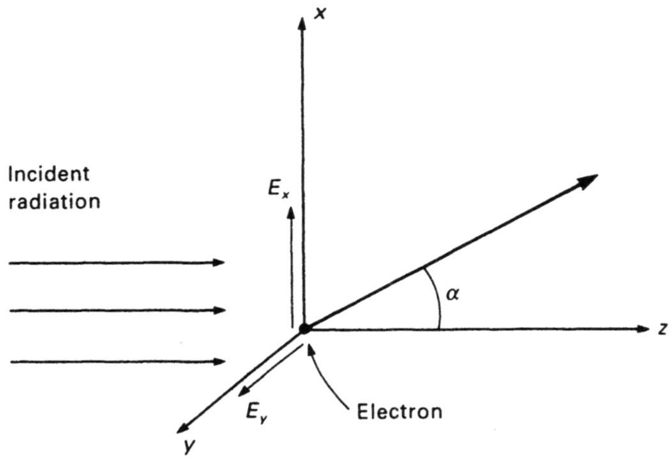

Fig. 2. Illustrating the geometry of Thomson scattering by a free electron of a beam of electromagnetic radiation.

We can therefore enter these accelerations into the radiation formula (9) which shows the angular dependance of the emitted radiation upon the polar angle θ. Let us treat the x-acceleration first. In this case, we can use the formula (Eq. 9) directly with the substitution $\alpha = \pi/2 - \theta$. Therefore, the intensity of radiation scattered through angle θ into the solid angle $\mathrm{d}\Omega$ is

$$-\left(\frac{\mathrm{d}E}{\mathrm{d}t}\right)_x \mathrm{d}\Omega = \frac{e^2|\ddot{r}_x|^2 \sin^2\theta}{16\pi^2\varepsilon_0 c^3}\,\mathrm{d}\Omega = \frac{e^4|E_x|^2}{16\pi^2 m_\mathrm{e}^2 \varepsilon_0 c^3}\cos^2\alpha\,\mathrm{d}\Omega. \tag{11}$$

We have to take time averages of E_x^2 and we find that $\overline{E_x^2} = E_{x0}^2/2$, where E_{x0} is the maximum field strength of the wave. We sum over all waves contributing to the E_x-component of radiation and express the result in terms of the incident energy per unit area upon the electron. The latter is given by Poynting's theorem, $\mathbf{S}_x = (\mathbf{E} \times \mathbf{H}) = c\varepsilon_0 E_x^2 \mathbf{i}_z$. Again, we take time averages and find that the contribution to the intensity in the direction α from the x-component of the acceleration is $S_x = \sum_i c\varepsilon_0 E_{x0}^2/2$. Therefore

$$-\left(\frac{\mathrm{d}E}{\mathrm{d}t}\right)_x \mathrm{d}\Omega = \frac{e^4\cos^2\alpha}{16\pi^2 m_\mathrm{e}^2\varepsilon_0 c^3}\sum_i \overline{E_x^2}\,\mathrm{d}\Omega = \frac{e^4\cos^2\alpha}{16\pi^2 m_\mathrm{e}^2\varepsilon_0^2 c^4} S_x\,\mathrm{d}\Omega. \tag{12}$$

Now let us look at the scattering of the E_y-component of the incident field. From the geometry of Fig. 2, it can be seen that the radiation in the $x-z$ plane from the acceleration of the electron in the y-direction corresponds to scattering at $\theta = 90°$ and so the scattered intensity in the α-direction is

$$-\left(\frac{\mathrm{d}E}{\mathrm{d}t}\right)_y \mathrm{d}\Omega = \frac{e^4}{16\pi^2 m_\mathrm{e}^2 \varepsilon_0^2 c^4} S_y \,\mathrm{d}\Omega. \tag{13}$$

The total scattered radiation into $\mathrm{d}\Omega$ is the sum of these components (notice that we add the intensities of the two independent field components).

$$-\left(\frac{\mathrm{d}E}{\mathrm{d}t}\right) \mathrm{d}\Omega = \frac{e^4}{16\pi^2 m_\mathrm{e}^2 \varepsilon_0^2 c^4}(1+\cos^2\alpha)\frac{S}{2}\,\mathrm{d}\Omega \tag{14}$$

where $S = S_x + S_y$ and $S_x = S_y$ for unpolarised radiation. We now express the scattered intensity in terms of a differential scattering cross section $\mathrm{d}\sigma_\mathrm{T}$ in the following way. We define the scattered intensity in direction α by the following relation

$$\frac{\mathrm{d}\sigma_\mathrm{T}(\alpha)}{\mathrm{d}\Omega} = \frac{\text{energy radiated per unit time per unit solid angle}}{\text{incident energy per unit time per unit area}}. \tag{15}$$

Since the total incident energy is S, the differential cross section for Thomson scattering is

$$\mathrm{d}\sigma_\mathrm{T}(\alpha) = \frac{e^4}{16\pi^2 \varepsilon_0^2 m_\mathrm{e}^2 c^4}\frac{(1+\cos^2\alpha)}{2}\,\mathrm{d}\Omega. \tag{16}$$

In terms of the *classical electron radius* $r_\mathrm{e} = e^2/4\pi\varepsilon_0 m_\mathrm{e} c^2$, this can be expressed

$$\mathrm{d}\sigma_\mathrm{T} = \frac{r_\mathrm{e}^2}{2}(1+\cos^2\alpha)\,\mathrm{d}\Omega. \tag{17}$$

To find the total cross section for scattering, we integrate over all angles α in the standard way,

$$\sigma_\mathrm{T} = \int_0^\pi \frac{r_\mathrm{e}^2}{2}(1+\cos^2\alpha)\,2\pi\sin\alpha\,\mathrm{d}\alpha \tag{18}$$

$$= \frac{8\pi}{3}r_\mathrm{e}^2 = \frac{e^4}{6\pi\varepsilon_0^2 m_\mathrm{e}^2 c^4} = 6.653\times 10^{-29}\,\mathrm{m}^2. \tag{19}$$

This is Thomson's famous result for the total cross section for scattering by stationary free electrons and is justly referred to as the *Thomson cross section*. It appears in all sorts of formulae involving radiation processes, as we will find as we proceed. Let us note some of the important properties of Thomson scattering.

- The scattering is symmetric with respect to the scattering of angle α. Thus as much radiation is scattered backwards as forwards.
- Another useful calculation is to work out the scattering cross section for 100% polarised emission. We can work this out by integrating the scattered intensity (Eq. 11) over all angles.

$$-\left(\frac{\mathrm{d}E}{\mathrm{d}t}\right)_x = \frac{e^2|\ddot{\mathbf{r}}_x|^2}{16\pi^2\varepsilon_0 c^3}\int \sin^2\theta\, 2\pi\sin\theta\,\mathrm{d}\theta \tag{20}$$

$$= \left(\frac{e^4}{6\pi\varepsilon_0^2 m_e^2 c^4}\right) S_x. \tag{21}$$

We find the same total cross section for scattering as before. This should not be surprising because it does not matter how the electron is forced to oscillate. The energy radiated is simply proportional to the sum of the incident intensities of the radiation field. Because of this last fact, the only important quantity so far as the electron is concerned is the total intensity incident upon it and it does not matter how anisotropic the radiation is. One convenient way of expressing this result is to write the formula for the scattered radiation in terms to the energy density of radiation $u_{\rm rad}$ in which the electron is located,

$$u_{\rm rad} = \sum_i u_i = \sum_i S_i/c, \tag{22}$$

and hence

$$-(\mathrm{d}E/\mathrm{d}t) = \sigma_{\rm T} c u_{\rm rad}. \tag{23}$$

- Thomson scattering is one of the most important processes which impedes the escape of photons from any region and it is useful to write Eq. (23) in terms of the scattering of photons out of a beam propagating in the positive x-direction. To do this, we write down the expression for the energy scattered by the electron in terms of the number density N of photons of frequency ν so that

$$-\frac{\mathrm{d}(Nh\nu)}{\mathrm{d}t} = \sigma_{\rm T} c N h\nu. \tag{24}$$

There is no change of energy of the photons in the scattering process and so, if we apply the above equation to the scattering of photons from the beam, and if there are N_e electrons per unit volume, the number density of photons decreases exponentially with distance

$$-\frac{\mathrm{d}N}{\mathrm{d}t} = \sigma_{\rm T} c N_e N \qquad -\frac{\mathrm{d}N}{\mathrm{d}x} = \sigma_{\rm T} N_e N$$

$$N = N_0 \exp\left(-\int \sigma_{\rm T} N_e\,\mathrm{d}x\right). \tag{25}$$

We can express this by stating that the *optical depth* of the medium to Thomson scattering is

$$\tau = \int \sigma_{\rm T} N_e\,\mathrm{d}x. \tag{26}$$

In this process, the photons are scattered in random directions and so they perform a random walk, each step corresponding to the *mean free path* $\lambda_{\rm T}$ of the photon through the electron gas where $\lambda_{\rm T} = (\sigma_{\rm T} N_e)^{-1}$.

Thus, there is a very real sense in which the Thomson cross section is the physical cross section of the electron for the scattering of electromagnetic waves.

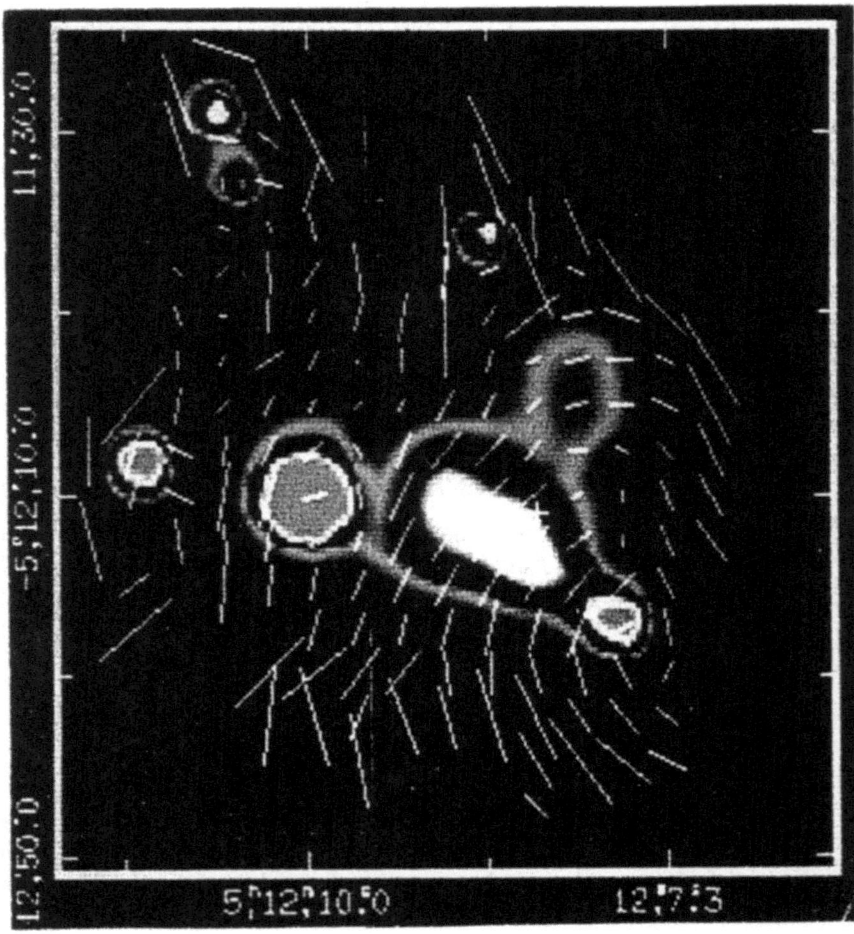

Fig. 3. An infrared polarisation image of an outflow source in the Orion Molecular Cloud observed at a wavelength of 2 μm. The compact object towards the bottom right is a very young star and is the source of the outflow which extends to the top left of the image. The polarisation vectors in the region of the outflow are roughly circularly symmetric about the young star, the typical signature of electron scattering. (Rayner & McLean 1987.)

- A distinctive feature of the process is that the scattered radiation is polarised, even if the incident beam of the radiation is unpolarised. This can be seen intuitively from Fig. 2 because all the E-vectors of the unpolarised beam lie in the $x - y$ plane. Therefore, in the case of observing the electron precisely in the $x - y$ plane, the scattered radiation is 100% polarised. On the other hand, if we look along the z-direction, we observe unpolarised radiation. If we define the degree of polarisation in the standard way,

$$\Pi = \frac{I_{\max} - I_{\min}}{I_{\max} + I_{\min}}, \tag{27}$$

we find by a simple calculation that the fractional polarisation of the radiation is

$$\Pi = \frac{1 - \cos^2 \alpha}{1 + \cos^2 \alpha}. \tag{28}$$

This is therefore a means of producing polarised radiation from an initially unpolarised beam. Fig. 3 is a pleasant example of this phenomenon, showing the infrared intensity and polarisation of an obscured source in Orion Molecular Cloud 1. The polarisation vectors are more or less circularly symmetric about the source to the right of the picture, showing that it is the source of excitation of the outflow which opens up towards the top left of the image, and that the radiation from the flow is scattered infrared radiation.

3.3 Radiation of an accelerated electron – improved version

We need to assemble some more tools before we get down to analysing specific radiation processes in more detail. Let us first describe briefly how the calculation given in Sect. 3.1 can be carried out starting from Maxwell's equations. We begin with Maxwell's equations in free space in standard form

$$\nabla \times \mathbf{E} = -\partial \mathbf{B}/\partial t \tag{29}$$

$$\nabla \times \mathbf{B} = \mu_0 \, \mathbf{J} + \frac{1}{c^2}\frac{\partial \mathbf{E}}{\partial t} \tag{30}$$

$$\nabla \cdot \mathbf{B} = 0 \tag{31}$$

$$\nabla \cdot \mathbf{E} = \rho_e/\varepsilon_0. \tag{32}$$

The standard procedure is to introduce the *scalar* and *vector potentials*, ϕ and $\mathbf{A}$ respectively, in order to simplify the evaluation of the vector fields $\mathbf{E}$ and $\mathbf{B}$ at distance $\mathbf{r}$ from the accelerated charge.

$$\mathbf{B} = \nabla \times \mathbf{A} \tag{33}$$

$$\mathbf{E} = \partial \mathbf{A}/\partial t - \nabla \phi. \tag{34}$$

From this point there follows a standard analysis in which Maxwell's equations are expressed in terms of $\mathbf{A}$ and ϕ. The freedom we have in choosing the precise form of the vector potential $\mathbf{A}$ is used to reduce these equations to the following pair of equations separately for $\mathbf{A}$ and ϕ:

$$\nabla^2 \mathbf{A} - \frac{1}{c^2}\frac{\partial^2 \mathbf{A}}{\partial t^2} = -\mu_0 \mathbf{J} \tag{35}$$

$$\nabla^2 \phi - \frac{1}{c^2}\frac{\partial^2 \phi}{\partial t^2} = -\frac{\rho_e}{\varepsilon_0}. \tag{36}$$

This process is known as selecting the gauge and this particular choice is known as the *Lorentz gauge* (see Jackson 1975 for more details). I have given some details of this analysis in *HEA1*, Sect. 3.3.3.

This pair of equations have standard forms of solutions. I have given a simple derivation of these in *Theoretical Concepts in Physics* (Longair 1984).

$$\mathbf{A}(\mathbf{r}) = \frac{\mu_0}{4\pi}\int \frac{\mathbf{J}(\mathbf{r}', t - |\mathbf{r}-\mathbf{r}'|/c)}{|\mathbf{r}-\mathbf{r}'|}\,\mathrm{d}^3\mathbf{r}' \tag{37}$$

$$\phi(\mathbf{r}) = \frac{1}{4\pi\varepsilon_0}\int \frac{\rho_e(\mathbf{r}', t - |\mathbf{r}-\mathbf{r}'|/c)}{|\mathbf{r}-\mathbf{r}'|}\,\mathrm{d}^3\mathbf{r}'. \tag{38}$$

The point at which the fields are measured is $\mathbf{r}$ and the integration is over the electric current and charge distributions throughout space. The terms in $|\mathbf{r}-\mathbf{r}'|/c$ take account of the fact that the current and charge distributions should be evaluated at *retarded times*. All this does is to take account of the fact that the electromagnetic waves take a finite time to propagate from the charged particle to the point of observation. We now make a number of simplifications in order to obtain the results we are seeking. First of all, we note that, in the case of an accelerated charged particle, the integral of the product of the current density $\mathbf{J}$ and the volume element $\mathrm{d}^3\mathbf{r}'$ is no more than the product of its charge times its velocity

$$\mathbf{J}\left(\mathbf{r}', t - \frac{|\mathbf{r}-\mathbf{r}'|}{c}\right)\,\mathrm{d}^3\mathbf{r}' = q\mathbf{v}\delta(\mathbf{r}) \tag{39}$$

where $\delta(\mathbf{r})$ is the Dirac delta function. The expression for the vector potential is therefore

$$\mathbf{A} = \frac{\mu_0}{4\pi}\frac{q\mathbf{v}}{r}. \tag{40}$$

We now take the time derivative of $\mathbf{A}$ in order to find $\mathbf{E}$

$$\mathbf{E} = -\frac{\partial \mathbf{A}}{\partial t} = -\frac{\mu_0}{4\pi}\frac{q\ddot{\mathbf{r}}}{r} = -\frac{q\ddot{\mathbf{r}}}{4\pi\varepsilon_0 c^2 r}. \tag{41}$$

This is exactly the same form as Eq. (5) which we derived by Thomson's argument and so there is no need to proceed further with this analysis. Notice, however, that Eqs (37) and (38) are very much more powerful tools than those adopted in that argument.

One important point about the use of these integrals is that they are correct provided the velocities of the charges are small. A more complete analysis results in the following expressions for the field potentials which are correct for all velocities – these are known as the *Liénard-Wiechert potentials* for the radiation of a moving charge q

$$\mathbf{A}(\mathbf{r},t)=\frac{\mu_0}{4\pi r}\left[\frac{q\mathbf{v}}{(1-\mathbf{v}\cdot\mathbf{n}/c)}\right]_{\text{ret}}\qquad \phi(\mathbf{r},t)=\frac{1}{4\pi\varepsilon r}\left[\frac{q}{(1-\mathbf{v}\cdot\mathbf{n}/c)}\right]_{\text{ret}} \tag{42}$$

$\mathbf{n}$ is the unit vector in the direction of the point of observation from the moving charge. In both cases, the potentials have to be evaluated at retarded times relative to the location of the observer. The reason for drawing attention to this more general form of the expressions for the potentials is that the factor $(1-\mathbf{v}\cdot\mathbf{n}/c)$ will reappear on a number of occasions in our treatment of the radiation of charges and sources of radiation moving at high velocities. In the case of particles moving highly relativistically, the factor represents the fact that the particle almost catches up with the radiation it has just emitted - this is one of the key aspects of synchrotron radiation, as we will discover.

3.4 A useful relativistic invariant

Let us establish next a useful relativistic invariant. Very often we have to transform the energy loss rate by radiation, $\mathrm{d}E/\mathrm{d}t$, from one inertial frame of reference to another. We can show that $\mathrm{d}E/\mathrm{d}t$ is a Lorentz invariant between inertial frames. To the expert in relativity, this is obvious. The total energy emitted in the form of radiation is the fourth component of the momentum four-vector $[\mathbf{p}, E/c^2]$ and $\mathrm{d}t$ is the fourth component of the displacement four-vector $[\mathrm{d}\mathbf{r}, \mathrm{d}t]$, using the four-vector notation of Rindler (1977), which we will use throughout these notes. Therefore, both the energy $\mathrm{d}E$ and the time interval $\mathrm{d}t$ transform in the same way between intertial frames of reference and their ratio $\mathrm{d}E/\mathrm{d}t$ is also an invariant.

Let us obtain this same result in a slightly gentler fashion. In the *instantaneous rest frame* of the accelerated charged particle, dipole radiation is emitted with zero net momentum, as may be seen from the polar diagram of dipole radiation (Fig. 1c), and therefore its four-momentum can be written $[0, \mathrm{d}E'/c^2]$. This radiation is emitted in the interval of proper time $\mathrm{d}t'$ which has four vector $[0, \mathrm{d}t']$. We may now use the inverse Lorentz transformation to relate $\mathrm{d}E'$ and $\mathrm{d}t'$ to $\mathrm{d}E$ and $\mathrm{d}t$.

$$\mathrm{d}E=\gamma\,\mathrm{d}E'\qquad \mathrm{d}t=\gamma\,\mathrm{d}t' \tag{43}$$

and hence

$$\frac{\mathrm{d}E}{\mathrm{d}t}=\frac{\mathrm{d}E'}{\mathrm{d}t'}. \tag{44}$$

It is a useful exercise to derive from this expression the corresponding result for the radiation rate as observed by the external observer who measures the velocity and acceleration of the electron to be $\mathbf{a}$ and $\mathbf{v}$ respectively, the proper acceleration measured in the instantaneous rest frame of the electron being $\mathbf{a}_0$. Then, from the above results, we find

$$\frac{\mathrm{d}E}{\mathrm{d}t}=\frac{\mathrm{d}E'}{\mathrm{d}t'}=\frac{e^2|\mathbf{a}_0|^2}{6\pi\varepsilon_0 c^3}. \tag{45}$$

To relate $\mathbf{a}_0$, $\mathbf{a}$ and $\mathbf{v}$, it is simplest to equate the norms of the four-accelerations of the accelerated electron in the frames S and S′. I leave it as an exercise to the reader to show that

$$\mathbf{a}_0^2 = \gamma^4 \left[\mathbf{a}^2 + \gamma^2 \left(\frac{\mathbf{v}\cdot\mathbf{a}}{c}\right)^2\right] \tag{46}$$

and so

$$\left(\frac{\mathrm{d}E}{\mathrm{d}t}\right)_{\text{in S}} = \frac{e^2\gamma^4}{6\pi\varepsilon_0 c^3}\left[\mathbf{a}^2 + \gamma^2 \left(\frac{\mathbf{v}\cdot\mathbf{a}}{c}\right)^2\right]. \tag{47}$$

Another useful exercise is to resolve $\mathbf{a}$ parallel and perpendicular to $\mathbf{v}$ so that

$$\mathbf{a} = a_{\parallel}\mathbf{i}_{\parallel} + a_{\perp}\mathbf{i}_{\perp} \tag{48}$$

and then to show that the radiation rate is

$$\left(\frac{\mathrm{d}E}{\mathrm{d}t}\right)_{\text{in S}} = \frac{e^2\gamma^4}{6\pi\varepsilon_0 c^3}\left(|a_{\perp}|^2 + \gamma^2|a_{\parallel}|^2\right). \tag{49}$$

I have shown how these relations are obtained in *HEA1*, Sect. 3.3.4.

3.5 Parseval's theorem and the spectral distribution of the radiation of an accelerated electron

The final tool we need before tackling the emission spectrum of accelerated electrons is the decomposition of the radiation field of the electron into its spectral components. *Parseval's theorem* provides an elegant method of relating the dynamical history of the particle to its radiation spectrum. We introduce the Fourier transform of the acceleration of the particle through the Fourier transform pair, which I write in symmetrical form:

$$\dot{\mathbf{v}}(t) = \frac{1}{(2\pi)^{1/2}}\int_{-\infty}^{\infty} \dot{\mathbf{v}}(\omega)\exp(-\mathrm{i}\omega t)\,\mathrm{d}\omega \tag{50}$$

$$\dot{\mathbf{v}}(\omega) = \frac{1}{(2\pi)^{1/2}}\int_{-\infty}^{\infty} \dot{\mathbf{v}}(t)\exp(\mathrm{i}\omega t)\,\mathrm{d}t. \tag{51}$$

According to Parseval's theorem, $\dot{\mathbf{v}}(\omega)$ and $\dot{\mathbf{v}}(t)$ are related by the following integrals:

$$\int_{-\infty}^{\infty} |\dot{\mathbf{v}}(\omega)|^2\,\mathrm{d}\omega = \int_{-\infty}^{\infty} |\dot{\mathbf{v}}(t)|^2\,\mathrm{d}t. \tag{52}$$

This is proved in all textbooks on Fourier anlaysis. We can therefore apply this relation to the energy radiated by a particle which has an acceleration history $\dot{\mathbf{v}}(t)$,

$$\int_{-\infty}^{\infty} \frac{\mathrm{d}E}{\mathrm{d}t} = \int_{-\infty}^{\infty} \frac{e^2}{6\pi\varepsilon_0 c^3}|\dot{\mathbf{v}}(t)|^2\,\mathrm{d}t \tag{53}$$

$$= \int_{-\infty}^{\infty} \frac{e^2}{6\pi\varepsilon_0 c^3}|\dot{\mathbf{v}}(\omega)|^2\,\mathrm{d}\omega. \tag{54}$$

Now, what we really want is $\int_0^\infty \ldots \mathrm{d}\omega$ rather than $\int_{-\infty}^\infty \ldots \mathrm{d}\omega$ Since the acceleration is a real function, there is another theorem in Fourier analysis which tells us that

$$\int_0^\infty |\dot{\mathbf{v}}(\omega)|^2 \,\mathrm{d}\omega = \int_{-\infty}^0 |\dot{\mathbf{v}}(\omega)|^2 \,\mathrm{d}\omega \tag{55}$$

and hence we find

$$\text{Total emitted radiation} = \int_0^\infty I(\omega)\,\mathrm{d}\omega = \int_0^\infty \frac{e^2}{3\pi\varepsilon_0 c^3} |\dot{\mathbf{v}}(\omega)|^2 \,\mathrm{d}\omega. \tag{56}$$

Therefore

$$I(\omega) = \frac{e^2}{3\pi\varepsilon_0 c^3} |\dot{\mathbf{v}}(\omega)|^2. \tag{57}$$

Note that this is the total energy per unit bandwidth emitted throughout the period during which the particle is accelerated. For a distribution of particles, this result must be integrated over all the particles contributing to the radiation at frequency ω.

4 Bremsstrahlung

Bremsstrahlung is the radiation associated with the acceleration of electrons in the electrostatic fields of ions and the nuclei of atoms. In X-ray and γ-ray astronomy, the most important cases are those in which bremsstrahlung is emitted by very hot plasmas at $T \geq 10^6$ K, at which temperatures the hydrogen and helium atoms are fully ionised. It is a useful exercise to use the tools introduced in Sect. 3 to derive classically the expressions for the bremsstrahlung emissivity of a hot plasma.

4.1 Encounters between charged particles

To begin with, let us study a slightly different problem from that needed for the analysis of bremsstrahlung, the collision of a high-energy proton or nucleus with the electrons of a fully ionised plasma. The interaction is illustrated in Fig. 4. The charge of the high-energy particle is ze, its mass M, and it is assumed that it is undeviated in the encounter with the electron; b, the distance of closest approach of the particle to the electron, is called the *collision parameter* of the interaction. The total *momentum impulse* given to the electron in the encounter is $\int F\,\mathrm{d}t$. By symmetry, the forces parallel to the line of flight of the high-energy particle cancel out and so we need only work out the component of force perpendicular to the line of flight. Then,

$$F_\perp = \frac{ze^2}{4\pi\varepsilon_0 r^2}\sin\theta \quad ; \quad \mathrm{d}t = \frac{\mathrm{d}x}{v}. \tag{58}$$

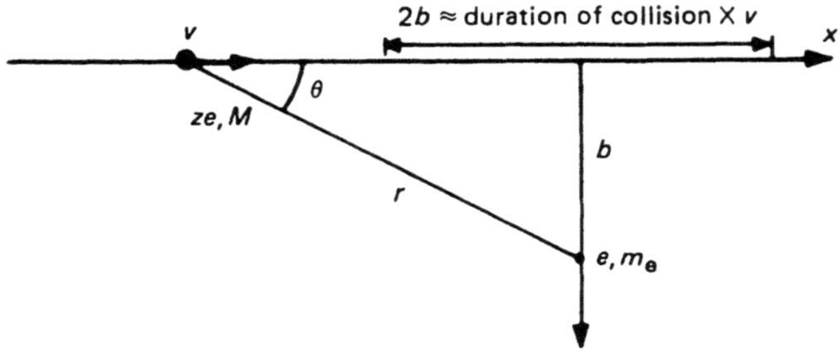

Fig. 4. Illustrating the geometry of an electrostatic encounter of a high-energy particle with a stationary electron and the definition of the collision parameter b.

Changing variables to θ, $b/x = \tan\theta$, $r = b/\sin\theta$ and therefore $\mathrm{d}x = (-b/\sin^2\theta)\,\mathrm{d}\theta$; v is effectively constant and therefore

$$\int_{-\infty}^{\infty} F_{\perp}\,\mathrm{d}t = -\int_0^{\pi} \frac{ze^2}{4\pi\varepsilon_0 b^2}\sin^2\theta\frac{b\sin\theta}{v\sin^2\theta}\,\mathrm{d}\theta = -\frac{ze^2}{4\pi\varepsilon_0 bv}\int_0^{\pi}\sin\theta\,\mathrm{d}\theta \quad (59)$$

Therefore

$$\text{momentum impulse } p = \frac{ze^2}{2\pi\varepsilon_0 bv} \quad (60)$$

and the kinetic energy transferred to the electron is

$$\frac{p^2}{2m_e} = \frac{z^2e^4}{8\pi^2\varepsilon_0^2 b^2 v^2 m_e} = \text{energy lost by high}-\text{energy particle.} \quad (61)$$

We now want the *average energy loss per unit length* and so we have to

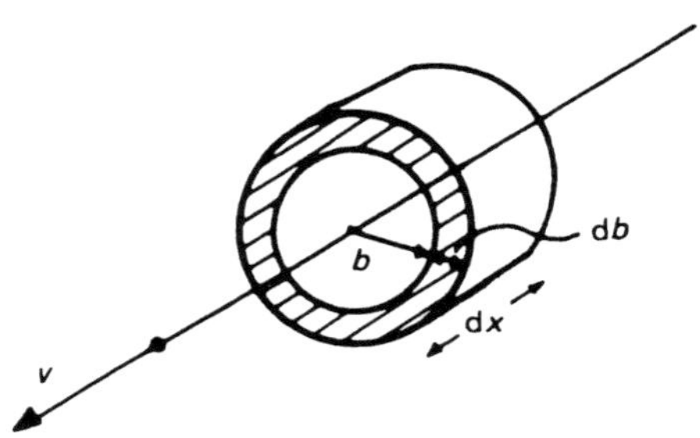

Fig. 5. Illustrating the integration of the energy loss rate over a range of collision parameters b.

work out the number of collisions with collision parameters in the range b to $b + \mathrm{d}b$ and integrate over collision parameters. From the geometry of Fig. 5, the total energy loss of the high-energy particle, $-\mathrm{d}E$, is:

(number of electrons in volume $2\pi b\,\mathrm{d}b\,\mathrm{d}x$) $\times$ (energy loss per interaction)

$$= \int_{b_{\min}}^{b_{\max}} N_{\mathrm{e}}\, 2\pi b\, \mathrm{d}b\, \frac{z^2 e^4}{8\pi^2 \varepsilon_0^2 b^2 v^2 m_{\mathrm{e}}}\, \mathrm{d}x \tag{62}$$

where N_{e} is the number density of electrons. Notice that I have included limits $b_{\max}$ and $b_{\min}$ to the range of collision parameters in this integral. Let us complete the integral:

$$-\frac{\mathrm{d}E}{\mathrm{d}x} = \frac{z^2 e^4 N_{\mathrm{e}}}{4\pi \varepsilon_0^2 v^2 m_{\mathrm{e}}} \ln\left(\frac{b_{\max}}{b_{\min}}\right). \tag{63}$$

Notice how the logarithmic dependence upon $b_{\max}/b_{\min}$ comes about. The closer the encounter, the greater the momentum impulse, $p \propto b^{-2}$. There are, however, more electrons at large distances ($\propto b\,\mathrm{d}b$) and hence, when we integrate, we obtain only a logarithmic dependence of the energy loss rate upon the range of collision parameters. You may well ask, 'Why introduce the limits $b_{\max}$ and $b_{\min}$, rather than work out the answer properly?' The reason is that the proper sum is very much more complicated and would take account of the acceleration of the electron by the high-energy particle and include a proper quantum mechanical treatment of the interaction. Our approximate methods give rather good answers, however, because the limits $b_{\max}$ and $b_{\min}$ only appear inside the logarithm and hence need not be known very precisely.

The reason for introducing this calculation is that it is the simplest example of the type of calculation which needs to be carried out in working out energy transfers and accelerations of electrons and protons in fully ionised plasmas. The logarithmic term $\ln(b_{\max}/b_{\min})$ appears in the guise of what are often referred to as *Gaunt factors* and care has to be taken to use the correct values of $b_{\max}$ and $b_{\min}$ in collision processes in different physical conditions. Similar forms of Gaunt factor appear in working out the radiation spectrum of bremsstrahlung and the electrical conductivity of a plasma.

4.2 The spectrum and energy loss rate of bremsstrahlung

In the classical limit, bremsstrahlung is the emission of an electron as it is accelerated in an electrostatic encounter with a nucleus. Electrons lose more energy in electron-electron collisions, but these do not result in the emission of dipole radiation since there is no net electric dipole moment associated with these encounters. We adopt here a classical approach, to which quantum mechanical parts are added as appropriate. First, we need an expression for the acceleration of an electron in the electrostatic field of the nucleus. Now the roles of the particles in the calculation of Sect. 4.1 are reversed – the electron is moving at a high speed past the stationary nucleus. Then, we take the Fourier transform of the acceleration of the electron and use Parseval's theorem to work out the spectrum of the emitted radiation. Next, we integrate this result over all collision parameters and we have to worry about suitable limits for $b_{\max}$ and $b_{\min}$. In the case in which the electron is relativistic, we have to perform a transformation back into the laboratory frame of reference.

Both the relativistic and non-relativistic cases begin in the same way and so let us treat both cases simultaneously. It is left as an exercise to the reader to show that the accelerations along the trajectory of the electron, $a_{\parallel}$, and perpendicular to it, $a_{\perp}$, in its rest-frame are given by

$$a_{\parallel} = \dot{v}_x = -\frac{eE_x}{m_e} = \frac{\gamma Z e^2 vt}{4\pi\varepsilon_0 m_e [b^2 + (\gamma vt)^2]^{3/2}} \tag{64}$$

$$a_{\perp} = \dot{v}_z = -\frac{eE_z}{m_e} = \frac{\gamma Z e^2 b}{4\pi\varepsilon_0 m_e [b^2 + (\gamma vt)^2]^{3/2}} \tag{65}$$

where Ze is the charge of the nucleus (see *HEA1*).

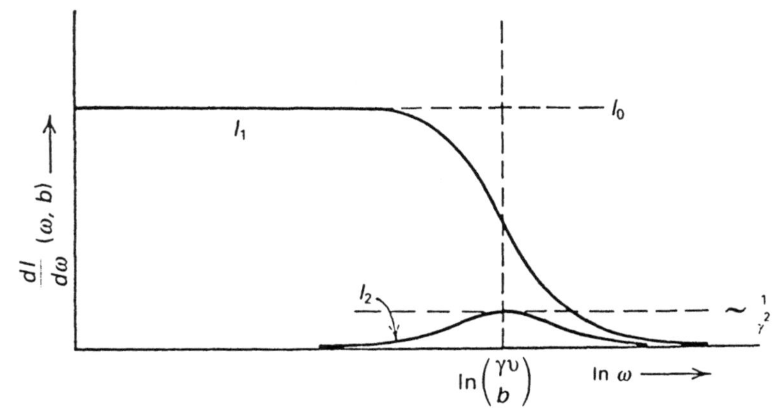

Fig. 6. The spectrum of bremsstrahlung resulting from the acceleration of the electron parallel and perpendicular to its initial direction of motion (Jackson 1975).

After some algebra we find that the radiation spectrum of the electron in an encounter with a charged nucleus is

$$I(\omega) = \frac{e^2}{3\pi\varepsilon_0 c^3}\left[|a_{\parallel}(\omega)|^2 + |a_{\perp}(\omega)|^2\right] \tag{66}$$

$$= \frac{e^2}{3\pi\varepsilon_0 c^3}\frac{1}{2\pi}\left(\frac{Ze^2}{4\pi\varepsilon_0 m_e b v}\right)^2\left[\frac{1}{\gamma^2} I_1^2(y) + I_2^2(y)\right] \tag{67}$$

$$= \frac{Z^2 e^6}{24\pi^4 \varepsilon_0^3 c^3 m_e^2}\frac{\omega^2}{\gamma^2 v^2}\left[\frac{1}{\gamma^2}K_0^2\left(\frac{\omega b}{\gamma v}\right) + K_1^2\left(\frac{\omega b}{\gamma v}\right)\right] \tag{68}$$

where $I_1(y) = 2iyK_0(y)$ and $I_2(y) = 2yK_1(y)$ and K_0 and K_1 are modified Bessel functions of order zero and one. This is the intensity spectrum which results from a single encounter between an electron and a nucleus with collision parameter b. It is interesting to plot the intensity spectrum, displaying separately the terms which arise from the accelerations parallel and perpendicular to the velocity vector of the electron (Fig. 6). It is apparent that the impulse perpendicular to the direction of travel contributes the greater intensity, even in the non-relativistic case $\gamma = 1$. In addition, this component results in significant radiation at low frequencies. When the particle is relativistic, the intensity due to acceleration along the trajectory of the particle decreases by a factor of γ^{-2}. Thus, in bremsstrahlung, the dominant contribution to the radiation spectrum results from the momentum impulse perpendicular to the line of flight of the electron.

It is instructive to study the asymptotic limits of $K_0(y)$ and $K_1(y)$. These are:

$$y \ll 1 \quad K_0(y) = -\ln y; \quad K_1(y) = 1/y \tag{69}$$

$$y \gg 1 \quad K_0(y) = K_1(y) = (\pi/2y)^{1/2}\exp(-y). \tag{70}$$

Thus, at high frequencies, there is an exponential cut-off to the spectrum

$$I(\omega) = \frac{Z^2e^6}{48\pi^3\varepsilon_0^3c^3m_e^2\gamma v^3}\left[\frac{1}{\gamma^2}+1\right]\exp\left(-\frac{2\omega b}{\gamma v}\right). \tag{71}$$

Notice the origin of the exponential cut-off. The duration of the relativistic collision is roughly $\tau = 2b/\gamma v$. Therefore, the dominant Fourier components in the radiation spectrum correspond to frequencies $\nu \approx 1/\tau = \gamma v/2b$ and hence to $\omega \approx \pi v\gamma/2b$, that is, to order of magnitude $\omega b/\gamma v \approx 1$. The exponential cut-off tells us that there is little power emitted at frequencies greater than $\omega \approx \gamma v/b$.

The low-frequency spectrum has the form

$$I(\omega) = \frac{Z^2e^6}{24\pi^4\varepsilon_0^3c^3m_e^2}\frac{1}{b^2v^2}\left[1-\frac{1}{\gamma^2}\left(\frac{\omega b}{\gamma v}\right)^2\ln^2\left(\frac{\omega b}{\gamma v}\right)\right]. \tag{72}$$

In the low-frequency limit, $\omega b/\gamma v \ll 1$, the second term in square brackets can be neglected and hence a good approximation for the low-frequency spectrum is

$$I(\omega) = \frac{Z^2e^6}{24\pi^4\varepsilon_0^3c^3m_e^2}\frac{1}{b^2v^2} = K. \tag{73}$$

This means that the low-frequency spectrum is almost entirely due to the momentum impulse perpendicular to the line of flight of the electron. In fact, we could have guessed that the low-frequency spectrum of the emission would be flat because, so far as these frequencies are concerned, the momentum impulse is a delta function, that is, the duration of the collision is very much less than the period of the waves. It is a standard result of Fourier analysis that the Fourier transform of the delta function is a flat spectrum, $I(\omega) =$

constant. To an excellent approximation, the low-frequency spectrum is flat up to frequency $\omega = \gamma v/b$, above which the spectrum falls off exponentially.

Finally, we have to integrate over all relevant collision parameters which contribute to the radiation at frequency ω. So far, we have performed a completely general analysis in the rest frame of the electron. If the electron is moving relativistically, the number density of nuclei it observes is enhanced by a factor γ because of relativistic length contraction. Hence, in the moving frame of the electron, $N' = \gamma N$ where N is the number density of nuclei in the laboratory frame of reference. The number of encounters per second is $N'v$ and since, properly speaking, all parameters are measured in the rest frame of the electron, let us add superscript dashes to all the relevant parameters. The radiation spectrum in frame of the electron is therefore

$$I(\omega') = \int_{b'_{\rm min}}^{b'_{\rm max}} 2\pi b' \gamma N v K' \, \mathrm{d}b' \tag{74}$$

$$= \frac{Z^2 e^6 \gamma N}{12\pi^3 \varepsilon_0^3 c^3 m_e^2} \frac{1}{v} \ln\left(\frac{b'_{\rm max}}{b'_{\rm min}}\right). \tag{75}$$

4.3 Non-relativistic and thermal bremsstrahlung

We are interested in two cases: in the first, we evaluate the total energy loss rate by bremsstrahlung of a high-energy but non-relativistic electron and, in the second, the continuum spectrum and radiation loss rate of hot ionised gas in which the velocity distribution of the electrons is Maxwellian at temperature T. In both cases, we can neglect the relativistic correction factors and hence obtain the low-frequency radiation spectrum from Eq. (75)

$$I(\omega) = \frac{Z^2 e^6 N}{12\pi^3 \varepsilon_0^3 c^3 m_e^2} \frac{1}{v} \ln \Lambda \tag{76}$$

where $\Lambda = b_{\rm max}/b_{\rm min}$. We have to make the correct choice of limiting collision parameters $b_{\rm max}$ and $b_{\rm min}$. For $b_{\rm max}$, we note that we should only integrate out to those values of b for which $\omega b/v = 1$. For larger values of b, the radiation at frequency ω lies on the exponential tail of the spectrum and we obtain a negligible contribution to the intensity. For $b_{\rm min}$, we have two options. At low velocities, $v \leq (Z/137)c$, we use the classical limit, $b_{\rm min} = Ze^2/8\pi\varepsilon_0 m_e v^2$, corresponding to the closest distance of approach before the expression for the acceleration of the electron breaks down. This value of $b_{\rm min}$ also corresponds to the electron giving up all its kinetic energy in the encounter and is appropriate for the bremsstrahlung of a region of ionised hydrogen at $T \approx 10^4 K$. At high velocities, $v \geq (Z/137)c$, the quantum restriction, $b_{\rm min} \approx \hbar/2m_e v$, is applicable and can be derived from Heisenberg's uncertainty principle (see *HEA1*). This is the appropriate limit to describe, for example, the X-ray bremsstrahlung of hot intergalactic gas in clusters of galaxies. Thus, the choices are

$$\Lambda = 8\pi\varepsilon_0 m_e v^3/Ze^2\omega \quad \text{for low velocities;} \tag{77}$$

$$\Lambda = 2m_e v^2/\hbar\omega \quad \text{for high velocities.} \tag{78}$$

Notice that we have simplified the algebra by restricting the analysis to the flat, low-frequency part of the radiation spectrum. There is, as usual, a cut-off at high frequencies $\omega \geq v/b$.

It is interesting to compare our result with the full answer which was derived by Bethe and Heitler using a full quantum mechanical treatment of the radiation process. The key aspect of their result is that the electron cannot give up more than its total kinetic energy in the radiation process and so no photons are radiated with energies greater than $\varepsilon = \hbar\omega = \frac{1}{2}m_e v^2$. In the same notation as above, the intensity of radiation from a single electron of energy $E = \frac{1}{2}m_e v^2$ in the non-relativistic limit is

$$I(\omega) = \frac{8}{3}Z^2\alpha r_e^2 \frac{m_e c^2}{E} vN \ln\left[\frac{1+(1-\varepsilon/E)^{1/2}}{1-(1-\varepsilon/E)^{1/2}}\right] \tag{79}$$

where $\alpha = e^2/4\pi\hbar\varepsilon_0 c \approx 1/137$ is the fine structure constant and $r_e = e^2/4\pi\varepsilon_0 m_e c^2$ is the *classical electron radius*. The constant in front of the logarithm in this expression is exactly the same as that in Eq. (76). In addition, in the limit of low energies $\varepsilon \ll E$, the term inside the logarithm reduces to $4E/\varepsilon$ which is exactly the same as Eq. (78).

In order to work out the bremsstrahlung, or free-free emission, of a gas at temperature T, we integrate the expression (76) over a Maxwellian distribution of electron velocities

$$N_e(v)\,dv = 4\pi N_e \left(\frac{m_e}{2\pi kT}\right)^{3/2} v^2 \exp\left(-\frac{m_e v^2}{2kT}\right) dv. \tag{80}$$

The algebra can become somewhat cumbersome at this stage. We can find the correct order-of-magnitude answer if we write $\frac{1}{2}m_e v^2 = \frac{3}{2}kT$ in expression (76). Then, the emissivity of a plasma having electron density N_e becomes in the low-frequency limit,

$$I(\omega) \approx \frac{Z^2 e^6 N N_e}{12\sqrt{3}\pi^3\varepsilon_0^3 c^3 m_e^2}\left(\frac{m_e}{kT}\right)^{1/2} g(\omega, T) \tag{81}$$

where $g(\omega, T)$ is a *Gaunt factor*, corresponding to $\ln \Lambda$, but now integrated over velocity. At high frequencies, the spectrum of thermal bremsstrahlung cuts off exponentially as $\exp(-\hbar\omega/kT)$, reflecting the population of electrons in the high-energy tail of a Maxwellian distribution at energies $\hbar\omega \gg kT$. Finally, the total energy loss rate of the plasma may be found by integrating the spectral emissivity over all frequencies. In practice, because of the exponential cut-off, we find the correct functional form by integrating Eq. (81) from 0 to $\omega = kT/\hbar$, that is,

$$-(dE/dt) = (\text{constant})\, Z^2 T^{1/2} \overline{g} N N_e \tag{82}$$

where $\bar{g}$ is a frequency averaged Gaunt factor.

Detailed calculations give the following answers: the spectral emissivity of the plasma is

$$\kappa_\nu = \frac{1}{3\pi^2}\left(\frac{\pi}{6}\right)^{1/2}\frac{Z^2e^6}{\varepsilon_0^3c^3m_e^2}\left(\frac{m_e}{kT}\right)^{1/2} g(\nu,T)NN_e\exp\left(-\frac{h\nu}{kT}\right) \tag{83}$$

$$= 6.8\times10^{-51}Z^2T^{-1/2}NN_e\,g(\nu,T)\exp(-h\nu/kT)\,\mathrm{W\,m^{-3}Hz^{-1}} \tag{84}$$

where the number densities of electrons N_e and of nuclei N are given in particles per cubic metre. At frequencies $\hbar\omega \ll kT$, the Gaunt factor has only a logarithmic dependence on frequency. Suitable forms at radio and X-ray wavelengths are:

$$\text{Radio}\quad g(\nu,T) = \frac{\sqrt{3}}{2\pi}\left[\ln\left(\frac{128\varepsilon_0^2k^3T^3}{m_ee^4\nu^2Z^2}\right)-\gamma^{1/2}\right] \tag{85}$$

$$\text{X}-\text{ray}\quad g(\nu,T) = \frac{\sqrt{3}}{\pi}\ln\left(\frac{kT}{h\nu}\right) \tag{86}$$

where $\gamma = 0.577...$ is Euler's constant. The functional forms of both logarithmic terms (85) and (86) can be readily derived from the corresponding expressions (77) and (78).

For frequencies $h\nu/kT \gg 1, g(\nu,T)$ is approximately $(h\nu/kT)^{1/2}$. The origin of this factor may be understood from the approximations for the Bessel functions used in deriving the relation (71). The high-frequency approximation differs from that at $h\nu \ll kT$ by a factor of roughly $y^{1/2}\exp(-y)$ and the form of bremsstrahlung emissivity given in Eq. (83) only takes account of the factor $\exp(-y)$ in the limit $y\to\infty$. Thus, the dominant term in the Gaunt factor is $y^{1/2} \approx (h\nu/kT)^{1/2}$.

The total loss rate of the plasma is

$$-\left(\frac{\mathrm{d}E}{\mathrm{d}t}\right)_{\text{brems}} = 1.435\times10^{-40}Z^2T^{1/2}\bar{g}NN_e \quad \mathrm{W\,m^{-3}}. \tag{87}$$

Detailed calculations show that the frequency averaged Gaunt factor $\bar{g}$ lies in the range $1.1-1.5$ and a good approximation is $\bar{g}=1.2$. A compilation of a large number of useful Gaunt factors for a wide range of physical conditions is given by Karzas & Latter (1961).

4.4 Non-relativistic and relativistic bremsstrahlung losses

To find the energy loss rate of a single high-energy electron, we integrate Eq. (76) over all frequencies. In practice, this means integrating from 0 to ω_{max} where ω_{max} corresponds to the cut-off, $b_{\text{min}} \approx \hbar/2m_ev$. This angular frequency is approximately

$$\omega_{\text{max}} = 2\pi/\tau \sim 2\pi v/b_{\text{min}} \approx 4\pi m_ev^2/\hbar, \tag{88}$$

that is, to order of magnitude $\hbar\omega \approx \frac{1}{2}m_e v^2$. This is just the kinetic energy of the electron and is obviously the maximum amount of energy which can be lost in a single encounter with the nucleus. We should therefore integrate Eq. (76) from $\omega = 0$ to $\omega_{\max} \approx m_e v^2/2\hbar$ and thus,

$$-\left(\frac{\mathrm{d}E}{\mathrm{d}t}\right)_{\text{brems}} \approx \int_0^{\omega_{\max}} \frac{Z^2 e^6 N}{12\pi^3\epsilon_0^3 c^3 m_e^2}\frac{1}{v}\ln\Lambda\,\mathrm{d}\omega \tag{89}$$

$$\approx \frac{Z^2 e^6 N v}{24\pi^3\epsilon_0^3 c^3 m_e \hbar}\ln\Lambda \tag{90}$$

$$= (\text{constant})Z^2 N v. \tag{91}$$

We note that the total energy loss rate of the electron is proportional to v, that is, to the square root of the kinetic energy E: $-\mathrm{d}E/\mathrm{d}t \propto E^{1/2}$. This is in contrast to the case of relativistic bremsstrahlung losses discussed below. In practical applications of this formula, it is necessary to integrate over the energy distribution of the particles. For example, the energy spectrum may well be of Maxwellian or power law form, $N(E)\,\mathrm{d}E \propto E^{-p}\,\mathrm{d}E$.

Let us look us briefly at the case of *relativistic bremsstrahlung*. The formulae we have derived are correct in the rest frame of the electron, namely,

$$I(\omega') = \int_{b'_{\min}}^{b'\max} 2\pi b'(\gamma N)vK'\,\mathrm{d}b' \tag{92}$$

where we have written the number density of nuclei γN because of length contraction. Since the collision parameters b' are perpendicular to the direction of motion, it follows that, since $y = y'$, the same collision parameters appropriate for the laboratory frame of reference can be used. I have given a discussion of the relevant collision parameters in *HEA1* and I will not repeat that discussion here. It suffices to note that we can write the emission spectrum in the frame of the electron

$$I(\omega') = \frac{Z^2 e^6 N\gamma}{12\pi^3\varepsilon_0^3 c^3 m_e^2}\frac{1}{v}\ln\Lambda. \tag{93}$$

Notice that there is at best a very weak dependence upon frequency ω and so we again obtain the characteristic flat bremsstrahlung intensity spectrum. On transforming this spectrum to the laboratory frame of reference, we note that the bandwidth changes as $\Delta\omega = \gamma\Delta\omega'$ and so the spectrum becomes

$$I(\omega) = \frac{Z^2 e^6 N}{12\pi^3\varepsilon_0^3 c^3 m_e^2}\frac{1}{v}\ln\Lambda \tag{94}$$

where the integral extends up to energies $E = \hbar\omega = \gamma m_e c^2$, where $\gamma \gg 1$. Thus, the rate of loss of energy of the relativistic electron is

$$-\left(\frac{\mathrm{d}E}{\mathrm{d}t}\right)_{\text{rel}} = \int_0^{E/\hbar} I(\omega)\,\mathrm{d}\omega = \frac{Z^2 e^6 N\bar{g}}{12\pi^3\varepsilon_0^3 c^4\hbar}E. \tag{95}$$

Notice that the dependence of the energy loss rate changes from $E^{1/2}$ to E between the non-relativistic and relativistic cases.

I leave it as an exercise to the reader to show that, if there is a non-thermal distribution of electron energies, $N(E)\,\mathrm{d}E \propto E^{-p}\,\mathrm{d}E$, the intensity spectrum of non-thermal bremsstrahlung in terms of the number of photons per unit energy interval is $N_\gamma(\omega) \propto \omega^{-p/2}$ and $N_\gamma(\omega) \propto \omega^{-p}$ in the non-relativistic and relativistic limits, respectively. The process of relativistic bremsstrahlung

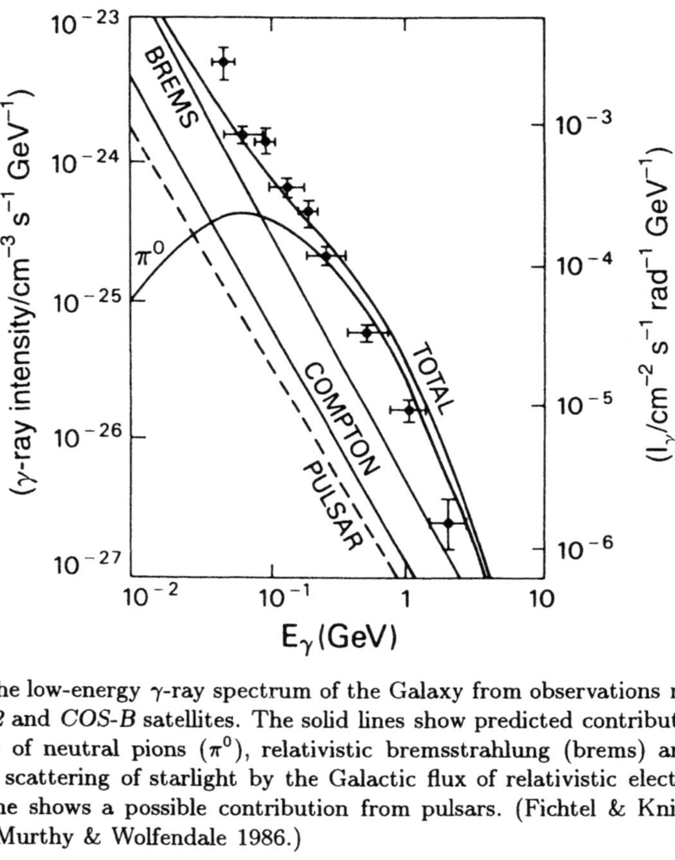

Fig. 7. The low-energy γ-ray spectrum of the Galaxy from observations made with the *SAS-2* and *COS-B* satellites. The solid lines show predicted contributions from the decay of neutral pions (π^0), relativistic bremsstrahlung (brems) and inverse Compton scattering of starlight by the Galactic flux of relativistic electrons. The dashed line shows a possible contribution from pulsars. (Fichtel & Kniffen 1981; Ramana Murthy & Wolfendale 1986.)

may well be important in accounting for the low-energy γ-ray emission of our Galaxy. Fig. 7 shows the γ-ray spectrum of our Galaxy, as presented by Fichtel & Kniffen (1981), as well as theoretical estimates of the emission by Stecker (1977). At energies $\varepsilon > 70$ MeV, the dominant emission mechanism is the decay of neutral pions created in collisions between cosmic rays and the nuclei of atoms and molecules of the interstellar gas. This spectrum peaks at about 70 MeV and so there must be another mechanism which contributes at the lower energies. It is quite conceivable that relativistic bremsstrahlung is the dominant source of emission at these energies. The spectrum labelled 'brems' in Fig. 7 is derived from an extrapolation of the relativistic electron spectrum in our Galaxy to energies $1 < E < 1000$ MeV (see the discussion of the synchrotron radiation of the interstellar medium in Sect. 6.7).

5 Hot Gas in Clusters of Galaxies

One of the most important discoveries of the *Uhuru* X-ray Observatory was the detection of intense X-ray emission from rich clusters of galaxies. The nature of the emission was soon identified as the bremsstrahlung of hot intracluster gas, the key observations being the extended nature of the emission and the detection of the highly ionised iron line Fe XXVI with the *Ariel-V* satellite. It was quickly appreciated that the X-ray emission of the gas provides a very powerful probe of the gravitational potential within the cluster. There are many ramifications of these observations. Let us summarise briefly the key properties of rich clusters of galaxies (for more details, see Longair 1998).

5.1 The properties of rich clusters of galaxies

Rich clusters of galaxies are the largest gravitationally bound systems in the Universe. Typically, the internal velocity dipersions of the galaxies in rich clusters are about $\langle v^2 \rangle^{1/2} \approx 1000$ km s^{-1} and their characteristic sizes $R_{\rm cl} \sim 1$ Mpc. Consequently, the *crossing time* for a galaxy in the centre of a cluster is $t_{\rm cr} \sim R_{\rm cl}/v \sim 10^9$ years, which is much less than cosmological time scales, and so the system must be gravitationally bound.

It might be thought that it would be easy to estimate the mass of the cluster from observations of the velocity dispersion of the galaxies and the characteristic size, but this is trickier than it seems. The straightforward approach is to use the *virial theorem* according to which the kinetic energy of the galaxies in the cluster should equal half its gravitational potential energy, that is

$$\sum_i \tfrac{1}{2} m_i v_i^2 = \tfrac{1}{2} \sum_{i \neq j} \frac{G m_i m_j}{r_{ij}} \tag{96}$$

If M is the total mass of the cluster and the velocity dispersion is independent of mass, Eq. (96) can be reduced to

$$\tfrac{1}{2} M \langle v^2 \rangle = \tfrac{1}{2} \frac{G M^2}{R_{\rm cl}} \tag{97}$$

where $R_{\rm cl}$ is a suitably-defined radius for the cluster. We can only measure the components of the velocities of the galaxies along the line of sight and so, if the velocity distribution is isotropic, $\langle v^2 \rangle = 3 \langle v_\parallel^2 \rangle$. The mass of the cluster would then be

$$M = \frac{3R_{\rm cl}\langle v_\parallel^2\rangle}{G} \tag{98}$$

It turns out that there are remarkably few clusters for which detailed analyses can be made. The reason is that it is essential to make a careful assessment of the galaxies which are truly cluster members and to measure accurate radial velocities for large enough samples of them. The Coma cluster is a good example of a regular rich cluster of galaxies which has been the subject of considerable study. Merritt (1987) considered a wide range of possible models for the mass distribution within the cluster. In the simplest reference model, with which the others can be compared, it is assumed that the mass distribution in the cluster follows the galaxy distribution, that is, that mass-to-luminosity ratio is a constant throughout the cluster, and that the velocity distribution is isotropic at each point in the cluster. With these assumptions, Merritt derived a mass for the Coma cluster of $1.79 \times 10^{15}h^{-1}$ $M_\odot$, assuming that the cluster extends to $16h^{-1}$ Mpc[1]. The mass within a radius of $1h^{-1}$ Mpc of the cluster centre is $6.1 \times 10^{14}\,M_\odot$. The corresponding value of the mass-to-blue luminosity ratio for the central regions of the Coma cluster is about $350h^{-1}\,M_\odot/L_\odot$.

The Coma cluster is a classic example of a rich regular cluster and the population is dominated in the central regions by elliptical and S0 galaxies for which the typical mass-to-luminosity ratios are about $15\,M_\odot/L_\odot$. There is therefore a discrepancy of about a factor of 20 between the mass which can be attributed to galaxies and the total mass which must be present. It is also where all the trouble begins. The *dark matter* dominates the mass of the cluster and there is no reason why it should have the same distribution as the visible matter. Likewise, there is no reason *a priori* why the velocity distribution of the galaxies should be isotropic. Merritt (1987) has provided a careful study of how the inferred mass-to-luminosity ratio would change for a wide range of different assumptions about the relative distributions of the visible and dark matter and the anisotropy of the velocity distribution. For the cluster as a whole, the mass-to-luminosity ratio varies from about 0.4 to at least three times the reference value, while the mass-to-luminosity ratio within $1\,h^{-1}$ Mpc is always very close to $350\,h^{-1}\,M_\odot/L_\odot$.

5.2 Hot gas in clusters of galaxies and isothermal gas spheres

Many rich clusters of galaxies are intense, extended sources of X-ray emission. It is natural that the gas should be hot since, otherwise, it could not form a stable extended atmosphere within the cluster. To a crude approximation, we require the thermal velocities of the particles of the plasma to be of the same

[1] $h = (H_0/100\,\mathrm{km\ s^{-1}\,Mpc^{-1}})$ is Hubble's constant measured relative to a reference value of 100 km s^{-1} Mpc^{-1}. Currently, estimates of h are settling down between about 0.5 and 0.8

order as the velocities of the galaxies, $v_{\rm th} \sim v_{\rm gal}$ and so, setting $\frac{1}{2}mv_{\rm th}^2 = \frac{3}{2}k_{\rm B}T$, we find $T \sim 10^{7-8}$ K. Let us carry out a better calculation.

In a regular cluster, the space density of galaxies increases towards the central regions, which is called the *core* of the cluster. Outside the core, the space density of galaxies decreases steadily until it becomes impossible to distinguish cluster galaxies from the background of unrelated objects. The regular structures of these clusters suggest that they can be represented by the distribution of mass in an *isothermal gas sphere*. These distributions are important both for the distribution of hot gas and galaxies in the discussion which follows, and so let us derive the relevant expressions for the density distribution of an isothermal gas sphere.

The term *isothermal* means that the temperature, or mean kinetic energy of the particles, is constant throughout the cluster. In physical terms, this means that the velocity distribution of the galaxies is Maxwellian with the same velocity dispersion (or temperature) throughout the cluster. If all the galaxies had the same mass, the velocity dispersion would be the same at all locations within the cluster.

To derive the structure of the cluster, we use the *Lane–Emden equation*, which describes the structure of a spherically symmetric object, normally a star, in hydrostatic equilibrium. The requirement of hydrostatic equilibrium is that, at all points in the system, the attractive gravitational force acting on a mass element $\varrho\,{\rm d}V$ at radial distance r from the centre of the system is balanced by the pressure gradient at that point.

$$\nabla p = \frac{{\rm d}p}{{\rm d}r} = -\frac{GM\varrho}{r^2}, \tag{99}$$

where M is the mass contained within radius r,

$$M = \int_0^r 4\pi r^2 \varrho(r)\,{\rm d}r \qquad {\rm d}M = 4\pi r^2 \varrho(r)\,{\rm d}r. \tag{100}$$

Reordering Eq. (99) and differentiating, we find

$$\frac{r^2}{\varrho}\frac{{\rm d}p}{{\rm d}r} = -GM \qquad \frac{\rm d}{{\rm d}r}\left(\frac{r^2}{\varrho}\frac{{\rm d}p}{{\rm d}r}\right) = -G\frac{{\rm d}M}{{\rm d}r} \tag{101}$$

$$\frac{\rm d}{{\rm d}r}\left(\frac{r^2}{\varrho}\frac{{\rm d}p}{{\rm d}r}\right) + 4\pi G r^2 \varrho = 0. \tag{102}$$

This is the *Lane–Emden equation*. We are interested in the simple case in which the pressure p and density ϱ are related by the perfect gas law at all radii r, $p = \varrho kT/\mu$, where μ is the mass of an atom, molecule or galaxy. In thermal equilibrium, $\frac{3}{2}kT = \frac{1}{2}\mu\langle v^2\rangle$, where $\langle v^2\rangle$ is the mean square velocity of the atoms, molecules or galaxies. Then, setting $p = \varrho kT/\mu$ in Eq. (102),

$$\frac{\rm d}{{\rm d}r}\left(\frac{r^2}{\varrho}\frac{{\rm d}\varrho}{{\rm d}r}\right) + \frac{4\pi G\mu}{kT} r^2 \varrho = 0. \tag{103}$$

Equation (103) is a non-linear differential equation and, in general, must be solved numerically. There is, however, an analytic solution for large values of r. If $\varrho(r)$ is expressed as a power series in r, $\varrho(r) = \sum A_n r^{-n}$, there is a solution for large r with $n = 2$. In this case,

$$\varrho(r) = \frac{2}{Ar^2} \qquad \text{where} \qquad A = \frac{4\pi G\mu}{kT}. \tag{104}$$

This mass distribution has two unfortunate properties: the density diverges at the origin and the total mass of the cluster diverges at large values of r,

$$\int_0^\infty 4\pi r^2 \varrho(r)\,\mathrm{d}r = \int_0^\infty \frac{8\pi}{A}\,\mathrm{d}r \to \infty. \tag{105}$$

There are, however, at least two reasons why there should be a cut-off at large radii. First, at very large distances, the particle densities become so low that the mean free path between collisions is very long. The thermalisation time scales are consequently very long indeed, longer than the time scale of the system. The radius at which this occurs is known as *Smoluchowski's envelope*. Second, in astrophysical systems, the outermost stars or galaxies are stripped from the system by tidal interactions with neighbouring systems. This process defines a *tidal radius* R_t for the cluster. Therefore, if clusters are modelled by isothermal gas spheres, it is perfectly permissible to introduce a cut-off at some suitably large radius, resulting in a finite total mass.

It is convenient to rewrite Eq. (103) in dimensionless form by writing $\varrho = \varrho_0 y$, where ϱ_0 is the central mass density, and introducing a *structural index* or *structural length* α, defined by the relation

$$\alpha = \frac{1}{(A\varrho_0)^{1/2}}. \tag{106}$$

Distances from the centre can then be measured in units of the structural length α by introducing a dimensionless distance $x = r/\alpha$. Then, Eq. (103) becomes

$$\frac{\mathrm{d}}{\mathrm{d}x}\left[x^2 \frac{\mathrm{d}(\log y)}{\mathrm{d}x}\right] + x^2 y = 0. \tag{107}$$

Two versions of the solution of Eq. (107) are illustrated in Fig. 8. What we observe on an image of the cluster is the distribution of galaxies projected onto the sky. It is a simple calculation to show that, if q is the projected distance from the centre of the cluster, the surface density $N(q)$ is related to $y(x)$ by the integral

$$N(q) = 2\int_q^\infty \frac{y(x)x}{(x^2 - q^2)^{1/2}}\,\mathrm{d}x. \tag{108}$$

Inspection of Fig. 8, shows that α is a measure of the size of the *core* of the cluster. It is convenient to fit the projected distribution $N(q)$ to the

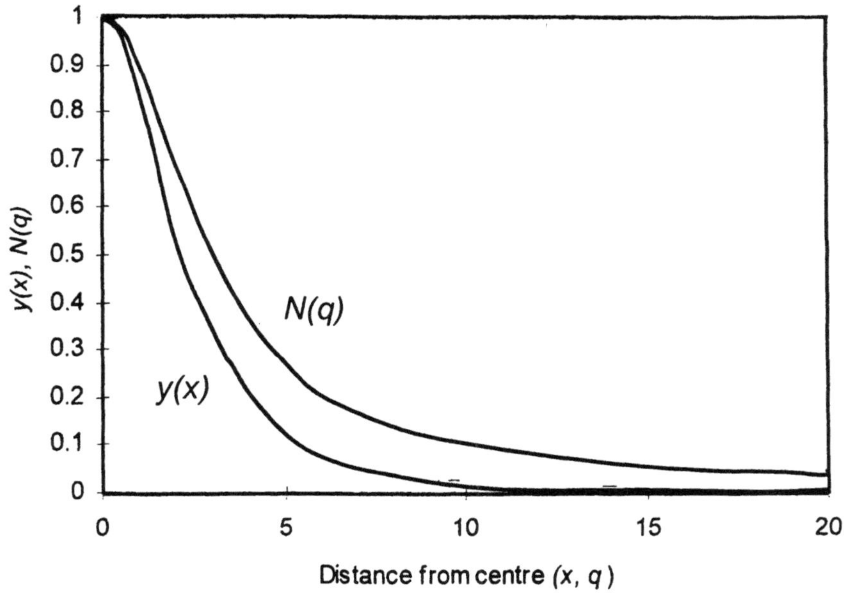

Fig. 8. The density distribution $y(x)$ and the projected density distribution $N(q)$ for an isothermal gas sphere.

distribution of stars or galaxies in a cluster and then a *core radius* for the cluster can be defined. It can be seen that the projected density falls to the value $N(q) = 1/2$ at $q = 3$, that is, at a core radius $R_{1/2} = 3\alpha$. $R_{1/2}$ is a convenient measure of the core radius of the cluster.

Having measured $R_{1/2}$, the central mass density of the cluster can be found if the velocity dispersion of the galaxies in this region is also known. From Maxwell's equipartition theorem, we know that $\frac{1}{2}\mu\langle v^2\rangle = \frac{3}{2}kT$ and therefore, from the definition of α,

$$\alpha^2 = \frac{1}{A\varrho_0} = \frac{kT}{4\pi G\mu\varrho_0} = \frac{\langle v^2\rangle}{12\pi G\varrho_0}. \tag{109}$$

Observationally, we can only measure the radial component of the galaxies' velocities $v_\parallel$. Assuming the velocity distribution of the galaxies in the cluster is isotropic,

$$\langle v^2\rangle = \langle v_x^2\rangle + \langle v_y^2\rangle + \langle v_z^2\rangle = 3\langle v_\parallel^2\rangle. \tag{110}$$

Expressing the central density ϱ_0 in terms of $R_{1/2}$ and $\langle v_\parallel^2\rangle$, we find

$$\varrho_0 = \frac{9\langle v_\parallel^2\rangle}{4\pi G R_{1/2}^2}. \tag{111}$$

Thus, assuming the central density of a cluster can be represented by an isothermal gas sphere, by measuring $\langle v_{\parallel}^2 \rangle$ and $R_{1/2}$, we can find the central mass density of the cluster.

It is remarkable how well isothermal gas sphere distributions can account for the observed distribution of galaxies and hot gas in clusters

5.3 X-ray observations of hot gas in clusters of galaxies

Let us repeat the simple calculation presented by Fabricant, Lecar & Gorenstein (1980) which shows how the mass distribution in clusters can be found from observations of the intensity and temperature distributions of its X-ray emission. For simplicity, we assume that the cluster is spherically symmetric so that the total gravitating mass within radius r is $M(< r)$. The gas is assumed to be in hydrostatic equilibrium within the gravitational potential defined by the total mass distribution in the cluster, that is, by the sum of the visible and dark matter, as well as the gaseous mass. If p is the pressure of the gas and ϱ its density, both of which vary with position within the cluster, the requirement of hydrostatic equilibrium is

$$\frac{\mathrm{d}p}{\mathrm{d}r} = -\frac{GM(< r)\varrho}{r^2}. \tag{112}$$

The pressure is related to the local gas density ϱ and temperature T by the perfect gas law

$$p = \frac{\varrho kT}{\mu m_{\mathrm{H}}}, \tag{113}$$

where m_{H} is the mass of the hydrogen atom and μ is the mean molecular weight of the gas. For a fully ionised gas with the standard cosmic abundance of the elements, a suitable value is $\mu = 0.6$. Differentiating Eq. (113) with respect to r and substituting into Eq. (112), we find

$$\frac{\varrho kT}{\mu m_{\mathrm{H}}}\left(\frac{1}{\varrho}\frac{\mathrm{d}\varrho}{\mathrm{d}r} + \frac{1}{T}\frac{\mathrm{d}T}{\mathrm{d}r}\right) = -\frac{GM(< r)\varrho}{r^2}. \tag{114}$$

Reorganising expression (114),

$$M(< r) = -\frac{kTr^2}{G\mu m_{\mathrm{H}}}\left[\frac{\mathrm{d}(\log\varrho)}{\mathrm{d}r} + \frac{\mathrm{d}(\log T)}{\mathrm{d}r}\right]. \tag{115}$$

Thus, the mass distribution within the cluster can be determined if the variation of the gas density and temperature with radius are known. Assuming the cluster is spherically symmetric, these can be derived from high-sensitivity X-ray intensity and spectral observations of the bremsstrahlung of the cluster gas. In practice, the spectral emissivity κ_ν has to be integrated along the line of sight through the cluster. Performing this integration and converting it into an intensity, the observed surface brightness at projected radius a from the cluster centre is

$$I_\nu(a) = \frac{1}{2\pi} \int_a^\infty \frac{\kappa_\nu(r) r}{(r^2 - a^2)^{1/2}}\, \mathrm{d}r. \tag{116}$$

Cavaliere (1980) noted that this is an Abel integral which can be inverted to find the emissivity of the gas as a function of radius

$$\kappa_\nu(r) = \frac{4}{r} \frac{\mathrm{d}}{\mathrm{d}r} \int_r^\infty \frac{I_\nu(a) a}{(a^2 - r^2)^{1/2}}\, \mathrm{d}a. \tag{117}$$

Ideally, one would like to measure precisely the spectrum of the X-ray emission along many lines of sight through the cluster. The problem is that, at present, X-ray telescopes either have high angular resolution and low spectral resolution, or good spectral resolution and low angular resolution. A further problem is that there is evidence that *cooling flows* are important in the central regions of a significant fraction of all clusters which are strong X-ray emitters (see below).

X-rays maps of about 200 clusters were made with the *Einstein* X-ray Observatory (Forman & Jones 1982) and more recently the *ROSAT* observatory has provided high-resolution maps of the X-ray emission of a number of nearby clusters (Böhringer 1995). A beautiful example of the quality of data now available is illustrated by the X-ray map of the central regions of the Virgo cluster obtained from the *ROSAT* All Sky Survey (Fig. 9). A number of galaxies belonging to the Virgo cluster have been detected as X-ray sources, as well as a few background clusters and active galaxies. In addition, the X-ray emission of the diffuse intergalactic gas can be seen centred on the massive galaxy M87. Evidence that the intergalactic gas traces the mass distribution of the cluster is provided by comparison of the contours of the X-ray surface brightness distribution with the surface distribution of galaxies as determined by the photometric survey of the Virgo cluster by Binggeli, Tammann & Sandage (1987). The distribution of galaxies in the cluster and the diffuse X-ray emission are remarkably similar (Fig. 9). This is an important result because it indicates that the dark matter, which defines the gravitational potential in the cluster and which is traced out by the distribution of hot gas, must be distributed like the galaxies. Within a radius of 1.8 Mpc, the total mass was found to lie in the range $1.5 - 5.5 \times 10^{14}\,\mathrm{M}_\odot$, the uncertainty in the mass resulting from uncertainties in the temperature profile. For comparison, the X-ray luminosity of the cluster is 8×10^{36} W in the energy band 0.1 to 2 keV, corresponding to a gaseous mass of $4 - 5.5 \times 10^{13}\,\mathrm{M}_\odot$.

Another beautiful example of the application of these procedures is to the Perseus cluster of galaxies (Böhringer 1995). In this case, the X-ray emission could be traced out to a radius $1.5h^{-1}$ Mpc. From the X-ray observations, it was possible to determine both the total gravitating mass, $M(< r)$, and the mass of gas within radius r, $M_{\mathrm{gas}}(< r)$, and to compare these with the mass in the visible parts of galaxies. This comparison is shown in Fig. 10, in which it can be seen that the mass of hot intra-cluster gas is about five times greater than the mass in galaxies, but that it is insufficient to account for all

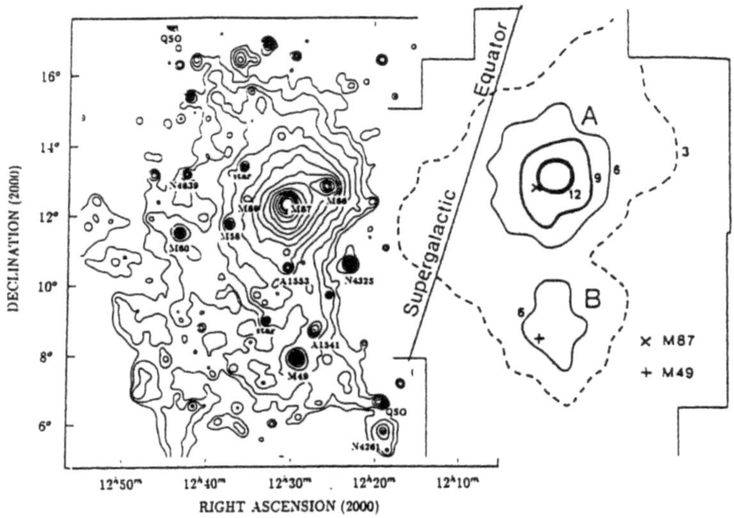

Fig. 9. Comparison of the X-ray emission and the surface density of luminous matter in galaxies in the Virgo cluster (Böhringer 1995). (**a**) The X-ray image of the Virgo cluster from the *ROSAT* All Sky Survey in the X-ray energy band 0.4 – 2 keV. The image has been smoothed with a variable Gaussian filter to enable low X-ray brightness regions to be detected. Some galaxies in the Virgo cluster have been detected as well as a few distant clusters and active galaxies. (**b**) The surface density of galaxies in the Virgo Cluster from the photometric survey of Binggeli, Tammann & Sandage (1987).

the gravitating mass which must be present. Some form of dark matter must be present to bind the cluster gravitationally.

Böhringer (1995) has summarised the typical masses found in rich clusters of galaxies. The typical total masses lie in the range 5×10^{14} to $5 \times 10^{15}\,M_{\odot}$, of which about 5% is attributable to the mass contained in the visible parts of galaxies and about 10 to 30% to hot gas. The remaining 60 to 85% of the mass is in some form of dark matter. Typically, the iron abundance of the hot gas is between about 20 and 50% of the solar value, indicating that the intergalactic gas has been enriched by the products of stellar nucleosynthesis.

5.4 Cooling flows in clusters of galaxies

In the dense central regions of clusters of galaxies, the gas density may become sufficiently high for the gas to cool by bremsstrahlung over cosmological time scales. The observation of peaks in the X-ray surface brightness distribution

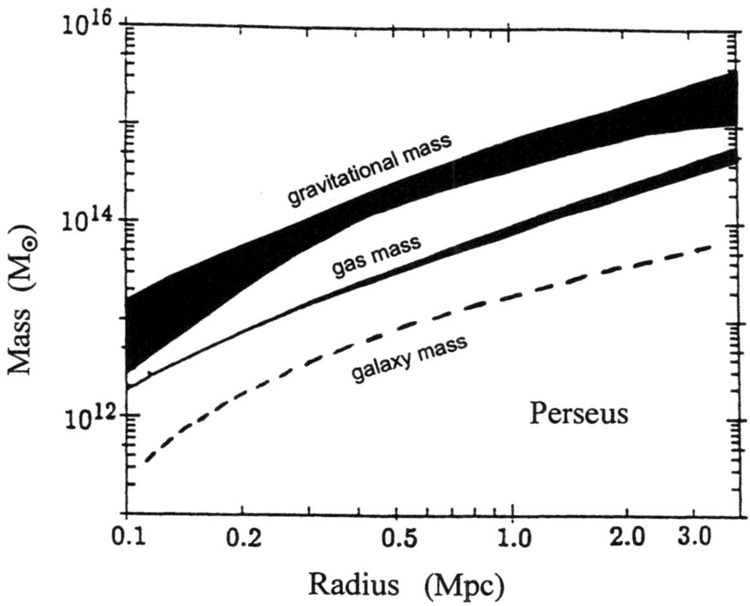

Fig. 10. Integrated radial profiles for the mass in the visible parts of galaxies, hot gas and total gravitating mass for the Perseus cluster of galaxies, as determined by observations with *ROSAT*. The upper band indicates the range of possible total masses and the central band the range of gaseous masses (Böhringer 1995).

in a number of clusters and the fact that the X-ray temperature is often lower in the centre than in the outer regions are convincing pieces of evidence for presence of cooling flows. At the high temperatures present in the intra-cluster gas, the principal cooling mechanism is bremsstrahlung, whose total energy loss rate is given by Eq. (87). Therefore, the cooling time scale τ_{cool} is

$$\tau_{\text{cool}} = \frac{E}{\left|\frac{\mathrm{d}E}{\mathrm{d}t}\right|} = \frac{3kT}{1.435 \times 10^{-40} Z^2 T^{1/2} \bar{g} N^2} = \frac{10^{10} T_8^{1/2}}{N_4} \text{ years} \tag{118}$$

where T_8 is the temperature in units of 10^8 K, N_4 is the number density of electrons measure in units of 10^4 m^{-3} and it is assumed that $Z = 1$ and $\bar{g} = 1$. Thus, it can be seen that, for the typical densities and temperatures found in the cores of clusters of galaxies, $T_8 \sim 1$, $N_e \geq 1$, the cooling time is likely to become less than the cosmological time scale and so a cooling flow develops in the central regions. A possible example of this phenomenon is seen in Fig. 11.

The observation of cooling flows is very important for many different aspects of the evolution of hot gas in clusters. First of all, the process adds mass to the central galaxy and may well be responsible for the formation of cool gas clouds in the central regions. In turn, this gas may well be responsible for fuelling the massive black holes present in the most massive galaxies in clusters. Notice that this is an example of a *thermal instability* in the sense that the gas cools, its pressure decreases and so the region is compressed by the overlying gas, which compresses the gas to a higher density, which increases the cooling rate .... This process can explain the origin of the intense X-ray emission observed in the very central regions of clusters, which would not be expected if the gas were distributed according to an isothermal density distribution.

5.5 The Sunyaev–Zeldovich effect in hot intra-cluster gas

A completely independent method of studying the hot gas in clusters of galaxies is to search for decrements in the intensity of the Cosmic Microwave Background Radiation in the directions of clusters of galaxies. As the photons of the background radiation pass through the gas cloud, a few of them suffer Compton scattering by the hot electrons. Although, to first order, the photons are just as likely to gain as lose energy in these Compton scatterings, to second order there is a net statistical gain of energy. Thus, the spectrum of the Cosmic Microwave Background Radiation is shifted to slightly higher energies and so, in the Rayleigh–Jeans region of the spectrum, there is expected to be a decrement in the intensity of the background radiation in the direction of the cluster of galaxies, while in the Wien region there should be a slight excess. These predictions were made by Sunyaev and Zeldovich as long ago as 1969 but it was almost 20 years before what has become known as the *Sunyaev–Zeldovich effect* was observed with confidence in the directions of clusters of galaxies, which were by then known to contain large masses of hot intra-cluster gas (Birkinshaw 1990).

The magnitude of the distortion is determined by the *Compton scattering optical depth* through the region of hot gas

$$y = \int \left(\frac{kT_{\mathrm{e}}}{m_{\mathrm{e}}c^2}\right) \sigma_{\mathrm{T}} N_{\mathrm{e}}\,\mathrm{d}l. \tag{119}$$

The resulting decrement in the Rayleigh–Jeans region of the spectrum is

$$\frac{\Delta I_\nu}{I_\nu} = -2y, \tag{120}$$

(for a discussion of the physical processes involved, see *HEA1*, Sect. 4.3.4). Thus, the magnitude of the decrement along any line of sight through the cluster provides a measure of the quantity $\int N_{\mathrm{e}} T_{\mathrm{e}}\,\mathrm{d}l$, in other words, the integral of the pressure of the hot gas along the line of sight. For the typical parameters of hot intra-cluster gas, the predicted decrement amounts to only $\Delta I/I \approx 10^{-4}$.

The most impressive maps of the decrements in the Cosmic Microwave Background Radiation have been obtained with the Ryle Telescope at Cambridge, the contour maps of the decrements observed in the Abell clusters A1413 and A1914 being shown in Figs. 11a and b, superimposed upon *ROSAT* maps of the X-ray surface brightness of the hot gas in these clusters (Courtesy of Drs. Michael Jones and Keith Grainge; see also, Jones et al. 1993). The contours defining the Sunyaev-Zeldovich decrements in both clusters follow closely the distribution of the X-ray emission of the hot cluster gas. In the case of Abell 1914, there is a region of high X-ray surface brightness to the left of the centre of the diffuse X-ray emission and this is interpreted as a cooling flow.

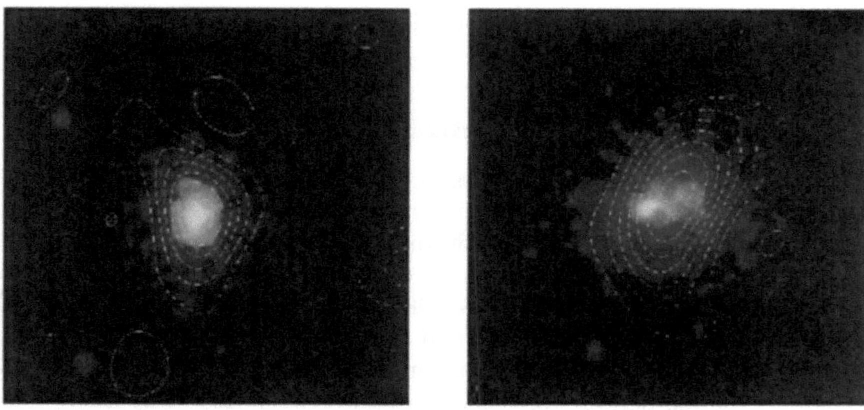

Abell 1413 **Abell 1914**

Fig. 11. Comparison of the decrements in the Cosmic Microwave Background Radiation with the distribution of X-ray emission for the rich Abell clusters (**a**) Abell 1413 and (**b**) Abell 1914 (Courtesy of Drs. Michael Jones and Keith Grainge).

In conjunction with X-ray bremsstrahlung measurements, these observations enable the physical conditions in the intra-cluster gas to be overdetermined and so the physical dimensions of the gas clouds can be estimated, without knowledge of the distance of the cluster. In conjunction with measurements of the angular sizes of the gas clouds, distances to the clusters can then be measured which are independent of their redshifts, enabling direct estimates of Hubble's constant to be made. Specifically, the X-ray surface brightness depends upon the electron density N_e and the electron temperature T_e through the relation $I_\nu \propto \int N_e^2 T_e^{-1/2} \mathrm{d}l$. The electron temperature T_e can be found from the shape of the bremsstrahlung spectrum. As discussed above, the magnitude of the decrement in the Cosmic Microwave Background Radiation in the direction of the cluster is proportional to the Compton optical depth $y = \int (kT_e/m_e c^2)\,\sigma_T N_e \mathrm{d}l \propto \int N_e T_e \mathrm{d}l$. Thus, the physical properties of the hot gas are over-determined and the physical dimensions of the X-ray emitting gas can be found. Myers et al. (1997) have estimated a value of $H_0 = 54 \pm 14$ km s^{-1} Mpc^{-1} from detailed studies of the Abell clusters A478, 2142 and 2256. Similar values are found from studies with the Ryle Telescope at Cambridge, which has now measured Sunyaev-Zeldovich decrements in 12 rich clusters. According to Dr. Richard Saunders, the clusters used in these studies must be selected with care. There are complications in the interpretation if the clusters are irregular, possess cooling flows or contain diffuse, non-thermal radio emission. If the clusters pass the selection criteria, he estimates that typically Hubble's constant can be measured to about 30% accuracy for an individual cluster, consistent with the findings of Myers et al. (1997).

5.6 The X-ray thermal bremsstrahlung of hot intergalactic gas

The diffuse intergalactic gas is almost fully ionised and one possibility is that it is very hot. For some years, there was the possibility that the isotropic X-ray background emission could be the X-ray bremsstrahlung of this gas. It was pointed out by Marshall et al. (1980) that the spectrum of the X-ray background emission in the 1 – 100 keV energy band can be very well described by the spectrum of thermal bremsstrahung at a temperature of $kT = 40$ keV (Fig. 12). If the emission originated from hot diffuse gas at redshift z, the temperature of this gas would be $kT = 40(1+z)$ keV.

There are, however, a number of problems with this seemingly simple explanation of the origin of the X-ray background emission. First of all, the number density of the hot gas can be evaluated using the formulae for the bremsstrahlung emissivity of the gas and it is found that the corresponding density parameter in baryons would be $\Omega_B h^2 \geq 0.23$. This value is significantly greater than the upper limit derived from studies of primordial nucleosynthesis, $\Omega_B h^2 \leq 0.036$. Second, in this picture, it is natural to attribute the high degree of ionisation of the gas to the same process responsible for heating the gas to a high temperature. Since the heating would have to take place at a large redshift, $z \geq 3$, the energy requirements for heating the gas would be very great. For example, the thermal energy density of the hot gas, if it had $\Omega_B = 0.23$ and temperature $kT = 160$ keV at a redshift $z = 3$, would be 7.6×10^7 eV m^{-3}, which is the same as the energy density of the Cosmic Microwave Background Radiation at that redshift. To put it in perspective, this energy corresponds to a density parameter $\Omega = 3.3 \times 10^{-3}$ at that redshift.

There is, however, a further serious concern. Such large quantities of hot gas would give rise to distortions of the Planck spectrum of the Cosmic Microwave Background Radiation. As first discussed by Sunyaev and Zeldovich,

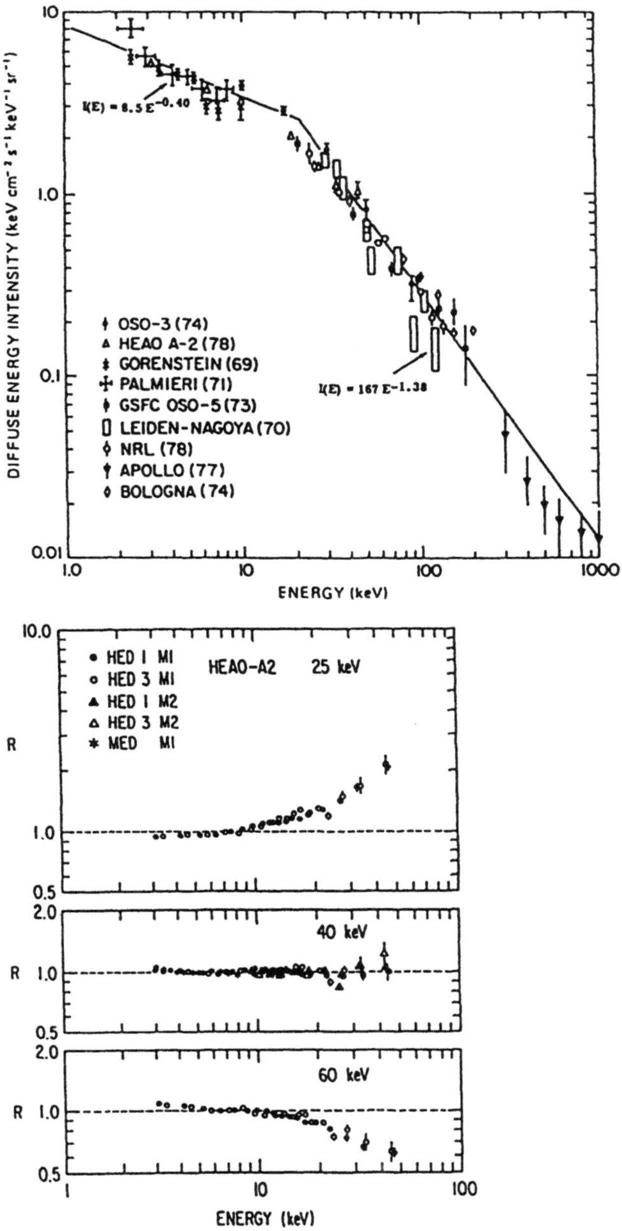

Fig. 12. Fits of bremsstrahlung spectra to the spectrum of the diffuse X-ray background in the energy range $1 < \varepsilon < 100$ keV. Adopting a temperature $kT = 40$ keV results in an excellent fit to the observations (Marshall et al. 1980).

Compton scattering of the photons of the background radiation leads to a characteristic distortion of the spectrum of the background radiation, by redistributing the photon energies. The perfect Planck spectrum of the background radiation as observed with *COBE* enabled Mather (1995) to set a powerful constraint to the Compton optical depth of hot diffuse intergalactic gas

$$y = \int \frac{kT_{\mathrm{e}}(z)}{m_{\mathrm{e}}c^2} \frac{\sigma_{\mathrm{T}} N_{\mathrm{e}}(z)}{(1+z)} \, \mathrm{d}r \leq 2.5 \times 10^{-5}. \tag{121}$$

The most conservative estimate we can make is to assume that the hot gas fills the intergalactic medium at the present epoch, in which case, we can make the approximation

$$y \approx \frac{c}{H_0} \sigma_{\mathrm{T}} N_{\mathrm{e}}(0) \frac{kT_{\mathrm{e}}(0)}{m_{\mathrm{e}}c^2}. \tag{122}$$

Assuming $\Omega_{\mathrm{B}} = 0.23$ and $kT_{\mathrm{e}} = 40$ keV, we find $y = 5 \times 10^{-4}$, already in significant conflict with expression (121). If we adopt a more realistic picture, in which the heating took place at a large redshift, the value of y would increase roughly as $(1+z)^{2.5}$, resulting in even greater conflict with the observations. More detailed models of the heating of the intergalactic gas, which can account for the observed X-ray background spectrum come to exactly the same conclusion (Taylor & Wright 1989).

Fabian & Barcons (1992) argue convincingly that, although the value of Ω_{B} can be reduced if the hot gas is clumped, the clumps would also result in large fluctuations in the Cosmic Microwave Background Radiation because of the thermal Sunyaev–Zeldovich effect, when the background radiation passes through the clumps. We can therefore exclude the possibility that the X-ray background is the diffuse emission of hot intergalactic gas. To complicate matters further, most of the X-ray background emission in the energy range $0.5 < \varepsilon < 2$ keV can be associated with discrete sources with energy spectra $I(\varepsilon) \propto \varepsilon^{-0.7}$ (Hasinger et al. 1993), and so this component of the background has to be subtracted from the X-ray background intensity at 1 keV. This has the effect of flattening the spectrum as compared with a thermal bremsstrahlung spectrum. Fabian & Barcons were led to conclude that 'the perfect bremsstrahlung shape of the X-ray background is just a cosmic conspiracy'.

5.7 The origin of the hard X-ray background

One obvious approach is to study the X-ray spectra of active galaxies in the energy range 1 – 20 keV in order to find out if any of them have spectra similar in form to that of the X-ray background. One possible clue comes from the spectra of a number of Seyfert-type galaxies observed with the Japanese *Ginga* satellite. Pounds et al. (1990) found evidence for significant distortions from a simple power law spectrum in this energy range when they added

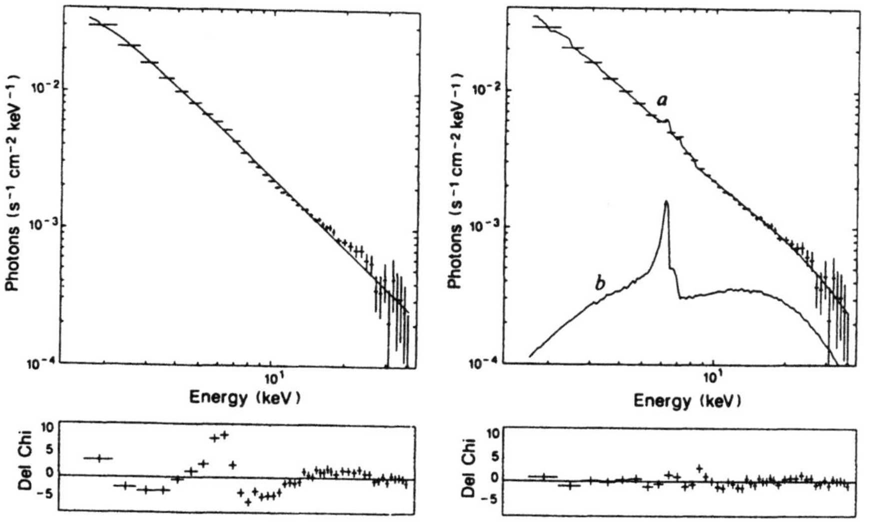

Fig. 13. (a) The summed X-ray spectra of 12 Seyfert galaxies compared with a best-fitting power law spectrum. The goodness-of-fit to this energy distribution is displayed in the lower panel in which it can be seen that there are significant deviations from a power law spectrum. (b) The addition of a reflection component to the power law distribution results in a much improved fit to the summed spectra (From Pounds et al. 1990).

together the spectra of a number of Seyfert galaxies (Fig. 13). Fig. 13(a) shows the summed spectrum for the 12 Seyfert galaxies and the residuals observed when a best-fitting power law is compared with the observations. Fig. 13(b) shows the improved fit obtained when a reflected X-ray component labelled b is added to the power law spectrum. Fabian et al. (1990) noted that the reflected component has a spectrum which could provide the type of spectral feature needed to account for the break in the spectrum of the background at about 30 keV.

The process of X-ray reflection involves illuminating a cool gas cloud with a power law X-ray energy spectrum and then working out the reflected X-ray spectrum. There are two competing processes, Compton scattering which is important at high energies and photoelectric absorption which becomes dominant at low X-ray energies. In a single Compton scattering of high-energy photons by stationary electrons, the average decrease in energy of each photon per Compton collision is

$$\left\langle \frac{\Delta\epsilon}{\epsilon} \right\rangle = \frac{h\nu}{m_e c^2}. \tag{123}$$

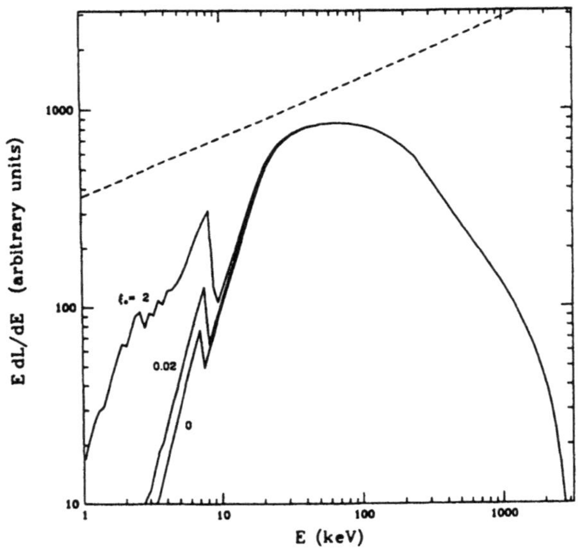

Fig. 14. The reflection spectra from a semi-infinite, plane-parallel cold medium assuming the cosmic abundance of the elements. The prominent feature at about 8 keV is associated with the K-absorption edge of iron. The input X-ray spectrum is of power law form $I(\varepsilon) \propto \varepsilon^{-0.7}$. The units on the ordinate are $\varepsilon I(\varepsilon)$ in our notation. ξ is the differential ionisation parameter which describes how the ionisation state of the cold cloud is modified by the incident X-ray spectrum. In the case $\xi = 0$, the matter remains cold. According to Lightman & White, the value $\xi = 2$ is at the upper limit of acceptable values for active galactic nuclei. (Lightman & White 1988).

Thus, the high-energy photons lose more energy per collision than the lower-energy photons, resulting in a progressive steepening of the spectrum with increasing energy. At low energies, photoelectric absorption is the dominant process and has the characteristic power law dependence of the absorption cross section $\sigma_{\rm ph}(\epsilon) \propto \epsilon^{-3}$ at energies greater than the characteristic absorption edges. The most important absorption edge in the 2 - 10 keV range is the K-edge of iron which occurs about 8 keV. Thus, at low X-ray energies there is strong absorption due to photoelectric absorption and a prominent feature at energies just greater than the K absorption edge of iron. The net result is that there is a maximum in the reflected spectrum at about 20 keV at which the combined energy losses due to Compton scattering and photoelectric absorption are at a minimum. To solve the problem properly, it is necessary to carry out a radiative-transfer analysis of the competing absorption and scattering processes and this has been carried out by Lightman & White (1988) whose results are shown in Fig. 14. It can be seen that qualitatively the reflected spectrum has the correct signature to account for the features observed in Seyfert galaxy spectra.

Another possibility for flattening the X-ray spectra in the $1 - 10$ keV waveband would be if the sources have strong photoelectric absorption within the nuclear regions. It is found that there there are about three times more sources in the $2 - 10$ keV waveband than expected from observations at $1 - 2$ keV and these could be sources with column depths of neutral hydrogen much greater than 10^{21} cm^{-2}. Some X-ray galaxies are known with very large column depths of neutral hydrogen, up to about $N_H \sim 10^{24}$ cm^{-2}, which means that the optical depth for photoelectric absorption would be about 1 at an energy of 10 keV (see, for example, Pounds 1990). Such sources would not be detected in sky surveys carried out at soft X-ray energies such as the *ROSAT* survey. The superposition of sources with different large column depths at different redshifts could lead to a flattening of the background relative to the typical X-ray spectra of active galaxies.

The problem of the origin of the hard X-ray background radiation is not resolved. Some of the many other possibilities are described in my review of all aspects of the extragalactic background radiation (Longair 1995).

6 Synchrotron Radiation

The synchrotron radiation of relativistic and ultra-relativistic electrons, the emission of very-high-energy electrons gyrating in a magnetic field, is the process which dominates high-energy astrophysics. It was originally observed in early betatron experiments in which electrons were first accelerated to ultra-relativistic energies. This process is responsible for the radio emission from the Galaxy, from supernova remnants and extragalactic radio sources. It is also responsible for the non-thermal optical emission observed in the Crab Nebula and possibly for the optical and X-ray continuum emission of quasars. The reasons for these assertions will become apparent in the course of this chapter.

The word *non-thermal* is used frequently in high-energy astrophysics to describe the emission of high-energy particles. I find this an unfortunate terminology since all emission mechanisms are 'thermal' in some sense. The word is conventionally taken to mean 'continuum radiation from particles, the energy spectrum of which is not Maxwellian'. In practice, continuum emission is often referred to as 'non-thermal' if it cannot be described by the spectrum of thermal bremsstrahlung or black-body radiation.

It is a very major undertaking to work out properly all the properties of synchrotron radiation. For details, I refer the enthusiast to the books by Bekefi (1966), Pacholczyk (1970) and Rybicki & Lightman (1979), and the review articles by Ginzburg and his colleagues. We will find that many of the most important results can be drived by simple arguments (see, for example,

Scheuer 1966). First of all, let us recall the motion of relativistic charged particles in magnetic fields.

6.1 Motion of an electron in a uniform, static magnetic field

We begin by writing down the equation of motion for a particle of rest mass m_0, charge ze and Lorentz factor $\gamma = (1 - v^2/c^2)^{-1/2}$ in a uniform static magnetic field $\mathbf{B}$.

$$\frac{\mathrm{d}}{\mathrm{d}t}(\gamma m_0 \mathbf{v}) = ze(\mathbf{v} \times \mathbf{B}) \tag{124}$$

We recall that the left-hand side of this equation can be expanded as follows:

$$m_0 \frac{\mathrm{d}}{\mathrm{d}t}(\gamma \mathbf{v}) = m_0 \gamma \frac{\mathrm{d}\mathbf{v}}{\mathrm{d}t} + m_0 \gamma^3 \mathbf{v} \frac{(\mathbf{v} \cdot \mathbf{a})}{c^2} \tag{125}$$

because the Lorentz factor γ should be written $\gamma = (1 - \mathbf{v} \cdot \mathbf{v}/c^2)^{-1/2}$. In a magnetic field, the three-acceleration $\mathbf{a} = \mathrm{d}\mathbf{v}/\mathrm{d}t$ is always perpendicular to $\mathbf{v}$ and consequently $\mathbf{v} \cdot \mathbf{a} = 0$. As a result,

$$\gamma m_0 \, \mathrm{d}\mathbf{v}/\mathrm{d}t = ze(\mathbf{v} \times \mathbf{B}) \tag{126}$$

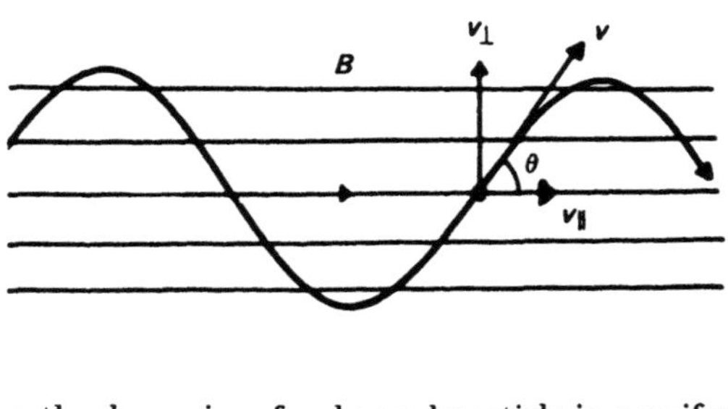

Fig. 15. Illustrating the dynamics of a charged particle in a uniform magnetic field.

We now split $\mathbf{v}$ into components parallel and perpendicular to the uniform magnetic field, $v_\parallel$ and $v_\perp$ respectively (Fig. 15). The pitch angle θ of the particle's path is given by $\tan\theta = v_\perp/v_\parallel$, that is, θ is the angle between the vectors $\mathbf{v}$ and $\mathbf{B}$. Since $v_\parallel \times \mathbf{B} = 0$, $v_\parallel =$ constant. The acceleration is perpendicular to the magnetic field direction and to $v_\parallel$.

$$\gamma m_0 \frac{\mathrm{d}\mathbf{v}}{\mathrm{d}t} = zev_\perp B(\mathbf{i}_\perp \times \mathbf{i_B}) = zevB(\mathbf{i_v} \times \mathbf{i_B}) \tag{127}$$

where $\mathbf{i_v}$ and $\mathbf{i_B}$ are unit vectors in the directions of $\mathbf{v}$ and $\mathbf{B}$ respectively.

Thus, the particle's acceleration vector is perpendicular to the plane containing both the instantaneous velocity vector $\mathbf{v}$ and the direction of the magnetic field **B**. Because the magnetic field is uniform, this constant acceleration perpendicular to the instantaneous velocity vector results in circular motion about the magnetic field. Equating this acceleration to the centripetal acceleration, we find

$$a_\perp = v_\perp^2/r = zevB\sin\theta/\gamma m_0, \quad \text{that is,} \quad r = \gamma m_0 v \sin\theta/zeB \qquad (128)$$

Thus, the motion of the particle consists of a constant velocity along the magnetic field direction and circular motion with radius r about it. This means that the particle moves in a *spiral path* with *constant pitch angle* θ. The radius r is known as the *gyroradius* of the particle. The angular frequency of the particle in its orbit ω_g is known as the *angular cyclotron frequency* or *angular gyrofrequency* and is given by

$$\omega_g = v_\perp/r = zeB/\gamma m_0 \qquad (129)$$

The corresponding *gyrofrequency* ν_g, that is, the number of times per second that the particle rotates about the magnetic field direction, is

$$\nu_g = \omega_g/2\pi = zeB/2\pi\gamma m_0 \qquad (130)$$

In the case of a non-relativistic particle, $\gamma = 1$ and hence $\nu_g = zeB/2\pi m_0$. A useful figure to remember is the non-relativistic gyrofrequency of an electron $\nu_g = eB/2\pi m_e = 28$ GHz T^{-1} where the magnetic field strength is measured in Tesla; alternatively, $\nu_g = 2.8$ MHz G^{-1} for those not yet converted from Gauss (G) to Teslas (T).

In this case, the axis of the particle's trajectory is parallel to the magnetic-field direction and this axis is known as the *guiding centre* of the particle's motion, that is, it is the mean direction of translation of the particle about which the gyration takes place. In more complicated magnetic field configurations, the particle follows the mean direction of the field, provided the field is slowly varying on the scale of a gyroradius, which is very often the case in high-energy astrophysical problems. The motion of the particle is then known as *guiding-centre motion* – more details of guiding-centre motion are given in *HEA1*, Chap. 11.

6.2 The total energy loss rate

Most of the essential tools needed in this analysis have already been derived in Sects. 3.1, 3.4 and 3.5. There are several ways of using them. First of all, we can go directly to Eq. (128) for the acceleration of the electron in its orbit and insert this into the expression for the radiation rate of a relativistic electron Eq. (9). In the present case, the acceleration is always perpendicular to the velocity vector of the particle and to **B** and hence $a_\parallel = 0$. Therefore, the total radiation loss rate of the electron is

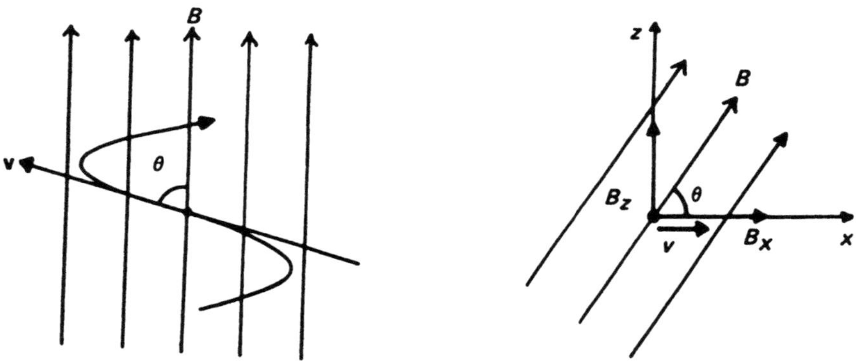

Fig. 16. The coordinates used in working out the total radiation rate due to synchrotron radiation.

$$-\left(\frac{\mathrm{d}E}{\mathrm{d}t}\right) = \frac{\gamma^4 e^2}{6\pi\varepsilon_0 c^3}|a_\perp|^2 = \frac{\gamma^4 e^2}{6\pi\varepsilon_0 c^3}\frac{e^2 v^2 B^2 \sin^2\theta}{\gamma^2 m_\mathrm{e}^2} \tag{131}$$

$$= \frac{e^4 B^2}{6\pi\varepsilon_0 c m_\mathrm{e}^2}\,\frac{v^2}{c^2}\,\gamma^2 \sin^2\theta \tag{132}$$

Another pleasant way of arriving at the same result, and revising some of the rules we have already derived, is to start from the fact that in the instantaneous rest frame of the electron, the acceleration of the particle is small and therefore in that frame we can use the non-relativistic expression for the radiation rate. Let us choose the coordinate system shown in Fig. 16. The instantaneous direction of motion of the electron in the laboratory frame, the frame in which **B** is fixed, is taken as the positive x-axis. Then, to find the force acting on the particle, we transform the field quantities into the instantaneous rest frame of the electron using the standard relativistic transformations for the magnetic flux density **B**. In S′, the force on the electron is

$$\mathbf{f}' = m_\mathrm{e}\dot{\mathbf{v}}' = e(\mathbf{E}' + \mathbf{v}' \times \mathbf{B}') = e\mathbf{E}' \tag{133}$$

since the particle is instantaneously at rest in S', $\mathbf{v}' = 0$. Therefore, in transforming the magnetic flux density **B** into S′, we need only consider the transformed components of the electric field **E**′.

$$E'_x = E_x \qquad\qquad E'_x = 0 \tag{134}$$

$$E'_y = \gamma(E_y - vB_z) \quad \text{and hence} \quad E'_y = -v\gamma B_z = -v\gamma B\sin\theta \tag{135}$$

$$E'_z = \gamma(E_z + vB_y) \qquad\qquad E'_z = 0 \tag{136}$$

Therefore

$$\dot{\mathbf{v}}' = -\frac{e\gamma v B\sin\theta}{m_\mathrm{e}} \tag{137}$$

Consequently, in the rest frame of the electron, the loss rate by radiation is

$$-\left(\frac{\mathrm{d}E}{\mathrm{d}t}\right)' = \frac{e^2|\dot{\mathbf{v}}'|^2}{6\pi\varepsilon_0 c^3} = \frac{e^4\gamma^2 B^2 v^2 \sin^2\theta}{6\pi\varepsilon_0 c^3 m_\mathrm{e}^2} \tag{138}$$

Since $(\mathrm{d}E/\mathrm{d}t)$ is a Lorentz invariant (Sect 3.4), we recover the formula (132). Let us rewrite this in the following way

$$-\left(\frac{\mathrm{d}E}{\mathrm{d}t}\right) = 2\left(\frac{e^4}{6\pi\varepsilon_0^2 c^4 m_\mathrm{e}^2}\right)\left(\frac{v}{c}\right)^2 c\,\frac{B^2}{2\mu_0}\,\gamma^2\,\sin^2\theta \tag{139}$$

The quantity in the first set of round brackets on the right-hand side of Eq. (139) is the Thomson cross section σ_T (see Sect. 3.2). We have also used the relation $c^2 = (\mu_0\varepsilon_0)^{-1}$. Therefore

$$-\left(\frac{\mathrm{d}E}{\mathrm{d}t}\right) = 2\sigma_\mathrm{T} c U_\mathrm{mag}\left(\frac{v}{c}\right)^2 \gamma^2 \sin^2\theta \tag{140}$$

where $U_\mathrm{mag} = B^2/2\mu_0$ is the energy density of the magnetic field. In the ultra-relativistic limit, $v \to c$, we may approximate this result by

$$-\left(\frac{\mathrm{d}E}{\mathrm{d}t}\right) = 2\sigma_\mathrm{T} c U_\mathrm{mag}\gamma^2 \sin^2\theta \tag{141}$$

These results apply for electrons of a specific pitch angle θ. Particles of a particular energy E, or Lorentz factor γ, are often expected to have an isotropic distribution of pitch angles and therefore we can work out their average energy loss rate by averaging over such a distribution of pitch angles $p(\theta)\,\mathrm{d}\theta = \frac{1}{2}\sin\theta\,\mathrm{d}\theta$

$$-\left(\frac{\mathrm{d}E}{\mathrm{d}t}\right) = 2\sigma_\mathrm{T} c U_\mathrm{mag}\gamma^2\left(\frac{v}{c}\right)^2 \frac{1}{2}\int_0^\pi \sin^3\theta\,\mathrm{d}\theta \tag{142}$$

$$= \tfrac{4}{3}\sigma_\mathrm{T} c U_\mathrm{mag}\left(\frac{v}{c}\right)^2\gamma^2 \tag{143}$$

Notice that there is a deeper sense in which expression (143) is the average loss rate for a particle of energy E. During its lifetime, it is likely that the high-energy particle is randomly scattered in pitch angle and then (143) is the correct expression for its average energy loss rate.

6.3 Non-relativistic gyroradiation and cyclotron radiation

Before tackling the spectral distribution of synchrotron radiation, let us make a short diversion into the non-relativistic and mildly relativistic cases which are of considerable interest astrophysically. Consider first of all the simplest case of non-relativistic gyroradiation, in which case $v \ll c$ and hence $\gamma = 1$. Then, the expression for the loss rate of the electron becomes

$$-\left(\frac{\mathrm{d}E}{\mathrm{d}t}\right) = 2\sigma_T c U_\mathrm{mag}\left(\frac{v}{c}\right)^2 \sin^2\theta = \frac{2\sigma_T}{c} U_\mathrm{mag} v_\perp^2 \tag{144}$$

and the radiation is emitted at the gyrofrequency of the electron $\nu_{\rm g} = eB/2\pi m_{\rm e}$.

One interesting aspect of this emission mechanism is the fact that its polarisation properties are quite distinctive. In the non-relativistic case, there are no beaming effects and thus what is observed by the distant observer can be derived from the simple rules given at the end of Sect. 3.1. When the magnetic field is perpendicular to the line of sight, *linearly polarised radiation* is observed because the acceleration vector is observed to perform simple harmonic motion in a plane perpendicular to the magnetic field by the distant observer. The electric field strength varies sinusoidally at the gyrofrequency as the dipole distribution of radiation sweeps past the observer. On the other hand, when the magnetic field is parallel to the line of sight, the acceleration vector is continually changing direction as the electron moves in a circular orbit about the magnetic field lines and therefore the radiation is observed to be 100% *circularly polarised*. When observed at an arbitrary angle θ to the magnetic field direction, the radiation is observed to be *elliptically polarised*, the ratio of axes of the polarisation ellipse being $\cos\theta$.

One of the most remarkable applications of gyroradiation is to the binary X-ray source Hercules X-1. Hard X-ray spectral observations have discovered what is referred to as a *cyclotron absorption feature* at 34 keV (Fig. 17). If this feature is attributed to absorption at the gyrofrequency of hot gas in the vicinity of the poles of the magnetised neutron star, an estimate of the magnetic flux density can be found. Inserting 34 keV into the formula for the gyrofrequency, we find a magnetic field strength of 3×10^8 T, a very strong field indeed but within the range of values found from studies of the rate of slow down of radio pulsars. Thus, it is entirely plausible that this is an example of absorption at the gyrofrequency in the intense magnetic fields present in magnetised neutron stars.

The other example concerns mildly relativistic cyclotron radiation in which account has to be taken of the beaming of the radiation. Even for slowly moving electrons, $v \ll c$, not all the radiation is emitted at the gyrofrequency because there are small aberration effects which slightly distort the observed angular distribution of the intensity from a simple $\cos^2\theta$ law. From the symmetry of these aberrations, it can be seen that the observed polar diagram of the radiation may be decomposed by Fourier analysis into a sum of equivalent dipoles radiating at harmonics of the relativistic gyrofrequency $\nu_{\rm r}$ where $\nu_{\rm r} = \nu_{\rm g}/\gamma$. These harmonics have frequencies

$$\nu_l = l\nu_{\rm r}/(1 - \frac{v_{\parallel}}{c}\cos\theta) \tag{145}$$

where l takes integral values, $l = 1, 2, 3, ..$ and the fundamental gyrofrequency has $l = 1$. The factor $[1-(v_{\parallel}/c)\cos\theta]$ in the denominator takes account of the Doppler shift of the radiation of the electron due to its translational velocity along the field lines $v_{\parallel}$, projected onto the line of sight to the observer. In the limit $lv/c \ll 1$, it can be shown that the total power emitted in a given

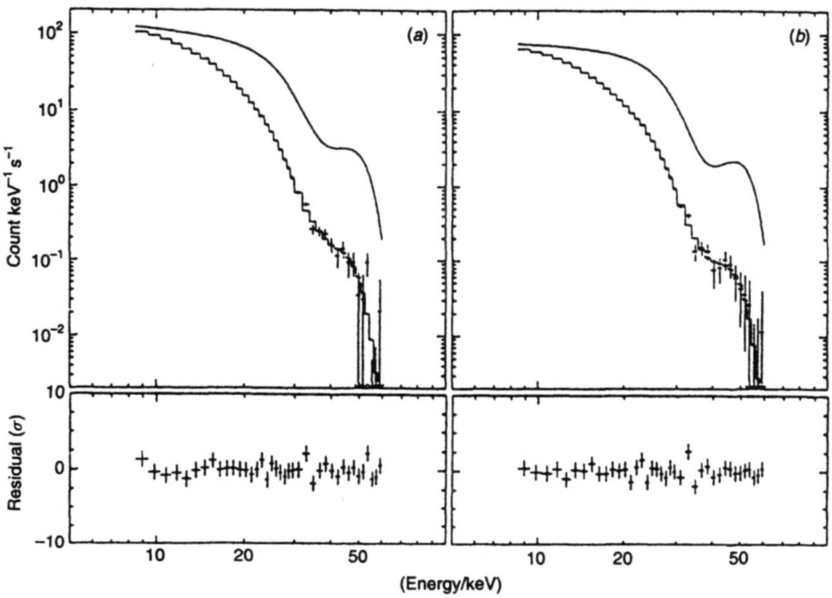

Fig. 17. The hard X-ray spectrum of the binary X-ray source Hercules X-1, as observed with the *Ginga* satellite, showing the cyclotron absorption feature at 34 keV. The X-ray source has a pulse period of 1.24 s, and in (**a**) the spectrum observed at pulse maximum is shown. In (**b**), the spectrum is derived from the difference between the spectra observed at pulse maximum and pulse minimum. The lower panels show the residuals obtained in model-fitting the observations. (Mihara et al. 1990.)

harmonic for the case $v_{\parallel} = 0$ is

$$-\left(\frac{\mathrm{d}E}{\mathrm{d}t}\right)_l = \frac{2\pi e^2 \nu_{\mathrm{g}}^2}{\varepsilon_0 c} \frac{(l+1)l^{2l+1}}{(2l+1)!} \left(\frac{v}{c}\right)^2 \tag{146}$$

and hence, to order of magnitude,

$$\left(\frac{\mathrm{d}E}{\mathrm{d}t}\right)_{l+1} \bigg/ \left(\frac{\mathrm{d}E}{\mathrm{d}t}\right)_l \approx \left(\frac{v}{c}\right)^2 \tag{147}$$

Thus, the energy radiated in high harmonics is small when the particle is non-relativistic. Notice that the loss rate (146) reduces to (144) for $l = 1$.

When the particle becomes significantly relativistic, $v/c \geq 0.1$, the energy radiated in the higher harmonics becomes important. The Doppler corrections to the observed frequency of the emitted radiation become significant and a wide spread of emitted frequencies is associated with the different pitch angles of an electron of energy $E = \gamma mc^2$. The result is broadening of the width of the emission line of a given harmonic and, for high harmonics, the lines are so broadened that the emission spectrum becomes continuous rather than consisting of a series of harmonics at well defined frequencies. The results of calculations for a relativistic plasma having $kT_e/m_ec^2 = 0.1$, corresponding to $\gamma = 1.1$ and $v/c \approx 0.4$, are shown in Fig. 18. The spectra of the first twenty harmonics are shown as well as the total emission spectrum found by summing the spectra of the individual harmonics. One way of looking at synchrotron radiation is to consider it as the relativistic limit of the process illustrated in Fig. 18 in which all the harmonics are washed out and a smooth continuum spectrum is observed.

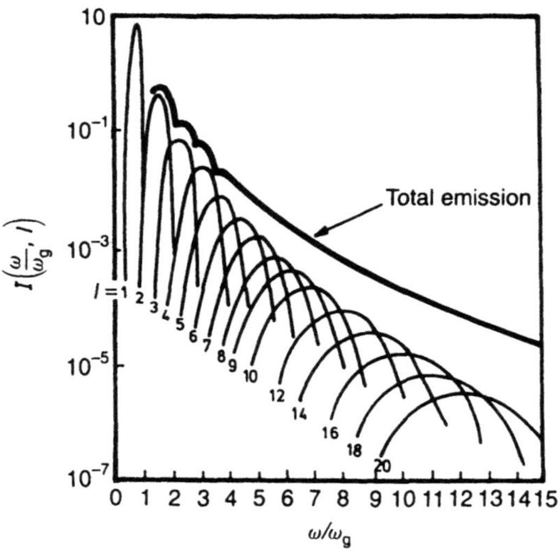

Fig. 18. The spectrum of emission of the first 20 harmonics of mildly relativistic cyclotron radiation. The electron has $v = 0.4c$ (Bekefi 1966).

Just as in the case of gyroradiation, the harmonics of cyclotron radiation are circularly polarised. This means that, if the elliptical polarisation of the radiation from a celestial object can be measured in detail, it is possible to learn a great deal, not only about the strength of the magnetic field but also about its orientation with respect to the line of sight. Circularly polarised optical emission has been discovered in the eclipsing magnetic binary stars known as *AM Herculis binaries*, or *polars*, circular polarisation percentages as large as 40% being observed. In these systems, a red dwarf star orbits a white dwarf with a very strong magnetic field. Accretion of matter from the surface of the red dwarf onto the magnetic poles of the white dwarf results in the heating of matter to temperatures in excess of 10^7 K. Thus, in addition to radiating X rays, these objects are strong sources of cyclotron radiation. Fields of order 2000 T have been found in these objects and hence the fundamental gyrofrequency is expected to correspond to a wavelength of about 5μm. Very often the individual harmonics of the cyclotron radiation are washed out but it has been possible to distinguish them in the X-ray source EXO 033319–2554.2 (Fig. 19). The frequency spacing between the harmonics has enabled an accurate estimate of the magnetic flux density to be made. This turns out to be 5600 T, the largest known magnetic flux density in an AM Herculis system.

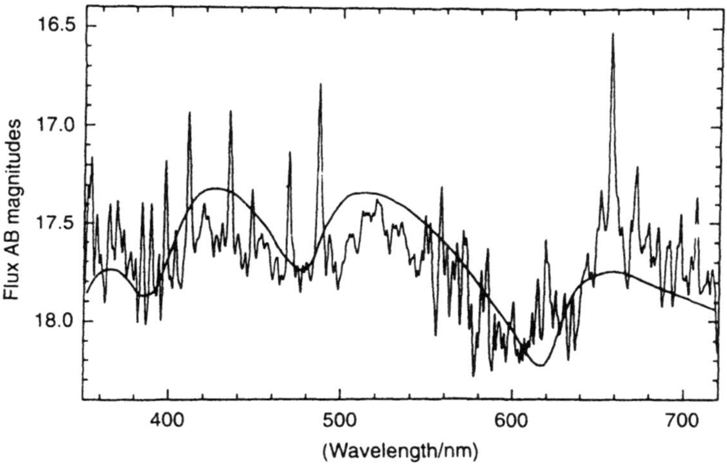

Fig. 19. A broad-band spectrum of the AM Herculis object EXO 033319–2554.2 which is a soft X-ray source. The presence of a strong magnetic field is inferred from the observation of strongly circularly polarised emission. The solid line shows a best fit of the cyclotron emission spectrum to the broad cyclotron harmonics at 420, 520 and 655 nm. The inferred strength of the magnetic field is 5600 T. (Ferrario et al. 1989.)

6.4 The spectral distribution of radiation from a single electron - physical arguments

The next step is to work out the spectral distribution of synchrotron radiation but an exact analysis requires very much more effort. Let us analyse first of all some basic aspects of radiation mechanisms involving relativistic particles which will prove to be invaluable in understanding where many of the exact results come from.

One of the basic features of the radiation of relativistic particles in general is the fact that the radiation is *beamed* in the direction of motion of the particle. Part of this effect is associated with the relativistic aberration formulae between the frame of reference of the particle and the observer's frame of reference. There are, however, subtleties about what is actually observed by the distant observer because, in addition to aberration, we have to consider carefully the time development of what is seen by the distant observer.

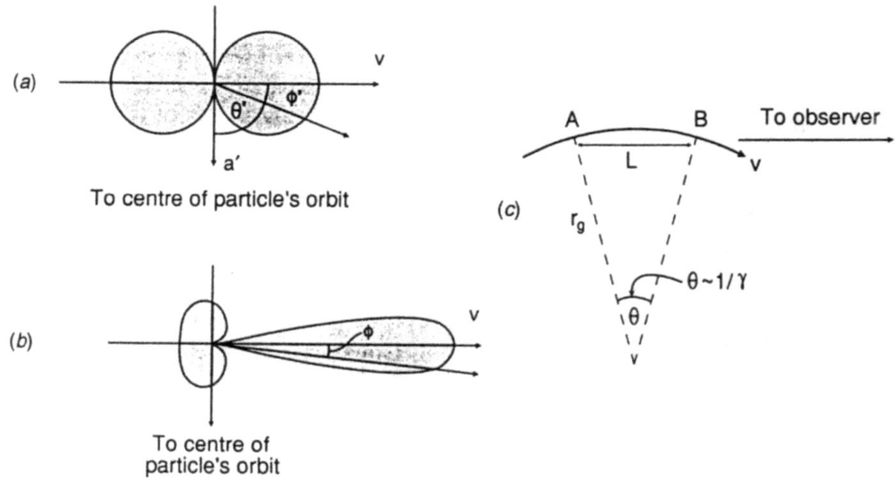

Fig. 20. Illustrating the relativistic beaming and Doppler effects associated with synchrotron radiation. (**a**) The polar diagram of the dipole radiation of the electron in its instantaneous rest frame. (**b**) The polar diagram of the radiation transformed into the laboratory frame of reference. (**c**) The geometry of the path of the electron when the radiation is observed by the distant observer.

Let us consider first the simple case of a particle gyrating about the magnetic field at a pitch angle of 90°. The electron is accelerated towards its guiding centre, that is, radially inwards, and in its instantaneous rest frame it emits the usual dipole pattern with respect to the acceleration vector. This is illustrated in Fig. 20a. We can therefore work out the radiation pattern in the laboratory frame of reference by applying the aberration formulae with the results illustrated schematically in Fig. 20b. As discussed in Sect. 3.1, the angular distribution of radiation with respect to the velocity vector in the frame S′ is $I_\nu \propto \sin^2\theta' = \cos^2\phi'$. We may think of this as being the probability distribution with which photons are emitted by the electron in its rest frame. The appropriate aberration formulae between the two frames are:

$$\sin\phi = \frac{1}{\gamma}\frac{\sin\phi'}{1+(v/c)\cos\phi'} \quad ; \quad \cos\phi = \frac{\cos\phi' + v/c}{1+(v/c)\cos\phi'} \tag{148}$$

To illustrate the beaming of the radiation, let us consider the angles $\phi' = \pm\pi/4$, these being the angles at which the intensity of radiation falls to half its maximum value in the instantaneous rest frame of the particle. The corresponding angles ϕ in the laboratory frame of reference are

$$\sin\phi \approx \phi \approx 1/\gamma \tag{149}$$

Thus, the radiation emitted within $-\pi/4 < \phi' < \pi/4$ is beamed in the direction of motion of the electron within an angle $-1/\gamma < \phi < 1/\gamma$. As observed in the frame S, the dipole beam pattern is very strongly distorted and the intensity of the radiation is strongly Doppler-shifted; a more detailed discussion of these aberration and redshift effects is given in Chap. 10. Only when this elongated beam pattern sweeps past the observer is a significant amount of radiation observed. A large 'spike' of radiation is observed every time the electron's velocity vector lies within an angle of about $1/\gamma$ to the line of sight to the observer. This analysis shows why the observed frequency of the radiation must be very much higher than the gyrofrequency of the electron in its orbit. The spectrum of the radiation received by the distant observer is the Fourier transform of this pulse once the effects of time retardation and aberration are taken into account. This phenomenon is illustrated in Fig. 20b which shows how the observed radiation pattern narrows as the velocity of the particle increases.

Thus, the observer sees significant radiation from only about $1/\gamma$ of the particle's orbit but the observed duration of the pulse is less than $1/\gamma$ times the period of the orbit because radiation emitted at the trailing edge of the pulse almost catches up with the radiation emitted at the leading edge. Let us illustrate this second effect by a simple calculation which is carried out entirely in the laboratory frame of reference and which concerns the time of arrival of signals at the distant observer. The portion of the particle's orbit from which radiation is received by the distant observer is shown in Fig. 20c. Consider the observer located at a distance R from the point A. The radiation from A reaches the observer at time R/c. Now consider the radiation emitted from B at time L/v later which then has to travel a distance $(R - L)$ at the speed of light to reach the observer. The trailing edge of the pulse therefore arrives at the observer at a time $L/v + (R - L)/c$. The duration

$$\Delta t = \left[\frac{L}{v} + \frac{(R-L)}{c}\right] - \frac{R}{c} = \frac{L}{v}\left(1 - \frac{v}{c}\right) \tag{150}$$

Notice that the observed duration of the pulse is much less than the value L/v which might naively have been expected. Only if light propagated at an infinite velocity would the duration of the pulse be L/v. The intriguing point about this analysis is that this factor $1 - (v/c)$ is exactly the same factor which appears in the Liénard-Weichert potentials (Eq. 42) and takes account of the fact that the source of radiation is not stationary but is moving towards the observer. In fact, the relativistic particle almost catches up with the radiation emitted at A since $v \approx c$, but not quite. Now we can rewrite the above expression noting that

$$\frac{L}{v} = \frac{r_g \theta}{v} \approx \frac{1}{\gamma \omega_r} = \frac{1}{\omega_g} \tag{151}$$

where ω_g is the non-relativistic angular gyrofrequency and $\omega_r = \omega_g/\gamma$. We also note that we can rewrite $(1 - v/c)$ as

$$\left(1 - \frac{v}{c}\right) = \frac{[1-(v/c)]\,[1+(v/c)]}{[1+(v/c)]} = \frac{(1 - v^2/c^2)}{1+(v/c)} \approx \frac{1}{2\gamma^2} \tag{152}$$

since $v \approx c$. Therefore, the observed duration of the pulse is roughly

$$\Delta t \approx \frac{1}{2\gamma^2 \omega_g} \tag{153}$$

This means that the duration of the pulse as observed by a distant observer in the laboratory frame of reference is roughly $1/\gamma^2$ times shorter than the non-relativistic gyroperiod $T_g = 2\pi/\omega_g$. The maximum Fourier component of the spectral decomposition of the observed pulse of radiation is expected to correspond to a frequency $\nu \sim \Delta t^{-1}$, that is,

$$\nu \sim \Delta t^{-1} \sim \gamma^2 \nu_g \tag{154}$$

where ν_g is the non-relativistic gyrofrequency.

In the above analysis, it has been assumed that the particle moves in a circle about the magnetic field lines, that is, the pitch angle θ is 90°. The same calculation can be performed for any pitch angle and then the result becomes

$$\nu \sim \gamma^2 \nu_g \sin\theta \tag{155}$$

The reason for performing this simple exercise in detail is that the beaming of the radiation of ultra-relativistic particles is a very general property and does not depend upon the nature of the force causing the acceleration.

Returning to an earlier part of the calculation, the observed frequency of the radiation can also be written

$$\nu \approx \gamma^2 \nu_g = \gamma^3 \nu_r = \frac{\gamma^3 v}{2\pi r_g} \tag{156}$$

where ν_r is the relativistic gyrofrequency and r_g is the radius of curvature of the particle's orbit. Notice that, in general, we may interpret r_g as the instantaneous radius of curvature of the particle's orbit and v/r_g is the angular frequency associated with it. This is a useful result because it enables us to work out the frequency at which most of the radiation is emitted, provided we know the radius of curvature of the particle's orbit. The frequency of the observed radiation is roughly γ^3 times the angular frequency v/r where r is the instantaneous radius of curvature of the particle in its orbit. This result is important in the study of *curvature radiation* which has important applications in the emission of radiation from the magnetic poles of pulsars.

For many order-of-magnitude calculations it is sufficient to know that the energy loss rate of the relativistic electron is given by Eq. (132) or (141) and that most of the radiation is emitted at a frequency which is roughly $\nu = \gamma^2 \nu_g$ where ν_g is the non-relativistic gyrofrequency. However, we often have to do somewhat better than this and that is the subject of the next section.

6.5 The spectrum of synchrotron radiation - improved version

I am not aware of any particularly simple way of deriving the spectral distribution of synchrotron radiation and I do not find the analysis particularly appealing. The analysis proceeds by the following steps:

1. Write down the expression for the energy emitted per unit bandwidth for an arbitrarily moving electron;
2. Select a suitable set of coordinates in which to work out the field components radiated by the electron spiralling in a magnetic field;
3. Then battle away at the algebra to obtain the spectral distribution of the field components.

This is a rather horrid calculation which is summarised by Rybicki & Lightman (1980) and in *HEA2*. Let me simply quote the results of these calculations. The emitted spectrum of a single electron, averaged over the particle's orbit is

$$j(\omega) = j_{\perp}(\omega) + j_{\parallel}(\omega) = \frac{\sqrt{3}e^3 B \sin\theta}{8\pi^2 \varepsilon_0 c m_e} F(x) \tag{157}$$

where

$$x = \omega/\omega_c, \quad \omega_c = \tfrac{3}{2}\left(\frac{c}{v}\right)\gamma^3 \omega_r \sin\theta \quad \text{and} \quad F(x) = x \int_x^{\infty} K_{5/3}(z)\,\mathrm{d}z. \tag{158}$$

ω_c is known as the critical angular frequency. In this expression, $K_{5/3}(z)$ is a modified Bessel function of order 5/3. In this form, the emissivity is expressed in terms of the angular frequency ω. The form of this spectrum is shown in linear and logarithmic form in Fig. 21. This form of spectrum confirms the physical arguments given in Sect. 6.4. It has a broad maximum centred roughly at the frequency $\nu \approx \nu_c$ with $\Delta\nu/\nu \sim 1$. The maximum of the emission spectrum in fact has value $\nu_{\rm max} = 0.29\nu_c$. The high-frequency emissivity of the electron is given by an expression of the form

$$j(\nu) \propto \nu^{1/2} e^{-\nu/\nu_c} \tag{159}$$

which is dominated by the exponential cut-off at frequencies $\nu \gg \nu_c$. This simply means that there is very little power at frequencies $\nu > \nu_c$ which can

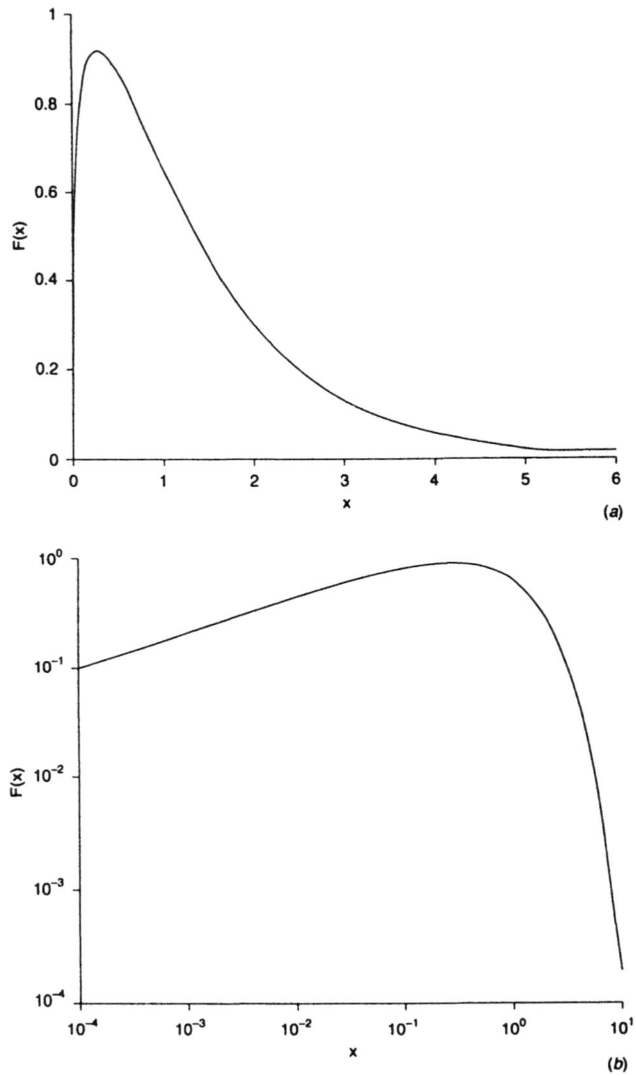

Fig. 21. The intensity spectrum of the synchrotron radiation of a single electron shown (**a**) with linear axes, and (**b**) with logarithmic axes. The function is plotted in terms of $x = \omega/\omega_c = \nu/\nu_c$, where ω_c is the critical angular frequency.

be understood on the basis of the physical arguments developed in Sect. 6.4 – there is very little structure in the observed polar diagram of the radiation emitted by the electron at angles $\theta \ll \gamma^{-1}$. At low frequencies, $\nu \ll \nu_c$, the spectrum is given by $j(\nu) \propto \nu^{1/3}$. A very pleasant argument given by Scheuer (1966) explains the origin of this dependence, which is repeated in *HEA2*.

The ratio of the powers emitted in the polarisations parallel and perpendicular to the magnetic field direction is

$$\frac{I_\perp}{I_\parallel} = 7. \tag{160}$$

To find the polarisation observed from a distribution of electrons at a particular observing frequency, however, we need to integrate over the energy spectrum of the emitting electrons and that is our next task.

6.6 The synchrotron radiation of a power law distribution of electron energies

We have shown that the emitted spectrum of electrons of energy E is quite sharply peaked near the critical frequency ν_c (Fig. 21) and is certainly very much narrower than the breadth of the electron energy spectrum. Therefore, to a good approximation, it may be assumed that all the radiation of an electron of energy E is radiated at the critical frequency ν_c which we may approximate by

$$\nu \approx \nu_c \approx \gamma^2 \nu_g = \left(\frac{E}{m_e c^2}\right)^2 \nu_g; \qquad \nu_g = \frac{eB}{2\pi m_e}. \tag{161}$$

Therefore, the energy radiated in the frequency range ν to $\nu + \mathrm{d}\nu$ can be attributed to electrons with energies in the range E to $E + \mathrm{d}E$, which we assume to have power law form $N(E) = \kappa E^{-p}$. We may therefore write

$$J(\nu)\,\mathrm{d}\nu = \left(-\frac{\mathrm{d}E}{\mathrm{d}t}\right) N(E)\,\mathrm{d}E. \tag{162}$$

Now

$$E = \gamma m_e c^2 = \left(\frac{\nu}{\nu_g}\right)^{1/2} m_e c^2, \tag{163}$$

$$\mathrm{d}E = \frac{m_e c^2}{2\nu_g^{1/2}} \nu^{-1/2} \mathrm{d}\nu, \tag{164}$$

$$-\left(\frac{\mathrm{d}E}{\mathrm{d}t}\right) = \tfrac{4}{3}\sigma_T c \left(\frac{E}{m_e c^2}\right)^2 \frac{B^2}{2\mu_0}. \tag{165}$$

Substituting these quantities into Eq. (162), we find the emissivity may be expressed in terms of κ, B, ν and fundamental constants.

$$J(\nu) = (\text{constants})\ \kappa B^{(p+1)/2} \nu^{-(p-1)/2}. \tag{166}$$

We note the important result that, if the electron energy spectrum has power law index p, the spectral index of the synchrotron emission of these electrons, defined by $J(\nu) \propto \nu^{-\alpha}$, is $\alpha = (p-1)/2$. The spectral shape is determined by the shape of the electron energy spectrum rather than by the shape of the emission spectrum of a single particle. The quadratic nature of the relation between emitted frequency and the energy of the electron accounts for the difference in slopes of the emission spectrum and the electron energy spectrum.

6.7 Why is synchrotron radiation taken so seriously?

Synchrotron radiation dominates a great deal of thinking in high-energy astrophysics and it is important to assess how convincing the evidence is.

- Perhaps the most important evidence comes from the comparison of the local flux of relativistic electrons measured at the top of the atmosphere with the predicted synchrotron radiation intensity if that flux of particles were present throughout the interstellar medium adopting the best estimate of the strength of the interstellar magnetic field. This turns out to be a much more difficult task than might be imagined. The big problem is that the observed electron spectrum is strongly modified at energies $E < 10$ GeV by the effects of *solar modulation*. By this is meant that the fluxes of low-energy electrons are strongly modified by scattering by irregularities in the interplanetary magnetic field. My own analysis of the comparison of the radio spectral emissivity of the interstellar medium with the predicted spectrum on the basis of the unmodulated part of the local electron spectrum and various estimates of the strength of the local magnetic field, normalised to a magnetic-flux density $B = 3 \times 10^{-10} x$ T, is shown in Fig. 22, for values of $x = 0.5$, 1 and 2. Another way of presenting this analysis is shown in Fig. 23, namely, in terms of the differential electron spectrum which is presented in the form $E^3 N(E)$. The hatched area shows the observed electron spectrum and it can be seen how the effects of solar modulation modify the intrinsic spectrum of the electrons in the energy range $0.5 < E < 10$ GeV. Also shown by a dashed line is the spectrum of low-energy electrons as inferred from the $1 - 50$ MeV γ-ray spectrum on the assumption that the emission is the bremsstrahlung of relativistic electrons. Thus, a convincing case can be made that the Galactic radio emission is the synchrotron radiation of ultra-relativistic electrons, adopting reasonable values for the local magnetic flux density.
- The next question is the origin of the high-energy electrons, and a convincing case can be made that they are accelerated in supernova remnants. These are observed to be very strong radio sources with power law intensity spectra and the radio emission is linearly polarised. These

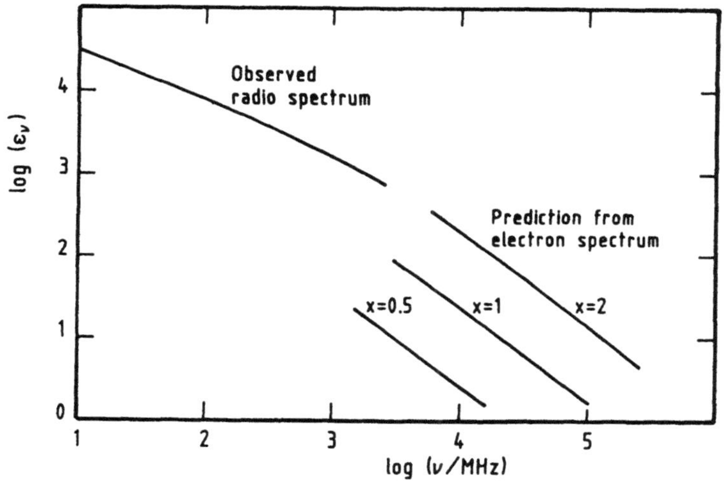

Fig. 22. Comparison of the radio emissivity of the interstellar medium as estimated from observations of the Galactic radio emission and the predicted emission on the basis of the local flux of high-energy electrons and estimates of the interstellar magnetic field strength, $B = 3 \times 10^{-10} x$ T (*HEA2*, Sect. 18.2).

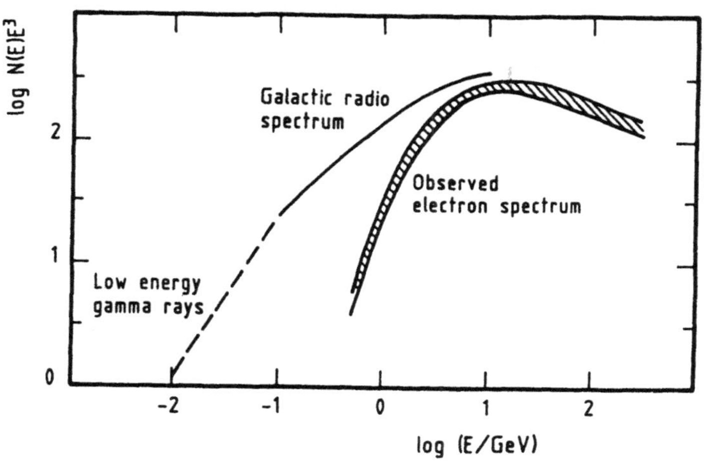

Fig. 23. The inferred interstellar electron spectrum from direct observations (shaded area), from the spectrum of the Galactic radio emission (solid line) and the low-energy γ-ray emission (dashed line) (*HEA2*, Sect. 18.2).

properties are similar to those of the diffuse radiation of the interstellar medium. Combining the frequency of occurrence of supernovae in our Galaxy with the typical energies they release in high-energy particles, it is quite feasible to account for the energy requirements of the Galactic radio emission.

- The logical extension of this argument is to the intense extragalactic radio sources, which have qualitatively the same properties of power law radio spectra and polarised radiation, but with intrinsic luminosities which are up to 10^8 greater than that of our own Galaxy. Moreover, the radio emission originates from enormous radio lobes rather than from the galaxy itself. Fig. 24 is a radio map of Cygnus A, the brightest extragalactic radio source in the northern sky, showing its extended double radio structure which has physical size about 200 kpc. The radio image shows jets extending from the nucleus of the radio galaxy to the outer radio components. The only reasonable way of accounting for the extended diffuse radio emission is that it is the synchrotron radiation of high-energy electrons gyrating in magnetic fields within the radio lobes and that the particles were accelerated in the interaction of the beams of high-energy particles from the nucleus of the active galaxy with the ambient intergalactic medium.

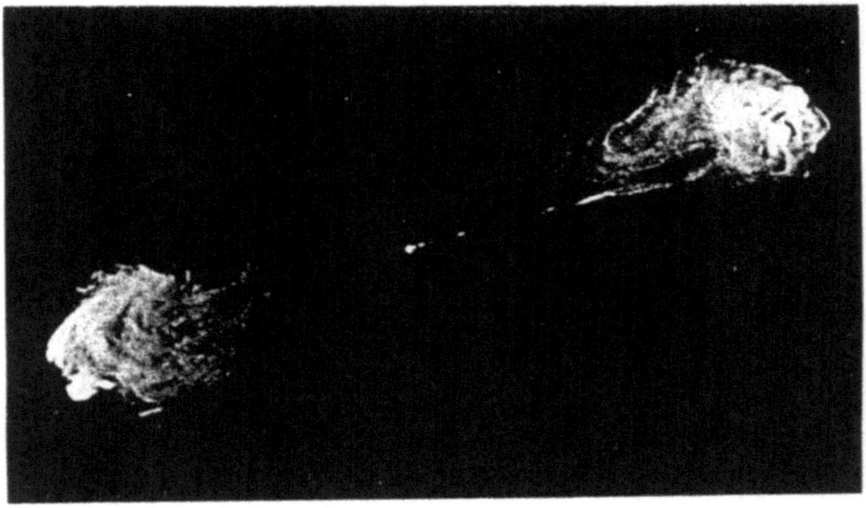

Fig. 24. The detailed radio structure of the brightest extragalactic radio source in the northern sky, Cygnus A. In addition to extensive radio lobes with remarakble fine structure, there are intense 'hot spots' towards the leading edges of the lobes, in which the energy densities of high-energy electrons and magnetic fields are high. There is a compact radio source in the nucleus of the radio galaxy as well as radio jets which channel energy from the nucleus to the hot spots. (Perley et al. 1984.)

- Direct evidence for relativistic particles in active galactic nuclei comes from the very high brightness temperatures observed in compact radio sources and, to understand this argument, we need to analyse the process of synchrotron self-absorption. That is our next task.

6.8 Synchrotron self absorption

According to the principle of detailed balance, to every emission process there is a corresponding absorption process - in the case of synchrotron radiation, this is known as *synchrotron self-absorption*. Let us begin with arguments which illustrate the physics of the full calculations, which are quite demanding algebraically.

Suppose a source of synchrotron radiation has a power law spectrum, $S_\nu \propto \nu^{-\alpha}$, where the spectral index $\alpha = (p-1)/2$. Its *brightness temperature* is defined to be $T_b = (\lambda^2/2k)(S_\nu/\Omega)$, and is proportional to $\nu^{-(2+\alpha)}$, where S_ν is its flux density and Ω is the solid angle it subtends at the observer at frequency ν. We recall that brightness temperature is the temperature of a black-body which would produce the observed surface brightness of the source at the frequency ν in the Rayleigh-Jeans limit, $h\nu \ll kT_e$. Thus, at low enough frequencies, the brightness temperature of the source may approach the kinetic temperature of the radiating electrons. When this occurs, self-absorption becomes important since thermodynamically the source cannot emit radiation of brightness temperature greater than its kinetic temperature.

We have derived expressions for the synchrotron radiation spectrum of a power law energy distribution of relativistic electrons, $N(E)\,\mathrm{d}E = \kappa E^{-p}\,\mathrm{d}E$, in Sect. 6.6. Now, this energy spectrum is *not* a thermal-equilibrium spectrum, which for relativistic particles would be a *relativistic Maxwellian distribution*. The concept of temperature can still be used, however, for particles of energy E for the following reasons. First of all, the spectrum of the radiation emitted by particles of energy E is peaked about the critical frequency $\nu \approx \nu_c \approx \gamma^2\nu_g$, where $\gamma = E/m_e c^2 \gg 1$ and $\nu_g = eB/2\pi m_e$ is the non-relativistic gyrofrequency. Thus, the emission and absorption processes at frequency ν are associated with electrons of roughly the same energy. Second, the characteristic time scale for the relativistic electron gas to relax to an equilibrium spectrum is very long indeed under typical cosmic conditions because the particle number densities are very small and all interaction times with matter are very long. Therefore, we can associate a temperature T_e with electrons of a given energy through the relativistic formula which relates particle energy to temperature,

$$\gamma m_e c^2 = 3kT_e. \tag{167}$$

One way of understanding the difference between this result and the standard result of kinetic theory, $E = \frac{3}{2}kT_e$ is to recall that the ratio of specific heats γ_{SH} is $\frac{4}{3}$ for a relativistic gas and $\frac{5}{3}$ for a non-relativistic gas. The internal thermal energy density of a gas is $u = NkT/(\gamma_{SH}-1)$, where N is the number density of particles. Setting $\gamma_{SH} = \frac{5}{3}$ we obtain the classical result and, setting $\gamma_{SH} = \frac{4}{3}$, we obtain (167) for the mean energy per particle.

The important point is that the *effective temperature* of the particles now becomes a function of their energies. Since $\gamma \approx (\nu/\nu_g)^{1/2}$,

$$T_e \approx (m_e c^2/3k)(\nu/\nu_g)^{1/2}. \tag{168}$$

For a self-absorbed source, the brightness temperature of the radiation must be equal to the kinetic temperature of the emitting particles, $T_b = T_e$, and therefore, in the Rayleigh-Jeans limit,

$$S_\nu = \frac{2kT_e}{\lambda^2}\Omega = \frac{2m_e}{3\nu_g^{1/2}}\Omega\,\nu^{5/2} \propto \frac{\theta^2\nu^{5/2}}{B^{1/2}}, \tag{169}$$

where Ω is the solid angle subtended by the source, $\Omega \approx \theta^2$.

This calculation shows the physical origin of the steep low-frequency spectrum expected in sources in which synchrotron self-absorption is important, $S_\nu \propto \nu^{5/2}$. It does not follow the standard Rayleigh-Jeans law because the effective kinetic temperature of the electrons varies with frequency. Notice that, in a self-absorbed source, the spectral form $S_\nu \propto \nu^{5/2}$ is independent of the spectrum of the emitting particles so long as the magnetic field is uniform. The typical spectrum of a self-absorbed radio source is shown in Fig. 25. Spectra of roughly this form are found at radio, centimetre and millimetre wavelengths from the nuclei of active galaxies and quasars and it is conventionally assumed that synchrotron self-absorption is the process responsible for the low-frequency cut-offs.

It is a straightforward, but long, calculation to work out the absorption coefficient $\chi(\nu)$ for synchrotron self-absorption and I show how this can be done in *HEA2*, Sect. 18.1.7. Let me simply quote the result for a randomly oriented magnetic field,

$$\chi_\nu = \frac{\sqrt{3}\pi e^3 \kappa B^{(p+2)/2} c}{64\pi^2\epsilon_0 m_e}\left(\frac{3e}{2\pi m_e^3 c^4}\right)^{p/2} \frac{\Gamma\left(\frac{3p+22}{12}\right)\Gamma\left(\frac{3p+2}{12}\right)\Gamma\left(\frac{p+6}{4}\right)}{\Gamma\left(\frac{p+8}{4}\right)}\nu^{-(p+4)/2}, \tag{170}$$

where the Γs are gamma functions.

To work out the emission spectrum from, say, a slab of thickness l, we write down the transfer equation

$$\frac{dI_\nu}{dx} = -\chi_\nu I_\nu + \frac{J(\nu)}{4\pi}. \tag{171}$$

The solution is

$$I_\nu = \frac{J(\nu)}{4\pi\chi_\nu}[1 - e^{-\chi_\nu l}]. \tag{172}$$

If the source is optically thin, $\chi(\nu)l \ll 1$, we obtain

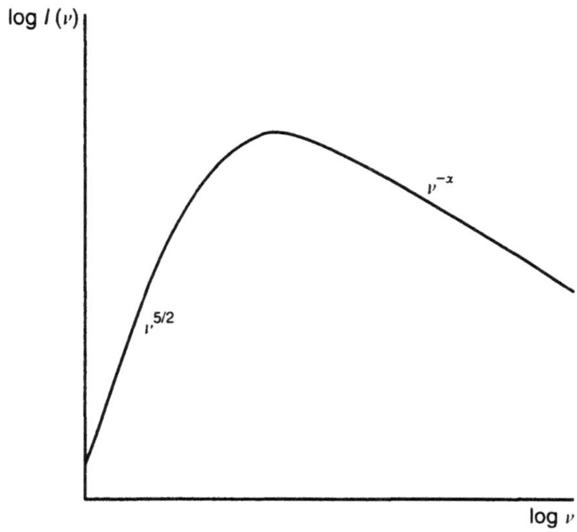

Fig. 25. The spectrum of a source of synchrotron radiation which exhibits synchrotron self-absorption.

$$I_\nu = \frac{J(\nu)l}{4\pi}. \tag{173}$$

If the source is optically thick, $\chi(\nu)l \gg 1$, we find

$$I_\nu = \frac{J(\nu)}{4\pi\chi_\nu}. \tag{174}$$

The quantity $J(\nu)/4\pi\chi_\nu$ is often referred to as the *source function*. Substituting for the absorption coefficient $\chi(\nu)$ from Eq. (170) and for J_ν from Eq. (166), we find

$$I_\nu = (\text{constant})\ \frac{m_e \nu^{5/2}}{B^{1/2}}, \tag{175}$$

where the constant involves numerous gamma functions. This is the same dependence as was found from the above physical arguments.

It is found that many of the most compact radio sources have spectra of roughly this form, in particular, their spectra are flat or inverted at centimetre wavelengths. VLBI observations show that the angular sizes of many of the synchrotron self-absorbed sources have angular sizes $\theta \approx 10^{-3}$ arcsec. For 1 Jy radio sources (1 Jy $= 10^{-26}$ W m^{-2} Hz^{-1}), the corresponding brightness temperatures are

$$T_b = \frac{\lambda^2}{2k_B}\frac{S_\nu}{\Omega} \sim 10^{11}\,\text{K}. \tag{176}$$

This is direct evidence for the presence of relativistic electrons within the source regions. The observation of synchtrotron self-absorption in compact radio sources can thus provide important information about physical conditions in active galactic nuclei.

6.9 Distortions of injection spectra of the electrons

In the optically thin regime of sources of synchrotron radiation, spectral breaks or cut-offs are often observed. In addition, different regions within individual sources may display spectral-index variations. Both of these phenomena can be attributed to the effects of ageing of the spectrum of the electrons within the source regions and so provide useful information about time scales. This is most simply appreciated from an estimate of the lifetimes τ of the electrons in the source regions,

$$\tau = \frac{E}{(\mathrm{d}E/\mathrm{d}t)} = \frac{m_{\mathrm{e}}c^2}{\frac{4}{3}\sigma_{\mathrm{T}} c U_{\mathrm{mag}} \gamma}. \tag{177}$$

For typical extended powerful radio sources, $\gamma \sim 10^3$ and $B \sim 10^{-9}$ T and so the lifetimes of the electrons are expected to be $\tau \leq 10^7 - 10^8$ years. In the case of X-ray sources, for example, the diffuse X-ray emission from the Crab Nebula and the jet of M87, the energies of the electrons are very much greater, the inferred magnetic field strengths are greater and so the relativistic electrons have correspondingly shorter lifetimes. In these specific cases, the inferred lifetimes of the electrons are significantly shorter than the light travel time across the sources and so the electrons must be continuously accelerated within these sources.

To obtain a quantitative description of the resulting distortions of synchrotron radiation spectra, it is convenient to introduce the *diffusion loss equation* for the electrons. I have given two derivations of this equation in *HEA2*, Chap. 19. If we write the loss rate of the electrons as

$$-\left(\frac{\mathrm{d}E}{\mathrm{d}t}\right) = b(E), \tag{178}$$

the diffusion loss equation is

$$\frac{\partial N(E)}{\partial t} = D\nabla^2 N(E) + \frac{\partial}{\partial E}[b(E)N(E)] + Q(E,t), \tag{179}$$

where D is a scalar diffusion coefficient and $Q(E)$ is a source term which describes the rate of injection of electrons and their injection spectra into the source region. We can obtain some useful results by inspection of a few special steady-state solutions.

Suppose, first of all, that there is an infinite, uniform distribution of sources, each injecting high-energy electrons with an injection spectrum given by $Q(E) = \kappa E^{-p}$. Then, diffusion is not important and the diffusion loss equation reduces to

$$\frac{\mathrm{d}}{\mathrm{d}E}[b(E)N(E)] = -Q(E); \tag{180}$$

$$\int \mathrm{d}[b(E)N(E)] = -\int Q(E)\,\mathrm{d}E. \tag{181}$$

We assume $N(E) \to 0$ as $E \to \infty$ and hence integrating we find

$$N(E) = \frac{\kappa E^{-(p-1)}}{(p-1)b(E)}. \tag{182}$$

We now write down $b(E)$ for high-energy electrons under interstellar conditions

$$b(E) = -\left(\frac{\mathrm{d}E}{\mathrm{d}t}\right) = A_1\left(\ln\frac{E}{m_e c^2} + 19.8\right) + A_2 E + A_3 E^2. \tag{183}$$

The first term on the right-hand side, containing the constant A_1, describes ionisation losses and depends only weakly upon energy; the second term, containing A_2, represents bremsstrahlung and adiabatic losses, and the last term, containing A_3, describes inverse Compton and synchrotron losses (see *HEA2*, Sect. 19.3.3). This analysis enables us to understand the effect of continuous energy losses upon the initial spectrum of the high-energy electrons. Thus, from Eq. (182),

- if ionisation losses dominate, $N(E) \propto E^{-(p-1)}$, that is, the energy spectrum is flatter by one power of E;
- if bremsstrahlung or adiabatic losses dominate, $N(E) \propto E^{-p}$, that is, the spectrum is unchanged;
- if inverse Compton or synchrotron losses dominate, $N(E) \propto E^{-(p+1)}$, that is, the spectrum is steeper by one power of E.

These are also the equilibrium spectra expected whenever the continuous injection of electrons takes place over a time scale longer than the lifetimes of the individual electrons involved. For example, if we inject electrons continuously with a spectrum E^{-p} into a source component for a time t and synchrotron radiation is the only important loss process, an electron of energy E_s loses all its energy in a time τ such that $-(\mathrm{d}E/\mathrm{d}t)\tau = E_s$. The electron energy spectrum in the source is different for electrons with energies greater and less than E_s; for lower energies, the electrons do not lose a significant fraction of their energy and therefore the spectrum is the same as the injection spectrum, $N(E) \propto E^{-p}$. For energies greater than E_s, the particles have lifetimes less than t and we only observe those produced during the previous synchrotron lifetime τ_s of the particles of energy E, that is, $\tau_s \propto 1/E$. Therefore, the spectrum of the electrons is one power of E steeper, $N(E) \propto E^{-(p+1)}$, in agreement with the analysis proceeding from the steady-state solution of the diffusion loss equation.

There are two useful analytic solutions for the electron energy distribution under continuous energy losses due to synchrotron radiation and inverse

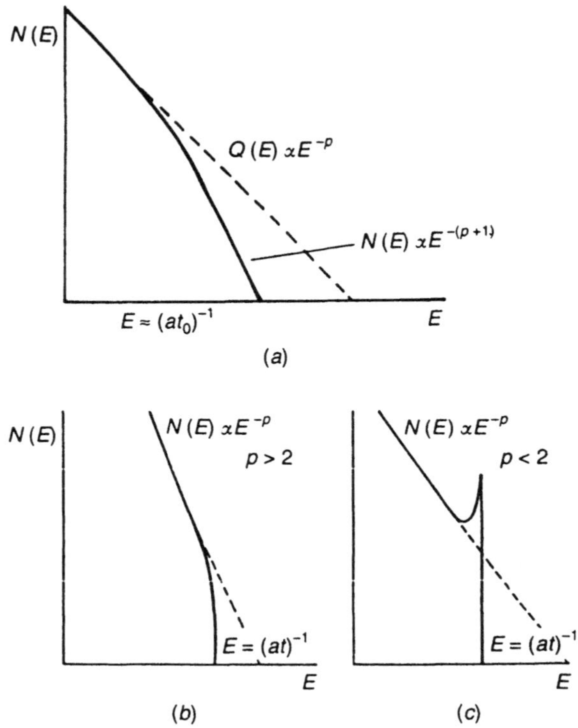

Fig. 26. (**a**) A solution of the diffusion loss equation for steady-state injection of electrons with a power law energy spectrum $Q(E) \propto E^{-p}$ in the presence of energy losses of the form $\mathrm{d}E/\mathrm{d}t = b(E) = -aE^2$. (**b**) The time evolution of a power law energy distribution injected at $t = 0$ with no subsequent injection of electrons for the case $p > 2$. (**c**) As in case (**b**) but for $p < 2$.

Compton scattering. In the first case, it is assumed that there is continuous injection of electrons with a power law energy spectrum $Q(E) = \kappa E^{-p}$ for a time t_0. If we write the loss rate of the electrons in the form $b(E) = aE^2$, the energy spectrum after time t_0 has the form

$$N(E) = \frac{\kappa E^{-(p+1)}}{a(p-1)}[1 - (1 - aEt)^{p-1}] \quad \text{if} \quad aEt_0 \leq 1; \tag{184}$$

$$N(E) = \frac{\kappa E^{-(p+1)}}{a(p-1)} \quad \text{if} \quad aEt_0 > 1. \tag{185}$$

This form of spectrum is shown in Fig. 26a and agrees with the physical arguments given in the last paragraph.

A second useful case is that of the injection of electrons with a power law energy spectrum at $t = 0$ with no subsequent injection of electrons. We can then write $Q(E) = \kappa E^{-p}\delta(t)$, where $\delta(t)$ is the Dirac delta function. It is straightforward to show that the solution of the diffusion loss equation, ignoring the diffusion term, is

$$N(E) = \kappa E^{-p}(1 - aEt)^{p-2}. \tag{186}$$

Thus, after time t, there are no electrons with energies greater than $(at)^{-1}$. Notice that, if $p > 2$, the spectrum steepens smoothly to zero at $E = (at)^{-1}$; if $p < 2$, there is a cusp in the energy spectrum at $E = (at)^{-1}$. The number of electrons, however, remains finite and constant. These spectra are illustrated in Figs. 26b and c.

These results find numerous applications in the study of extragalactic radio sources. For example, in extended radio sources it is commonly found that the 'hot spots' found towards the advancing edges of the extended source components have flatter spectra than the extended radio lobes. Since there are good grounds to believe that the electrons are accelerated in the hot spots, they are likely to be younger in these regions than in the extended lobes and the steepening of the spectra can be attributed to synchrotron losses in the extended source regions, thus enabling time scales to be determined for the radio source.

Another possible example of the use of these results is in accounting for the form of the electron energy spectrum in the local interstellar medium. In Fig. 27, a schematic representation of the electron energy spectrum in the local interstellar medium is shown as derived from direct observations of the electrons themselves, from interpretation of the spectrum of the diffuse Galactic radio emission and from the low-energy γ-ray emission from the general direction of the Galactic centre (see Fig. 23). Interpreting the energy spectrum by power laws of the form $N(E) \propto E^{-p}$, the spectrum steepens with increasing energy from $p \approx 1.6$ at ~ 30 MeV to 2.8 between 1 and 10 GeV and to 3.3 at the highest energies $10 - 100$ GeV.

The simplest way of interpreting the observations is to study first those energy ranges in which the time scale for energy losses is less than the escape time $\tau(E)$. The observed abundances of ^{10}Be nuclei in the cosmic rays suggest that cosmic-ray nuclei have escape times from the Galaxy of about $(1-3)\times 10^7$ years (see *HEA2*, Sect. 20.3.1). If the electrons have similar escape times, the time scales for energy losses are less than this value for the lowest-energy electrons due to ionisation losses, and for those with the highest energies by a combination of synchrotron and inverse Compton losses. Taking the ionisation loss rate to be $10^{-5}N$ eV per year and $N = 10^6$ m^{-3}, we find a lifetime of 3×10^7 years for 300 MeV electrons; taking $B = 6 \times 10^{-10}$ T, we find the lifetime of 10 GeV electrons to be about 3×10^7 years due to synchrotron losses. Thus, if electrons are continuously injected uniformly into the interstellar medium, the electron spectra in these spectral regions should

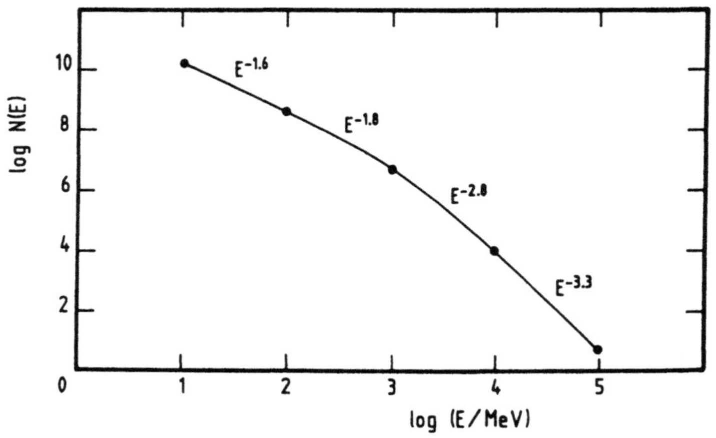

Fig. 27. A schematic representation of the electron energy spectrum in the local interstellar medium. This spectrum has been subject to energy losses at high and low energies during propagation of the electrons through the interstellar medium. The units on the ordinate are relative units.

reach a steady state under losses. From the above analysis, we find that, in the low-energy regions in which ionisation losses dominate, $N(E) \propto E^{-(p-1)}$ and so the injection spectrum would be $Q(E) \propto E^{-2.6}$. In the high-energy region, in which synchrotron losses dominate, $N(E) \propto E^{-(p+1)}$ and hence the injection spectrum is $Q(E) \propto E^{-2.3}$. These values are not too different, suggesting that the injection electron spectrum might be quite close to $E^{-2.5}$ throughout the energy range 100 MeV to 100 GeV.

6.10 The energetics of sources of synchrotron radiation

An important calculation involving sources of synchrotron radiation is the estimation of the minimum energy requirements in relativistic electrons and magnetic fields to account for the observed synchrotron emission. Suppose a source has luminosity L_ν at frequency ν and its volume is V. The spectrum of the radiation is of power law form, $L_\nu \propto \nu^{-\alpha}$, and the emission mechanism is assumed to be synchrotron radiation. The following arguments can be applied to the synchrotron radiation emitted by the source at any frequency, be it radio, optical or X-ray wavelengths. The luminosity can be related to the energy spectrum of the ultra-relativistic electrons and the magnetic field B present in the source through the expression (166) for synchrotron radiation

$$L_\nu = A(\alpha) V \kappa B^{1+\alpha} \nu^{-\alpha}, \tag{187}$$

where the electron energy spectrum per unit volume is $N(E)\,\mathrm{d}E = \kappa E^{-p}\,\mathrm{d}E$, $p = 2\alpha + 1$ and $A(\alpha)$ is a constant which depends only weakly on the spectral index α; detailed numerical values for $A(\alpha)$ are given in *HEA2*, Sect. 18.1.8. Writing the energy density in relativistic electrons as ε_e, the total energy present in the source is

$$W_{\text{total}} = V\varepsilon_e + V\frac{B^2}{2\mu_0} \tag{188}$$

$$= V\int \kappa E N(E)\,\mathrm{d}E + V\frac{B^2}{2\mu_0}. \tag{189}$$

From Eq. (185), it can be seen that the luminosity of the source L_ν determines only the product $V\kappa B^{1+\alpha}$. If V is assumed to be known, the luminosity may either be produced by a large flux of relativistic electrons in a weak magnetic field, or *vice versa*. There is no way of deciding which combination of ε_e and B is appropriate from observations of L_ν. Between the extremes of dominant magnetic field and dominant particle energy, there is, however, a minimum total energy requirement.

Before proceeding to that calculation, we should consider the problem of how much energy might also be present in the form of relativistic protons, which presumably must also be present in the source. There are, unfortunately, very few sources for which estimates of both the electron and proton fluxes are known. On the one hand, in our own Galaxy, there seems to be about 100 times as much energy in relativistic protons as there is in electrons, whereas in the Crab Nebula, the energy in relativistic protons cannot be much greater than the energy in the electrons from dynamical arguments. Therefore, to take account of the protons, it is customary to assume that they have energy β times that of the electrons, that is,

$$\varepsilon_{\text{protons}} = \beta\varepsilon_e, \tag{190}$$

$$\varepsilon_{\text{total}} = (1+\beta)\varepsilon_e = \eta\varepsilon_e. \tag{191}$$

We therefore write

$$W_{\text{total}} = \eta V\int_{E_{\min}}^{E_{\max}} \kappa E N(E)\,\mathrm{d}E + V\frac{B^2}{2\mu_0}. \tag{192}$$

The energy requirements as expressed in Eq. (192) depend upon the unknown quantities κ and B, but they are related through Eq. (187) for the observed luminosity of the source L_ν. We also require the relation between the frequency of emission of an ultra-relativistic electron of energy $E = \gamma m_e c^2 \gg m_e c^2$ in a magnetic field of strength B. We use the result that the maximum intensity of synchrotron radiation occurs at a frequency

$$\nu = \nu_{\max} = 0.29\nu_c = 0.29\,\tfrac{3}{2}\gamma^2\nu_g \tag{193}$$

$$= 0.29 \times 4.199 \times 10^{10}\gamma^2 B = CE^2B, \tag{194}$$

where ν_g is the non-relativistic gyrofrequency and $C = 1.22 \times 10^{10}/(m_e c^2)^2$. Therefore, the relevant range of electron energies in the integral (192) is related to the range of observable frequencies through

$$E_{\max} = \left(\frac{\nu_{\max}}{CB}\right)^{1/2} \qquad E_{\min} = \left(\frac{\nu_{\min}}{CB}\right)^{1/2}. \tag{195}$$

$\nu_{\max}$ and $\nu_{\min}$ are the maximum and minimum frequencies for which the spectrum is known or the range of frequencies relevant to the problem at hand. Then

$$W_{\text{particles}} = \eta V \int_{E_{\min}}^{E_{\max}} E\kappa E^{-p}\,\mathrm{d}E \tag{196}$$

$$= \frac{\eta V \kappa}{(p-2)}(CB)^{(p-2)/2}\left[\nu_{\min}^{(2-p)/2} - \nu_{\max}^{(2-p)/2}\right]. \tag{197}$$

Substituting for κ in terms of L_ν and B from Eq. (187),

$$W_{\text{particles}} = \frac{\eta V}{(p-2)}\left[\frac{L_\nu}{A(\nu)VB^{1+\alpha}\nu^{-\alpha}}\right](CB)^{(p-2)/2}[\nu_{\min}^{(2-p)/2} - \nu_{\max}^{(2-p)/2}]. \tag{198}$$

Preserving only the essential dependences,

$$W_{\text{particles}} = G(\alpha)\eta L_\nu B^{-3/2}, \tag{199}$$

where $G(\alpha)$ is a constant which depends weakly on α, $\nu_{\max}$ and $\nu_{\min}$ if $\alpha \approx 1$. Therefore

$$W_{\text{total}} = G(\alpha)\eta L_\nu B^{-3/2} + V\frac{B^2}{2\mu_0}. \tag{200}$$

The variations of the energies in particles and magnetic field are shown in Fig. 28 as a function of B. There is a minimum total energy which can be found by minimising expression (200) with respect to B.

$$B_{\min} = \left[\frac{3\mu_0}{2}\frac{G(\alpha)\eta L_\nu}{V}\right]^{2/7}. \tag{201}$$

This magnetic field strength $B_{\min}$ corresponds to approximate equality of the energies in the relativistic particles and magnetic field. Substituting $B_{\min}$ into Eq. (199), we find

$$W_{\text{mag}} = V\frac{B_{\min}^2}{2\mu_0} = \tfrac{3}{4}W_{\text{particles}} \tag{202}$$

Thus, the condition for minimum-energy requirements corresponds closely to the condition that there are equal energies in the relativistic particles and the magnetic field. This condition is often referred to as *equipartition*. The minimum total energy is

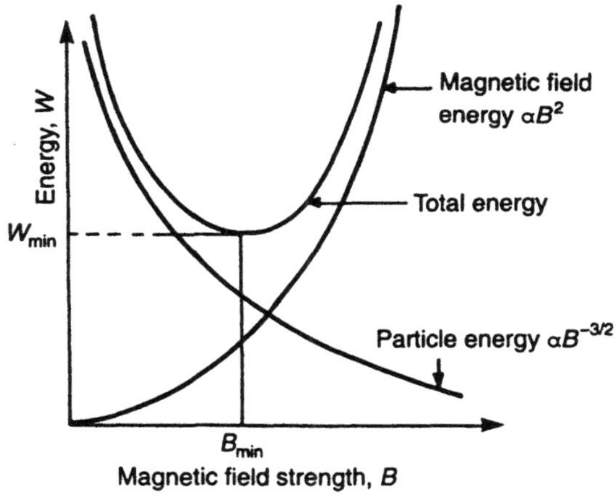

Fig. 28. Illustrating the origin of the minimum energy requirements of a source of synchrotron radiation as a function of magnetic-flux density B.

$$W_{\mathrm{total}}(\mathrm{min}) = \frac{7}{6\mu_0} V^{3/7} \left[\frac{3\mu_0}{2} G(\alpha)\eta L_\nu \right]^{4/7} . \tag{203}$$

The expressions (201) and (203) are the answers we have been seeking. These are the magnetic-field strength and minimum total energy needed to account for the observed luminosity of the source. These results are frequently used in the study of the synchrotron radiation from radio, optical and X-ray sources but their limitations should be appreciated.

1. There is no physical justification for the source components being close to equipartition. It might be that the particle and magnetic-field energies in the source components tend towards equipartition but there is no proof that this must be so. For example, it has been conjectured that the magnetic field in the source components may be stretched and tangled by motions in the plasma and so there might be rough equipartition between the magnetic-energy density and the energy density in turbulent motions. The turbulent motions might also be responsible for accelerating the high-energy particles and these particles might come into equipartition with the turbulent-energy density if the acceleration mechanism were very efficient. In this way, it is possible that there might be a physical justification for the source components being close to equipartition, but this is really no more than a conjecture.
2. The amount of energy present in the source is sensitive to the value of η, that is, the amount of energy present in the form of relativistic protons and nuclei.
3. The total amount of energy in relativistic particles is dependent upon the limits assumed to the energy spectrum of the particles. It can be seen

that, if $\alpha = 1$, we need only consider the dependence upon $\nu_{\min}$ which is quite weak, $W_{\min} \propto \nu_{\min}^{-0.5}$. However, there might be large fluxes of low-energy relativistic electrons present in the source components with a quite different energy spectrum and we would have no way of knowing that they are present from the radio observations.

4. Even more important is the fact that the energy requirements depend upon the volume of the source. The calculation has been carried out assuming that the particles and magnetic field fill the source volume uniformly. In fact, the emitting regions might occupy only a small fraction of the apparent volume of the source, for example, if the synchrotron emission originated in filaments or subcomponents within the overall volume V. In this case, the volume which should be used in the expressions (201) and (203) should be smaller than V. Often, a *filling factor* f is used to describe the fraction of the volume occupied by radio emitting material. Clearly, the energy requirements are reduced if f is small.
5. On the other hand, we can obtain a firm lower limit to the *energy density* within the source components since

$$U_{\min} = \frac{W_{\text{total}}(\min)}{V} = \frac{7}{6\mu_0} V^{-4/7} \left[\frac{3\mu_0}{2} G(\alpha)\eta L_\nu\right]^{4/7} . \qquad (204)$$

For dynamical purposes, the energy density is more important than the total energy since it is directly related to the pressure within the source components $p = (\gamma - 1)U$ where γ is the ratio of specific heats. In the case of an ultra-relativistic gas, $\gamma = 4/3$ and so $p = \frac{1}{3}U$ as usual.

For these reasons, the values of the magnetic-field strength and minimum energy which come out of these arguments should be considered only order of magnitude estimates. Obviously, if the source components depart radically from the equipartition values, the energy requirements are increased and this can pose problems for some of the most luminous sources.

It is often cumbersome to have to go through the procedure of working out $G(\alpha)$ to estimate the minimum-energy requirements and magnetic-field strengths. A simplified calculation can be performed in the following way. If we assume that the spectral index $\alpha = 0.75$, which is a good approximation for many galactic and extragalactic radio sources, we can neglect the upper limit $\nu_{\max}$ in comparison with $\nu_{\min}$ in evaluating $G(\alpha)$. Then, if we know the luminosity $L(\nu)$ at a certain frequency ν, we obtain a lower limit to the energy requirements if we set $\nu = \nu_{\min}$. Making these simplifications, we find that the minimum-energy requirement is:

$$W_{\min} \approx 3.0 \times 10^6 \, \eta^{4/7} V^{3/7} \nu^{2/7} L_\nu^{4/7} \quad \text{J}, \qquad (205)$$

where the volume of the source V is measured in m^3, the luminosity $L(\nu)$ in W Hz^{-1} and the frequency ν in Hz. In the same units, the minimum magnetic- field strength is:

$$B_{\rm min} = 1.8 \left(\frac{\eta L_\nu}{V}\right)^{2/7} \nu^{1/7} \quad {\rm T}. \tag{206}$$

This line of reasoning was very important in the late 1950s when the energy requirements of extragalactic radio sources were first estimated by Burbidge (1956). A good example was provided by the radio source Cygnus A (Fig. 24). At that time, it was thought that the source consisted of two components roughly 100 kpc in diameter. The source had luminosity roughly 8×10^{28} W Hz^{-1} at 178 MHz. Inserting these values into Eq. (205), we find that the minimum total energy is $2 \times 10^{52}\eta^{4/7}$ J, which corresponds to the rest mass energy of $3 \times 10^5\eta^{4/7}\,M_\odot$ of matter. The realisation of the enormous energy demands of extragalactic radio sources in the form of relativistic particles and magnetic fields was one of the most important problems which stimulated the very rapid growth of high-energy astrophysics in the 1960s. Evidently a very considerable amount of mass has to be converted into relativistic particle energy and ejected from the nucleus of the galaxy into enormous lobes well outside the body of the galaxy.

Another important example is the supernova Cassiopeia A. Performing the same calculation, the magnetic-flux density corresponding to the minimum-energy requirements is $B = 10\eta^{2/7}$ nT and the minimum total energy is $W_{\rm min} = 2 \times 10^{41}\eta^{4/7}$ J. The latter figure can be compared with the kinetic energy of the filaments which amounts to about 2×10^{44} J.

7 Inverse Compton Scattering

Comptonisation is a vast subject and some aspects of it will be covered in Dr. Liedahl's lectures (see also Pozdnyakov, Sobol & Sunyaev 1983). Inverse Compton scattering involves the scattering of low-energy photons to high energies by ultra-relativistic electrons so that the photons gain and the electrons lose energy. The process is called *inverse Compton scattering* because the electrons lose energy rather than the photons, the opposite of the standard Compton effect. We will treat the case in which the energy of the photon in the centre of momentum frame of the interaction is much less that $m_e c^2$, and consequently the Thomson scattering cross section can be used to describe the probability of scattering.

Many of the most important results can be worked out using simple physical arguments (see, for example, Blumenthal & Gould 1970, and Rybicki & Lightman 1979). We consider the geometry of inverse Compton scattering shown in Fig. 29, which shows the collision between a photon and a relativistic electron as seen in the laboratory frame of reference S and in the rest frame of the electron S′. Since $\gamma\hbar\omega \ll m_e c^2$, the centre of momentum frame is very closely that of the relativistic electron. If the energy of the photon is $\hbar\omega$ and the angle of incidence θ in S, its energy in the frame S′ is

$$\hbar\omega' = \gamma\hbar\omega[1 + (v/c)\cos\theta] \tag{207}$$

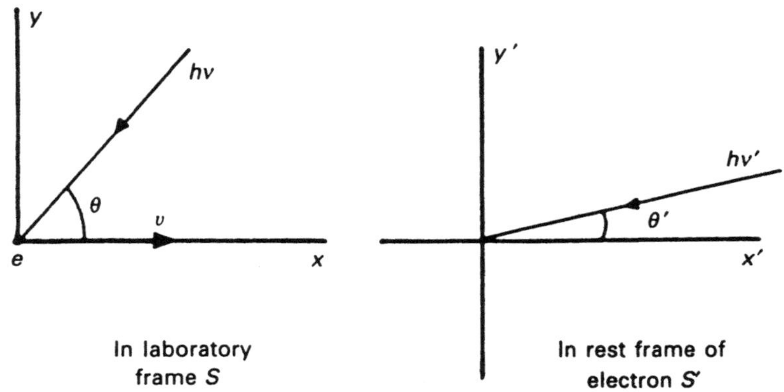

Fig. 29. The geometry of inverse Compton scattering in the laboratory frame of reference S and that in which the electron is at rest S′.

according to the standard relativistic Doppler shift formula. Similarly, the angle of incidence θ' in the frame S′ is related to θ by the formulae

$$\sin\theta' = \frac{\sin\theta}{\gamma[1+(v/c)\cos\theta]} \quad ; \quad \cos\theta' = \frac{\cos\theta + v/c}{1+(v/c)\cos\theta}. \tag{208}$$

Now, provided $\hbar\omega' \ll m_e c^2$, the Compton interaction in the rest frame of the electron is simply Thomson scattering and hence the energy loss rate of the electron in S′ is just the rate at which energy is reradiated by the electron. According to Eq. (23), this loss rate is

$$-(\mathrm{d}E/\mathrm{d}t)' = \sigma_\mathrm{T} c U'_\mathrm{rad}, \tag{209}$$

where U_rad is the energy density of radiation in the rest frame of the electron. As shown in Sect. 3.2, it is of no importance whether or not the radiation is isotropic. The free electron oscillates in response to any incident radiation field. Our strategy is threfore to work out U'_rad in the frame of the electron S′ and then to use Eq. (209) to work out $(\mathrm{d}E/\mathrm{d}t)'$. Using the result obtained in Section 3.4, this is also the loss rate $(\mathrm{d}E/\mathrm{d}t)$ in the observer's frame S.

Suppose the number density of photons in a beam of radiation incident at angle θ to the x-axis is N. Then, the energy density of these photons in S is $N\hbar\omega$. The flux density of photons incident upon an electron stationary in S is $U_\mathrm{rad} c = N\hbar\omega c$. Now let us work out the flux density of this beam in the frame of reference of the electron S′. We need two things, the energy of each photon in S′ and the rate of arrival of these photons at the electron in S′. The first of these is easy and is given by Eq. (207). The second factor requires a little bit of care, although the answer is obvious in the end. The beam of photons incident at angle θ in S arrives at an angle θ' in S′ according to the

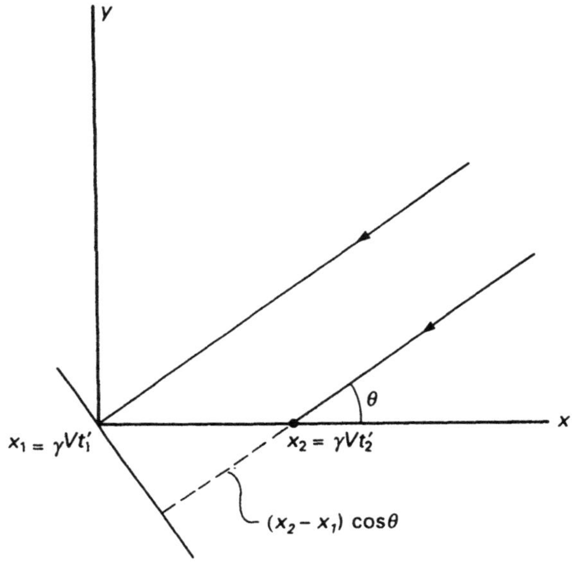

Fig. 30. Illustrating the rate of arrival of photons at the observer in the laboratory frame of reference (see text).

aberration formulae (208). We are interested in the rate of arrival of photons at the origin of S′ and so let us consider two photons which arrive there at times t'_1 and t'_2. The coordinates of these events in S are

$$[x_1, 0, 0, t_1] = [\gamma V t'_1, 0, 0, \gamma t'_1] \quad \text{and} \quad [x_2, 0, 0, t_2] = [\gamma V t'_2, 0, 0, \gamma t'_2] \qquad (210)$$

respectively. This calculation makes the important point that the photons in the beam are propagated along parallel but separate trajectories in S as illustrated by Fig. 30. From the geometry of the figure, it is apparent that the time difference when the photons arrive at a plane perpendicular to their direction of propagation in S is

$$\Delta t = t_2 + \frac{(x_2 - x_1)}{c} \cos\theta - t_1 = (t'_2 - t'_1)\gamma[1 + (v/c)\cos\theta], \qquad (211)$$

that is, the time interval between the arrival of photons from the direction θ is shorter by a factor $\gamma[1 + (v/c)\cos\theta]$ in S′ than it is in S. Thus, the rate of arrival of photons, and correspondingly their number density, is greater by this factor $\gamma[1 + (v/c)\cos\theta]$ in S′ as compared with S. This is exactly the same factor by which the energy of the photon has increased (207). On reflection, we should not be surprised by this result because these are two different aspects of the same relativistic transformation between the frames S and S′.

Thus, as observed in S′, the energy density of the beam is

$$U'_{\rm rad} = [\gamma(1 + (v/c)\cos\theta)]^2\, U_{\rm rad}. \tag{212}$$

Now, we may think of this as the energy density associated with the photons incident at angle θ in the frame S and consequently arriving within solid angle $2\pi \sin\theta\, \mathrm{d}\theta$. We assume that the radiation field in S is isotropic and therefore we can now work out the total energy density seen by the electron in S′ by integrating over solid angle in S, that is,

$$U'_{\rm rad} = U_{\rm rad} \int_0^\pi \gamma^2 [1 + (v/c)\cos\theta]^2\, \tfrac{1}{2} \sin\theta\, \mathrm{d}\theta. \tag{213}$$

Integrating, we find

$$U'_{\rm rad} = \tfrac{4}{3} U_{\rm rad} (\gamma^2 - \tfrac{1}{4}). \tag{214}$$

Therefore, substituting into (209), we find

$$(\mathrm{d}E/\mathrm{d}t)' = \tfrac{4}{3}\sigma_{\rm T} c U_{\rm rad} (\gamma^2 - \tfrac{1}{4}). \tag{215}$$

Because $(\mathrm{d}E/\mathrm{d}t) = (\mathrm{d}E/\mathrm{d}t)'$, we find

$$\mathrm{d}E/\mathrm{d}t = \tfrac{4}{3}\sigma_{\rm T} c U_{\rm rad} (\gamma^2 - \tfrac{1}{4}). \tag{216}$$

Now, this is the energy gained by the photon field due to the scattering of the low-energy photons. We have therefore to subtract the energy of these photons to find the total energy gain to the photon field in S. The rate at which energy is removed from the low-energy photon field is just $\sigma_{\rm T} c U_{\rm rad}$ and therefore, subtracting, we find

$$\mathrm{d}E/\mathrm{d}t = \tfrac{4}{3}\sigma_{\rm T} c U_{\rm rad} (\gamma^2 - \tfrac{1}{4}) - \sigma_{\rm T} c U_{\rm rad} = \tfrac{4}{3}\sigma_{\rm T} c U_{\rm rad} (\gamma^2 - 1). \tag{217}$$

We now use the identity $(\gamma^2 - 1) = (v^2/c^2)\gamma^2$ to write the loss rate in its final form

$$\mathrm{d}E/\mathrm{d}t = \tfrac{4}{3}\sigma_{\rm T} c U_{\rm rad} \left(\frac{v^2}{c^2}\right) \gamma^2. \tag{218}$$

This is the remarkably elegant result we have been seeking. It is exact so long as $\gamma\hbar\omega \ll m_{\rm e}c^2$.

Notice the remarkable similarity between the expressions for the loss rates by synchrotron radiation (143) and by inverse Compton scattering (218), even down to the factor of $\frac{4}{3}$ in front of the two expressions. This is not an accident. The reason for the similarity is that, in both cases, the electron is accelerated by the electric field which it observes in its instantaneous rest-frame. The electron does not really care about the origin of the electric field. In the case of synchrotron radiation, the constant accelerating electric field is associated with the motion of the electron through the magnetic field $\mathbf{B}$, $\mathbf{E}' = \mathbf{v} \times \mathbf{B}$, and, in the case of inverse Compton scattering, it is the sum of all the electric fields of the incident waves. Notice that, in the latter case, the fields of the waves add incoherently and it is the sum of the squares of the

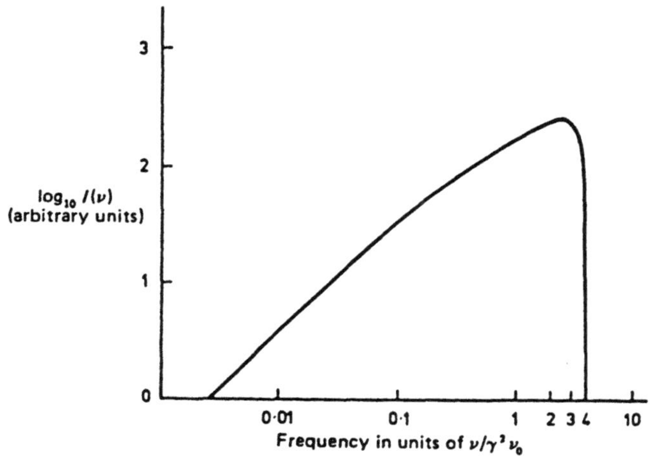

Fig. 31. The emission spectrum of inverse Compton scattering; ν_0 is the frequency of the unscattered radiation. (Blumenthal & Gould 1970.)

electric field strengths of the waves which appears in the formulae. Another way of understanding this similarity, discussed by Jackson (1975) (Chap. 15), is that synchrotron radiation can be considered to be the scattering of 'virtual photons' observed by the electron as it gyrates about the magnetic field.

The next calculation is the determination of the spectrum of the scattered radiation. This can be found by performing two successive Lorentz transformations, first transforming the photon distribution into the frame S′ and then transforming the scattered radiation back into the laboratory frame of reference S. This is not a trivial calculation, but the exact result is given by Blumenthal & Gould (1970) for an incident isotropic photon field at a single frequency ν_0 (Fig. 31). They show that the spectral emissivity $I(\nu)$ may be written

$$I(\nu)\,\mathrm{d}\nu = \frac{3\sigma_{\mathrm{T}}c}{16\gamma^4}\frac{N(\nu_0)}{\nu_0^2}\nu\left[2\nu\ln\left(\frac{\nu}{4\gamma^2\nu_0}\right)+\nu+4\gamma^2\nu_0-\frac{\nu^2}{2\gamma^2\nu_0}\right]\mathrm{d}\nu, \quad (219)$$

where the radiation field is assumed to be monochromatic with frequency ν_0; $N(\nu_0)$ is the number density of photons. At low frequencies, the term in square brackets in Eq. (219) is a constant and hence the scattered radiation has the form $I(\nu) \propto \nu$.

It is an easy calculation to show that the maximum energy which the photon can acquire corresponds to a head-on collision in which the photon is sent back along its original path. It is a useful exercise to show that the maximum energy of the photon is

$$(\hbar\omega)_{\max} = \hbar\omega\gamma^2(1+v/c)^2 \approx 4\gamma^2\hbar\omega_0. \quad (220)$$

Another interesting result comes out of formula (218) for the total energy loss rate of the electron. The number of photons scattered per unit time is $\sigma_{\mathrm{T}} c U_{\mathrm{rad}}/\hbar\omega_0$ and hence the average energy of the scattered photons is

$$\hbar\omega = \tfrac{4}{3}\gamma^2 (v/c)^2 \hbar\omega_0 \approx \tfrac{4}{3}\gamma^2 \hbar\omega_0. \tag{221}$$

This result gives substance to the hand-waving argument that the photon gains one factor of γ in transforming into S′ and then gains another on transforming back to S.

The general result that the frequency of the scattered photons is $\nu \approx \gamma^2 \nu_0$ is of profound importance in high-energy astrophysics. We know that there are electrons with Lorentz factors $\gamma \sim 100 - 1000$ in various types of astronomical source and consequently they scatter any low-energy photons to very much higher energies. To give some simple examples of how this might apply, consider radio, infrared and optical photons scattered by electrons with $\gamma = 1000$. The scattered radiation has frequency (or energy) 10^6 times that of the incoming photons. Thus, radio photons with $\nu_0 = 10^9$ Hz become ultraviolet photons with $\nu = 10^{15}$ Hz ($\lambda = 300\,\mathrm{nm}$); far-infrared photons with $\nu_0 = 3 \times 10^{12}$ Hz, typical of the photons seen in galaxies which are powerful far-infrared emitters, produce X-rays with frequency 3×10^{18} Hz, that is, about 10 keV; optical photons with $\nu_0 = 4 \times 10^{14}$ Hz become γ-rays with frequency 4×10^{20} Hz, that is, about 1.6 MeV. It is apparent that the inverse Compton scattering process is a means of producing very-high-energy photons indeed. It also becomes an inevitable drain of energy for high-energy electrons whenever they pass through a region in which there is a large energy density of photons.

When these formulae are used in astrophysical calculations, it is necessary to integrate over both the spectrum of the incident radiation and the spectrum of the relativistic electrons. The enthusiast is urged to consult the excellent review paper by Blumenthal & Gould (1970). Some of the results are, however, immediately apparent from the close analogy between the inverse Compton scattering and synchrotron radiation processes. For example, we can understand immediately the spectrum of the inverse Compton scattering of photons of energy $h\nu$ by a power law distribution of electron energies

$$\mathrm{d}N \propto E^{-p}\,\mathrm{d}E. \tag{222}$$

By analogy with the results of the calculation which resulted in Eq. (166), the intensity spectrum of the scattered radiation is

$$I(\nu) \propto \nu^{-(p-1)/2}, \tag{223}$$

because of the γ^2 dependence of the energy loss rate by inverse Compton scattering and the fact that the frequency of the scattered radiation is $\nu \approx \gamma^2 \nu_0$.

One of the key questions is whether or not this process is important in extragalactic radio sources. Evidently, the ratio of the total amount of energy liberated by synchrotron radiation process and by inverse Compton scattering by the same distribution of electrons is

$$\frac{(\mathrm{d}E/\mathrm{d}t)_{\mathrm{sync}}}{(\mathrm{d}E/\mathrm{d}t)_{\mathrm{IC}}} = \frac{\int I_\nu \, \mathrm{d}\nu \text{ (radio)}}{\int I_X \, \mathrm{d}\nu_X \text{ (X-ray)}} = \frac{B^2/2\mu_0}{U_{\mathrm{rad}}}, \tag{224}$$

where U_{rad} is the energy density of radiation and B the magnetic-flux density in the source region. Thus, if we measure the radio and X-ray flux densities from a source region and we know U_{rad}, we can find the magnetic flux density in the source. This type of phenomenon has been sought for in the case of extended radio sources in which it is likely that the dominant source of low-energy photons is the Cosmic Microwave Background Radiation. Diffuse X-ray emission from the extended radio lobes has been searched for in the cases of the bright radio sources Cygnus A, Centaurus A and Fornax A. It has proved difficult to find convincing evidence for the inverse Compton scattered X rays from the same population of electrons responsible for the radio emission. Probably the most convincing case is that of Fornax A in which the X-ray emission is coincident with the radio lobes (Feigelson et al. 1995). The inferred magnetic field strength is $B \approx 2 - 3 \times 10^{-10}$ T, a value close to that derived from equipartition arguments.

Another important piece of astrophysics involving the Cosmic Microwave Background Radiation is that relativistic electrons can never escape from it since it permeates all space. The energy density of the Cosmic Microwave Background Radiation is $U_0 = aT^4 = 2.6 \times 10^5$ eV m^{-3}. Therefore, the maximum lifetime τ of any electron is

$$\tau = \frac{E}{|\mathrm{d}E/\mathrm{d}t|} = \frac{E}{\frac{4}{3}\sigma_{\mathrm{T}} c \gamma^2 U_0} = \frac{2.3 \times 10^{12}}{\gamma} \text{ years} \tag{225}$$

For example, we observe 100 GeV electrons at the top of the atmosphere and so they must have lifetimes $\tau \leq 10^7$ years.

8 Synchro-Compton Radiation and the Inverse Compton Catastrophe

Inverse Compton scattering is likely to be an important source of X rays and γ rays, for example, in the intense extragalactic γ-ray sources. Wherever there are large number densities of soft photons, the presence of ultra-relativistic electrons must result in the production of high-energy photons, X rays and γ rays. The case of special interest in this section is that in which the same relativistic electrons which are the source of the soft photons are also responsible for scattering these photons to X-ray and γ-ray energies – this is the process known as *synchro-Compton radiation*. One case of special importance is that in which the number density of low-energy photons is so great that most of the energy of the electrons is lost by synchro-Compton radiation rather then by synchotron radiation. This line of reasoning leads to what is known as the *inverse Compton catastrophe*.

We can derive the essential results from the formulae we have already derived. The ratio, η, of the rates of loss of energy of an ultra-relativistic electron by inverse Compton and synchrotron radiation in the presence of a photon energy density $U_{\rm rad}$ and a magnetic field of magnetic-flux density B is

$$\eta = \frac{(\mathrm{d}E/\mathrm{d}t)_{\rm IC}}{(\mathrm{d}E/\mathrm{d}t)_{\rm sync}} = \frac{U_{\rm photon}}{B^2/2\mu_0}. \tag{226}$$

The synchro-Compton catastrophe occurs if this ratio is greater than 1. In that case, low-energy photons, say, radio photons produced by synchrotron radiation, are scattered to X-ray energies by the same flux of relativistic electrons. Since η is greater than 1, the energy density of the X rays is greater than that of the radio photons and so the electrons suffer an even greater rate of loss of energy by scattering these X rays to γ-ray energies. In turn, these γ rays have a greater energy density than the X rays ...and so on. It can be seen that as soon as the ratio (226) becomes greater than one, all the energy of the electrons is lost at the very highest energies and so the radio source should instead be a very powerful source of X rays and γ rays. Before considering the higher-order scatterings, let us study the first stage of the process for the case of compact synchrotron self-absorbed radio sources.

We need to determine the energy density of radiation within a synchrotron self-absorbed radio source. As shown in Sect. 6.8, the flux density of such a source is

$$S_\nu = \frac{2k_{\rm B}T_{\rm e}}{\lambda^2}\Omega \quad \text{where} \quad \Omega \approx \theta^2 = \frac{r^2}{D^2}. \tag{227}$$

Ω is the solid angle subtended by the source, r is the size of the source and D its distance. We recall that, for a synchrotron self-absorbed source, the electron temperature of the relativistic electrons is the same as its brightness temperature $T_{\rm e} = T_{\rm b}$. The radio luminosity of the source in W Hz^{-1} is

$$L_\nu = 4\pi D^2 S_\nu = \frac{8\pi k_{\rm B} T_{\rm e}}{\lambda^2} r^2. \tag{228}$$

Therefore, the energy density of the radio emission $U_{\rm photon}$ is

$$U_{\rm photon} \sim \frac{L_\nu \nu}{4\pi r^2 c} = \frac{2k_{\rm B}T_{\rm e}\nu}{\lambda^2 c}. \tag{229}$$

Notice that L_ν is the luminosity per unit bandwidth, and so the bolometric luminosity is roughly νL_ν. Therefore,

$$\eta = \frac{\left(\dfrac{2k_{\rm B}T_{\rm e}\nu}{\lambda^2 c}\right)}{\left(\dfrac{B^2}{2\mu_0}\right)} = \frac{4k_{\rm B}T_{\rm e}\nu\mu_0}{\lambda^2 c B^2}. \tag{230}$$

We can now use the theory of self-absorbed radio sources to express the magnetic-flux density B in terms of observables. Repeating the calculations carried out in Sect. 6.8,

$$\nu_g = \nu/\gamma^2 \quad \text{and} \quad 3k_B T_b = 3k_B T_e = \gamma m_e c^2, \tag{231}$$

where T_b is the brightness temperature of the source. Reorganising these relations, we find

$$B = \frac{2\pi m_e}{e}\left(\frac{m_e c^2}{3k_B T_e}\right)^2 \nu. \tag{232}$$

Therefore, the ratio of the loss rates, η, is

$$\eta = \frac{(dE/dt)_{IC}}{(dE/dt)_{sync}} = \left(\frac{81 e^2 \mu_0 k_B^5}{\pi^2 m_e^6 c^{11}}\right) \nu T_e^5. \tag{233}$$

This is the key result. It can be seen that the ratio of the loss rates depends very strongly upon the brightness temperature of the radio source. Putting in the values of the constants, we find that the critical brightness temperature is

$$T_b = T_e = 10^{12}\,\nu_9^{-1/5}\ \text{K}, \tag{234}$$

where ν_9 is the frequency at which the brightness temperature is measured in units of 10^9 Hz, that is, in GHz. Thus, according to this calculation, no compact radio source should have brightness temperature greater than $T_b \approx 10^{12}$ K, if the emission is incoherent synchrotron radiation.

The most compact sources, which have been studied by VLBI at centimetre wavelengths, have brightness temperatures which are less than the synchro-Compton limit, typically, the values found being $T_b \approx 10^{11}$ K, which is reassuring. Notice that this is direct evidence that the radiation is the emission of relativistic electrons since the temperature of the emitting electrons must be at least 10^{11} K. This is not, however, the whole story. If the time scales of variability τ of the compact sources are used to estimate their physical sizes, $l \sim c\tau$, the source regions must be considerably smaller than those inferred from VLBI, and then values of T_b exceeding 10^{11} K are found. It is likely that relativistic beaming is the cause of this discrepancy, a topic which we take up in Chap. 10.

There is no definite evidence that synchro-Compton radiation has been observed in any of the X-ray and γ-ray sources, but it would certainly be no surprise if it were. the origin of the emission in some of these sources. Examples of the expected spectra of sources of synchro-Compton radiation have been evaluated by Band & Grindlay (1985). They take into account the transfer of radiation within the self-absorbed source, as well as considering both homogeneous and inhomogeneous sources. A number of important refinements are included in their computations. Of particular importance is that they take account of the fact that, at relativistic energies $h\nu \geq 0.5$ MeV, the Klein-Nishina cross section rather than the Thomson cross section should

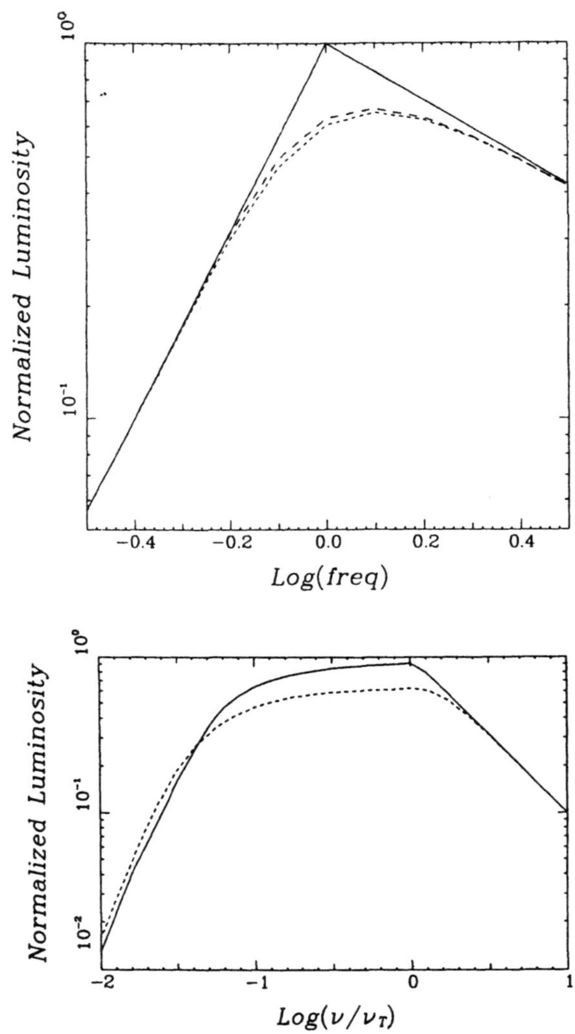

Fig. 32. Examples of the spectra of synchro-Compton radiation of compact radio sources. (**a**) The radio spectrum of the homogeneous source and (**b**) the inhomogeneous source. (Band & Grindlay 1985.)

be used for photon-electron scattering. In the ultra-relativistic limit, the cross section is

$$\sigma_{\mathrm{KN}} = \frac{\pi^2 r_e^2}{h\nu} \left[\ln\,(2h\nu) + \tfrac{1}{2}\right], \tag{235}$$

and so the cross section decreases as $(h\nu)^{-1}$ at high energies. Consequently, higher-order scatterings result in much reduced luminosities as compared with the non-relativistic calculation. Many features of these computations

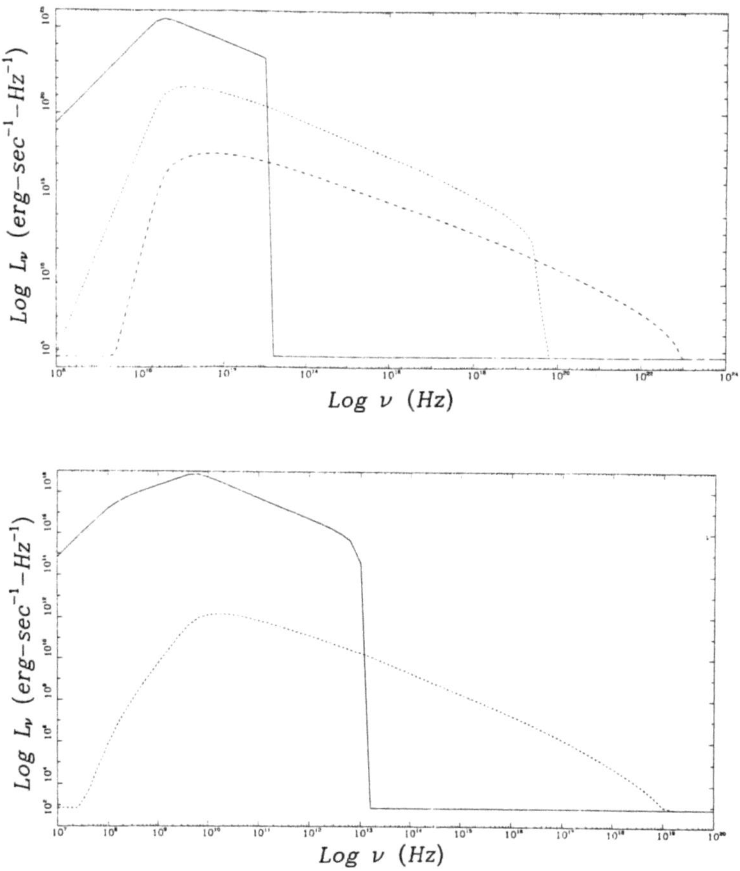

Fig. 33. Examples of the spectra of synchro-Compton radiation of compact radio sources. (**a**) The synchro-Compton spectrum of the homogeneous source and (**b**) the inhomogeneous source. (Band & Grindlay 1985.)

can be understood from the results of the detailed calculations by Band & Grindlay. Figs. 32a and b show the radio spectra of the homogenous and inhomogeneous sources. The homogeneous source has the standard form of spectrum, namely, a power law distribution in the optically thin spectral region $L_\nu \propto \nu^{-\alpha}$, while, in the optically thick region, the spectrum has the form $L_\nu \propto \nu^{5/2}$. In the case of the inhomogeneous source, the magnetic-field strength and number density of relativistic electrons decrease outwards as power laws, resulting in a much broader 'synchrotron-peak'.

Figs. 33a and b show the X-ray and γ-ray spectra of the homogeneous and inhomogeneous sources. In Fig. 33a, the relativistic boosting of the spectrum of the radio emission from the compact radio source is clearly seen. Both the low- and high-frequency spectral features of the radio source spectrum follow the relativistic 'boosting' relations

$$\nu_{\rm g} \to \gamma^2 \nu_{\rm g} \to \gamma^4 \nu_{\rm g} \to \gamma^6 \nu_{\rm g} \ldots \tag{236}$$

These features are most apparent in the case of the homogeneous source. The higher-order scatterings for photon energies $h\nu \gg m_{\rm e}c^2$ are much reduced because of the use of the Klein-Nishina cross section at high energies. In the case of the inhomogeneous model, shown in Fig. 33b, only one Compton scattering is apparent because of the wide range of photon energies produced by the radio source.

It is not clear what role this process plays in the astrophysics of the X-ray and γ-ray spectra of active galaxies. There is evidence that radio quasars have greater X-ray luminosities than radio quiet quasars and synchro-Compton radiation may well be involved, but the situation is far from clear.

9 γ-ray Processes, Photon-photon Interactions and the Compactness Parameter

The processes of synchrotron radiation, inverse Compton scattering and relativistic bremsstrahlung are effective means of creating high-energy γ-ray photons, but there are other mechanisms. One of the most important is the decay of neutral pions created in collisions between relativistic protons and nuclei of atoms and ions of the interstellar gas.

$$\rm p + p \to \pi^+, \pi^-, \pi^0 . \tag{237}$$

The charged pions decay into muons and neutrinos

$$\pi^+ \to \mu^+ + \nu_\mu \quad ; \quad \pi^- \to \mu^- + \bar{\nu}_\mu \tag{238}$$

with a mean lifetime of 2.551×10^{-8} s. The charged muons then decay with mean lifetime of 2.2001×10^{-6} s

$$\mu^+ \to {\rm e}^+ + \nu_{\rm e} + \bar{\nu}_\mu \quad ; \quad \mu^- \to {\rm e}^- + \bar{\nu}_{\rm e} + \nu_\mu . \tag{239}$$

In contrast, the neutral pions decay into pairs of γ-rays, $\pi^0 \to \gamma + \gamma$, in only 1.78×10^{-16} s. The cross section for this process is $\sigma_{\rm pp \to \gamma\gamma} \approx 10^{-30}$ m^2 and the emitted spectrum of γ rays has a broad maximum centred on a γ-ray energy of about 70 MeV (see *HEA2*, Sect. 20.1). This is the process responsible for the continuum emission of the interstellar gas at energies $\varepsilon \geq 100$ MeV. A simple calculation shows that, if the mean number density of the interstellar gas is $N \sim 10^6$ m^{-3} and the average energy density of cosmic-ray protons with energies greater than 1 GeV about 10^6 eV m^{-3}, the γ-ray luminosity of the disc of our Galaxy is about 10^{32} W, as observed.

9.1 Electron-positron annihilation

Perhaps the most extreme form of energy loss mechanism for electrons is annihilation with their antiparticles, the positrons, resulting in the production of pairs of γ-rays. Electron-positron annihilation is of particular interest because definite evidence for this process has been found in the central regions of our Galaxy from observations of the 511 MeV annihilation line (Fig. 34).

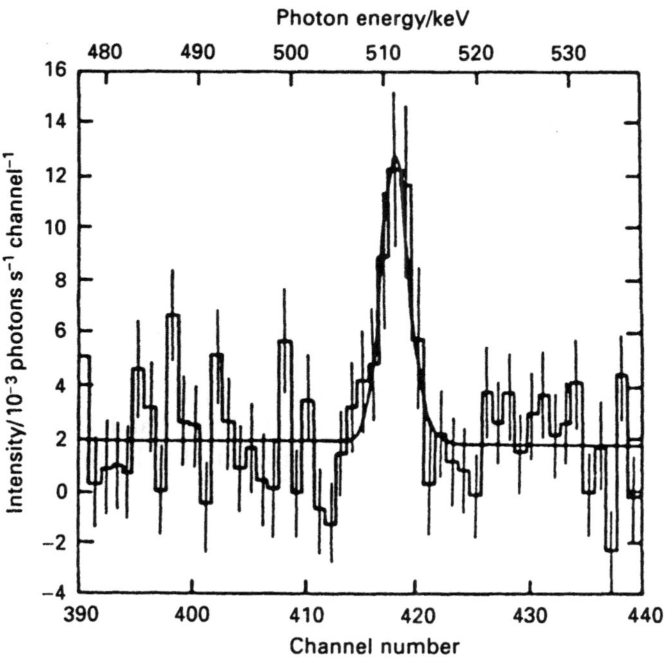

Fig. 34. *HEAO-2* observations of the 0.511 MeV electron-positron annihilation line from the general direction of the Galactic Centre. (Riegler et al. 1981.)

Electron-positron annihilation can proceed in two ways. In the first case, the electrons and positrons annihilate at rest or in flight through the interaction $e^+ + e^- \to 2\gamma$. When emitted at rest, the photons both have energy 0.511 MeV. When the particles annihilate 'in flight', meaning that they suffer a fast collision, there is a dispersion in the photon energies. It is a pleasant exercise in relativity to show that, if the positron is moving with velocity v with corresponding Lorentz factor γ, the centre of momentum frame of the collision has velocity $V = \gamma v(1+\gamma)$ and that the energies of the pair of photons ejected in the direction of the line of flight of the positron and in the backward direction are

$$E = \frac{m_e c^2(1+\gamma)}{2}\left(1 \pm \frac{V}{c}\right). \tag{240}$$

From this result, it can be seen that the photon which moves off in the direction of the incoming positron carries away most of the energy of the positron and that there is a lower limit to the energy of the photon ejected in the opposite direction of $m_e c^2/2$.

If the velocity of the positron is small, *positronium atoms*, that is, bound states consisting of an electron and a positron, can form by radiative recombination: 25% of the positronium atoms form in the singlet $^1S_0$ state and 75% of them in the triplet $^3S_1$ state. The modes of decay from these states are different. The singlet $^1S_0$ state has a lifetime of 1.25×10^{-10} s and the atom decays into two γ-rays, each with energy 0.511 MeV. The majority triplet $^3S_1$ states have a mean lifetime of 1.5×10^{-7} s and three γ-rays are emitted, the maximum energy being 0.511 MeV in the centre of momentum frame. In this case, the decay of positronium results in a continuum spectrum to the low-energy side of the 0.511 MeV line. If the positronium is formed from positrons and electrons with significant velocity dispersion, the line at 0.511 MeV is broadened, both because of the velocities of the particles and because of the low-energy wing due to continuum three-photon emission. This is a useful diagnostic tool in understanding the origin of the 0.511 MeV line. If the annihilations take place in a neutral medium with particle density less than $10^{21}\mathrm{m}^{-3}$, positronium atoms are formed. On the other hand, if the positrons collide in a gas at temperature greater than about 10^6 K, the annihilation takes place directly without the formation of positronium.

The cross section for electron-positron annihilation in the extreme relativistic limit is

$$\sigma = \frac{\pi r_e^2}{\gamma}[\ln 2\gamma - 1]. \tag{241}$$

For thermal electrons and positrons, the cross section becomes

$$\sigma \approx \frac{\pi r_e^2}{(v/c)}. \tag{242}$$

There are several sources of positrons in astronomical environments. Perhaps the simplest is the decay of positively charged pions, π^+, which are created in collisions between cosmic-ray protons and nuclei and the interstellar gas, roughly equal numbers of positive, negative and neutral pions being created. Since the π^0s decay into γ rays, the flux of interstellar positrons created by this process can be estimated from the γ-ray luminosity of the interstellar gas. A second process is the decay of long-lived radioactive isotopes created by nucleosynthesis in supernova explosions. For example, the β^+ decay of ^{26}Al has a mean lifetime of 1.1×10^6 years. This element is formed in supernova explosions and so it is ejected into the interstellar gas where the decay results in a flux of interstellar positrons. A third process is the creation of electron-positron pairs through *photon-photon collisions*, a process of considerable importance in compact γ-ray sources.

9.2 Photon-photon collisions

Let us work out the threshold energy for this process. If $\mathbf{P}_1$ and $\mathbf{P}_2$ are the momentum four-vectors of the photons before the collision

$$\mathbf{P}_1 = [(\varepsilon_1/c)\mathbf{i}_1, \varepsilon_1/c^2] \quad ; \quad \mathbf{P}_2 = [(\varepsilon_2/c)\mathbf{i}_2/c^2], \tag{243}$$

then conservation of four-momentum requires

$$\mathbf{P}_1 + \mathbf{P}_2 = \mathbf{P}_3 + \mathbf{P}_4 \tag{244}$$

where $\mathbf{P}_3$ and $\mathbf{P}_4$ are the four-vectors of the created particles. To find the threshold for pair production, we require that the particles be created at rest and therefore

$$\mathbf{P}_3 = [0, m_\mathrm{e}] \quad ; \quad \mathbf{P}_4 = [0, m_\mathrm{e}]. \tag{245}$$

Squaring both sides of Eq. (244) and noting that $\mathbf{P}_1 \cdot \mathbf{P}_1 = \mathbf{P}_2 \cdot \mathbf{P}_2 = 0$ and that $\mathbf{P}_3 \cdot \mathbf{P}_3 = \mathbf{P}_4 \cdot \mathbf{P}_4 = \mathbf{P}_3 \cdot \mathbf{P}_4 = m_\mathrm{e}^2 c^2$,

$$\mathbf{P}_1 \cdot \mathbf{P}_1 + 2\mathbf{P}_1 \cdot \mathbf{P}_2 + \mathbf{P}_2 \cdot \mathbf{P}_2 = \mathbf{P}_3 \cdot \mathbf{P}_3 + 2\mathbf{P}_3 \cdot \mathbf{P}_4 + \mathbf{P}_4 \cdot \mathbf{P}_4, \tag{246}$$

$$2\left(\frac{\varepsilon_1\varepsilon_2}{c^2} - \frac{\varepsilon_1\varepsilon_2}{c^2}\cos\theta\right) = 4m_\mathrm{e}^2 c^2, \tag{247}$$

$$\varepsilon_2 = \frac{2m_\mathrm{e}^2 c^4}{\varepsilon_1(1-\cos\theta)}, \tag{248}$$

where θ is the angle between the incident directions of the photons. Thus, if electron-positron pairs are created, the threshold for the process occurs for head-on collisions, $\theta = \pi$ and hence,

$$\varepsilon_2 \geq \frac{m_\mathrm{e}^2 c^4}{\varepsilon_1} = \frac{0.26 \times 10^{12}}{\varepsilon_1}\,\mathrm{eV}, \tag{249}$$

where ε_1 is measured in electron volts. This process thus provides not only a means for creating electron-positron pairs, but also results an important source of opacity for very-high-energy γ rays. Table 1 shows some important examples of combinations of ε_1 and ε_2. Photons with energies greater than those in the last column are expected to suffer some degree of absorption when they traverse regions with high energy densities of photons with energies listed in the first column.

The cross section for this process for head-on colisions in the ultra-relativistic limit is

$$\sigma = \pi r_\mathrm{e}^2 \frac{m_\mathrm{e}^2 c^4}{\varepsilon_1\varepsilon_2}\left[2\ln\left(\frac{2\omega}{m_\mathrm{e}c^2}\right) - 1\right] \tag{250}$$

where $\omega = (\varepsilon_1\varepsilon_2)^{1/2}$ and r_e is the classical electron radius. In the limit $\hbar\omega \approx m_\mathrm{e}c^2$, the cross section is

$$\sigma = \pi r_\mathrm{e}^2 \left(1 - \frac{m_\mathrm{e}^2 c^4}{\omega^2}\right)^{1/2} \tag{251}$$

Table 1. The energies of ultra-high energy photons (ε_2) which give rise to electron-positron pairs in collision with photons of different energies (ε_1).

	ε_1(eV)	ε_2(eV)
Microwave Background Radiation	6×10^{-4}	4×10^{14}
Starlight	2	10^{11}
X-ray	10^3	3×10^8

(see Ramama Murthy & Wolfendale 1986). Thus, near threshold, the cross section for the interaction $\gamma\gamma \to e^+e^-$ is

$$\sigma \sim \pi r_e^2 \sim 0.2\sigma_T. \tag{252}$$

These cross sections enable the opacity of the interstellar and intergalactic medium to be evaluated as well as providing a mechanism by which large fluxes of positrons could be generated in the vicinity of active galactic nuclei.

9.3 The compactness parameter

These considerations are particularly important in the case of the extremely luminous and highly variable extragalactic γ-ray sources discovered with the Compton Gamma-Ray Observatory (*CGRO*). A key role is played by the *compactness parameter*, which arises in considerations of whether or not a γ-ray source is opaque for $\gamma\gamma$ collisions because of pair production. Let us carry out a simple calculation which indicates how the compactness parameter arises. We will carry out the calculation for the flux of γ rays at threshold, $\varepsilon \sim m_e c^2$, for simplicity. The mean free path of the γ ray for $\gamma\gamma$ collisions is $\lambda = (N_\gamma \sigma)^{-1}$ where N_γ is the number density of photons with energies $\varepsilon = h\nu \sim m_e c^2$. If the source has luminosity L_γ and radius r, the number density of photons within the source region is

$$N_\gamma = \frac{L_\gamma}{4\pi r^2 c\varepsilon} \tag{253}$$

The condition for the source to be opaque is $r \approx \lambda$, that is,

$$r \sim \frac{4\pi r^2 c m_e c^2}{L_\nu \sigma}, \quad \text{that is,} \quad \frac{L_\nu \sigma}{4\pi m_e c^3 r} \sim 1 \tag{254}$$

The compactness factor C is defined to be the quantity

$$C = \frac{L_\nu \sigma}{4\pi m_e c^3 r} \tag{255}$$

Notice that sometimes the compactness parameter is defined without the factor of 4π in the denominator. If the compactness parameter is very much greater than unity, the γ rays are all destroyed by electron-positron pair production, resulting in a huge flux of electrons and positrons within the source region. Consequently, the source would no longer be a hard γ-ray source. The significance of the compactness parameter can be appreciated from observations of some of the intense γ-ray sources observed by the *CGRO*. These have enormous luminosities, $L_\gamma \sim 10^{41}$ W and have been observed to vary significantly in intensity over time scales of the order of days. Inserting these values into Eq. (255), it is found that $C \gg 1$ and so there is a problem in understanding why these sources exist. Fortunately, an answer is at hand since all the ultra-luminous γ-ray sources are associated with compact radio sources, which exhibit synchrotron self-absorption and many of which display superluminal motions. The inference is that the luminosities of the γ-ray sources and the time scales of variation have been significantly changed by the relativistic motion of the source region. This is a topic which we address in more detail in the next section.

10 Relativistic Beaming

We have been gradually assembling evidence that some high-energy extragalactic sources seem to involve relativistic motion of the emitting regions. The most direct evidence comes from the superluminal motions observed in the compact radio jets found in VLBI observations of active galactic nuclei. In the classic case of the radio core of the radio source 3C 273, one of the radio components appeared to move a distance of 25 light-years in only three years, corresponding to an observed transverse velocity of about eight times the speed of light (Fig. 35). This is a common phenomenon in the compact, variable radio sources which often have spectra which are synchrotron self-absorbed. The phenomenon has also been observed in Galactic radio sources, for example, the source GRS 1915+105, which is a binary X-ray source in which the compact X-ray source is associated with a stellar-mass black hole (Mirabel & Rodriguez 1998).

The observation of compact radio sources with brightness temperatures exceeding the critical value of 10^{12} K on the basis of their temporal variability has already been discussed as evidence that relativistic beaming may be required to overcome the inverse Compton catastrophe. A convincing case can be made that relativistic beaming is the origin of the very rapid variations in intensity observed in some of the most extreme active galactic nuclei, the BL Lac objects. Very rapid radio variability is observed in these sources and so, if dimensions are estimated using the causality relation $l = c\tau$, where τ is the time scale of the variability, the brightness temperature would exceed the critical value of 10^{12} K. A beautiful piece of evidence which supports this view of the BL Lac phenomenon is the radio map of the 3C 371 made with

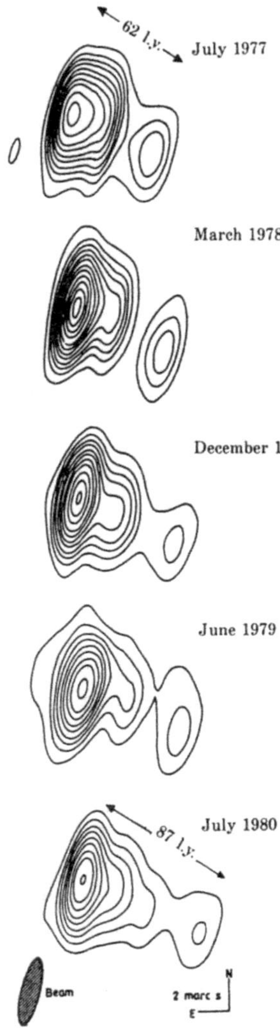

Fig. 35. The superluminal radio source 3C 273 as observed by VLBI over the period July 1977 to July 1980 (Pearson et al. 1981).

the VLA, which had very large dynamic range (Fig. 36). The central compact radio source is the 'blazar' and is extended in the direction of a jet leading to one side of a classical double radio source. According to the consensus view, this source is similar to a classical double radio source, such as Cygnus A (Fig. 24), but now observed at a small angle to the axis of the relativistic jet.

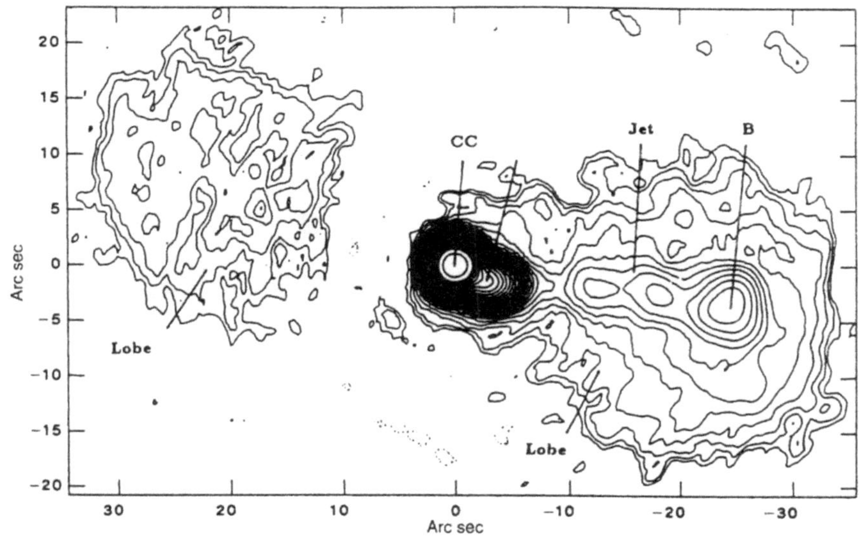

Fig. 36. The highly variable radio source 3C 371 observed with high dynamic range with the VLA (Wrobel & Lind 1990).

Another striking piece of evidence for superluminal motions has already been discussed – the extreme luminosities of the γ-ray sources observed with the Compton Gamma-Ray Observatory. It is wholly persuasive that relativistic motion is involved in these γ-ray sources since essentially all of them are associated with compact self-absorbed radio sources and many of these show direct evidence for relativistic motion.

Let us begin with the simplest, and most popular, model for superluminal sources, what is commonly referred to as the *relativistic ballistic model*. Let us carry out first the simplest part of the calculation, the determination of the *kinematics* of relativistically moving source components. The aim is to determine the observed transverse speed of a component ejected at some angle θ to the line of sight at a high velocity v (Fig. 37). The observer is located at a distance D from the source. The source component is ejected from the origin O at some time t_0 and the signal from that event sets off towards the observer, where it arrives at time $t = D/c$ later. After time t_1, the component is located at a distance vt_1 from the origin and so is observed at a projected distance $vt_1 \sin\theta$ according to the distant observer. The light signal bearing this information arrives at the observer at time

$$t_2 = t_1 + \frac{D - vt_1 \cos\theta}{c}, \tag{256}$$

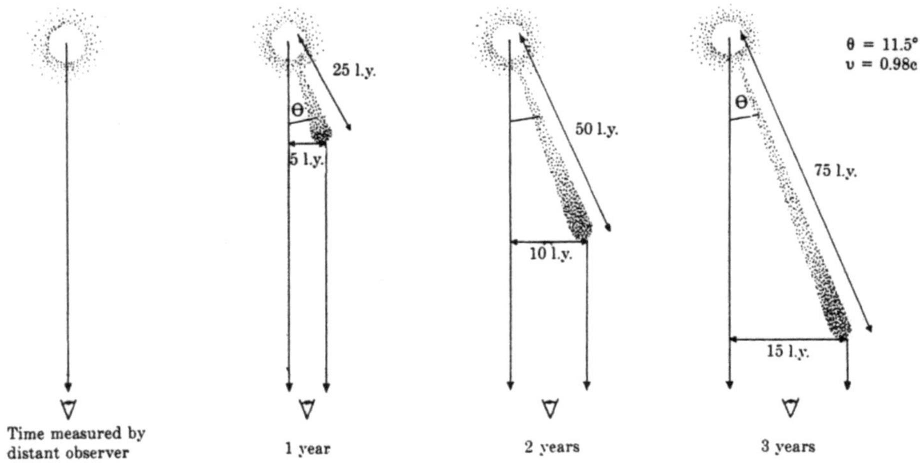

Fig. 37. The motion of a relativistically moving source component as viewed from above. In this example, the parameters have been chosen so that the component is observed to move at five times the speed of light.

since the signals have to travel a slightly shorter distance $D - vt_1 \cos\theta$ to reach the observer. Therefore, according to the distant observer, the transverse speed of the component is

$$v_\perp = \frac{vt_1 \sin\theta}{t_2 - t} = \frac{vt_1 \sin\theta}{t_1 - \dfrac{vt_1 \cos\theta}{c}} = \frac{v \sin\theta}{1 - \dfrac{v\cos\theta}{c}}. \tag{257}$$

It is a simple sum to show that the maximum observed transverse speed occurs at an angle $\cos\theta = v/c$ and is $v_\perp = \gamma v$, where $\gamma = (1 - v^2/c^2)^{-1/2}$ is the Lorentz factor. Thus, provided the source component moves at a speed close enough to the speed of light, apparent motions on the sky $v_\perp > c$ can be observed without violating causality and the postulates of special relativity. For example, if the source component were ejected at a speed $0.98c$, transverse velocities up to $\gamma c = 5c$ are perfectly feasible, the case illustrated in Fig. 37.

This is the easy bit of the story. The trickier bit is to understand the effects of what is loosely referred to as 'relativistic beaming' upon the observed intensities of the source components. Let us consider first a classical undergraduate problem in relativity:

- *A rocket travels towards the Sun at speed $v = 0.8c$. Work out the luminosity, colour, angular size and brightness of the Sun as observed from the spaceship when it crosses the orbit of the Earth. It may be assumed that the Sun radiates like a uniform disc with a black-body spectrum at temperature T_0.*

This problem includes many of the effects found in relativistic beaming problems. Let us work out the separate effects involved in evaluating the intensity of radiation observed in the moving frame of reference. Consider the radiation from an annulus of angular width $\Delta\theta$ at angle θ with respect to the centre of the Sun (Fig. 38).

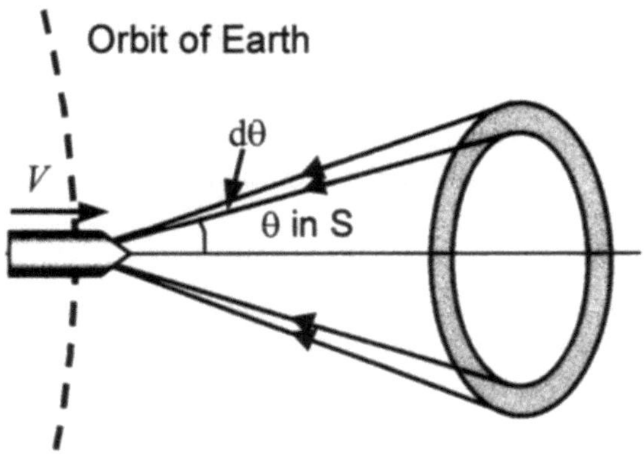

Fig. 38. Illustrating the geometry of the propagation of light from the Sun to the observer in a spaceship moving radially towards the Sun at speed v.

- *The frequency shift of the radiation* It is simplest to use four-vectors to work out the frequency shifts and aberrations. The frequency four-vector in the frame of the Solar System S in Rindler's notation[2] is

$$\mathbf{K} = \left[-k_0 \cos\theta, -k_0 \sin\theta, 0, \frac{\omega_0}{c^2}\right], \tag{258}$$

where the light rays are assumed to propagate towards the observer at the orbit of the Earth, as illustrated in Fig. 38. The frequency four-vector in the frame of reference of the spaceship S′ is

$$\mathbf{K}' = \left[-k_0' \cos\theta', -k_0' \sin\theta', 0, \frac{\omega_0'}{c^2}\right]. \tag{259}$$

We use the time transform to relate the 'time'-components of the four-vectors:

---

[2] In Rindler's notation, the components of the four-vectors transform exactly as $[x, y, z, t]$ according to the standard Lorentz transformation $x' = \gamma(x - Vt), y' = y, z' = z, t' = \gamma(t - Vx/c^2)$. The invariant norm of the four-vector is $|R|^2 = c^2t^2 - x^2 - y^2 - z^2$.

$$t' = \gamma\left(t - \frac{Vx}{c^2}\right), \tag{260}$$

and so

$$\frac{\omega'}{c^2} = \gamma\left(\frac{\omega_0}{c^2} + \frac{Vk_0\cos\theta}{c^2}\right). \tag{261}$$

Since $k_0 = \omega_0/c$,

$$\nu' = \gamma\nu_0\left(1 + \frac{V}{c}\cos\theta\right) = \kappa\nu_0. \tag{262}$$

This is the expression for the 'blueshift' of the frequency of the radiation due to the motion of the spacecraft.

- The *waveband* $\Delta\nu$, in which the radiation is observed, is blueshifted by the same factor
$$\Delta\nu' = \kappa\Delta\nu_0. \tag{263}$$
- *Time intervals* are also different in the stationary and moving frames. This can be appreciated by comparing the periods of the waves as observed in S and S′
$$\nu' = \frac{1}{T'} \quad ; \quad \nu_0 = \frac{1}{T_0}, \tag{264}$$
and so
$$\frac{T'}{T} = \frac{\nu_0}{\nu'}. \tag{265}$$
Since the periods T and T' can be considered to be the times measured on clocks, the radiation emitted in the time interval Δt is observed in the time interval $\Delta t'$ by the observer in S′ such that $\Delta t' = \Delta t/\kappa$.
- *Solid Angles* Finally, we need to work out how the solid angle subtended by the annulus shown in Fig. 38 changes between the two frames of reference. It is simplest to begin with the cosine transform, which is derived from the 'x' Lorentz transformation of the frequency four-vector:
$$\cos\theta' = \frac{\cos\theta + \dfrac{V}{c}}{1 + \dfrac{V}{c}\cos\theta}. \tag{266}$$
Now, differentiating with θ on both sides of this relation, we find
$$\sin\theta'\,\mathrm{d}\theta' = \frac{\sin\theta\,\mathrm{d}\theta}{\gamma^2\left(1 + \dfrac{V}{c}\cos\theta\right)^2} = \frac{\sin\theta\,\mathrm{d}\theta}{\kappa^2}. \tag{267}$$
This result has been derived for an annular solid angle with respect to the x-axis, but we can readily generalise to any solid angle since $\mathrm{d}\phi' = \mathrm{d}\phi$ and so
$$\sin\theta'\,\mathrm{d}\theta'\,\mathrm{d}\phi' = \frac{\sin\theta\,\mathrm{d}\theta\,\mathrm{d}\phi}{\kappa^2} \qquad \mathrm{d}\Omega' = \frac{\mathrm{d}\Omega}{\kappa^2}. \tag{268}$$

Thus, the solid angle in S$'$ is smaller by a factor κ^2 as compared with that observed in S. This is a key aspect of the derivation of the aberration formulae. Exactly the same form of beaming occurred in the derivation of the formulae for synchrotron radiation.

We can now put these results together to work out how the intensity of radiation from the region of the Sun within solid angle $\mathrm{d}\Omega$ changes between the two frames of reference. First of all, the intensity $I(\nu)$ is defined to be the power arriving at the observer per unit frequency interval per unit solid angle from the direction θ. The observer in the spacecraft observes the radiation arriving in the solid angle $\mathrm{d}\Omega'$ about the angle θ' and we need to transform its other properties to those observed in S$'$. Let us enumerate how the factors change the observed intensity. The energy $h\nu N(\nu)$ received in S in the time interval Δt, in the frequency interval $\Delta\nu$ and in solid angle $\Delta\Omega$ is observed in S$'$ as an energy $h\nu' N(\nu')$ in the time interval $\Delta t'$, in the frequency interval $\Delta\nu'$ and in solid angle $\Delta\Omega'$, where $N(\nu) = N(\nu')$ is the invariant number of photons. Therefore, the intensity observed in in S$'$ is

$$I(\nu') = I(\nu) \times \frac{\kappa \times \kappa \times \kappa^2}{\kappa} = I(\nu)\kappa^3. \tag{269}$$

Now, let us apply this result to the spectrum of black-body radiation, for which

$$I(\nu) = \frac{2h\nu^3}{c^2}\left(\mathrm{e}^{h\nu/kT} - 1\right)^{-1}. \tag{270}$$

Then,

$$I(\nu') = \frac{2h\nu^3\kappa^3}{c^2}\left(\mathrm{e}^{h\nu/kT} - 1\right)^{-1} = \frac{2h\nu'^3}{c^2}\left(\mathrm{e}^{h\nu'/kT'} - 1\right)^{-1}, \tag{271}$$

where $T' = \kappa T$. In other words, the observer in S$'$ observes a black-body radiation spectrum with temperature $T' = \kappa T$. A number of useful results follow from this analysis. For example, Eq. (271) describes the temperature distribution of the Cosmic Microwave Background Radiation over the sky as observed from the Solar System which is moving through the frame of reference in which the sky would be perfectly isotropic on the large scale at a velocity of about 600 km s^{-1}. Since $V/c \approx 2 \times 10^{-3}$ and $\gamma \approx 1$, the temperature distribution is rather precisely a dipole distribution, $T = T_0[1 + (V/c)\cos\theta]$ with respect to the direction of motion of the Solar System through the Cosmic Microwave Background Radiation.

In the example of the spacecraft travelling at $v = 0.8c$ towards the Sun, we can illustrate a number of the features of relativistic beaming. In this case, $\gamma = 5/3$ and the angle at which there is no change of temperature, corresponding to $\gamma[1 + (V/c)\cos\theta] = 1$, is $\theta = 60°$.

Let us now turn to the case of relativistically moving source components. Evidently, all we need do is determine the value of κ for the source component moving at velocity V at an angle θ with respect to the line of sight from the observer to the distant quasar as illustrated in Fig. 37. In this case, a straightforward calculation shows that the value of κ is

$$\kappa = \frac{1}{\gamma\left(1 - \dfrac{V\cos\theta}{c}\right)}, \tag{272}$$

where the source is moving towards the observer as illustrated in the figure. Just as in the above example, the observed flux density of the source is therefore

$$S(\nu_{\rm obs}) = \frac{L(\nu_0)}{4\pi D^2} \times \kappa^3, \tag{273}$$

where $\nu_{\rm obs} = \kappa\nu_0$. In the case of superluminal sources, the spectra can often be described by a power law $L(\nu_0) \propto \nu_0^{-\alpha}$ and so

$$S(\nu_0) = \frac{L(\nu_0)}{4\pi D^2} \times \kappa^{3+\alpha}. \tag{274}$$

It is interesting to compare this result with the expressions used in the case in which the sources are at cosmological distances. In that case, we have to use the appropriate distance measure D to describe how the radiation spreads out over a sphere in the curved geometry of space (see, Longair 1998). In that case,

$$S(\nu_0) = \frac{L(\nu_0)}{4\pi D^2} \times \kappa^{1+\alpha}, \tag{275}$$

where $\kappa = (1+z)^{-1}$. Notice the difference of a factor of κ^2 between these formulae. The reason is that, in the cosmological case, the radiation is emitted isotropically over a sphere centred on the source, whereas in the case of the moving source component, the radiation of beamed towards the observer, the change of solid angle being given by Eq. (268).

Thus, if the superluminal sources consisted of identical components ejected from the radio source at the same angle in opposite directions, the relative intensities of the two components would be in the ratio

$$\frac{S_1}{S_2} = \left(\frac{1 + \dfrac{v}{c}\cos\theta}{1 - \dfrac{v}{c}\cos\theta}\right)^{3+\alpha}. \tag{276}$$

It is therefore expected that there should be large differences in the observed intensities of the jets. For example, if we adopt the largest observed velocities for a given value of γ, $\cos\theta = v/c$, then in the limit $v \approx c$,

$$\frac{S_1}{S_2} = \left(2\gamma^2\right)^{3+\alpha}. \tag{277}$$

Thus, since values of $\gamma \sim 10$ are quite plausible and $\alpha \sim 0 - 1$, it follows that the advancing component would be very much more luminous than the receding component. It is, therefore, not at all unexpected that the sources should be one-sided.

Another complication is the fact that the emission is often assumed to be associated with jets. Care has to be taken because, if the jet as a whole is moving at velocity v, then the time dilation formula (265) shows that the advancing component is observed in a different proper time interval as compared with the receding component, the time which has passed in the frame of the source being $\Delta t_1 = \kappa \Delta t_0$ where Δt_0 is the time measured in the observer's frame of reference. If the jet consisted of a stream of components ejected at a constant rate from the active galactic nucleus, the observed intensity of the jet would be enhanced by a factor of only $\kappa^{2+\alpha}$. Thus, the precise form of the relativistic beaming factor is model dependent and care needs to be taken about the assumptions made.

Let us apply these considerations to sources exceeding the limiting surface brightness $T_b = 10^{12}$ K discussed in Sect. 8, and the *compactness parameter* discussed in Sect. 9.3. In the case of the Inverse Compton Catastrophe, Eq. (233) shows that the ratio of the loss rates for inverse Compton scattering and synchrotron radiation depends upon the product νT_b^5. Since the brightness temperature $T_{obs} = \kappa^5 T_0$ and $\nu_{obs} = \kappa \nu_0$, it follows that $\eta \propto \kappa^6$ and so the observed value of T_b can exceed 10^{12} K if the source is moving at such a high velocity that $\kappa \gg 1$. In the case of the compactness parameter,

$$C = \frac{L_\nu \sigma_T}{4\pi m_e c^3 \times ct}, \tag{278}$$

the relativistic beaming factors enable us to understand why these sources should exist. In Eq. (278), it is assumed that the dimensions of the source are $l \approx ct$ from its rapid time variability. The observed luminosity is enhanced by a factor $\kappa^{3+\alpha}$ and, in addition, because the time scale of variability appears on the denominator of Eq. (278), the observed value is shorter by a factor κ and so the compactness parameter is increased by relativistic beaming by a factor of roughly $\kappa^{4+\alpha}$. Since $\alpha \approx 1$, it can be seen that $C \propto \kappa^5$ and so, in the frame of the source components themselves, the value of the compactness parameter can be reduced below the critical value.

11 The Acceleration of Charged Particles

This is a huge subject of importance for many aspects of high-energy astrophysics and there is only space to give a brief impression of the mechanism which now dominates much of the thinking in the field. I have given some more details in *HEA2*, Chap. 21. The preferred mechanism involves first-order Fermi acceleration of particles in strong shock waves. Let us begin with a simple general formulation of the acceleration process in which the average energy of the particle after one collision is $E = \beta E_0$ and the probability that the particle remains within the accelerating region after one collision is P. Then, after k collisions, there are $N = N_0 P^k$ particles with energies $E = E_0 \beta^k$. Eliminating k between these quantities,

$$\frac{\ln(N/N_0)}{\ln(E/E_0)} = \frac{\ln P}{\ln \beta}, \tag{279}$$

and hence

$$\frac{N}{N_0} = \left(\frac{E}{E_0}\right)^{\ln P/\ln \beta}. \tag{280}$$

In fact, this value of N is $N(\geq E)$, since this number reach energy E and some fraction of them is accelerated to higher energies. Therefore,

$$N(E)\,\mathrm{d}E = \text{constant} \times E^{-1+\ln P/\ln \beta}\,\mathrm{d}E. \tag{281}$$

Notice that we have obtained a power law energy spectrum of the particles, exactly what is required to account for the non-thermal spectra of many different classes of high-energy astrophysical sources.

In Fermi's original version of the *Fermi mechanism*, α was proportional to $(V/c)^2$, because of the decelerating effect of the following collisions. The original version of Fermi's theory is therefore known as *second-order Fermi acceleration* and is a very slow process. We would do much better if there were only head-on collisions, in which case the energy increase would be $\Delta E/E \propto V/c$, that is, first-order in V/c and, appropriately, this is called *first-order Fermi acceleration*.

A very attractive version of first-order Fermi acceleration in the presence of strong shock waves was discovered independently by a number of workers in the late 1970s. The papers by Axford, Leer & Skadron (1977), Krymsky (1977), Bell (1978) and Blandford & Ostriker (1978) stimulated an enormous amount of interest in this process for the many environments in which high-energy particles are found in astrophysics. There are two different ways of tackling the problem, one starting from the diffusion equation for the evolution of the momentum distribution of high-energy particles in the vicinity of strong shock waves (for example, Blandford & Ostriker 1978) and the second, a more physical approach in which the behaviour of individual particles is followed (for example, Bell 1978). Let us adopt Bell's version of the theory which makes the essential physics clear and indicates why this version of first-order Fermi acceleration results remarkably naturally in a power law energy spectrum of high-energy particles.

To illustrate the basic physics of the acceleration process, let us consider the case of a strong shock propagating through the interstellar medium. A flux of high-energy particles is assumed to be present both in front of and behind the shock front. The particles are considered to be of very high energy and so the velocity of the shock is very much less than those of the high-energy particles. The key point about the acceleration mechanism is that the high-energy particles hardly notice the shock at all since its thickness is normally very much smaller than the gyroradius of a high-energy particle. Because of

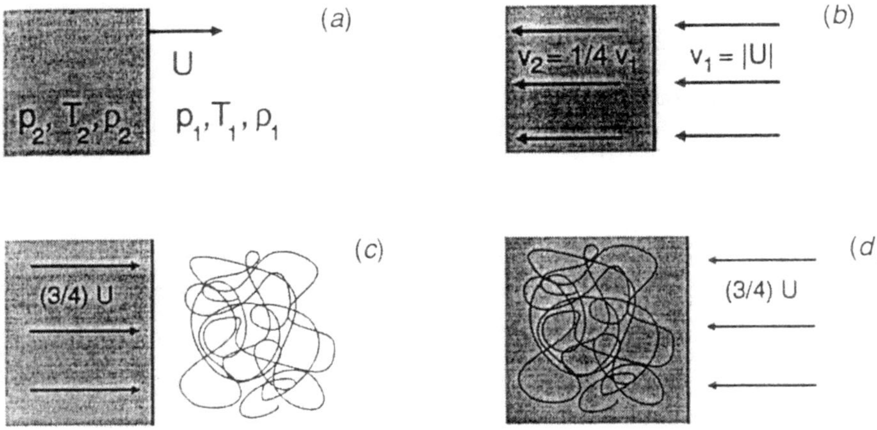

Fig. 39. The dynamics of high-energy particles in the vicinity of a strong shock wave. (**a**) A strong shock wave propagating at a supersonic velocity U through stationary interstellar gas with density ρ_1, pressure p_1 and temperature T_1. The density, pressure and temperature behind the shock are ρ_2, p_2 and T_2, respectively. (**b**) The flow of interstellar gas in the vicinity of the shock front in the reference frame in which the shock front is at rest. In this frame of reference, the ratio of the upstream to the downstream velocity is $v_1/v_2 = (\gamma+1)/(\gamma-1)$. For a fully ionised plasma, $\gamma = 5/3$ and the ratio of these velocities is $v_1/v_2 = 4$ as shown in the figure. (**c**) The flow of gas as observed in the frame of reference in which the upstream gas is stationary and the velocity distribution of the high-energy particles is isotropic. (**d**) The flow of gas as observed in the frame of reference in which the downstream gas is stationary and the velocity distribution of high-energy particles is isotropic.

turbulence behind the shock front and irregularities ahead of it, when the particles pass though the shock in either direction, they are scattered so that their velocity distribution rapidly becomes isotropic on either side of the shock front. The key point is that the distributions are isotropic with respect to the frames of reference in which the fluid is at rest on either side of the shock.

In the case of the material ejected in supernova explosions, the shock velocities can be up to about $U = 10^4$ km s^{-1}, compared with the sound and Alfvén speeds of the interstellar medium which are at most about 10 km s^{-1}. Thus, these are certainly strong shock waves with Mach numbers $\mathcal{M} = U/c_s \gg 1$, where c_s is the sound speed in the ambient medium. It is often convenient to transform into the frame of reference in which the shock front is at rest and then the upstream gas flows into the shock front at velocity $v_1 = U$ and leaves the shock with a downstream velocity v_2 (Fig. 39b). The equation of continuity requires mass to be conserved through the shock and so

$$\rho_1 v_1 = \rho_2 v_2 .$$

In the case of a strong shock, $\rho_2/\rho_1 = (\gamma+1)/(\gamma-1)$ where γ is the ratio of specific heats of the gas. Taking $\gamma = 5/3$ for a monatomic or fully ionised gas, we find $\rho_2/\rho_1 = 4$ and so $v_2 = (1/4)v_1$.

Now let us consider the high-energy particles ahead of the shock. Scattering ensures that the particle distribution is isotropic in the frame of reference in which the gas is at rest. It is instructive to draw diagrams illustrating the dynamical situation so far as typical high-energy particles upstream and downstream of the shock are concerned. Let us consider the upstream particles first. The shock advances through the medium at velocity U but the gas behind the shock travels at a velocity $(3/4)U$ relative to the upstream gas (Fig. 39c). When a high-energy particle crosses the shock front, it obtains a small increase in energy, of the order $\Delta E/E \sim U/c$. The particles are then scattered by the turbulence behind the shock front so that their velocity distributions become isotropic with respect to that flow.

Now let us consider the opposite process of the particle diffusing from behind the shock to the upstream region in front of the shock (Fig. 39d). Now the velocity distribution of the particles is isotropic behind the shock and, when they cross the shock front, they encounter gas moving towards the shock front again with the same velocity $(3/4)U$. In other words, the particle undergoes exactly the same process of receiving a small increase in energy ΔE on crossing the shock from downstream to upstream as it did in travelling from upstream to downstream. This is the clever aspect of this acceleration mechanism. Every time the particle crosses the shock front it receives an increase of energy, there are never crossings in which the particles lose energy, and the increment in energy is the same going in both directions. Thus, unlike the standard Fermi mechanism in which there are both head-on and following collisions, in the case of strong shock fronts, the collisions are always head-on and energy is transferred to the particles. The beauty of the mechanism is the complete symmetry between the passage of the particles from upstream to downstream and from downstream to upstream through the shock wave.

I will simply quote the result of some simple calculations which are described in detail in Sect. 21.4 of *HEA2*. The average energy gain when particles cross from one side of the shock to the other is

$$\left\langle \frac{\Delta E}{E} \right\rangle = \frac{2}{3}\frac{V}{c}, \tag{282}$$

the factor of $\frac{2}{3}$ coming from averaging over all angles of incidence of the particles with respect to the shock wave. $V = \frac{3}{4}U$ is the speed of the material behind the shock. Thus, in one round trip, the fractional energy gain is

$$\left\langle \frac{\Delta E}{E} \right\rangle = \frac{4}{3}\frac{V}{c} = \frac{U}{c}, \tag{283}$$

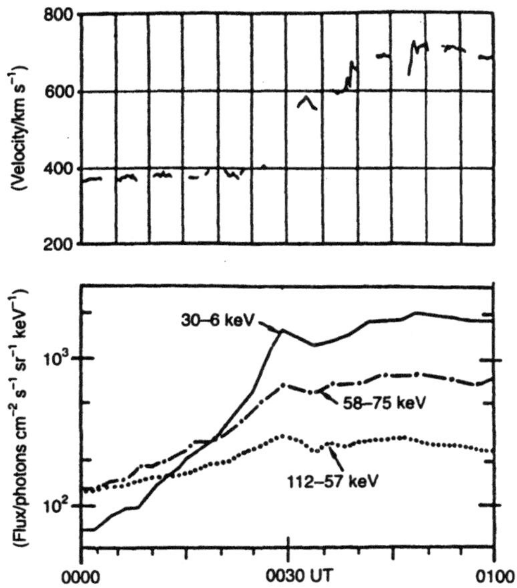

Fig. 40. The distribution of energetic particles in the vicinity of an interplanetary shock wave observed by the *ISEE-3* satellite on November 12, 1978. The upper diagram shows the solar-wind velocity, which suddenly increases as the shock passes the spacecraft. The energetic proton flux increases roughly exponentially ahead of the shock, and the length scale increases with increasing particle energy. After the passage of the shock, the fluxes of energetic particles remain roughly constant on the downstream side of the shock. (From Kennel et al. 1986.)

The other factor we need is the fraction of the particles which are lost per cycle. This can be found from the approach developed by Bell (1978) in which he noted that particles are lost by being 'advected' downstream by the flow of gas behind the shock, the downstream flux being $\frac{1}{4}UN$, whereas the number of particles crossing the shock is $\frac{1}{4}Nc$. Thus, the loss probability is the ratio of these fluxes U/c and the probability of the particles remaining within the accelerating region is

$$P = 1 - \frac{U}{c}. \tag{284}$$

Therefore,

$$\ln\beta = \ln\left(1 + \frac{4V}{3c}\right) = \frac{4V}{3c} = \frac{U}{c} \qquad \ln P = \ln\left(1 - \frac{U}{c}\right) = -\frac{U}{c}. \tag{285}$$

Inserting these values into Eq. (279), we find

$$\ln P/\ln\beta = -1, \tag{286}$$

and so the differential spectrum of the accelerated electrons is

$$N(E)\,\mathrm{d}E \propto E^{-1+\ln P/\ln\beta}\,\mathrm{d}E = E^{-2}\,\mathrm{d}E. \tag{287}$$

This is the remarkable result of this version of first-order Fermi acceleration. The predicted spectrum is of power law form with spectral index -2, corresponding to a synchrotron emission spectrum $\alpha = 0.5$. It may be argued that this is a somewhat flatter spectrum than that of many non-thermal galactic and extragalactic sources. Nonetheless, it is a remarkable result that roughly the correct form of spectrum is found, particularly when it is appreciated that the result depends only upon the assumption that the particles diffuse back and forth across a strong shock wave.

The obvious question is whether or not this actually takes place in strong shock waves. A pleasant direct piece of evidence that it does has been provided by direct observations of particle fluxes in the interplanetary medium on either side of a strong shock, which was observed with the *ISEE* satellite on 12 November 1978 (Fig. 40). The passage of the shock wave past the spacecraft was determined by the abrupt increase in the solar-wind velocity from just less than 400 km s^{-1} to about 700 km s^{-1}. The fluxes of energetic protons increased roughly exponentially as the spacecraft approached the shock and then remained at roughly a constant level in the downstream region. The more energetic protons had a longer length scale in front of the shock than the lower-energy particles. These observations are exactly what would be expected according to the theory of first-order particle acceleration in shocks. According to Völk (1987), the shock mechanism of acceleration can give a good account of the distributions of fast particles observed in shocks which propagate along the magnetic-field direction, but is less successful for oblique shocks.

The subject of particle accelertion in strong shocks has developed dramatically since the results derived in the last section were established. Detailed reviews of these developments have been presented by Drury (1983), Blandford & Eichler (1987) and Völk (1987). The most interesting question concerns how the spectral index of the particle energy distribution x changes as the properties of the shock wave change. The following summary is based upon a survey kindly provided by Dr. Alan Heavens.

Fig. 41 summarises how the spectral index of the radio emission of optically thin synchrotron radiation changes for different assumptions about the physical conditions in non-relativistic shock waves. The results discussed above apply only for strong non-relativistic shocks in which the pressure of the accelerated particles can be neglected. If there is a magnetic field present, the standard results are applicable when the field is uniform and the field direction is perpendicular to the shock. Bell (1978) showed that the predicted electron energy spectral index remains the same, $x = 2$, $\alpha = 0.5$, in the case in which the strong shock propagates at an angle to the magnetic-field direction. If the shock is weak, the compression ratio $\rho_2/\rho_1 < 4$, the velocity

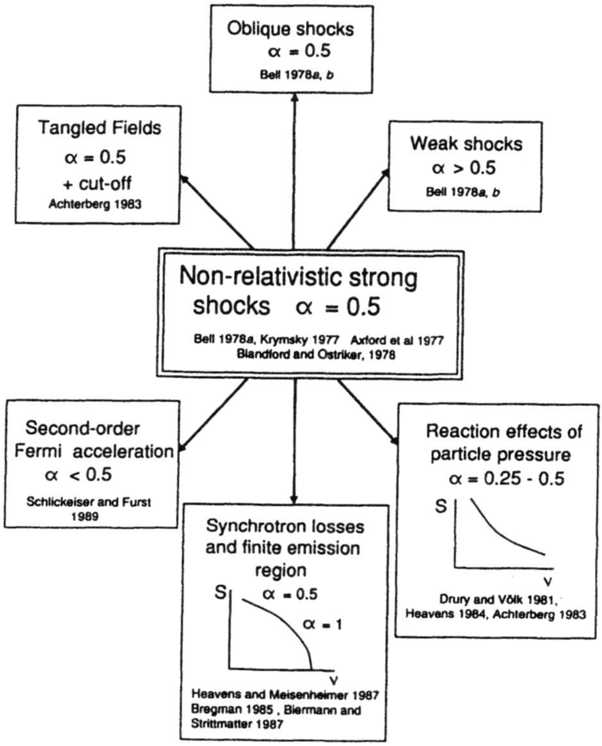

Fig. 41. Modifications to the standard model of shock acceleration for non-relativistic shock waves (Courtesy of Dr. Alan Heavens; for references, see *HEA2*, Chap. 21.)

discontinuity is smaller $r = v_1/v_2 < 4$ and the spectral index of the synchrotron radiation becomes $\alpha = 3/[2(r-1)]$. Therefore, steeper spectra are expected for weak shock waves. The effect of the relativistic particle pressure is to flatten the spectra.

One way of steepening the electron energy spectra is to invoke synchrotron losses of the accelerated particles. For example, in the model of Heavens & Meisenheimer (1987), the accelerated electrons are swept downstream where they suffer synchrotron losses. If the synchrotron loss time of the electrons is less than the age of the source, the steady-state spectrum steepens by $\Delta\alpha = 0.5$, as shown in Sect. 6.9. Notice that, in this case, the energy spectrum of the particles at a given distance behind the shock front has an abrupt cut-off at that energy at which the lifetime of the particles to synchrotron losses is equal to the time since they were accelerated in the shock front. However, when the spectra of all the particles at different distances behind the shock are summed, the standard result is obtained. This steepening may account for the forms of spectra observed in some of the 'hot spots' in extragalactic radio sources which have also been observed in the near-infrared and optical wavebands (Fig. 42). At the very highest energies, there is also expected to be an abrupt cut-off for those particles for which the synchrotron radiation loss time is equal to the characteristic time for acceleration of the particles in the vicinity of the shock front. At low frequencies, the standard spectral index $\alpha = 0.5$ is expected.

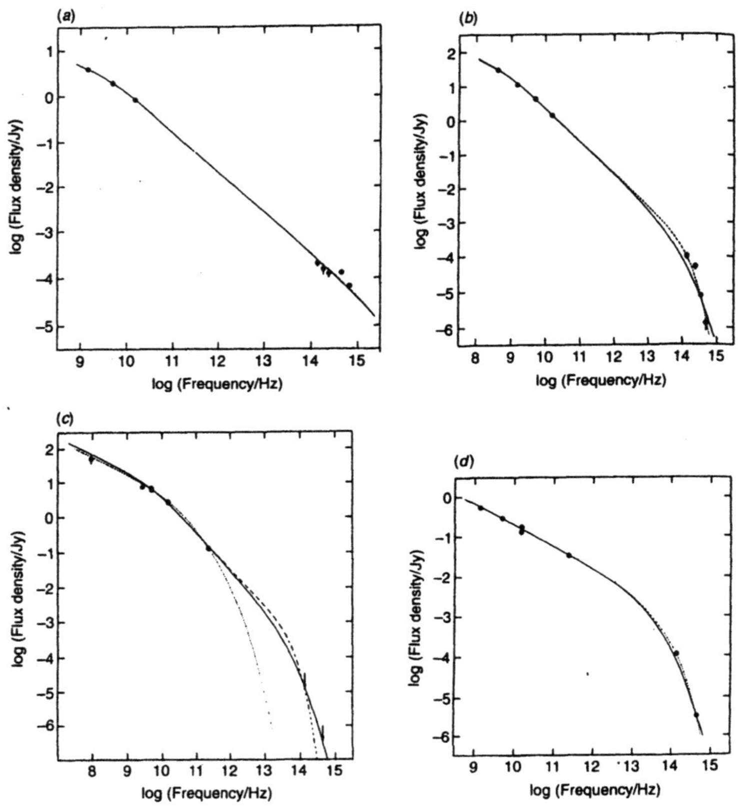

Fig. 42. The spectra of the hot spots in the radio sources Pictor A (west), 3C 273, 3C 123 (east) and 3C33 (south). These forms of spectra can be accounted for by the standard theory of acceleration including the effects of synchrotron losses. (From Meisenheimer et al. 1989.)

The case of relativistic shocks is more complicated and the results of various analyses are displayed in Fig. 43. The relativistic equivalents of the calculations described above result in radio spectral indices in the range $\alpha = 0.35 - 0.6$, still somewhat flatter than the typical spectra of cosmic rays and extragalactic radio sources. As in the case of non-relativistic shocks, steeper spectra are found in weak relativistic shocks. The case of oblique shocks becomes more complicated because, at small enough angles between the shock normal and the magnetic field direction, the shock propagates along the magnetic field lines superluminally and so the particles gyrating about the field lines cannot recross the shock. Kirk & Heavens (1989) find values of the radio spectral index in the range $\alpha = 0 - 0.5$.

It is now possible to test some of these models directly by following the trajectories of individual particles as they propagate back and forth across the shock front. For example, Ballard & Heavens (1992) have studied the acceleration of particles in relativistic shocks in which it is assumed that the magnetic field is tangled on either side of the shock. By averaging over large numbers of particles, the energy spectra of the particles can be found. In their computations, they find the standard value $\alpha \sim 0.5$ for shock velocities $v \leq 0.5c$ but the spectra steepen to $\alpha \sim 0.6 - 1.1$ for greater shock velocities. This result may be relevant to the spectra of the jets in extragalactic radio sources in which the most luminous sources have the steepest radio spectra.

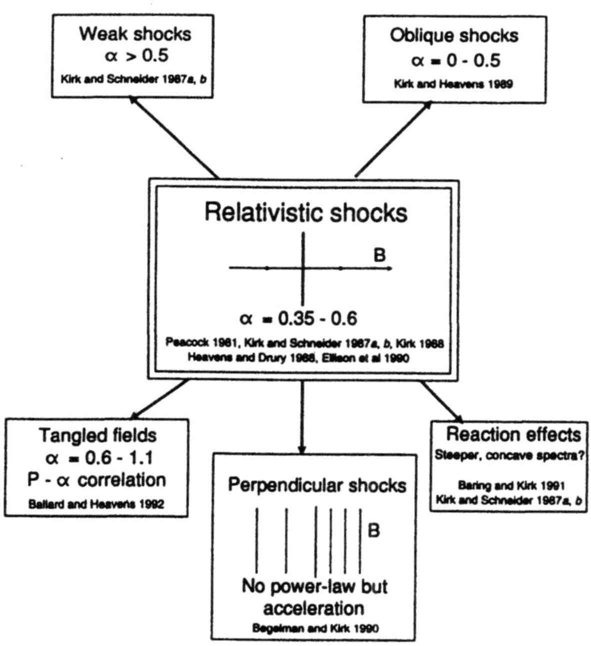

Fig. 43. Modifications to the standard model of shock acceleration for relativistic shock waves (Courtesy of Dr. Alan Heavens; for references, see *HEA2, Chap. 21*)

References

Axford, W.I., Leer, E. & Skadron, G. (1977): *Proc. 15th Intl. Cosmic ray Conf.*, 11, 132
Band, D.L. & Grindlay, J.E. (1985): ApJ 298, 128
Ballard, K.R. & Heavens, A.F. (1992): MNRAS 259, 89
Bekefi, G. (1966): *Radiation Processes in Plasmas*, New York: John Wiley and Sons, Inc.
Bell, A.R. (1978): MNRAS 182, 147
Binggeli, B., Tammann, G.A. & Sandage, A.R. (1987): AJ 94, 251
Birkinshaw, M. (1990): in *The Cosmic Microwave Background: 25 Years Later*, Eds N. Mandolesi & N. Vittorio, Dordrecht: Kluwer Academic Publishers, p. 77
Blandford, R.D. & Eichler, D.(1987): Phys. Rep. 54, 1
Blandford, R.D., Netzer, H. & Woltjer, L. (1990): *Active Galactic Nuclei*, Berlin: Springer-Verlag
Blandford, R.D. & Ostriker, J.P. (1978): ApJ 221, L29
Blumenthal, G.R. & Gould, R.J. (1970): Rev. Mod. Phys. 42, 237
Böhringer, H. (1995): Ann. New York Acad. Sci. 759. 67
Burbidge, G.R. (1956): Phys. rev. 103, 264
Cavaliere, A. (1980): in *X-ray Astronomy*, R. Giacconi & G. Setti (Eds), NAT ASI Vol. C60, Reidel Publ. Cy, p. 217
Drury, L.O'C. (1983): Rep. Prog. Phys. 154, 973
Fabian, A.C., George, I.M., Miyoshi, S. & Rees, M.J. (1990): MNRAS 242, 14
Fabian, A.C. & Barcons, X. (1992): ARA&A 30, 429
Fabricant, D., Lecar, M. & Gorenstein, P. (1980): ApJ 241, 552
Feigelson, E.P., Laurent-Muehleisen, S.A., Kollgaard, R.I. & Fomalont, E.B. (1996): ApJ 449, 149
Ferrario, L., Wickramsinghe, D.T., Bailey, J., Tuohy, I.R. & Hough, J.H. (1989): ApJ 337, 832
Forman, W. & Jones, C. (1982): ARA&A 20, 547
Ginzburg, V.L., Sasonov, V.N. & Syrovarskii, S.I. (1968): Uspekhi Fiz. Nauk 94, 60
Hasinger, G., Burg, R., Giacconi, R., Hartner, G., Schmidt, M., Trümper, J. & Zamorani, G. (1993): A&A 275, 1
Heavens, A.F. & Meisenheimer, K. (1987): MNRAS 225, 335
Jackson, J.D. (1975): *Classical Electrodynamics*, New York: Wiley and Sons, Inc.
Jones, M., Saunders, R. et al. (1993): Nature 365, 320
Karzas, W.J. & Latter, R. (1961): ApJS 6, 167
Kennel, C.F., Coroniti, F.V., Scarf, F.L. et al. (1986): J. geophys. Res. 91 (11), 917
Kirk, J.G. & Heavens, A.F. (1989): MNRAS 239, 995
Kniffen, D.A. & Fichtel, C.E. (1981): ApJ 250, 389
Krymsky, G.F. (1977): Dok. Acad. Nauk. USSR 234, 1306
Lightman, A.P. & White, T.R. (1988): ApJ 335, 57
Longair, M.S. (1984) (revised edition 1992): *Theoretical Concepts in Physics*. Cambridge: Cambridge University Press
Longair, M.S. (1992) (revised edition 1997): *High Energy Astrophysics. Vol. 1.*. Cambridge: Cambridge University Press (*HEA1*)
Longair, M.S. (1994) (revised edition 1997): *High Energy Astrophysics. Vol. 2.*. Cambridge: Cambridge University Press (*HEA2*)

Longair, M.S. (1995): in *The Deep Universe*, Eds A. Sandage, R.G. Kron & M.S. Longair, Berlin: Springer-Verlag
Longair, M.S. (1998): *Galaxy Formation*, Berlin: Springer-Verlag (in press)
Marshall, F.E., Boldt et al. (1980): ApJ 235, 4
Mather, J. (1995): in *The Extragalactic Background Radiation*, Eds D. Calzetti, M. Livio & P. Madau, Cambridge: Cambridge University Press, p. 169
Meisenheimer, K., Röser et al. (1989): A&A 219, 63
Merritt, D. (1987): ApJ 313, 121
Mihara, T., Makashima, K. et al. (1990): Nature 346, 250
Mirabel, I.F. & Rodriguez, L.F. (1998): Nature 392, 673
Myers, S.T., Baker, J.E., Readhead, A.C.S, Leitch, E.M. & Herbig, T. (1997): ApJ 485, 1
Pacholczyk, A.G. (1970): *Radio Astrophysics*, San Francisco: W.H. Freeman and Co.
Pearson, T.J., Unwin, S.C. et al. (1981): Nature 290, 365
Perley, R.A., Dreher, J.W. & Cowan, J.J. (1984): ApJ 285, L35
Pounds, K.A. (1990): MNRAS 242, 20P
Pounds, K.A., Nandra, K., Stewart, G.C., George, I.M. & Fabian, A.C. (1990): Nature 344, 132
Pozdnyakov, L.A., Sobel, I.M. & Sunyaev, R.A. (1983): Astrophys. Sp. Phys. Reviews 2, 263
Ramana Murthy, P. & Wolfendale, A.W. (1986, 1993): *Gamma-ray Astronomy*, Cambridge: Cambridge University Press
Rayner, J. & McLean, I.S. (1987): in *Infrared Astronomy with Arrays*, Eds C.G. Wynn-Williams & E.E. Becklin, Hawaii: University of Hawaii Publications, p. 277
Riegler, G.R., Ling, J.C. et al. (1981): ApJ 248, L113
Rindler, W. (1977): *Essential Relativity*, New York: Springer-Verlag
Rybicki, G.B. & Lightman, A.P. (1979): *Radiative Processes in Astrophysics*, New York: Interscience Publishers
Scheuer, P.A.G. (1966): in *Plasma Astrophysics*, Ed P.A. Sturrock, Proceedings of the International School of Physics 'Enrico Fermi', 39, 289.
Stecker, F.W. (1977): ApJ 212, 60
Taylor, G.B. & Wright, E.L. (1989): ApJ 405, 125
Völk, H.J. (1987): *Proc. 20th Intl. Cosmic Ray Conf.*, Moscow USSR
Wrobel, J. & Lind, K. (1990): ApJ 348, 135

Atomic Physics of Hot Plasmas

Rolf Mewe

Space Research Organization Netherlands (SRON), Sorbonnelaan 2,
NL-3584 CA Utrecht, The Netherlands

Abstract. Plasma with temperatures above a million Kelvin is common in the Universe. It exists for a large part in the intergalactic space, but is also found in a variety of sources which range from optically thin to optically thick. X-ray spectral modeling is an important tool to study the physical parameters of such hot plasmas. In this paper the general procedure of such a modeling is briefly considered and various processes that generate X rays in hot cosmic plasmas are reviewed. Several plasma models, such as the coronal, nebular, and optically thick models are discussed with emphasis on the optically thin coronal model. Various effects of relaxing the restrictions of this model such as those arising from high density, optical depth, and transient ionization are discussed. Most of this paper is dedicated to the atomic physics of calculating X-ray spectra from optically thin plasmas: the ionization balance, the rate coefficients for ionization and recombination, and line excitation, including the formation of dielectronic recombination satellite lines. Finally, the diagnostics of plasma properties such as electron temperature and density, ionization balance, differential emission measure, and non-Maxwellian electron distributions is discussed.

1 Introduction

During the past decades, the results from a sequence of X-ray space missions with ever increasing sensitivity and spectral and spatial resolutions have demonstrated the existence of thermal X-ray emission from hot plasmas above a million Kelvin in a large variety of astrophysical sources including optically thin sources like solar and stellar coronae, clusters of galaxies, supernova remnants, the tenuous intergalactic and interstellar space, and at higher densities also accretion powered sources such as compact X-ray binaries, cataclysmic variables, and active galactic nuclei, where a central X-ray emitting region is surrounded by a cooler medium that is partially photo-ionized.

In modeling these various types of sources, X-ray spectroscopy has proven an invaluable tool. Motivated by the X-ray observations, various authors have developed computer codes in the past in order to explain the observed X-ray emission and to understand the physics of the emitting objects, for optically thin plasmas [e.g., Raymond & Smith 1977 (RS); Mewe & Gronenschild 1981 (MG); Masai 1984; Mewe et al. 1985a; Landini & Monsignori Fossi 1990] or for photo-ionized plasmas (e.g., Kallman& McCray 1982; McCray 1984; see also the Chapter by D. Liedahl).

In part I I consider briefly several thermal and non-thermal processes that can generate X rays in hot plasmas, compare different plasma models, and discuss the general procedure of spectral modeling. I concentrate on the coronal model of a tenuous plasma applicable to hot, optically thin, thermal plasmas in equilibrium. In part II I deal with the basic atomic physics underlying the computation of the ionization balance in a coronal plasma, i.e. the rate coefficients for ionization and recombination. In part III the formation of line and continuum X-ray spectra of optically thin plasmas is considered, in particular the line excitation processes. Finally, part IV deals with the diagnostics of plasma parameters such as electron temperature and density, differential emission measure, and the diagnostics of dielectronic satellite lines including non-Maxwellian electron distributions.[1] Recent reviews on plasma spectroscopy have been given by Kahn & Liedahl (1995), Raymond & Brickhouse (1996), and Griem (1997).

I. X-ray Spectral Modeling of Hot Plasmas

2 Radiation Processes and Plasma Models

In nature, X rays are produced by a variety of processes which may be classified roughly as *thermal* or *non-thermal* processes. The emission mechanisms include blackbody radiation, bremsstrahlung (thermal or nonthermal), line emission, recombination radiation, synchrotron radiation, and inverse Compton radiation (for more details and literature, e.g., Mewe 1992). Observations of certain features of the X-ray flux, such as its spectral energy distribution, emission lines, absorption edges, and degree of polarization, may reveal the dominant emission mechanism in the source. X-ray emission usually results from an electron-photon process. If the energy of the generating electron is thermal in nature, i.e. the electrons are described by a Maxwellian energy distribution characterized by a certain temperature T, we speak of *thermal* processes. Temperatures of a million Kelvin or more are required. The most important processes are line and continuum emission from optically thin plasmas and blackbody radiation from optically thick plasmas. *Non-thermal* radiation is produced when the electrons are non-thermal in nature (e.g., occur in beams). The most important processes for non-thermal X-ray emission in cosmic sources involve the acceleration of (relativistic) electrons in magnetic fields (synchrotron radiation or magnetic bremsstrahlung), and

[1] A few useful textbooks for background studies are: Herzberg (1944); Condon & Shortley (1970) [general atomic spectroscopy]; Griem (1997) [plasma spectroscopy]; Kahn & Liedahl (1995); Raymond & Brickhouse (1996), and Peacock (1996) [review articles on plasma spectroscopy]; and Pal'chikov & Shevelko (1995) [atomic data on highly ionized atoms].

the interaction of energetic electrons with visible, infrared, or microwave photons (inverse Compton radiation), or the production of bremsstrahlung and line emission by the deceleration of electron beams as a result of Coulomb interaction with ions in solar and stellar flare plasmas (cf. the Chapters by Longair, and Liedahl).

All processes are given in Table 1 of Mewe (1992), together with examples of the cosmic sources in which they occur, but here we summarize only the radiation processes in optically thin plasmas (such as stellar coronae, supernova remnants, hot interstellar matter, intra-cluster gas, normal galaxies and galactic halos). Note that the first two processes are usually the most important ones.

- Thermal bremsstrahlung continuum produced by a transition of a free plasma electron between two continuum states above the ionization limit of the ion: energy spectrum $\propto T^{-\frac{1}{2}}e^{-E/kT}$. Dominant at $T \gtrsim 10^8$ K.
- Discrete line emission (electronic transition between two bound levels of the ion). Dominant at $T \lesssim 5\ 10^7$ K. Collisionally excited spectral lines from highly ionized atoms are signatures of the thermal nature of hot plasmas.
- Radiative recombination continuum (capture of the electron into a bound state of the ion) with emission edges that fade out for $T \gtrsim 10^7$ K, when bremsstrahlung dominates.
- Dielectronic recombination lines (capture of a free electron into a doubly excited ion state through simultaneous excitation of a bound electron of the ion).
- Two-photon continuum (simultaneous emission of two photons from a metastable state).

Astrophysical plasmas are usually discussed in terms of the three thermal models presented in Table 1 or in terms of combinations thereof. Since the discovery in 1948 of X rays from the solar corona. Elwert (1952) started his pioneering work by applying the optically thin model to the solar corona, and this is therefore designated as the *coronal model*. The *nebular model* is the X-ray analogon of a planetary nebula, in which a central continuum source photoionizes the surrounding gas. This, and the optically thick *atmosphere model* can be applied to the important classes of X-ray binary sources containing a compact object, such as a white dwarf (WD), neutron star (NS), or black hole (BH). In practice, we may encounter a mixture of all models because in many cases the observed sources are not spatially resolved, which impedes the disentanglement of emission regions with different model properties. For example, low-resolution spectral observations with the Objective Grating Spectrometer on the *Einstein* observatory of twenty bright galactic X-ray binaries revealed blended line emission (and absorption) features from highly ionized atoms which can neither be explained by photoionization nor by coronal models (Vrtilek et al. 1991). Probably the features are observed from a mixture of various distinct - and spatially unresolved - "coronal" and "nebular" regions.

How can we discriminate between different models such as the coronal and nebular models? For this we need spectroscopy with very high resolution ($\lesssim$0.05 Å) such as provided by NASA's Advanced X-ray Astrophysics Facility *AXAF* and ESA's X-ray Multi-Mirror Mission *XMM*, space missions that are expected to fly near the end of this century.[2] First, it is possible to disentangle the different line emission regions by the Doppler measurement of line velocities as a function of binary phase and, second, to resolve the detailed differences in the spectra due to the different line formation processes in the two models.

Table 1. Comparison between plasma models

	Coronal Model	Nebular Model	Atmosphere Model
Assumptions	Optically thin; collisional ionization & collisional excitation	Optically thin/thick; photo-ionization & -exc., and el. scatt. by radiat. from ext. X-ray source	Optically thick; collisional ionization
Examples	-stellar coronae & flares -SNRs (transient) -clusters of galaxies	-compact objects (WD, NS,BH in X-r.binaries) -stellar winds -AGNs	-photosphere of hot O stars -hot DA WDs -NS
Model parameters	-$EM = \int n_e N_H dV$ or -$DEM = n_e N_H dV/d\log T$ -T (controlled externally) -n_e (dens.-sensitive lines) -abundances	-$\xi = L_X/nr^2$ (ioniz. par.) -$(L_X n)^{1/2}$ (continuum opt. depth) -n_e -abundances	-T_{eff} -g -abundances
Characteristics	Spectra with emiss.lines excited by electron coll.; He-like singl./tripl.ratio & Fe XVII-XIX lines discriminate between coronal/nebular model	Spectra with emission & absorption lines formed by recombination; *many* ion stages at low T	Spectra with ionization absorption edges/lines and continuum
Problems	-atomic rates -non-stationary ionization balance	T is not free parameter, but is determined by local energy equations	-NS comptonization (Monte Carlo calc.) -layered atmosphere -non-LTE effects

In the coronal model the heat input is coupled directly to the ions and free electrons, and its characteristic parameters are (electron) temperature T, ele-

[2] See also the Chapter by Paerels.

ment abundances, and (differential) emission measure $(D)EM$ (cf. Sect. 14). At given T only one or two ionization stages of a given element are abundant. In the nebular model T is not a free parameter, but instead is determined by absorption and emission in the gas of the incident radiation from an external X-ray source embedded in the gas. The gas is primarily ionized by inner-shell photoionization. As a result, a wider range of ionization stages of a given element can simultaneously occur and the elements are more highly ionized ("overionized") at a given temperature than they would be in the coronal model. In a photo-ionized plasma the excitation of lines is dominated by radiative recombination (with cascades) and photo-excitation as opposed to collisional excitation from the ground state in a coronal plasma. Therefore, high-resolution spectra of X-ray photoionized plasmas can appreciably differ from those of collisionally ionized plasmas with similar ion concentrations.

Liedahl et al. (1991) have developed a useful diagnostic to distinguish between coronal and nebular models using high-resolution Fe L-shell $3 \rightarrow 2$ spectra (from Fe XVII–XIX ions) at ~1 keV on the basis of the different line formation processes in a coronal plasma (collisionally excited $3d$ lines at T a few MK) and a nebular plasma (recombination-cascade-populated $3s$ lines at T about 0.1 MK).

3 Spectral Modeling of Optically Thin Plasmas

3.1 General scheme

We need to assume a model to infer from the observations the relevant physical parameters including electron temperature, emission measure, density distributions, ion and elemental abundances, mass motions, and the nature of the ambient radiation field. The usual procedure is to apply a forward modeling technique by convolving theoretical model spectra with the instrumental response and to vary the model parameters in order to optimize the fit of the model to the observational data. A common approach is to consider first a simplified plasma model for the X-ray source, neglecting much of the complexity of the temperature and density structure and of the effects of opacity, and to synthesize such models into successively more sophisticated approximations of the source model. The processing flow diagram in Fig. 1 illustrates the spectral modeling.

3.2 Spectral fitting with SPEX

The advent of the new series of X-ray satellites with high sensitivity and spectral resolution such as *EUVE*, *ASCA*, *BeppoSAX*, *AXAF* (1998), *XMM* (1999), *Spectrum X-γ* (1999), and *ASTRO-E* (2000) strongly demands the availability of spectral codes with higher accuracy and more detail than before. Since 1992 a completely new software package **SPEX** (**SPE**ctral **X**-ray

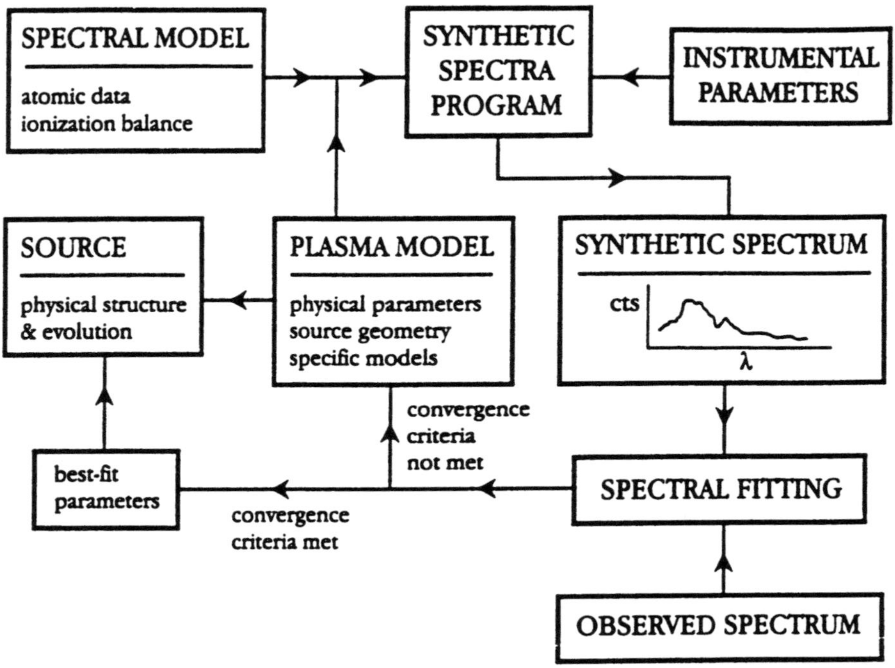

Fig. 1. Processing flow diagram of spectral modeling of optically thin plasmas (from Mewe & Kaastra 1994).

and UV modeling & analysis software package) was designed and developed at our SRON institute to model and fit complicated spectra of hot astrophysical plasmas (Kaastra, Mewe & Nieuwenhuijzen 1996). It encompasses a number of modules for the computation of physical parameters and associated emergent spectra of optically thin plasmas in *collisional ionization equilibrium* **(CIE)** (Mewe et al. 1985a, 1986; Kaastra & Mewe 1993; Mewe & Kaastra 1994) ranging from simple one-temperature to more complicated models, such as stellar coronal loop structures and supernova remnants [also including transient ionization effects, so-called *non-equilibrium-ionization* **(NEI)** models], photoionized plasmas, and optically thick plasmas. Moreover, it contains a variety of differential emission measure (DEM) analysis methods for optically thin plasmas (cf. Sect. 14).

The model spectra can be convolved with instrumental response functions. Different spectral fitting procedures are available. Also simple models (e.g., power law, blackbody, delta functions, and Gaussian lines, and in the future also photoionization models) are included. SPEX has been installed at the

High Energy Astrophysics Science Archive Research Center (HEASARC) of NASA since 1995, and in a more updated version at our institute since 1996. Calculations for the Fe-L complexes have been *preliminarily* updated using results from the HULLAC code at Livermore (Liedahl, Osterheld & Goldstein 1995), and various other improvements have been made (Mewe, Kaastra & Liedahl 1995a).

Several earlier versions of our optically thin plasma code have been implemented as the "Mewe" code at various institutes and in the widely distributed spectral-fitting package XSPEC. The latest version of the code – which contains the updates mentioned above – was installed in 1995 under the name "MEKAL" (Mewe–Kaastra–Liedahl) in XSPEC (version 8.3) of HEASARC.

4 The Coronal Model

The coronal model is a familiar standard model that was first applied to the *solar corona* (Elwert 1952). It implies the following assumptions:

- The plasma is optically thin, so that the X rays are not attenuated by the interaction with the atoms or ions in the plasma, and do not affect the populations in the bound atomic levels.
- The gas density is sufficiently low so that the excited state populations are negligible compared to the ground state population.
- Radiation losses are balanced by non-radiative (mechanical) heating.
- The plasma electrons and the ions are relaxed to Maxwellian energy distributions with a common temperature, T, a free parameter controlled by external processes.
- The gas is assumed to be in a steady state of statistical equilibrium both for the bound atomic states and for the ionization balance.

Assumption (1) implies that the emergent X-ray spectrum faithfully represents the microscopic emission processes in the plasma and therefore is directly linked to the physical conditions in the plasma. It further implies that *photo-excitation* and *photo-ionization* (processes that are very important in nebular-type plasmas) are neglected here.

Assumption (2) may break down in the case of metastable levels which then should be taken into account separately.

Assumption (3) is valid for cosmic sources such as stellar coronae and supernova remnants which are heated by non-radiative processes that include e.g., magnetic-field annihilation, MHD waves, and shock waves.

Assumption (4) implies that the electron-electron relaxation time $t_{ee} \approx 0.01T^{3/2}n_e^{-1}$ (s) [Spitzer 1962; T in K, electron density, n_e, in cm^{-3}] is short enough to ensure a Maxwellian velocity distribution for the electrons, unless the time scales for energy loss or gain, or particle containment are smaller than t_{ee}. If the mechanisms of energy supply to the plasma preferentially heat one kind of particles (e.g., heavy ions in shocks or electrons in microturbulent plasmas) ion and electron temperatures may differ significantly, if the Coulomb collision equilibration time (Spitzer 1962) $t_{ei} \approx 10\ T^{3/2} n_e^{-1}$ (s) is too long, unless plasma instabilities reduce the equilibration time scale (e.g., in the turbulent shock front in supernova remnants; Mewe 1984).

In order to keep the plasma in a steady state, the heating is balanced by the cooling due to line and continuum emission and the ionization equilibrium must be reached (assumption 5). Line emission is mainly due to electron impact excitation followed by spontaneous decay of a bound level within a highly ionized atom, while continuum emission is produced by the interaction of a free electron with an ion either by a free-free transition between two continuum states, a free-bound transition, or a two-photon process.

In the other extreme case of a *high-density* plasma in Local Thermodynamic Equilibrium (*LTE*) every atomic process is as frequent as its inverse process: we call this *the Principle of Detailed Balancing (PDB).*[3]

In an optically thin plasma the radiation which originates in the interior escapes the plasma so that each collision process cannot be balanced by its inverse collision process as in the LTE model. A simple description is then only possible if we assume that the electron density and the radiation field intensity are so small that an excited atom will decay by spontaneous radiation and an ionized atom will recombine by radiative or dielectronic recombination. The stationary ionization state in the coronal model is established by a balance between electron impact ionization and excitation-autoionization, and radiative and dielectronic recombination; the electron density is too low for *3-body recombination* (the inverse process of collisional ionization) with dielectronic or radiative recombination. Such a plasma is said to be in *collisional ionization equilibrium* (CIE).

Examples of steady coronal plasmas:

- stellar coronae;
- hot gas in interstellar and intergalactic medium;
- clusters of galaxies.

A transient, but optically thin plasma such as a supernova remnant is an example of *non-ionization equilibrium* (NEI) (cf. Sect. 4.1.3).[4]

High-resolution soft X-ray (5–140 Å) spectroscopy (e.g., with *AXAF* and *XMM*) will allow detection and identification of a multitude of prominent spectral features from nearly all ion stages of many abundant elements, including the K-shell transitions of C, N, O, Ne, Mg, and Si, and the L-shell transitions of Si, S, Ar, Ca, Ni, and Fe. The interpretation of such an X-ray

[3] We can use the PDB to derive the relation between the atomic rate coefficients of two inverse processes which is independent of the equilibrium (cf. Sections 7.1, 7.2, and 8.1.2).

[4] Though strictly speaking assumption (5) excludes this category of transient plasmas from the coronal model, all rate coefficients valid for optically thin plasmas can also be used in this case.

spectrum requires a detailed knowledge of the ionization, recombination and excitation rates involved.

It is obvious that the physical parameters can be inferred only at the accuracy level of the available rates of ionization, recombination and excitation. Later on I discuss the expected accuracy of these data, but I may also refer to Raymond (1988) and Mewe (1990) who have discussed the atomic physics of the ionization balance and the line excitation.

4.1 Deviations from the coronal CIE model approximation

Apart from the uncertainty in the basic atomic parameters other uncertainties can arise from the simplifying assumptions we have made in the coronal model. Mewe (1990) and Raymond (1988) have discussed the effects of relaxing the restrictions for the simple steady optically thin model by considering effects of photo-ionization, optical depth, resonance line scattering, high density, non-Maxwellian velocity distributions, and non-equilibrium ionization (NEI) in a transient plasma. I consider the effects of optical depth, high density, and of NEI and mention briefly the effect of a non-Maxwellian velocity distribution. For more details I refer to the above papers.

4.1.1 Optical-depth effects: criteria for the optically thin approximation. In the coronal model we assume that the plasma is optically thin, so that the observed intensities and plasma emissivities are directly proportional. Once some radiation is absorbed, radiative transfer must be considered.

In the extreme case of very density plasmas, e.g., those present in the inner regions of accretion flows on compact objects such as hot white dwarfs and neutron stars, the source is optically thick to both continuum and line radiation. The spectrum will resemble, at very high optical depths, blackbody emission.

At intermediate optical depths, the spectral formation is influenced by complicated radiation transfer effects as well as by fundamental atomic processes. Discrete spectral structure is expected which can provide much information about the source (Ross 1979). For X-ray emitting plasmas, Compton scattering plays a significant role as well, and transfer through the scattering plasma will broaden and shift line profiles and alter the continuum distribution, depending on temperature and column density (Lightman et al. 1981).

For the optically thin approximation we apply the criterion that the intensity of a given type of radiation should not differ from the value obtained from the optically thin approximation by more than 10%, which can be expressed as (e.g., Cooper 1966):

$$\tau_\lambda(D) \lesssim 0.2 \ , \qquad (1)$$

where $\tau_\lambda = \alpha_\lambda D$ is the optical depth at wavelength λ, α_λ the linear absorption coefficient (cm^{-1}) and D the typical dimension (cm) of the (homogeneous) plasma. I evaluate criterion (1) for the following radiation processes which may contribute to the optical depth (ignoring stimulated emission or stimulated recombinations) [cf. Cooper 1966; Wilson 1962]:

(*i*) photo-absorption of line radiation,
(*ii*) photo-absorption of recombination radiation,
(*iii*) photo-absorption of bremsstrahlung,
(*iv*) scattering by free electrons.

(*i*). For line radiation, resonance line absorption is most serious. For a Doppler broadened line profile, the application of criterion (1) to the central wavelength λ (Å) gives [expressing the density $N_{Z,z}$ of the absorbing ion Z^{+z} of element Z in terms of the electron density n_e (cm^{-3}), the ion fraction $\eta_{Z,z}$ and the element abundance $A_Z = N_Z/N_H$, i.e. $N_{Z,z} = 0.85 n_e \eta_{Z,z} A_Z$ for a plasma with cosmic abundances]:

$$Dn_e \lesssim 2\ 10^{13} \frac{\sqrt{T_i/M_i}}{\lambda f \eta_{Z,z} A_Z}\ (\mathrm{cm}^{-2})\ , \qquad (2)$$

where f is the absorption oscillator strength of the line, T_i(K) the ion temperature and M_i the ion mass number. For example, for λ=20Å, f=0.5, η=0.5, A_Z=10^{-4}, T_i=3 10^6 K, and D=10^8 cm, we have $n_e \lesssim 2\ 10^{11}$ cm^{-3}, which emphasizes the importance of considering very carefully opacity effects in solar and stellar flare and active region conditions. This criterion will be relaxed when additional line broadening from Stark [$\Delta\lambda_S \sim 4\ 10^{-18}\lambda^2 n_e^{2/3}(z+1)^{-1}$] or Zeeman effect [$\Delta\lambda_Z \sim 10^{-12}\lambda^2 B$] becomes comparable to the Doppler broadening ($\Delta\lambda_D \sim 7\ 10^{-7}\lambda\sqrt{T_i/M_i}$) at sufficiently large plasma densities or magnetic fields (B in Gauss, line widths $\Delta\lambda$ and wavelengths in Å).

When the plasma becomes optically thick ($\tau \gtrsim 1$) the effect on the intensity of the resonance line is determined by the processes competing with the spontaneous radiative decay to the ground level. In a high-density plasma a resonance photon will be completely destroyed after an absorption, but in a low-density plasma (with $\tau > 1$) the situation can exist that each time a photon is absorbed, only a small fraction b ($\ll$1) is destroyed. Since, on average, a photon is absorbed and re-emitted ("scattered") $\sim \max(\tau, \tau^2)$ times before escaping, the plasma can still be considered effectively optically thin as long as $b\tau \ll$1 (e.g., Hearn 1966). E.g., for Lyman α, where the loss occurs through electron excitation to the next higher level, $b \sim$2 $10^{-23}\lambda^3 T^{-1/2} n_e$, but for Lyman β the situation is completely different because here the main process competing with Lyman β emission is Balmer α emission, hence $b \sim$0.44, independent of temperature and density.

If the emitting region is not spherical (e.g., a long thin coronal magnetic flux tube), the effects of resonant scattering can drastically enhance the intensity in the direction of the shortest plasma dimension where the re-emitted photons (from absorption in the longitudinal direction) more readily escape (e.g., Acton & Brown 1978; Sylwester et al. 1986, and discussion by Raymond

1988). In principle, one can extract from the spectra information about the geometry of the source, although the interpretation can be quite complicated.

(*ii*). For photoexcitation we obtain approximately for transitions from the ground state to the absorption edge (where the absorption is maximum):

$$Dn_e \lesssim 5\ 10^{16} \frac{(z+1)^2}{\eta_{Z,z} A_Z}\ (\mathrm{cm}^{-2})\ . \tag{3}$$

Breakdown of criterion (3) occurs in an important class of cosmic X-ray sources involving *photoionized nebulae*, such as accretion-powered sources like X-ray binaries, cataclysmic variables, and active galactic nuclei, where a central X-ray emitting region is surrounded by a cooler, partially ionized medium, and hot shocks in the winds of early-type stars.

In all these cases photoelectric absorption edges will substantially modify the emergent X-ray spectrum particularly at longer wavelengths. High-resolution spectral measurements of the strength of K-shell absorption edges in combination with emission lines produced by recombination, provide information on the geometry of the medium surrounding the source along the line of sight. The nebular model (e.g., Holt & McCray 1982; McCray 1984; Kallman & McCray 1982) is the X-ray analogue of a planetary nebula, in which a central continuum source ionizes the surrounding gas. The gas may be optically thick to photoabsorption but not to electron scattering. The ionization and temperature structure of the gas are established by a stationary balance between photoionization (collisional ionization can be neglected) and heating due to the central X-ray source and, on the other hand, (radiative plus dielectronic) recombination and charge exchange and cooling of the gas. When the gas is optically thin, the local radiation field is determined by geometrical dilution of the source spectrum. Then the local state of the gas (at radius R from the central X-ray source) can be parametrized in terms of the scaling parameter $\xi = L/nR^2$ [L is total luminosity of the central source, n is the local gas density; when electron scattering also is important, the ionization parameter is defined as $\xi = L/nR^2\tau$, where τ is the electron scattering optical depth (Ross 1979; Fabian & Ross 1981)].

The model can be applied to a wide variety of astrophysical X-ray sources and ranges from optically thin to optically thick in the photoionization continuum of abundant elements. In the latter case, the transfer of continuum radiation should be taken into account which yields one additional parameter $(Ln)^{1/2}$ which characterizes the continuum optical depth at a given value of ξ. The photoionization model is treated in detail in the Chapter by D. Liedahl.

(*iii*). Using the free-free absorption coefficient (e.g., Spitzer 1962; Cooper 1966), criterion (1) reduces to

$$Dn_e^2 \lesssim 1.5\ 10^{46}\, T^{1/2} \lambda^3 \bar{G}_c^{-1}\ (\mathrm{cm}^{-5})\ , \tag{4}$$

where T is the electron temperature (K) and $\bar{G}_c$ the average Gaunt factor (Mewe et al. 1986). The criterion is best evaluated for the region of maximum bremsstrahlung emission, i.e. for $h\nu \sim 2kT$, or $\lambda T \sim 10^8$Å K [here $G_c \sim 1$ for a plasma with cosmic abundances (Mewe et al. 1986)].

(*iv*). For electron scattering to produce no appreciable optical depth, the Thomson cross section ($\sigma_T = 6.65\ 10^{-25}$ cm^2) gives

$$Dn_e \lesssim 3\ 10^{23}\ (\mathrm{cm}^{-2}). \tag{5}$$

For the case of plasmas in which Compton scattering dominates I refer to the Chapter by D. Liedahl.

With the above criteria (*i*)–(*iv*) we can estimate when the intensity at a given wavelength differs by more than 10% from the optically thin value. Examination of these criteria indicates that criterion (*i*) is often most severe and criterion (*ii*) will be so in photoionized plasmas, while the criteria (*iii*) and (*iv*) will break down only for very dense plasmas in compact X-ray sources (powered by accretion onto a neutron star or black hole) where Compton scattering plays a significant role.

4.1.2 High-density effects: transition from coronal to thermal model. The ionization distribution in a low-density coronal plasma is very different from that described by the *Saha equation* (e.g., Unsöld 1955; Allen 1973; Griem 1964, 1997) in a very dense plasma in thermodynamic equilibrium which, e.g., for a hydrogenic plasma reads (Z^{+z} is hydrogenic ion with ionization energy χ_z, $Z^{+(z+1)}$ is bare nucleus of charge number $z+1=Z$), omitting subscript Z:

$$\frac{N_{z+1}n_e}{N_z} = \frac{2w_{z+1}}{w_z}\left(\frac{2\pi m_e kT}{h^2}\right)^{3/2} \exp\left(-\frac{\chi_z}{kT}\right) =$$
$$4.8294\ 10^{15}\ \frac{w_{z+1}}{w_z}\ T^{3/2} \exp\left(-\frac{1.5789\ 10^5 Z^2}{T}\right)\ , \tag{6}$$

where T is in K and n_e in cm^{-3}, and the usual symbols are used for the fundamental physical constants. For the ratio of the statistical weights (or more generally partition functions) we can take $w_{z+1}/w_z = 1/2$ in the hydrogenic case. Whereas in the coronal approximation N_{z+1}/N_z is independent of n_e, this ratio varies as $\sim n_e^{-1}$ in the high-density limit. Clearly, in the latter case Eq. (6) predicts for low-Z ions at lower densities a much higher degree of ionization (i.e. larger N_{z+1}/N_z) compared to the coronal model (cf. Eq. 13) in which the ions of a given ionization stage can exist at much higher temperatures. From approximate hydrogenic formulae (Wilson 1962; Griem 1964, 1997; McWhirter 1965) it can be estimated whether coronal conditions are valid or not. The coronal domain is roughly bounded by the condition (deviations from coronal ionization balance less than about a factor of two):

$$n_e \lesssim 4\ 10^4\,(z+1)^2\,T^2\ (\mathrm{cm}^{-3}), \tag{7}$$

whereas the extreme high-density limit is reached at densities approximately

$$n_e \gtrsim 1.4\ 10^{15}\,(z+1)^6\,T^{1/2}\ (\mathrm{cm}^{-3}), \qquad (8)$$

where T is the electron temperature in K. The latter condition for complete Local Thermal Equilibrium (LTE) may be relaxed by about two orders of magnitude if the resonance transitions are self-absorbed and become optically thick, because this effectively reduces the radiative population rate (e.g., Mewe 1967, 1970).

In the region intermediate between the coronal and thermal domains, the situation is complicated, and the problem is to solve the differential equations describing the population and depopulation of many bound levels [e.g., reviews by Cooper (1966) and Wilson (1962)]. Moreover, if the plasma is neither optically thin nor thick towards the resonance lines, the rate coefficients depend also on the optical depths and an exact computation would require a solution of the level population rate equations coupled with the equations of radiative transfer.

Because collisional rates between bound levels increase with principal quantum number, and corresponding radiative decay rates decrease, the upper bound states near the ionization limit are strongly collisionally coupled to the continuum of free electrons and weakly coupled to the ground state, whereas for the lower levels the reverse is the case. From the work by Wilson (1962) the physical picture emerged that the thermal equilibrium in the continuum extends down to the upper bound levels owing to the high collisional rates between the upper bound levels and the continuum. Since this is imposed on the upper levels by the free electrons, the thermal equilibrium of the bound levels is linked to the continuum and their populations are given by Saha-Boltzmann equations. There is therefore a certain level n_t in the ion, for which upward collisional rates balance the downward radiative decay rates. It is known as the *thermal limit* (TL) because it defines the limit above which the levels are in thermal equilibrium with the continuum and below which the level distributions are approximately coronal. For low n_e, the TL is very close to the ionization limit (coronal approximation valid), for increasing n_e the TL drops until, at sufficiently high densities it reaches the ground level (n_t=1, LTE). For a hydrogenic ion Z^{+z} in an optically thin plasma the thermal limit is given by (Wilson 1962):

$$n_t^7 = 1.4\ 10^{15}\,(z+1)^6\,T^{1/2} n_e^{-1}. \qquad (9)$$

For n_t=1 this equation indeed corresponds to Eq. (8).

4.1.3 Non-equilibrium ionization (NEI) in transient plasmas. When processes such as plasma instabilities, shock compression, rapid expansion, heating or cooling of the gas, etc., cause a change in the physical plasma parameters (like electron temperature T and electron density n_e) on a plasma time scale, t_{pl}, much shorter than the time t_{rel} in which the plasma relaxes to ionization equilibrium, the assumptions of a steady-state equilibrium break down: the plasma is in a *transient* state. The establishment of

Table 2. Parameters for transient plasmas

Source		$t_{\rm pl}$[1]	n_e[2]	T[3]	$n_e t_{\rm pl}$	η[4]	$Z^{+z} \rightleftharpoons Z^{+(z+1)}$	T[3]
Hot ISM		10^{13}-10^{14}	10^{-3}-0.1	0.5-2	10^{10}-10^{13}	2 10^{11}	C V VI	1
						2 10^{12}	O VII VIII	1
SNR	old	10^{12}-10^{13}	0.1-10	1	10^{11}-10^{14}	4 10^{11}	O VII VIII	2
						4 10^{10}	Fe XVIII XIX	2
	young	10^{10}-10^{11}	0.1-10	5-10	10^{9}-10^{12}	5 10^{12}	Si XIII XIV	6[5]
						3 10^{10}	Fe XXIV XXV	6[5]
				10-100		10^{12}	Fe XXV XXVI	50[6]
Impulsive solar flare		1-100	10^{9}-10^{12}	20-60	10^{9}-10^{13}	10^{11}	Fe XXIV XXV	30
						10^{12}	Fe XXV XXVI	30
Tokamak		10^{-3}-10^{-2}	10^{14}-10^{15}	$\gtrsim 10$	10^{11}-10^{13}	2 10^{11}	Fe XXIV XXV	15
Θ Pinch		10^{-6}-10^{-5}	10^{15}-10^{16}	$\gtrsim 10$	10^{10}-10^{11}	3 10^{11}	Fe XXV XXVI	15
Laser pl.		10^{-9}-10^{-8}	10^{19}-10^{20}	$\gtrsim 10$	10^{10}-10^{11}			

[1] Characteristic plasma time scale or age in s.
[2] Electron density in cm^{-3}.
[3] Electron temperature in MK.
[4] Ionization parameter $\equiv n_e t_{\rm rel}$ (cm^{-3} s^{-1}), where $t_{\rm rel}$ is relaxation time for ionization equilibrium.
[5] Reverse shock.
[6] Main shock.

the ionization balance then lags the temperature changes, and this can have dramatic effects on the emergent X-ray spectrum. As $t_{\rm rel} \propto 1/n_{\rm e}$ the parameter that characterizes this *non-equilibrium ionization* (NEI) is the *ionization parameter* $n_{\rm e}t$ or more generally $\int n_{\rm e}dt$ when $n_{\rm e}$ varies with time (Kaastra & Jansen 1993). X-ray emitting, hot plasmas are out of equilibrium typically for $n_{\rm e}t \lesssim 10^{10} - 10^{13}$ cm^{-3} s. As a result, such a non-equilibrium gas initially radiates thermal X-ray bremsstrahlung characteristic of the high temperature just after the heating and virtually independent of the ionization balance, together with line plus two-photon and free-bound emission characteristic of the non-equilibrium ionization structure corresponding to the pre-heating state. The line spectrum is therefore much softer than the bremsstrahlung continuum, and as the gas ionizes out, this NEI enhancement of the soft X rays will eventually disappear, with spectral hardening occurring along the way (for examples, e.g., Shapiro & Moore 1977).

To evaluate in which circumstances transient conditions apply, we estimate the ionization equilibrium relaxation time scale $t_{\rm rel} \simeq \min(t_{\rm ion}, t_{\rm rec})$ with approximate formulae (Mewe 1984) for the time scales for ionization ($t_{\rm ion}$) and recombination ($t_{\rm rec}$)

$$n_e t_{\rm ion} \approx 10^{10}(z+1)^4 n_z^{-4} \zeta_z{}^{-1} T^{-1/2} e^y \,({\rm cm^{-3}s}) \ , \qquad (10)$$

$$n_e t_{\rm rec} \approx 10^{11}(z+1)^{-2} n_z^{5/2} \xi_z{}^{-1} T^{1/2} \,({\rm cm^{-3}s}) \ , \qquad (11)$$

where $y = 1.58\ 10^5 (z+1)^2 n_z^{-2} T^{-1}$, ζ_z (or ξ_z) the number of valence electrons (or empty spaces) in the outer shell with principal quantum number n_z of the ionizing (or recombining) ion Z^{+z} (or $Z^{+(z+1)}$), and compare this to the relevant dynamical plasma time scale $t_{\rm pl}$ (for more details and examples of NEI plasmas, see the Chapter by D. Liedahl).

In Table 2 I have presented a number of representative cases. It is seen that notwithstanding the widely different categories of optically thin astrophysical plasmas, such as young supernova remnants (SNR), hot interstellar media of galaxies, solar or stellar flares, and fusion-oriented laboratory plasmas, that span a wide range of dynamical time scales ($t_{\rm pl}$ ~ 10^{-9} to 10^{14} s), these cases cover only a restricted range in $n_e t_{\rm pl}$ (~ 10^9 to 10^{14} cm^{-3} s). Because the ionization equilibrium time scales $n_e t_{\rm rel}$ range from ~ 10^{10} to 10^{13} cm^{-3} s, the transient situation is believed to apply to many of these cases (with probably the only exception of quiet stellar coronae and the hot gas in clusters of galaxies which both have $n_e t_{\rm pl} \approx 10^{13}$–$10^{14}$ cm^{-3}; Mewe 1984).

For a highly transient plasma like a SNR the equilibrium ionization model completely fails and we must introduce the NEI model for an adequate description. However, in the early 1980s astronomers first misinterpreted the *Einstein* X-ray spectra of SNRs by applying a CIE model and assumed special evolution effects to explain the derived artificially anomalous abundances. This is a typical example of a more complex plasma model in which the transient nature of the plasma and its spatial structure must be simultaneously taken into account for a proper interpretation of the observed X-ray spectrum; the X-ray observations of young SNR clearly confirm that non-equilibrium effects in the ionization balance are very important (e.g., Gronenschild & Mewe 1982; Hamilton & Sarazin 1983; Itoh 1984a; Masai 1984, 1994a). The spectral modeling is quite complicated. To generate a SNR model the following steps can be distinguished (e.g., Kaastra & Jansen 1993). First, the hydrodynamical model describing the blast and the reverse shock waves is calculated yielding the distribution of density, velocity and temperature as a function of position and time. Then the ion densities are evaluated by integrating the T- and n_e-dependent equations for the ionization balance over time and the X-ray spectrum is computed for the known set of ion concentrations for each position in the SNR. Finally, we include Doppler shifts due to the plasma motion and project the spectrum onto the sky and integrate it over the relevant area of the sky.

4.1.4 Non-Maxwellian electron velocity distributions. If departures from a Maxwellian electron velocity distribution occur this may significantly influence the calculation of the ionization balance and of the excitation rates. In Sections 15.4.3 and 15.4.4 we will consider various situations in which

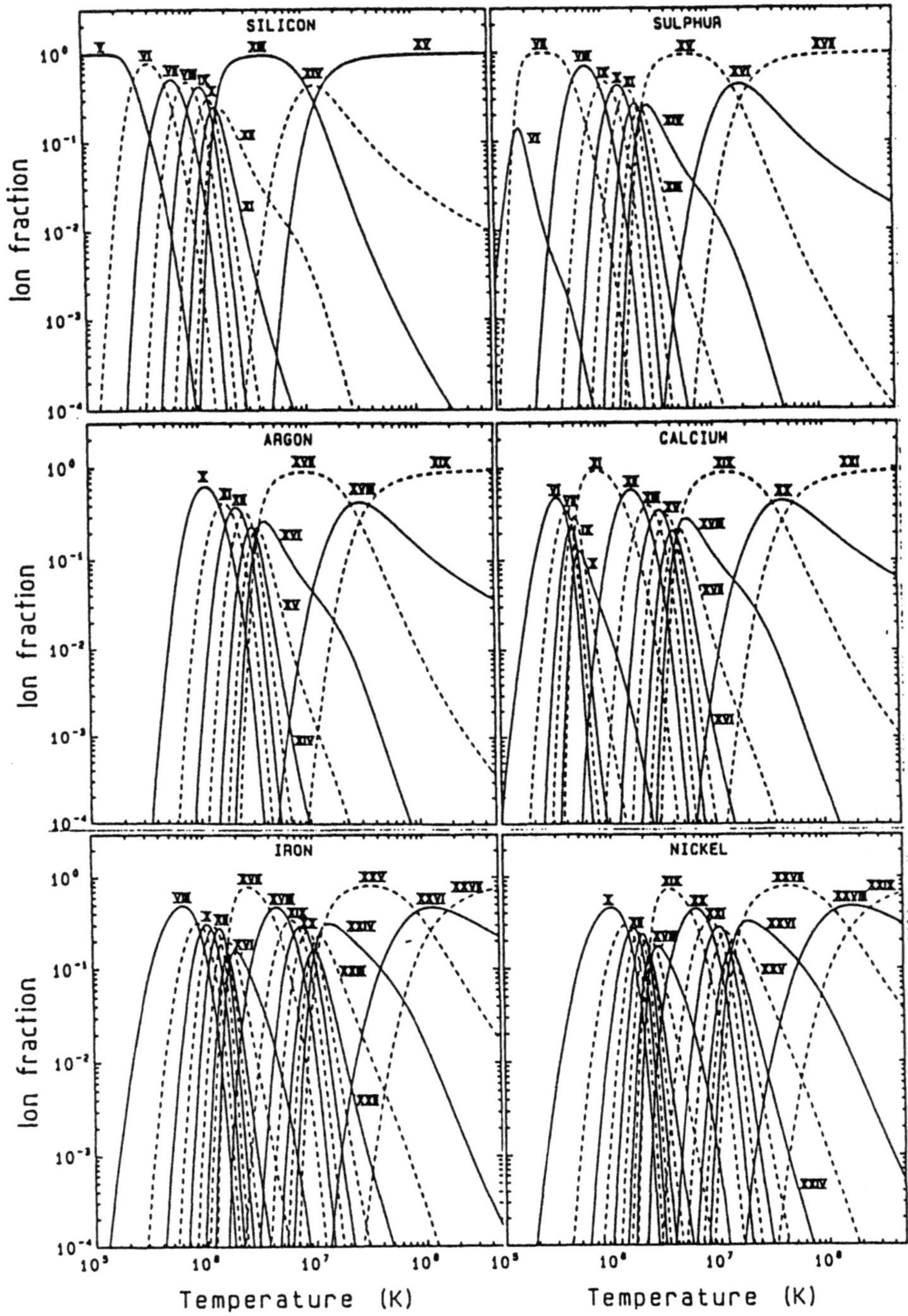

Fig. 2. Ion fractions of Si, S, Ar, Ca, Fe, and Ni as a function of electron temperature (from Mewe 1988).

such deviations can occur and how we can use these as a diagnostics. In these sections the line and continuum emission is calculated for a power law and a mono-energetic electron distribution.

II. Ionization and Recombination in a Coronal Plasma

5 Ionization Balance

Much of the temperature sensitivity of the soft X-ray spectrum is associated with the ionization structure. Under the assumptions of the coronal model, all ions can be taken to be in their ground states. The ionization state is controlled by electron impact ionization (including sometimes a contribution from excitation-autoionization) and by radiative plus dielectronic recombination. The rate of change of the population density $N_{Z,z}$ (in cm^{-3}) of ion Z^{+z} from element of atomic number Z is given by

$$\frac{1}{n_e}\frac{\mathrm{d}N_{Z,z}}{\mathrm{d}t} = N_{Z,z-1}S_{Z,z-1} - N_{Z,z}(S_{Z,z} + \alpha_{Z,z}) + N_{Z,z+1}\alpha_{Z,z+1} \ , \qquad (12)$$

where $S_{Z,z}$ and $\alpha_{Z,z}$ are the total ionization ($z \to z+1$) and recombination ($z \to z-1$) rate coefficients (in cm^3 s^{-1}) of ion Z^{+z}, etc. ($z = 0, 1, \ldots, Z$).[5] This equation shows that, in an evolving plasma, the characteristic time scale to approach to a steady state is $\approx [n_e(S_{Z,z} + \alpha_{Z,z})]^{-1}$ and one can expect solutions to contain terms of the form $\exp[n_e(S_{Z,z} + \alpha_{Z,z})t]$ (see also the Chapter by D. Liedahl). In Eq. (12) we neglect *multiple ionizations.*[6] The ionization and recombination rate coefficients will be dealt with in Sections 6 and 7, respectively. The ionization structure can be derived by solving for each element Z a set of $Z+1$ coupled rate equations. Eq. (12) can be written in the form of a matrix equation where the ion concentrations are represented by vectors of length $Z+1$ and the matrix is a tridiagonal $(Z+1) \times (Z+1)$ matrix composed of the ionization and recombination rate coefficients (Masai 1994a). The solution can be obtained using a matrix inversion technique.

The population of stage z depends on the four rates which connect it with the neighbouring ionization stages $z-1$ and $z+1$. If ion $z-k$ is most abundant, then in principle all the rates connecting stages between $z-k$ and z would enter. If, for example, all ionization rates would be systematically too high by a factor of two, then the predicted population of ion z would be

[5] Because we focus on the atomic rates we omit in Eq. (12) transport terms relating to the dynamic effects of mass motions (plasma expansion, diffusion, etc.). These must be appropriately taken into account for each particular plasma model if they have a non-negligible effect, e.g., in a laboratory plasma of finite size (see Düchs & Griem 1966; Peacock 1996).

[6] For the lower ions multiple ionization can be sometimes important (cf. Pal'chikov & Shevelko 1995).

off by 2^k. Fortunately, if k is more than say 1 or 2, the population is generally to small too matter.

For a *steady-state equilibrium* (collisional ionization equilibrium, *CIE*) $\mathrm{d}N_{Z,z}/\mathrm{d}t = 0$ in Eq. (12), and the population density ratio $N_{Z,z+1}/N_{Z,z}$ of two adjacent ionization stages $Z^{+(z+1)}$ and Z^{+z} can be expressed by

$$\frac{N_{Z,z+1}}{N_{Z,z}} = \frac{n_e S_{Z,z}(T)}{n_e \alpha_{Z,z+1}(T)} = \frac{S_{Z,z}(T)}{\alpha_{Z,z+1}(T)} \, , \tag{13}$$

which is, to first order, only dependent on T and not on n_e, as long as stepwise ionization in $S_{Z,z}$ and collisional coupling to the continuum in $\alpha_{Z,z+1}$ can be neglected (Mewe 1970; Wilson 1962; Mewe & Schrijver 1978a; Mewe & Gronenschild 1981).

For each pair of subsequent ions the stationary ionization balance (Eq. 13) can be solved. With $\sum_{i=0}^{i=Z} N_{Z,z} = N_Z$ the total density of species Z, the solution for the fraction of ions in a specific ionization stage z can be expressed as

$$\eta_{Z,z} = \frac{N_{Z,z}}{N_Z} = \left[1 + \sum_{i=1}^{i=z} \prod_{k=i}^{k=z} \frac{N_{Z,k-1}}{N_{Z,k}} + \sum_{i=1}^{i=Z-z} \prod_{k=z+1}^{k=z+i} \frac{N_{Z,k}}{N_{Z,k-1}} \right]^{-1} \, . \tag{14}$$

Many results of such calculations have been reported in the literature (references up to 1984 are cited by Mewe 1984). As an example, in Fig. 2 I plot the ion fractions of six cosmically abundant elements from Si to Ni as functions of electron temperature, as calculated with the rate coefficients given by Mewe & Gronenschild (1981); for plots for He to Mg, see Mewe (1988). These results are very similar to those reported by Arnaud & Rothenflug (1985) which are often applied in the analysis of tenuous, astrophysical plasmas.

5.1 Accuracy of atomic physics for the ionization balance

Most spectroscopists agree that the ionization balance is one of the major uncertainties entering the comparison of models. In both ionization and recombination rates there are uncertainties typically ~20–50%, but sometimes even more, say a factor ~2 to 4. E.g., for the low Fe ions having $3d$ electrons the ionization rates as given by Arnaud & Rothenflug (1985) (ARo) are probably underestimated by factors 3–6, while for the Be- and Li-like and Ne- and F-like Fe ions the dielectronic recombination rates may be underestimated by factors 2–4, as was pointed out by Arnaud & Raymond (1992) (ARa). So the predicted ratio of ion concentrations of adjacent ionization stages may be off sometimes by a factor ~2 or more. Fortunately, the ionization and recombination rates for the H- and He-like ions, which emit the lines that are among the strongest from hot astrophysical plasmas, are known more accurately than most of the other rates. In the case of ions which have a closed outer electron shell (e.g., He- and Ne-like ions) it doesn't matter very much because such ions cover a broad plateau in temperature. This is because this ionization stage can easily be reached from the preceding Li- or Na-like stage (with only one outer electron with low binding energy) and persists long, since the next ionization step towards the H- or F-like stage needs a much (~4 times) higher ionization energy (it is much more difficult to take out an electron from a closed shell). The adjacent ionization stages (e.g., Li-like) depend more critically on temperature.

The effect of a systematic error can be readily estimated. Suppose that all ionization rates are underestimated by say a factor of 4. Since the rates typically vary as $\exp(-\chi/kT)$, and since the concentrations of ions with ionization energy χ typically peak at $\chi/kT \sim 3$, the ion peaks will be shifted to $\chi/kT \sim 4.4$, so that all ions will be shifted to lower temperatures by about 0.17 in $\log T$ (i.e. 50% in T).

Mewe (1990) has compared the results from various ionization balance calculations (RS, ARo, MG) and concludes that the overall shapes of the ion fraction $\eta_z(T)$ curves are quite similar and that the shifts of the ion peaks are generally limited within about $\Delta \log T \approx 0.1$-$0.2$, but that the ion peak values may differ by about 10–30% and sometimes up to a factor of two, especially for the ions adjacent to the He-like and Ne-like "plateaus".

The differences mainly result from different dielectronic recombination rates (for inaccuracies see Sections 7.2 and 7.2.2). For many diagnostic purposes the ARo ion balance is used, and for Fe the updated ARa calculations.

Recently Masai (1997) has discussed the effect of the ionization balance on the X-ray spectral analysis by comparing the ionization balance for iron based on ARo, ARa, and Masai (1984), respectively.

5.2 Update of the ionization balance by improved calculations for the rate coefficients

Arnaud & Raymond (1992) have made an attempt to update the ionization balance for the special case of Fe by introducing improved dielectronic recombination rates, but we are still lacking the revision of the Arnaud & Rothenflug (1985) ionization balance for the other elements. Moreover, a more general revision is desirable for the rates of radiative and dielectronic recombination.

5.2.1 Ionization rates. The most recent work has been done by Kaastra & Mewe (1998) who have fitted the ionization cross sections of all shells of ions from H to Zn (atomic number $Z = 1$–30). Apart from direct ionization, they include excitation-autoionization, resonance excitation, double autoionization, and direct multiple ionization. Further, an extended compilation of ionization rate coefficients has been made by Kato et al. (1991).

5.2.2 Recombination rates. Verner et al. (1993) and Verner & Yakovlev (1995) have calculated the partial photoionization cross sections of atoms and ions from He to Zn for all subshells $n\ell$ of ground-state species with the Hartree-Dirac-Slater method and have fitted their results by simple analytic functions. By use of the Milne relation (Eq. 34) one can obtain radiative recombination rates, and recently Verner & Ferland (1996) have followed such a procedure. They have calculated the rates for H-like, He-like, Li-like, and Na-like ions of all elements from H through Zn. These are the ionic species for which radiative recombination is not dominated by dielectronic recombination (at least at low temperature $\lesssim 10^4$ K and low density) because these ions do not have low-lying autoionization levels and therefore are not subject to low-temperature dielectronic recombination.

Recently, Huaguo & Zhizhan (1996) have fitted analytic formulae to dielectronic recombination rate coefficients for the H-like isoelectronic sequence, calculated with an intermediate-coupling multi-configuration Hartree-Fock method; Huaguo et al. (1994a,b,c,d) have done this for the Li-, Ne-, He-, and F-like sequences, respectively. Some further references are Karim & Bhalla [1988 (H), 1989 (He)] and Romanik (1988) (He, Li, Be, Ne) (cf. also Itikawa et al. 1995 for more references).

6 Rate Coefficients for Ionization

6.1 Collisional ionization

Under the conditions of the coronal model valid for low-density plasmas (see Sect. 4) the ionization equilibrium is determined by the balance between electron impact ionization and radiative and dielectronic recombination. Here I consider the ionization processes which, apart from *direct ionization* from the ground state (with rate coefficient $S^{\rm d}_{Z,z}$),

$$Z^{+z} + e^- \rightarrow Z^{+(z+1)} + 2e^- \tag{15}$$

may also may contain a contribution from *excitation-autoionization*, i.e., electron impact excitation from the ground state to an autoionizing level (above the first ionization energy) followed by autoionization (rate coefficient $S^{\rm a}_{Z,z}$):

$$Z^{+z} + e^- \rightarrow (Z^{+z})^* + e^- \rightarrow Z^{+(z+1)} + 2e^- \ . \tag{16}$$

The total rate coefficient (in cm^{+3} s^{-1}) is

$$S_{Z,z} = S^{\rm d}_{Z,z} + S^{\rm a}_{Z,z} \ , \tag{17}$$

hence the number of ionizations (cm^{-3} s^{-1}) of ion Z^{+z} is: $n_{\rm e} N_{Z,z} S_{Z,z}$, where $n_{\rm e}$ and $N_{Z,z}$ are the electron and ion densities (in cm^{-3}). The inelastic process of ionization (or excitation) is most effective when the relative velocity of the impacting particle is of the order of the orbital velocity of the bound electron in the ion to be ionized (or excited). Thus if electrons and ions have comparable kinetic energies the electrons are much more effective, unless we consider population redistribution between nearby levels where e.g., proton collisions can be efficient. In the following I neglect ionization by ion impact (cf. also Sect. 8.1).

For the impact of an impinging electron with velocity v relative to the ion Z^{+z} to be ionized (or excited) the number of ionizations (excitations) per second reads:

$$N_{Z,z}\ v\ Q(v)\ , \tag{18}$$

where $Q(v)$ is the collision cross section (cm^2) for the particular process at the given relative velocity v (which is practically the velocity of the light electron itself). We obtain the rate coefficient S by averaging over the velocity distribution $f(v)$ of the electrons in the plasma:

$$S = \langle vQ(v)\rangle = \int_{v_0}^{\infty} vQ(v)f(v)\,\mathrm{d}v = \int_{E_0}^{\infty} \sqrt{2E/m_\mathrm{e}}Q(E)f(E)\,\mathrm{d}E\ , \tag{19}$$

where $E = \frac{1}{2}m_\mathrm{e}v^2$ is the electron energy and v_0 (or E_0) is the threshold velocity (or energy) of the collision process. Under the assumptions of the coronal model the electrons have a normalized[7] Maxwellian velocity (or energy) distribution with a temperature T:

$$f_\mathrm{M}(v)\mathrm{d}v = 4\pi\left(\frac{m_\mathrm{e}}{2\pi kT}\right)^{3/2} v^2 \exp\left(-\frac{m_\mathrm{e}v^2}{2kT}\right)\mathrm{d}v,\ \text{or}$$

$$f_\mathrm{M}(E)\mathrm{d}E = \frac{2}{\sqrt{\pi}}\left(\frac{E}{kT}\right)^{1/2} \exp\left(-\frac{E}{kT}\right)\mathrm{d}\left(\frac{E}{kT}\right)\ . \tag{20}$$

Inserting Eq. (20) into Eq. (19) yields

$$S = \sqrt{\frac{8kT}{\pi m_\mathrm{e}}}y^2 \int_{E_0}^{\infty} Q(U)\mathrm{e}^{-yU}\,U\,\mathrm{d}U\ , \tag{21}$$

where $U = E/E_0$ is the initial energy E of the impinging electron in terms of the threshold energy E_0, and $y = E_0/kT$ the reduced threshold energy. It turns out that the cross sections for excitation or ionization obey a scaling law by which the cross section can be written in a form

$$Q(U) = \alpha(E_\mathrm{H}/E_0)^2 Q_\mathrm{red}(U)\,\pi a_0^2\ , \tag{22}$$

where the reduced cross section $Q_\mathrm{red}(U)$ is a scaled dimensionless cross section that for all ions in a given isoelectronic sequence has approximately the same form and is not sensitively dependent of the atomic parameters; a_0 is the first

[7] $\int_0^\infty f_\mathrm{M}(E)\mathrm{d}E = 1$.

H atom Bohr radius and $E_{\rm H}$ the ionization energy of a hydrogen atom,[8] and α is a numerical factor of the order unity.

Writing $E_0 = E_{\rm H} z_{\rm e}^2/n^2$ ($z_{\rm e}$ is the effective charge, n the principal quantum number of the state from which ionization takes place), Eq. (22) implies that the ionization cross section scales as $Q \propto (n/z_{\rm e})^4$. We may explain this as follows. At first sight we would expect $Q \sim \pi a_n^2 \sim (n^4/z_{\rm e}^2)\pi a_0^2$ because the radius for the orbit (n) is $a_n = (n^2/z_{\rm e})a_0$. However, the electron-electron interaction has to compete with the electron-nucleus interaction (which goes as $\sim z_{\rm e}/r$ at distance r). On a crude classical picture one may argue the e-e interaction to dominate only over a *fraction* $1/z_{\rm e}^2$ of the total effective target area πa_n^2. Taking this into account we obtain indeed the *scaling* $Q \sim (n^4/z_{\rm e}^2)(1/z_{\rm e}^2) = (n/z_{\rm e})^4$.

6.1.1 Direct ionization. Numerous authors have attempted to represent the functional dependence of the direct electron-impact ionization cross section by a semi-empirical formula (e.g., Elwert 1952; Drawin 1961; Lotz 1968; Arnaud & Rothenflug 1985), which correctly represents both the behaviour at threshold ($U = 1$), $Q \propto (U - 1)$ as well as the quantum-mechanical Born-Bethe dependence at high energies, $Q \propto U^{-1} \ln U$ [cf. also discussion of more complicated formulae by Arnaud & Rothenflug (1985), and Kaastra & Mewe (1998)].

In astrophysics the rates derived from the semi-empirical formula of Lotz (1968) for the cross section for direct impact ionization from the ground state have been widely used. The latter reads

$$Q(U) = \pi a_0^2 \sum_{k=m}^{N} C_k \xi_k W(U_k) \left(\frac{E_{\rm H}}{\chi_k}\right)^2 \frac{\ln(U_k)}{U_k} \ , \tag{23}$$

where the summation is over (sub)shells m of the initial ion Z^{+z} with ionization energy χ_m, $U_m = E/\chi_m$, and where ξ_m is the effective number of equivalent electrons in the shell with principal quantum number m. In most cases it is sufficient to set N equal to 1 (outermost shell) or 2 (next inner subshell). The function $W(U) = 1 - b_m \exp[-c_m(U - 1)]$ represents the deviation from linear behaviour near threshold (only significant for ions of low charge and for low temperature). From fits to experimental data or extrapolations Lotz (1968) found that one can take for ions ionized more than four times $W(U_m)$=1, and for the constant C_m=2.76. Then the "Lotz" rate coefficient for direct ionization (d) [for one shell only and omitting subscript m] follows from Eqs. (21) and (22) (with $\alpha = 2.76\xi_m$, $E_0 = \chi$, and $Q_{\rm red}(U) = U^{-1} \ln U$):

$$S_{Z,z}^{\rm d} = 3.244\ 10^{-4} \xi \chi_{\rm [eV]}^{-1} T_{\rm [K]}^{-1/2} E_1(\chi/kT)\ ({\rm cm^3 s^{-1}}) \ , \tag{24}$$

[8] $a_0 = 5.2918\ 10^{-9}$ cm, $\pi a_0^2 = 8.797\ 10^{-17}$ cm^2, and $E_{\rm H} = 13.6057$ eV = 1 *Rydberg* (Ry) = $\frac{1}{2}$ *atomic unit* (a.u.).

where $E_1(y) = \int_1^\infty U^{-1}\exp(-yU)\,dU \simeq e^{-y}/(y+1) \approx e^{-y}/y$ $(y \gg 1)$ is the first exponential integral [cf. also Eq. (71)], since ion concentrations typically peak at $y = \chi/kT \sim 2$–5. Hence direct ionization rates scale with temperature as $S^d \propto T^{1/2}\chi^{-1}\exp(-\chi/kT)$.

Since threshold energies for ionization (and for $\Delta n \neq 0$ excitation) scale as $\chi \propto z_e^2$, where z_e is the effective charge number of the nucleus acting on the electrons in the shell (in particular for hydrogenic ions Z^{+z} : $z_e = z + 1$, for non-hydrogenic ions z_e is defined from $\chi_m = E_H z_e^2/m^2$), the cross section scales as $Q \propto z_e^{-4}$ (Eqs 22 and 23) and the rate coefficient, expressed as a function of *reduced* temperature $\Theta = T/z_e^2$, as $S^d \propto z_e^{-3}$ (Eq. 24). Calculations based on the Exchange Classical Impact Parameter (ECIP) method (Burgess et al. 1977; Summers 1974) predict rates typically about half those given by the Lotz formula, whereas the ionization rates now coming into widest use, those from Distorted Wave calculations (e.g., Younger 1981), appear to lie in between [see discussion by Raymond (1988)]. Many ionization cross sections have been measured for the lower ions with 10–20% accuracy (e.g., Gregory et al. 1987), while for the higher ions many atomic *rate* coefficients for ionization, recombination, and excitation have been obtained from plasma measurements (e.g., Wang et al. 1986, 1987, 1988; for reviews see Kunze 1972; Griem 1988). For cases in which experimental or theoretical data are not available, Burgess & Chidichimo (1983) suggest to use the Lotz formula (23) with $C_{(m)} \simeq 2.3$ and with $\xi_{(m)}$ and $\chi_{(m)}$ properly assigned. Arnaud & Rothenflug (1985) present an extensive compilation of ionization rates for 15 cosmically abundant elements (H, He, C, N, O, Ne, Na, Mg, Al, Si, S, Ar, Ca, Fe, and Ni) based on fits to available experimental cross sections and on Younger's theoretical results. Arnaud & Raymond (1992) have updated the rate coefficients for the special case of Fe. The most complete work has been done by Kaastra & Mewe (1998) who have fitted the ionization cross sections of all shells of ions from H to Zn (atomic number $Z = 1$–30).

6.1.2 Excitation-Auto-ionization. An additional ionization process via collisional excitation of an inner-shell electron is most important for ions which have a large number of electrons in the first inner subshell compared with the number in the outermost shell. An example is the sodium isoelectronic sequence ($1s^2 2s^2 2p^6 3s$, e.g., Fe XVI), where eight L-shell electrons can contribute to auto-ionization (AI) (e.g., $2p$-$3p$ and $2p$-$3d$ transitions), while there is only a single valence $3s$ electron that contributes readily to direct ionization (DI). Here an inner electron can be excited to a bound level above the first ionization threshold and subsequently can either autoionize (probability A^a), or can decay radiatively (probability A^r) to a bound state below the ionization threshold (e.g., Bely & van Regemorter 1970; Mewe 1988). This is illustrated in Fig. 3 for the Li-like case. In the total ionization cross section the AI contribution is recognized as a steep bump above the DI cross section. The net contribution to the ionization cross section can be

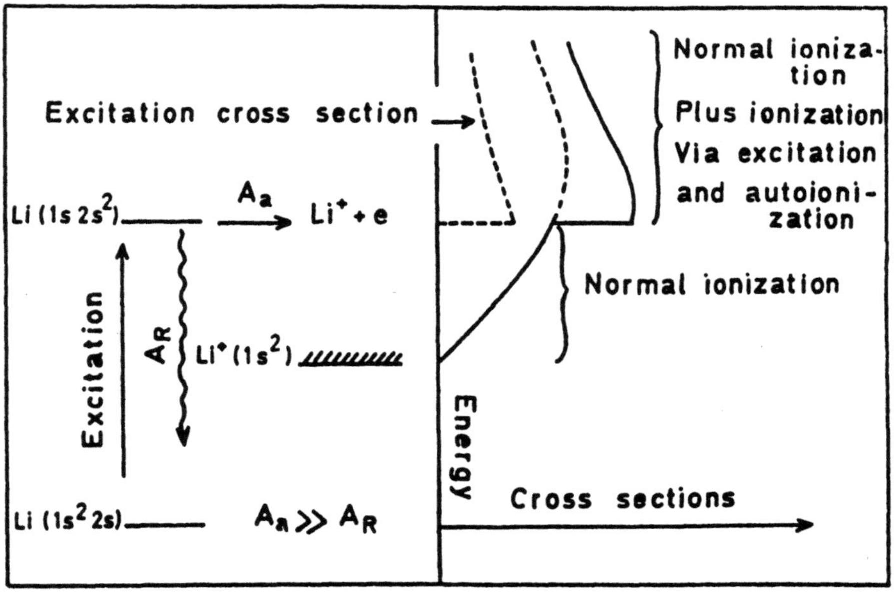

Fig. 3. Ionization via excitation and auto-ionization (after Bely & van Regemorter 1970).

written as a sum $\sum_s Q_{s,\mathrm{exc}} BR_s$ over all target bound states (s) lying above the first ionization limit and where the branching ratio $BR_s = A_s^{\mathrm{a}}/[A_s^{\mathrm{a}} + A_s^{\mathrm{r}}]$. Generally $A^{\mathrm{r}} \propto Z^4$ along an iso-electronic sequence, whereas A^{a} is approximately constant, so that B_i strongly varies along the sequence, while it can change very suddenly along successive ionization stages of one element. Typically $Q \propto E_0^{-2} U^{-1} \propto Z^{-4}$ (Eq. 65), so that the rate coefficient scales as (see Eq. 68) $S^{\mathrm{a}} \propto T^{-1/2} E_0^{-1} \exp(-E_0/kT)$ for $E_0/kT \gg 1$, where E_0 is the excitation energy of the autoionizing level. Arnaud & Rothenflug (1985) present a compilation based on semi-empirical fits to the results of calculations by Sampson (1982).

Resonant ionization. Other indirect processes which lead to ionization are multi-step resonant processes leading to the appearance of narrow resonances in the total ionization cross section. The first step in resonant ionization (RI) is the capture of a free electron (the same as in DR, cf. Sect. 7.2) and the creation of a doubly excited autoionizing state in the target ion which can decay by the emission of two electrons; the net result is a single ionization. Various combinations are possible such as "resonant-excitation-double-autoionization" (REDA) which, e.g., for the ionization of Fe XVI contributes up to 30% to the total ionization cross section [cf. Pal'chikov & Shevelko (1995), and Moores (1988)].

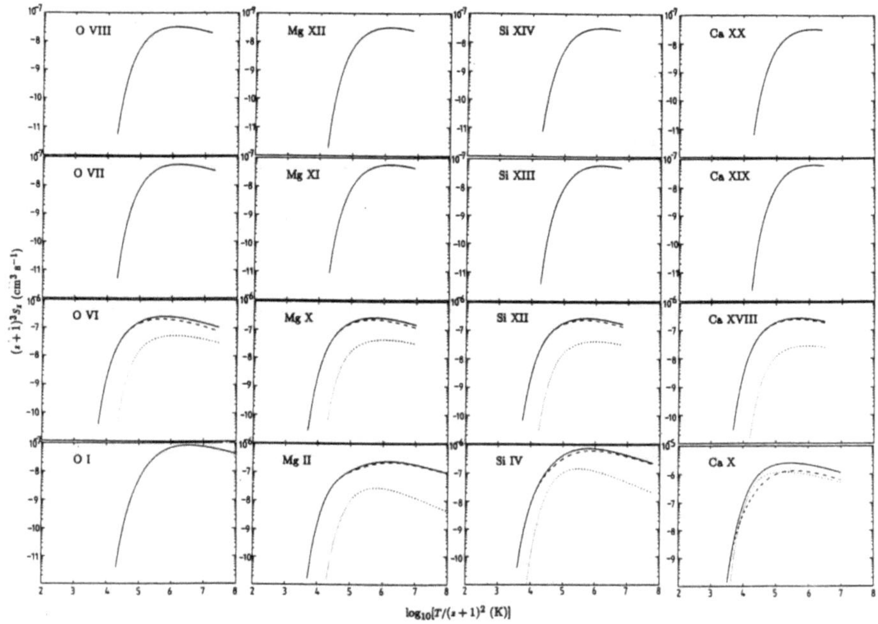

Fig. 4. Reduced rate coefficients $(z+1)^3 S_z$ (in cm^3 s^{-1}) for direct ionization (d) (dashed line) plus autoionization (a) (dotted) (solid line indicates total rate) of ion Z^{+z} as a function of reduced electron temperature $T/(z+1)^2$ from Arnaud & Rothenflug (1985) (e.g., z=7 for ionization of O VIII, etc.) (from Mewe 1990).

In Figs. 4 and 5 I have presented, for a number of representative cases covering both low- and high-Z ions, the results of Arnaud & Rothenflug (which are often used in the present spectral codes). I show the scaled rate coefficient $(z+1)^3 S_z$ for ionization of ion Z^{+z} as a function of the reduced temperature $\Theta = T/(z+1)^2$. In cases where the contribution from autoionization becomes also important, I have indicated this in the figure. I estimate that in many cases the rates are good to 10–20% and that the overall uncertainty will be at the level of up to ~40% (cf. also Raymond 1988).

6.1.3 High-density effects on the ionization rates. Any transition to levels below the TL (see Sect. 4.1.2) is equivalent to recombination, since the hole that is left is immediately populated by a collisional transition from the continuum; similarly, any excitation from low levels to the thermal levels above the TL is equivalent to ionization, since further excitation and ionization dominates downward radiative decay. If the TL is low enough [say $n_t \lesssim 5$, corresponding to $n_e \gtrsim 10^{12}(z+1)^{7.5}$] effects of density (e.g., stepwise ionization) on the ionization and recombination rate coefficients should be

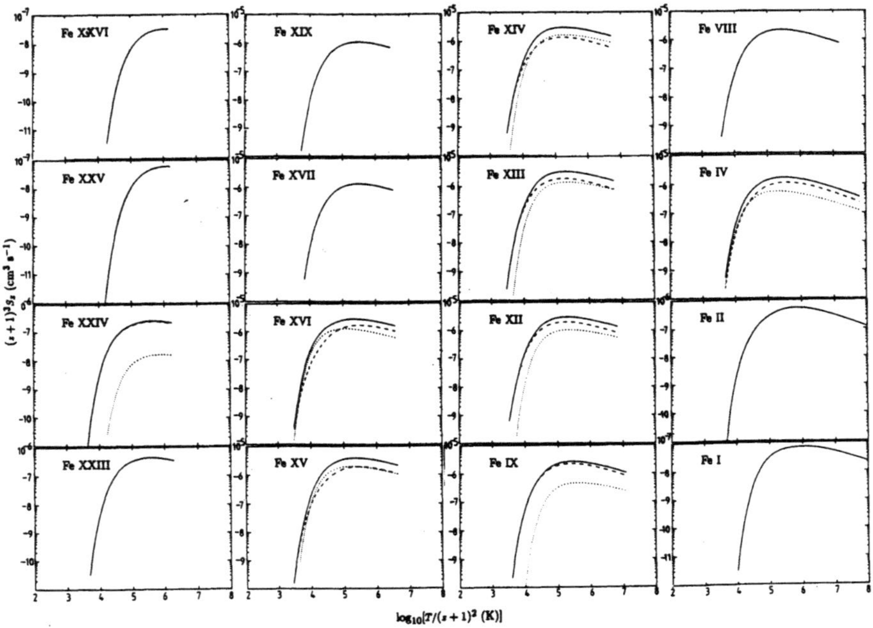

Fig. 5. Reduced ionization rate coefficients for a number of ionization stages of iron. For explanation see caption to Fig. 4.

taken into account. In certain cases where metastable levels are important (e.g., low-Z Li-, Be-, and B-like ions) stepwise excitation-ionization can occur already at much lower densities (e.g., Vernazza & Raymond 1979).

Bates et al. (1962a,b) were the first to make calculations fully taking into account the combined effects of collisions and radiative decay for the case of an optically thin *or* thick *hydrogenic* plasma. They expressed the net rates of ionization and radiative recombination in terms of binary coefficients which they called *"collisional-radiative"* ionization or recombination coefficients. For the complicated case in which also trapping of resonance radiation becomes important, I have made some rough estimates for these rate coefficients with semi-empirical approximation formulae (based on the results of Bates et al.) for a highly simplified hydrogenic two-level + continuum scheme (Mewe 1970, also 1988). The transition between low- and high-density cases occurs at different densities, roughly at the plasma dimension $D \sim 3\ 10^{27}(z+1)^{10.5} n_e^{-2}$ cm, which represents the effect of resonance radiation trapping in the case of Doppler broadening.

In the high-density limit the effective rate coefficient (cm^3 s^{-1}) for ionization of hydrogenic ions $X^{+(Z-1)}$ approaches (cf. Mewe 1970, 1988) (T in K):

$$S_{Z-1}(n_e \to \infty) = 1.3\ 10^{-5} Z^{-2} T^{-1/2} \exp(-1.184\ 10^5 Z^2/T) \ , \qquad (25)$$

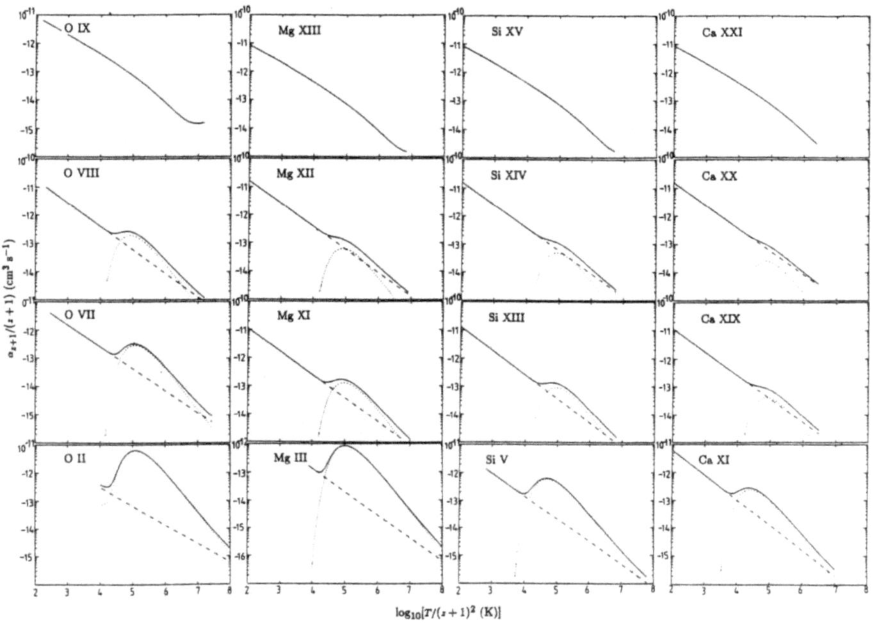

Fig. 6. Reduced rate coefficients $\alpha_{z+1}/(z+1)$ (in cm^3 s^{-1}) for radiative recombination (RR; dashed line) plus dielectronic recombination (DR; dotted line) of ion $Z^{+(z+1)}$ as a function of reduced electron temperature $T/(z+1)^2$ from Arnaud & Rothenflug (1985) E.g., z=7 for recombination of O IX, etc. The solid line indicates the total rate. (from Mewe 1990).

which equals the excitation rate to the first excited level for this simplified model.

7 Rate Coefficients for Recombination

In coronal equilibrium, each ionization is balanced by a recombination process, either radiative or dielectronic. For hot plasmas we can neglect processes of charge transfer, which can be very important for the ionization structure of cooler plasmas, such as photoionized nebulae. The most important charge transfer process is generally the capture of an electron from neutral hydrogen by a charged ion, resulting in recombination of the ion. For temperatures below ~0.01 MK this becomes important [see, e.g., Arnaud & Rothenflug (1985), and the review by Butler & Dalgarno (1980)]. Further, I neglect 3–body recombination which is only important at very high densities ($n_e \gtrsim 10^{15} Z^2$ cm^{-3}, cf. Sect. 7.1.1).

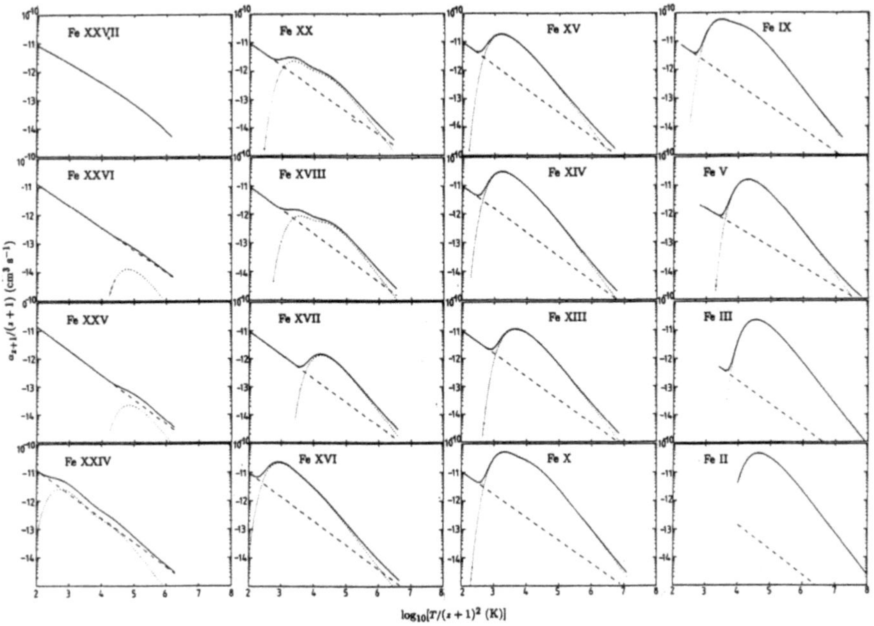

Fig. 7. Reduced recombination rate coefficients for a number of ionization stages of iron. For explanation see caption to Fig. 6.

For low-density plasmas we have (a) the process of *radiative recombination*,

$$Z^{+(z+1)} + e^- \to Z^{+z}(m) + h\nu \ , \tag{26}$$

where m denotes the m-th state of the recombined ion and (b) the process of *dielectronic recombination* which occurs in the following steps. First an energetic free electron is captured and its kinetic energy is used for the simultaneous excitation of a bound electron into a doubly excited state of ion Z^{+z}:

$$Z^{+(z+1)} + e^- \to (Z^{+z})^{**} \ . \tag{27}$$

If auto-ionization follows (the reverse of the capture process), the system returns to its original state and no recombination takes place. Alternatively a fraction of the ions in the autoionizing state decays by spontaneous radiative transition of the inner excited electron to a state below the first ionization limit (stabilization process):

$$(Z^{+z})^{**} \to (Z^{+z})^{*} + h\nu \ , \tag{28}$$

where the stabilizing transition in the recombined ion Z^{+z} results in the emission of a dielectronic satellite line of the parent transition in the recombining ion $Z^{+(z+1)}$. Eventually, the singly excited state cascades down to the ground state with the subsequent emission of photons:

$$(Z^{+z})^* \rightarrow Z^{+z} + h\nu' + h\nu'' + \ldots \tag{29}$$

Though dielectronic recombination (DR) was already recognized by Massey & Bates (1942) as a recombination process wherein radiationless capture of a free electron can occur, its importance was not emphasized until it was shown by Burgess (1964, 1965) that it provides an important additional recombination mechanism in a hot dilute plasma like the solar corona. Taking into account this process he pointed out that as a consequence, a considerably higher temperature is needed to maintain a certain state of ionization, thus solving the outstanding problem of the discrepancy between the ionization temperature and the two to three times higher Doppler line width temperature. Another aspect is, as was first shown by Gabriel (1972), that this process is responsible for the formation of spectral lines appearing as (mostly) long-wavelength satellites to the resonance lines of highly ionized atoms in hot plasmas [see review by Dubau & Volonté (1980) and Sect. 15]. In fact, such lines constitute the only observable effect by which dielectronic recombination manifests itself.

If we write for the total radiative and dielectronic recombination rate coefficients $\alpha^{\mathrm{r}}_{Z,z+1}$ and $\alpha^{\mathrm{d}}_{Z,z+1}$, respectively, then the total recombination rate coefficient (in cm^3 s^{-1}) is given by

$$\alpha_{Z,z+1} = \alpha^{\mathrm{r}}_{Z,z+1} + \alpha^{\mathrm{d}}_{Z,z+1} \quad , \tag{30}$$

so that the number of recombinations (cm^{-3} s^{-1}) of ion $Z^{+(z+1)}$ is given by $n_e N_{Z,z+1} \alpha_{Z,z+1}$, where n_e and $N_{Z,z+1}$ are the electron and ion densities (in cm^{-3}), respectively. Though we will concentrate on a tenuous plasma, at the end I will briefly describe the influence of stepwise processes on the radiative recombination rate in a hydrogenic plasma of high density.

7.1 Radiative recombination; the Milne equation

Since radiative recombination is the capture of a plasma electron into a bound state of an ion with emission of a photon, this process is inverse to photoionization so that there is a connection between the rates of recombination and photoionization, the so-called Milne relation (cf. Milne 1924; Elwert 1952; Bates & Dalgarno 1962). This relation can be derived by applying the principle of detailed balance (see Sect. 4) in thermal or thermodynamic equilibrium (cf. Rybicki & Lightman 1979), hence assuming a thermal Maxwell electron velocity distribution (Eq. 20) and a Planck radiation field. Since the rate coefficients refer to atomic properties, the relation is generally valid. The power per unit area (perpendicular to the propagation direction), frequency, and solid angle of the blackbody radiation field is given by:

$$B_\nu = \frac{2h\nu^3}{c^2}\frac{1}{e^{h\nu/kT}-1} \ . \tag{31}$$

Let $Q^{\mathrm{r}}_{z+1,m}$ be the cross section for radiative recombination of ion $Z^{+(z+1)}$ in the ground state with statistical weight w_{z+1} and density N_{z+1} (cm^{-3}) with an electron of velocity v (or kinetic energy $E = \frac{1}{2}m_e v^2$) towards an ion $Z^{+z}(m)$ in state m with statistical weight $w_{z,m}$. Then the number of recombinations per second and per cm^3 with thermal electrons per velocity interval dv is

$$N_{z+1} n_e Q^{\mathrm{r}}_{z+1,m} f_M(v) v \, \mathrm{d}v \ , \tag{32}$$

where $f_M(v)$ is given by (Eq. 20). On the other hand, the number of photoionizations from state m (in ion Z^{+z} with density $N_{z,m}$) per second and per cm^3 in an isotropic blackbody radiation field B_ν per frequency interval $d\nu$ is

$$\frac{4\pi}{h\nu} N_{z,m} \kappa_{z,m}(\nu) \left(1 - \mathrm{e}^{-h\nu/kT}\right) B_\nu \, \mathrm{d}\nu \ , \tag{33}$$

where $\kappa_{z,m}(\nu)$ is the absorption cross section for photo-ionization from state m in ion Z^{+z} and the factor $\left(1 - e^{-h\nu/kT}\right)$ is the correction on the absorption for stimulated emission (which follows immediately from the well-known relations between the Einstein radiation coefficients). Then equating expressions (32) and (33), using Eq. (20), Eq. (31), the Saha-Boltzmann equation [i.e., Eq. (6) with $N_z \to N_{z,m}$, $\chi_z \to \chi_{z,m}$, and $w_z \to w_{z,m}$], using energy conservation for non-relativistic electrons [$\mathrm{d}(h\nu) = \mathrm{d}(\frac{1}{2}m_e v^2)$ or $h\mathrm{d}\nu = m_e v \mathrm{d}v$] and $h\nu = E + \chi_{z,m}$ is the energy of the emitted photon ($\chi_{z,m} = h\nu_m$ is the ionization energy from state m, ν_m is the frequency of the absorption edge), we finally obtain the *Milne relation*:

$$\frac{Q^{\mathrm{r}}_{z+1,m}}{\kappa_{z,m}(\nu)} = \frac{(h\nu)^2}{2Em_e c^2}\frac{w_{z,m}}{w_{z+1}} = 1.3313 \ 10^{-5} \ \frac{(h\nu)^2}{EE_{\mathrm{H}}}\frac{w_{z,m}}{w_{z+1}} \ , \tag{34}$$

where E_{H} is the ionization energy of hydrogen. This is an important relation that is used to calculate recombination rates from photoionization data.

For a hydrogenic ion of nuclear charge Z, in a state with principal quantum number m the absorption cross section (in cm^2) is given by the Kramers-Gaunt formula (Elwert 1952; Kramers 1923; Gaunt 1930)

$$\kappa_m(\nu) = 7.9 \ 10^{-18} \ mZ^{-2} f_1 (\nu_m/\nu)^3 \ (\nu \geq \nu_m) \ , \tag{35}$$

where f_1 is the Gaunt correction factor of order unity ($f_1 = 0.8$ for $m = 1$, 0.9–1 for $m > 1$); $\kappa_m(\nu) = 0$ for $\nu < \nu_m$. Combining Eqs (34) and (35) and omitting subscripts we obtain for the recombination cross section (in cm^2)

$$Q^{\mathrm{r}}(U) = 2.10 \ 10^{-22} \ f_1 m[U(U+1)]^{-1} \ , \tag{36}$$

where $U = E/\chi_m$. The recombination rate coefficient α^{r} (in $\mathrm{cm}^3 \ \mathrm{s}^{-1}$) can be obtained by integrating vQ^{r} over a Maxwellian electron distribution (see Eq. 19), but taking the electron threshold energy $E_0 = 0$ and $y = \chi_m/kT$):

$$\alpha^{\mathrm{r}} = \langle vQ^{\mathrm{r}}\rangle = \int_0^\infty v f_M(v) Q^{\mathrm{r}}(v)\,\mathrm{d}v =$$
$$6.21\ 10^5\ T^{1/2} y^2 \int_0^\infty Q^{\mathrm{r}}(U) e^{-yU} U\,\mathrm{d}U\ , \tag{37}$$

or inserting Eq. (36)

$$\alpha^{\mathrm{r}}(m) = 2.06\ 10^{-11}\ f_1 m^{-1} Z^2 T^{-1/2} G_1(\chi_m/kT)\ , \tag{38}$$

with T in K and $G_1(y) = ye^y E_1(y) \simeq y\ln[(y+1)/y] \approx 1$ ($y \gg 1$) (see Eq. 72). To obtain the total rate we have to sum over all states m.

Seaton (1959) derived a formula for the total radiative recombination rate towards hydrogenic ions which is based on an expansion of the Kramers-Gaunt factor:

$$\alpha^{\mathrm{r}} = 5.197\ 10^{-14} Z\lambda^{1/2}[0.4288 + 0.5\ln(\lambda) + 0.469\lambda^{-1/3}]\ , \tag{39}$$

where $\lambda = 157890 Z^2/T$ (T in K). This expression which is accurate to 3% for $T \leq 10^6 Z^2$, is used in the calculations by Arnaud & Rothenflug (1985).

For recombination towards non-hydrogenic ions in excited states ($m > m_0$) one can also use the hydrogenic approximation since with increasing m excited states readily approach hydrogenic ones, but for the recombination towards the ground state (m_0) we can evaluate the recombination coefficient from the cross section of photoionization from the ground state (e.g., Reilman & Manson 1979; also Pradhan 1987) using the Milne relation (Eq. 34). A number of authors, e.g., Aldrovandi & Péquignot (1973) and Arnaud & Rothenflug (1985) applied this and made simple power law fits in temperature to the resulting recombination rates. For references, see Verner & Ferland (1996) who have made more sophisticated fits to the photoionization cross sections and recombination rates. As an example, I present in Figs. 6 and 7 some representative results of Arnaud & Rothenflug (1985). The estimated overall uncertainties are probably at the level of 30–40% (see discussion by Raymond 1988).

For the general case of non-hydrogenic ions we can use the Milne relation when photoionization rates are available (see Sect. 5.2.2), but for making first-order estimates we may use the approach of Elwert (1952) who replaced in this case Z by the effective nuclear charge z_e, defined by $\chi_{m_0} = E_{\mathrm{H}} z_{\mathrm{e}}^2/m_0^2$, where χ_{m_0} is the energy of ionization of the recombined ion Z^{+z} in the ground state with principal quantum number m_0 (e.g., for the hydrogenic case $z_e = Z$ and $m_0 = 1$), and introduced a correction factor G to account for the recombination to excited states. Then (omitting subscript Z) we obtain for the total radiative recombination rate coefficient ($\mathrm{cm}^3\ \mathrm{s}^{-1}$) :

$$\alpha^{\mathrm{r}}_{z+1} = 2.06\ 10^{-11}(G/m_0)G_1(\chi_{m_0}/kT) z_{\mathrm{e}}^2 T^{-1/2} \approx 3\ 10^{-11}\ z_{\mathrm{e}}^2 T^{-1/2}\ , \tag{40}$$

where $G \simeq 1 + 0.37 m_0 + 0.25 m_0^{1.36} (\chi_{m_0}/kT)^{0.43}$ and T is in K. This expression shows that radiative recombination rates in dependence of reduced temperature $\Theta = T/z_{\mathrm{e}}^2$ scale $\propto z_{\mathrm{e}}$. For a global estimate we can use the right-hand approximation within about 15% for $\chi_{m_0}/kT \gg 1$.

7.1.1 High-density effects on the radiative recombination rates. Analogously to the calculation of the collisional–radiative ionization rate coefficient (see Sect. 6.1.3) we consider the effect of high density on recombination. In the high-density limit the rate coefficient ($\mathrm{cm^3\,s^{-1}}$) for recombination towards hydrogenic ions $Z^{+(Z-1)}$ approaches (n_e in $\mathrm{cm^{-3}}$, T in K) (Mewe 1970, 1988):

$$\alpha_Z(n_e \to \infty) = 5.37\ 10^{-21}\ Z^{-2}T^{-2}n_e \exp(3.946\ 10^4 Z^2/T)\ , \tag{41}$$

which is $\propto n_e$ and equal to the 3-body recombination rate. The latter is directly related to the inverse process of ionization (Eq. 25) by the principle of detailed balancing (Sect. 4) so that the ratio of Eq. (25) and Eq. (41) yields indeed the Saha ionization balance equation (Eq. 6).

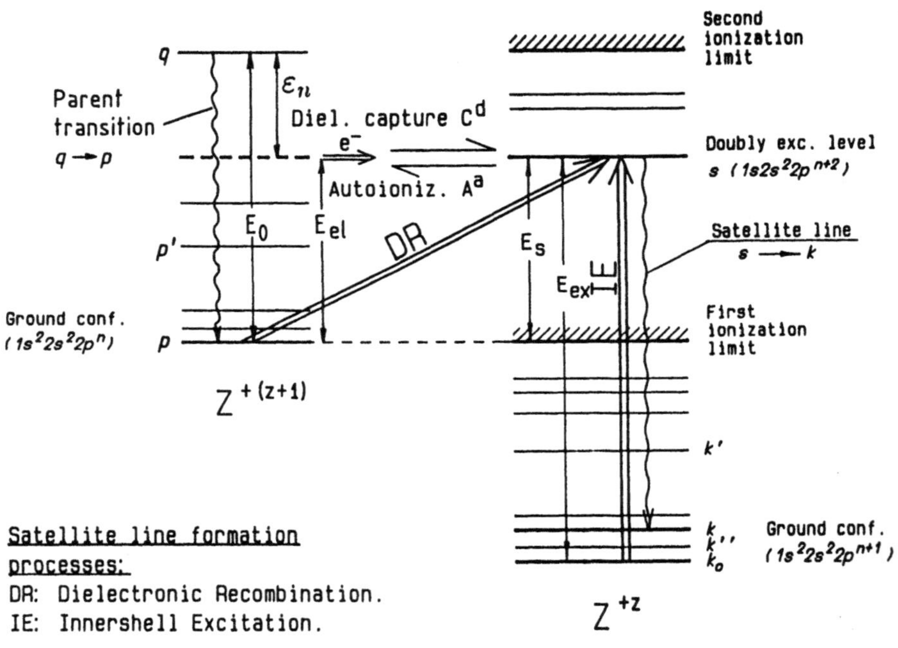

Fig. 8. Energy level diagram showing the dielectronic capture of an electron into a doubly excited level $s = q, n\ell$ and the subsequent formation of a dielectronic recombination (DR) satellite line $s \to k$ in ion Z^{+z}. Here the doubly excited level s is formed by the DR of ion $Z^{+(z+1)}$ in one of its ground configuration levels p (= ground state 1 for a low-density plasma). An alternative population process is electron impact inner-shell excitation (IE) (energy $E_{ex} \simeq E_0$) of ion Z^{+z} from one of its ground configuration levels k'' (= presumably the ground state k_0). De-excitation of state s occurs either radiatively towards levels $k' < s$ (e.g., $k = p, n\ell$) to form the DR satellite line $s \to k$) or by auto-ionization towards levels $p' < s$.

7.2 Dielectronic recombination

The recombination of many coronal ions, however, is dominated by dielectronic recombination (for reviews, e.g., Hahn 1985; Hahn & LaGattuta 1988; Bell & Seaton 1985).

Dielectronic recombination (DR) can be considered as a two-step process illustrated in Fig. 8.[9] It is initiated by the resonant radiationless capture of an energetic plasma electron by an ion $Z^{+(z+1)}$ into a high Rydberg state $n\ell$ of ion Z^{+z}, accompanied by the excitation $p \to q$ of one of the bound electrons in the core of the recombining ion $Z^{+(z+1)}$, thus forming a doubly excited state in ion Z^{+z}:

$$Z^{+(z+1)}(p) + e^- \to Z^{+z}(q, n\ell) \ . \tag{42}$$

The initial state p of the recombining ion $Z^{+(z+1)}$ is usually the ground state in the coronal approximation. The high Rydberg electron $(n\ell)$ is just a "spectator" as far as the radiation of the core electron is concerned. Thus, an electron is incident on ion $Z^{+(z+1)}$ with an energy $E_{\rm el} = E_s = E_0 - \varepsilon_n$ just ε_n less than the threshold energy E_0 needed to excite the core electron. As it approaches the ion, it gains kinetic energy in the Coulomb field of the ion, so that close in it has sufficient energy to excite the bound inner core electron, but if it does so, it has not enough energy to escape, and is bound in Rydberg level $(n\ell)$ with binding energy $\varepsilon_n = E_{\rm H} z^2/n^2$. Since there is an infinite number of Rydberg states, the DR capture cross section versus electron energy consists of a corresponding series of narrow resonances located just below the threshold for core-electron excitation (cf. also Sect. 8.1).

Once the ion is in the doubly excited state $s = q, n\ell$ (which is above the first ionization limit), having one electron removed from an inner shell, it has *two options.* It may return to a state of lower energy by ejecting one of its own electrons without the emission of radiation. This internal conversion process in which the energy of the excited core electron is again transferred to the Rydberg electron (the reverse of the capture process) is called the *Auger effect.* Then the electron leaves and *autoionization* occurs. Alternatively, the vacancy in the inner shell is filled by an electron from a less tightly bound state (this process may be successively repeated until all the excess energy of the ion has been radiated). The excited core electron radiates the excess energy away as fluorescence radiation and is captured so that *dielectronic recombination* occurs. The doubly excited state $s = q, n\ell$ undergoes a stabilizing radiative transition (satellite line $s \to k$) directly (or eventually via several cascades) towards a final state $k = p, n\ell$ that lies below the first ionization limit of the recombined ion:

$$Z^{+z}(q, n\ell) \to Z^{+z}(p, n\ell) + h\nu \ . \tag{43}$$

[9] The ion $Z^{+(z+1)}$ in this diagram may represent one of the Fe XVIII–Fe XXIII ions for which the parent resonance line $2p \to 1s$ occurs in Fe XXV.

The *satellite* line $(h\nu)$ $s \rightarrow k$ in ion Z^{+z} is slightly shifted to the long-wavelength side of the corresponding parent $q \rightarrow p$ transition in the recombining ion $Z^{+(z+1)}$ due to the electrostatic shielding by the spectator electron.

Finally, the singly excited state $k = p, n\ell$ eventually cascades down to the ground state $k_0 = p, n_0\ell_0$ [i.e. the spectator electron makes the transition $n\ell \rightarrow n_0\ell_0$ and $Z^{+z}(p, n\ell) \rightarrow Z^{+z}(p, n_0\ell_0) + h\nu'$].

The dielectronic capture rate C_s^{d} (in $\mathrm{cm^3\ s^{-1}}$) to level s with statistical weight $w_s = 2(2\ell + 1)$ is directly related to the inverse process of autoionization from level s (rate A_s^{a} in $\mathrm{s^{-1}}$) towards the ground state (statistical weight w_1) of the recombining ion $Z^{+(z+1)}$ by applying the principle of *detailed balancing* in thermodynamical equilibrium (cf. Sect. 4):

$$N_{z+1}^{*} n_{\mathrm{e}} C_s^{\mathrm{d}} = N_{z,s}^{*} A^{\mathrm{a}} \ , \tag{44}$$

where $N_{z,s}^{*}$ and N_{z+1}^{*} are the *fictitious* densities of the satellite level (s in the recombined ion Z^{+z}) and of the recombining ion ($Z^{+(z+1)}$) given by the Saha-Boltzmann equation (cf. Eq. 6) with w_{z+1}/w_z replaced by w_1/w_s). Then we immediately obtain for the dielectronic capture rate (cf. also Gabriel 1972; Dubau & Volonté 1980):

$$C_s^{\mathrm{d}} = 2.071\ 10^{-16} T_{[\mathrm{K}]}^{-3/2} (w_s/w_1) A^{\mathrm{a}}{}_{[\mathrm{s}^{-1}]} \exp(-E_s/kT)\ (\mathrm{cm^3\ s^{-1}}) \ , \tag{45}$$

where E_s is the energy difference between the autoionizing state s and the ground state of the recombining ion which is just the kinetic energy $E_0 - \varepsilon_n$ of the plasma electron being captured. This relation is valid as long as the plasma electrons have a Maxwellian energy distribution, since C_s^{d} and A_s^{a} are atomic rate coefficients relating two exactly inverse processes. For the dielectronic rate coefficient we can write:

$$\alpha^{\mathrm{d}}(s) = C_s^{\mathrm{d}}\ BR_s \ , \tag{46}$$

where

$$BR_s = \frac{A_s^{\mathrm{r}}}{A_s^{\mathrm{a}} + \sum A_s^{\mathrm{r}}} \ , \tag{47}$$

the branching ratio for the dielectronic recombination channel where A_s^{r} is the radiative decay rate for the satellite line $s \rightarrow k$ with the summation over all possible radiative transitions from the satellite level s. By summing over all possible resonant autoionizing Rydberg states $s = n\ell$ we obtain the total DR rate coefficient ($\mathrm{cm^3\ s^{-1}}$) (e.g., Bely-Dubau et al. 1979):

$$\alpha^{\mathrm{d}} = \sum_s \alpha^{\mathrm{d}}(s) = 2.071\ 10^{-16} T^{-3/2} \sum_s B_s \mathrm{e}^{-E_s/kT} \ , \tag{48}$$

where

$$B_s = \frac{w_s A_s^{\mathrm{a}} A_s^{\mathrm{r}}}{w_1 (A_s^{\mathrm{a}} + \sum A_s^{\mathrm{r}})} \ . \tag{49}$$

The behaviour of the DR rate depends sensitively on the branching ratio BR, i.e. on the relative magnitudes of $A^{\rm a}$ and $A^{\rm r}$. The $A^{\rm a}$ can be found by extrapolating the $p \to q$ excitation cross section below threshold. They vary as n^{-3} (e.g., Hahn & LaGattuta 1988). For small n, $A^{\rm a}$ is likely to be much larger than $A^{\rm r}$ for the values of ℓ which contribute strongly, and $\alpha_{\rm DR}$ is then $\propto A^{\rm r}$. As along an isoelectronic sequence the radiative transition probability scales as $A^{\rm r} \propto Z^4$ (for $\Delta n \neq 0$ transitions), it turns out that for low-Z ions (say $Z \lesssim 10$), $A^{\rm a}$ still exceeds $A^{\rm r}$ for large n values up to several hundred (and ℓ values up to ~ 6), so that many resonant Rydberg states will contribute. However, as Z increases (say $Z \gtrsim 20$), $A^{\rm a}$ already becomes smaller than $A^{\rm r}$ for smaller n, and DR from lower states becomes increasingly important. For instance, recombination of Fe XXV ($Z = 26$) occurs primarily through the $n = 2$ shell (Bely-Dubau et al. 1979). For a given (n, ℓ), $A^{\rm r}(n, \ell)$ is nearly constant with n and ℓ (because the Rydberg electron is effectively only a spectator as far as the core-excited electron is concerned), whereas $A^{\rm a}(n, \ell) \propto n^{-3}$. For low n and l, $A^{\rm a} \gg A^{\rm r}$, then $BR \propto A^{\rm r} n^3 \propto n^3$, and $\alpha_{\rm DR}(n, \ell) \propto g_s A^{\rm a} BR \propto (2\ell + 1) A^{\rm r}$. Furthermore, $A^{\rm a}(n, \ell)$ decreases rapidly with ℓ [e.g., $\propto \exp(-0.25\ell^2)$], effectively cutting off the contribution from large ℓ (say above $\ell = \ell_c$ between 5 and 10), since for $A^{\rm a} \ll A^{\rm r}$, $\alpha_{\rm DR} \propto A^{\rm a}$. Thus, though (especially for low-Z ions) there is a large number of resonances which *could* contribute to DR (*if* their autoionization rates were larger) $\propto \sum_{\ell=0}^{n-1} 2(2\ell + 1) = 2n^2$, only ℓ's for $\ell \lesssim \ell_c$ (i.e., $A^{\rm a} \gtrsim A^{\rm r}$) will contribute $\propto 2(\ell_c + 1)^2$ (e.g., for n=20 and ℓ_c=10 only 200 states out of the possible 800). This is due to the strong decrease of $A^{\rm a}$ with ℓ.

Burgess (1964, 1965) was the first to compute DR rates and fit the results to a general semi-empirical formula, similar to the expression (48), later corrected by Burgess & Tworkowski (1976). The expected accuracy was ~30% and the formula is probably valid only near the temperature $T_{\rm m}$ at which the ion concentration peaks. For $T \gg T_{\rm m}$ DR rarely matters, but for very low temperatures ($T \ll T_{\rm m}$), it can be quite important in photoionized plasmas or in rapidly cooling gas. For low temperatures Nussbaumer & Storey (1983) have computed DR rates for abundant elements up to Si. The Burgess approximation is best for $\Delta n = 0$ inner-electron transitions and modified versions of the formula have been proposed for $\Delta n \neq 0$ transitions (Merts et al. 1976; Hahn 1985). Another modification, as was first pointed out by Jacobs et al. (1977), is the inclusion of additional autoionizing decay channels of the doubly excited state into *excited* states of the recombining ion (auto-ionization not always occurs only via the true inverse radiationless capture, i.e., into the channel associated with the ground state of the recombining ion). For certain ions this can drastically reduce the DR rate. Though the Jacobs rates are good for low-Z ions, the Jacobs correction may seriously underestimate the rates for high-Z ions [e.g., a factor ~2–3 for recombination of Fe XVII (Smith et al. 1985)]. It appears that the disagreement among various computations allows an overall accuracy of only ~40% [see discussion by Raymond (1988)].

Unfortunately, for dielectronic recombination rates there is less experimental guidance than for ionization rates, because the crossed-beam experimental results suffer from the influence of electric fields experienced by ion beams crossing a magnetic field (e.g., Hahn 1985). This causes field ionization that limits the value of n to ~ 30, but more importantly, Stark mixing of different l levels of a given n state, which may cause a dramatic enhancement of the experimental cross section for low-Z ions of up to an order of magnitude compared to the computed zero-field case (see also Sect. 7.2.2).

Arnaud & Rothenflug (1985) have evaluated many theoretical data and updated them in a number of cases. Figs. 6 and 7 present their results for the radiative recombination (RR) and dielectronic recombination (DR) rate coefficients in a reduced form, i.e. $\alpha_{z+1}/(z+1)$, as a function of the reduced temperature $\Theta = T/(z+1)^2$. This scaling is appropriate for RR (see Eq. 40), but seems also reasonably valid for DR (Summers 1974; Hahn et al. 1980). It is seen that dielectronic recombination becomes increasingly important for lower Z and at higher temperatures, whereas for the higher ionization stages of high-Z ions radiative recombination becomes dominant.

7.2.1 High-density effects on the dielectronic recombination rates. Dielectronic recombination rates can be affected by the density: at increasing density the DR rate will be reduced by the ionization of the highly excited $n\ell$ states. The magnitude of this suppression has a fairly weak n_e dependence and roughly scales as $\sim (n_e/(z+1)^7)^{0.25}$ (e.g., Burgess & Summers 1969; Summers 1974). It will be most important for the lower-Z ions which emit UV emission lines rather than for the X-ray emitting ions (Raymond 1988). The effects set in roughly for $n_e \gtrsim (10^3\text{–}10^4)(z+1)^7$ cm^{-3}.

7.2.2 Inaccuracies: effects of electric fields. Several crossed-beam experiments for singly ionized atoms have shown that dielectronic recombination (DR) is very sensitive to electric fields. The results show that the measured DR rates are a factor of $\sim$5–10 larger than the predicted ones for lower ions (e.g., Müller et al. 1987). This strong discrepancy was explained in terms of the effect of an electric field $E_{\rm el}$ of $\sim$ 10–100 V/cm that cuts off the high $n\ell$ Rydberg states by field ionization but causes Stark mixing of different ℓ levels for lower n, resulting in a net enhancement of the DR rate by nearly an order of magnitude for lower ions. In the experiment, a magnetic field B $\sim$200 Gauss was used to focus the electron beam. The ions which cross the electron beam with velocities v $\sim$$10^7$ cm/s then experience in their rest frame a Lorentz field $E_{\rm el} = 10^{-8} vB \approx 20$ V/cm, sufficient to produce full Stark mixing. As $A^{\rm a}$ decreases with ℓ the result of mixing high- and low-ℓ levels is that $A^{\rm a}$ is increased for high ℓ and decreased for low ℓ, thus flattening the $A^{\rm a}(\ell)$ versus ℓ curve, so that more states effectively participate in the recombination process, and the DR cross section is increased.

In the actual plasma environment the mean thermal drift $v_{\perp} \sim \sqrt{2kT_{\rm i}/m_{\rm i}}$ of ions (i) across a magnetic field B can indeed generate such Lorentz electric fields. For $E_{\rm el} \gtrsim 20$ V/cm this reduces to $B_{[\rm G]}\sqrt{T_{\rm i[MK]}/M_{\rm i}} \gtrsim 90$, which, e.g., for $M_{\rm i} \sim 20$ and $T_{\rm i} \sim 1$ MK implies $B \gtrsim 400$ G. Magnetic fields on the order of this value can exist in solar or stellar coronal loop structures and much higher values (~1000 G) in the lower chromosphere. As the effect is only important for the lowly ionized atoms, it can play a role in cool ($\lesssim 2$ MK) coronal or photoionized plasmas.

Mewe (1990) made an attempt to visualize this effect using a simple schematic model as given by Müller et al. (1987). The typical enhancement factor for the DR recombination of ion $Z^{+(z+1)}$ due to full Stark mixing was approximated by $f = 1 + 10\ (z+1)^{-1.15}$. The results show that the effects are most noticeable for the lower ions (where many n states take part in the DR process) formed at low ($\lesssim 2$ MK) temperatures (cf. his Fig. 4 in which a comparison is made for ion fractions of oxygen and iron calculated with and without correction for ℓ-level Stark mixing).

III. Formation of X-ray Spectra in a Coronal Plasma

The overall appearance of the X-ray spectra will be dominated by the ionization structure which varies dramatically with temperature throughout the range 0.1–100 MK.

In extremely hot ($T > 100$ MK) plasmas all abundant elements are nearly fully ionized and the X-ray emission is dominated by the free-free continuum from hydrogen and helium. With decreasing temperature, the heavier trace elements are only partly ionized, beginning with iron, and spectral emission lines excited by electron collisions show up and will begin to dominate the X-ray spectrum. This is especially true for the Fe L-shell lines around 10 Å, the Fe K-shell lines at ~ 2 Å and the Fe $2s$–$2p$ lines around 100 Å. In a wide temperature range (0.01–10 MK) the X-ray spectra of optically thin sources are rich in emission lines from many ions. The spectral lines are broadened mainly by Doppler broadening.

The wavelength band 1–140 Å contains a multitude of prominent lines from nearly all ionization stages of cosmically abundant elements, including the K-shell transitions of carbon through iron, and the L-shell transitions from silicon through iron. If suitably resolved, these lines are powerful diagnostics of plasma parameters because the line strengths are generally very sensitive to the electron temperature, the elemental abundances, and in some cases to the electron density. In certain circumstances (e.g., transient plasmas) the line intensities are dependent on deviations from the ionization equilibrium or on deviations from a Maxwellian electron energy distribution (e.g., Mewe 1990).

The emission line spectra and continua from optically thin plasmas are fairly well known from calculations (for reviews, e.g., Raymond 1988; Mewe

1990, 1991; for examples of measured and simulated spectra see also the Chapters by Paerels and by Kaastra). Comparison of measured line fluxes with known theoretical line emissivities yields the differential emission measure distribution as a function of temperature of the source, which is an essential first step in building a model for a stellar corona and assessing the important terms in the coronal energy balance. Therefore, high-resolution X-ray spectroscopy has its most obvious application in diagnosing optically thin sources such as the coronae of late-type stars.

I consider the processes that produce the line and continuum radiation, in particular the electron impact excitation of spectral lines and the formation of so-called satellite lines. The possibilities for the diagnostics of temperature, density, differential emission measure, abundances, and velocities have been described in more detail elsewhere (Mewe 1990, 1991).

8 Line Radiation

The spectral lines that dominate the soft X-ray spectrum and the cooling of astrophysical plasmas at temperatures up to ~10 MK are mainly excited by electron collisional excitation from the ground state. Though this is often a good approximation to calculate the stronger (resonance) lines in a low-density plasma, we have to follow a more complicated scheme of processes connecting many more levels (e.g., also fine structure levels) when we consider the effects of electron density on the line intensities (see also the Chapter by Liedahl).

The volume emissivity P_{ji} (phot cm^{-3} s^{-1}) of a particular spectral line transition $j \to i$ in ion Z^{+z} from element of atomic number Z is written as (omitting subscript Z, z):

$$P_{ji} = N_j A_{ji} \ , \tag{50}$$

where N_j is the population density (cm^{-3}) of the ions in the upper line level j and A_{ji} is the probability (s^{-1}) of a spontaneous radiative transition from the upper line level j towards the lower level i. In the general case the level population density is solved from a set of rate equations which read for level j [neglecting processes of photo-excitation and photo-ionization and also transport terms which can be important in a laboratory plasma of finite size (e.g., Düchs & Griem 1966; Peacock 1996)]:

$$\frac{\mathrm{d}N_j}{\mathrm{d}t} = \sum_{h<j} N_h n_e S_{hj} + \sum_{k>j} N_k (A_{kj} + n_e S'_{kj}) + N_{z+1} n_e \alpha_j^{\mathrm{r}} + N_{z-1} n_e \beta_j^{\mathrm{II}}$$
$$- N_j \left[\sum_{h<j} (A_{jh} + n_e S'_{jh}) + \sum_{k>j} n_e S_{jk} + n_e S_j^{\mathrm{i}} \right] \ , \tag{51}$$

where S_{hj} is the rate coefficient for electron impact excitation $h \to j$ (h denotes the lower level) and S'_{kj} the rate coefficient for collisional de-excitation

$k \rightarrow j$ (k the upper level). Furthermore, α_j^{r} and β_j^{II} are the rate coefficients for recombination and collisional inner-shell ionization (the latter is important in the formation of the He-like forbidden line in a transient plasma, see Mewe & Schrijver 1978b, and Sect. 15.3), which may give a contribution to the population of level j, while S_j^{i} denotes the rate of collisional ionization from level j. If we assume a *steady state*, i.e. $\mathrm{d}N_j/\mathrm{d}t = 0$, Eqs (51) reduce for each element to a set of linear algebraic equations describing the statistical equilibrium. Because we neglected the transport terms we have the condition that $\sum_{j=1}^{N} N_j = N_{Z,z}$, the total ion density in charge state z of element Z. The equation of charge conservation can be added to the system, which couples the elements, but in practice, we are interested primarily in the trace elements, which have practically no effect on the charge balance. For each element, the system of equations can be solved for all level populations N_j by using a technique of matrix inversion [the ion densities N_{z+1} and N_{z-1} can be expressed into $N_z \simeq N_g$ via the ionization balance (Eq. 14)].

Even for a transient plasma we can often make the assumption of a steady state because it turns out that in most cases the excited levels come very quickly, i.e. effectively instantaneously, into equilibrium with a particular value of the ion density in the ground state (N_g) corresponding to the current plasma state. Such a state is called a *quasi-equilibrium* or quasi-steady-state (QSS) (cf. McWhirter & Hearn 1963, and the Chapter by Liedahl). I note that a metastable state takes more time to approach a QSS. An additional condition for the quasi-equilibrium is that the excited-level population densities are negligible with regard to the ground state density, so that the densities of the free electrons and the ions remain essentially constant during the time in which the quasi-steady state is established.

So we deal with the steady-state equations and, for convenience, we make the following assumptions (number in parentheses is the number of the corresponding term in the r.h.s. of Eq. 51): we neglect recombination (3), inner-shell ionization (4), ionization (7), collisional de-excitation processes (5) and excitations to higher levels (6); we assume low plasma density, i.e., excited level populations are so low that we can neglect cascades from levels above level j (2), and all excitations from excited levels (i.e., we consider in the first term only excitation from the ground level g and neglect possible effects from metastable levels); $N_g \simeq N_{Z,z}$, i.e. practically all ions are in the ground state. Then Eq. (51) reduces to

$$0 = N_g n_e S_{gj} - N_j \sum_{h<j} A_{jh} \ , \tag{52}$$

which in many situations is a suitable approximation for a coronal plasma. However, in the case where we study density effects on spectral lines, we should consider the more complete set of Equations (51) including sometimes transitions between fine-structure levels (cf. Sect. 13). Equations (50) and (52) yield for the line emissivity (phot $\mathrm{cm}^{-3}\ \mathrm{s}^{-1}$)

$$P_{ji} = N_j A_{ji} = N_g n_e S_{gj} BR_{ji} = n_e N_H A_Z BR_{ji} G_{ji}(T) \ , \qquad (53)$$

where $BR_{ji} = A_{ji}/\sum_{h<j} A_{jh}$ is the radiative branching ratio with respect to all lower levels, N_g the ground state population (cm^{-3}) of ion Z^{+z}, and $A_Z = N_Z/N_H$ the abundance of element Z relative to hydrogen (H). Finally, $G_{ji}(T) = \eta_{Z,z}(T) S_{gj}(T)$, where $\eta_{Z,z} = N_{Z,z}/N_Z \simeq N_g/N_Z$ is the fraction of ions from element Z in ionization stage z and $S_{gj}(T)$ the rate coefficient (cm^3 s^{-1}) for electron impact excitation $g \to j$. The function $G_{ji}(T)$ represents the temperature dependence due to the combination of ionization and excitation, and the temperature T_m at which it peaks is called the temperature of *maximum formation* of the given line. For a fully ionized (in H and He) plasma with cosmic abundances (Anders & Grevesse 1989) the ratio N_H/n_e is fixed primarily by the abundances of H and He: $N_H \simeq 0.77 n_e \simeq n_e$, hence $P_{ji} \propto n_e^2$.

8.1 Excitation processes

The formation of line and continuum radiation mainly occurs by the *electron impact* of atoms or ions. The excitation becomes effective when the impinging electron has a speed comparable to or somewhat larger than the orbital velocity of the bound electron in the atom or ion that is to be excited, i.e., this occurs when the energy of the impinging electron $E \gtrsim \Delta E$, the threshold energy for excitation of the ion. For the excitation of an ion the excitation cross section is finite (and generally largest) at threshold owing to the Coulomb attraction that accelerates the impinging electron.

For an ion (e.g., a *proton*) to be effective in exciting an atom or ion, its speed should also be comparable to the orbital speed of the bound electron which implies that its energy should be higher than for an electron by the ratio of proton/electron mass $R_{pe} = m_p/m_e \simeq 1800$. The excitation cross section for proton excitation begins to rise slowly at threshold and attains its maximum roughly at energies of the impinging proton of $\sim R_{pe}\Delta E \gg \Delta E$. In a plasma with equal electron and ion temperatures $T_e = T_i = T$, those ions that are formed in collisional ionization equilibrium (CIE) (Sect. 5) have ionization energies $\chi \sim \Delta E \sim kT$. Hence, in this situation, electrons are by far more effective in the excitation (because $kT_e \sim \Delta E$) than protons (or heavier ions) which rather would need the condition $kT_i \sim R_{pe}\Delta E \gg kT_e$.

However, there are cases where proton excitation is quite important. For instance, in a NEI plasma (cf. Sect. 4.1.3) the situation can occur in which the plasma temperature T suddenly rises and the ionization balance is still frozen in so that kT becomes $\gg \Delta E$, and the protons are more efficient in exciting the ions rather than the electrons for which the excitation cross section drops at least as fast as $\log E/E$. This is, for example, the case in very fast shock waves for which conditions are created where proton excitation is important: behind the shock low-ionization species are immersed in very hot shocked gas so $kT_i \gg \Delta E$.

The condition $kT \gg \Delta E$ is also fulfilled in CIE for the excitation of transitions between closely spaced energy levels in fine structures which affects the density sensitivity of line intensities. Then proton impact can be important, e.g., by initiating $2s$–$2p$ transitions in H-like ions (Lyman α doublet line) (Zygelman & Dalgarno 1987) or $2s\ ^3S$–$2p\ ^3P$ transitions in He-like ions (Blaha 1971) (cf. Fig. 13 and Sect. 13).

For further literature, I refer to Raymond & Brickhouse (1996) and to Pal'chikov & Shevelko (1995). The latter authors distinguish in the excitation of ions by electrons,

$$Z^{+z} + e^- \rightarrow (Z^{+z})^* + e^- \ , \tag{54}$$

the following processes:

(i) excitation of outer-shell electrons (cf. Sect. 8.1.1)

$$Z^{+z}(\zeta n\ell^q) + e^- \rightarrow (Z^{+z})^*(\zeta n\ell^{q-1} n'\ell') + e^- \ , \tag{55}$$

(ii) excitation of inner-shell electrons (cf. Sect. 15.3)

$$Z^{+z}(\zeta n\ell^q \xi) + e^- \rightarrow (Z^{+z})^*(\zeta n\ell^{q-1} \xi n'\ell') + e^- \ , \tag{56}$$

and

(iii) resonant excitation (see below)

$$Z^{+z}(\xi_0) + e^- \rightarrow (Z^{+(z-1)})^{**}(\xi n\ell) \rightarrow (Z^{+z})^*(\xi_1) + e^- \ , \tag{57}$$

where ζ and ξ denote a part of the electron configuration and q denotes the number of equivalent electrons of the $n\ell$ shell.

Resonant excitation

Process (57) is a two-step process that is illustrated in Fig. 9. Because of the long-range attractive Coulomb force, a multi-charged ion Z^{+z} can capture a free electron of energy E and create a doubly excited ion $(Z^{+(z-1)})^{**}$:

$$Z^{+z}(\xi_0) + e^- \rightarrow (Z^{+(z-1)})^{**}(\xi n\ell) \ , \tag{58}$$

where the free electron is captured into a high Rydberg state $n\ell$ and an inner core electron of $Z^{+(z-1)}$ is simultaneously excited to state ξ. The excited ion $(Z^{+(z-1)})^{**}$ is unstable and can decay by two competitive channels:

auto-ionization (AI)

$$(Z^{+(z-1)})^{**}(\xi n\ell) \rightarrow (Z^{+z})^*(\xi_1) + e^- \ , \tag{59}$$

or radiative decay (dielectronic recombination, cf. Sect. 7)

$$(Z^{+(z-1)})^{**}(\xi n\ell) \rightarrow (Z^{+(z-1)})^*(\xi_1 n\ell) + h\nu \ . \tag{60}$$

via the emission of a satellite line ($h\nu$) and subsequent cascading ($h\nu'$, $h\nu''$, ...) to the ground state ξ_0.

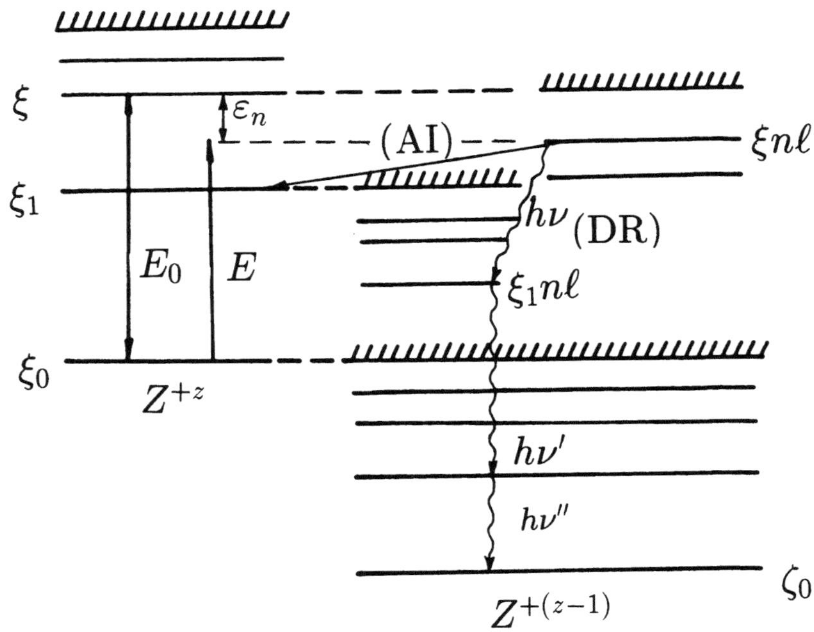

Fig. 9. Energy level scheme and processes for electron resonance capture. AI and DR denote auto-ionization and dielectronic recombination, respectively (from Pal'chikov & Shevelko 1995, with some modifications).

The auto-ionization leads to excitation $\xi_0 \rightarrow \xi_1 (\neq \xi_0)$ of the ion Z^{+z} via the intermediate state $\zeta = \xi n\ell$. This excitation process in addition to the usual one is called *resonant excitation* (if $\xi_1 = \xi_0$ it is called resonant elastic scattering).

The resonant capture of an electron is only possible if its energy

$$E = E_0 - \varepsilon_n \simeq E_0 - E_{\rm H}(z-1)^2/n^2 \ , \qquad (61)$$

where E_0 is the threshold energy to excite the core electron in ion Z^{+z} and ε_n is the binding energy of the captured electron in a high Rydberg state $n\ell$.

This leads to the appearance of a series of narrow resonance structures (for different values of n) on top of the "normal" excitation cross section just below the threshold for the core-electron excitation and just above the threshold energy of the normal excitation (cf. Fig. 10). The width of such a resonance is given by the *Heisenberg relation*:

$$\delta E_\zeta = \frac{h}{2\pi}[A_{\rm a} + A_{\rm r}] \ , \qquad (62)$$

where the sum of the autoionization ($A_{\rm a}$) and radiative decay ($A_{\rm r}$) probabilities determines the total decay probability of the doubly-excited state ζ.

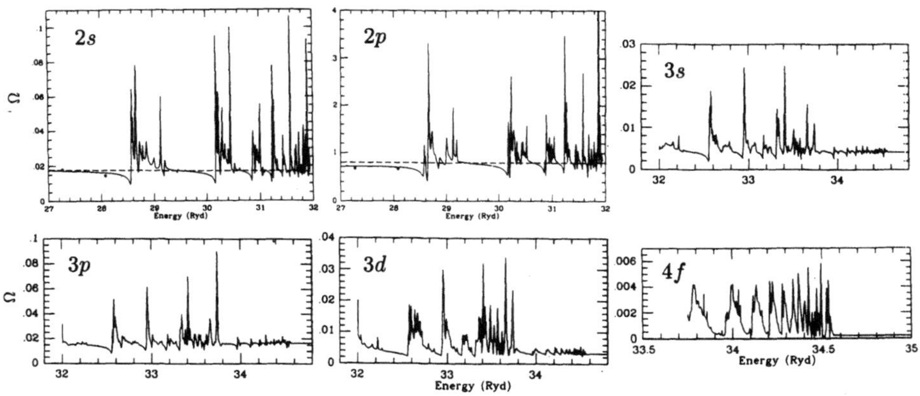

Fig. 10. Collision strengths for excitation of the transitions from the ground state $1s\ ^2S$ in the H-like ion C VI to the $n = 2$ states $2s\ ^2S$ and $2p\ ^2P$, the $n = 3$ states $3s\ ^2S$, $3p\ ^2P$, and $3d\ ^2D$, and to $4f\ ^2F$ (from Aggarwal & Kingston 1991).

Typically, for $[A_a + A_r] \lesssim 10^{14}\ s^{-1}$, $\delta E_\zeta \lesssim 0.0048$ Rydberg. The contribution of these resonances (smoothed over an electron Maxwellian distribution) should be taken into account for plasmas at relatively low temperatures, but can often be neglected for hot plasmas. Moreover, as the charge of the ion increases, the radiative transition probability ($A_r \propto Z^4$) becomes larger than the autoionization probability ($A_a \sim$ const. in an isoelectronic seq.) so that the branching ratio $A_a/(A_a + A_r) \propto Z^{-4}$ and the resonances will disappear and dielectronic recombination prevails (see discussion by Moores 1988).

8.1.1 Electron impact excitation rate coefficients. The cross section $Q_{ij}(U)$ for electron impact excitation from level i to level j can be written in terms of the *collision strength* $\Omega(U)$:

$$Q_{ij}(U) = \frac{\pi a_0^2}{w_i}\left(\frac{E_H}{E_{ij}}\right)\frac{\Omega(U)}{U} = \frac{\pi a_0^2}{w_i}\left(\frac{E_H}{E}\right)\Omega(E) , \tag{63}$$

where $U = E/E_{ij}$, the impact energy E of the electron expressed in terms of the line excitation energy E_{ij}, and w_i is the statistical weight of the initial lower (often the ground) level. This expression is similar to Eq. (22) with $\alpha = E_{ij}/(w_i E_H)$ and $Q_{red}(U) = \Omega(U)/U$. Note that E_{ij}/E_H is the excitation energy expressed in so-called *Ry(dbergs)* (cf. Sect. 6.1). For optically allowed (electric dipole) transitions we can introduce the concept of the excitation *Gaunt factor* $g(U)$, first introduced by Seaton (1962) and by van Regemorter (1962), which is related to the collision strength by

$$\frac{\Omega(U)}{w_i} = \frac{8\pi}{\sqrt{3}} \frac{E_{\rm H}}{E_{ij}} f\, g(U) = 197.3\, f\, g(U)\ E_{ij[\rm eV]}^{-1}$$
$$= 0.01592\, [\lambda_{[\rm\AA]}/a]\, f\, g(U)\ , \qquad (64)$$

where $f_{(ij)}$ is the absorption oscillator strength and λ the wavelength of the line, and $a = E_{ij}/h\nu$, the ratio of line excitation and photon energy ($a > 1$ for a line transition not ending on the ground level). Then the excitation cross section reads as

$$Q_{ij}(U) = \frac{8\pi}{\sqrt{3}} \left(\frac{E_{\rm H}}{E_{ij}}\right)^2 f\, \frac{g(U)}{U}\, \pi a_0^2 = 14.51 \left(\frac{E_{\rm H}}{E_{ij}}\right)^2 f\, \frac{g(U)}{U}\, \pi a_0^2\ . \qquad (65)$$

Note that if g for a given transition is approximately constant in an iso-electronic sequence, the collision strength Ω scales as f/E_{ij}, or $\propto Z^{-2}$, and $Q_{ij} \propto Z^{-4}$ because $E_{ij} \propto Z^2$ [cf. Eqs (64) and (65)].

Integration of vQ_{ij} over a Maxwellian electron velocity distribution (Eq. 20) leads to the *excitation rate* S_{ij} [same as Eq. (21)]:

$$S_{ij} = \sqrt{\frac{8kT}{\pi m_e}}\, y^2 \int_1^\infty Q_{ij} {\rm e}^{-yU} U {\rm d}U\ , \qquad (66)$$

where $y = E_{ij}/kT$. By defining the Maxwellian-averaged collision strength $\bar{\Omega}(y)$ as

$$\bar{\Omega}(y) = y{\rm e}^y \int_1^\infty \Omega(U) {\rm e}^{-yU} {\rm d}U\ , \qquad (67)$$

we substitute Eq. (63) into Eq. (66), use Eq. (67) and obtain numerically for S_{ij} in units of cm^3 s^{-1}:

$$S_{ij} = 8.63\ 10^{-6}\ \frac{\bar{\Omega}(y)}{w_i} T^{-1/2} e^{-y} \ =\ 1.703\ 10^{-3}\, f\, \bar{g}(y) E_{ij[\rm eV]}^{-1}\ T^{-1/2} e^{-y}\ , \qquad (68)$$

where $y = E_{ij}/kT = 1.1604\ 10^4\ E_{ij}/T = 1.4388\ 10^8\ a/[\lambda T]$, with E_{ij} the line excitation energy in eV, T in K, and λ in Å. Finally, $\bar{g}(y)$ is the excitation Gaunt factor averaged over the Maxwell distribution.

When no data are available one can make a *first-order estimate* of the excitation rate for optically allowed dipole (and sometimes also forbidden) transitions by using the Gaunt factor approximation. With an accuracy of a factor of two or so we may use near threshold ($U \gtrsim 1$) typical values of $g \sim 0.2$ (or 1) for $\Delta n \neq 0$ (or $= 0$) transitions, and at $U \gg 1$ the asymptotic Born-Bethe approximation $g \approx (\sqrt{3}/2\pi) \ln U$. Usually spectroscopists have used as a crude approximation a *constant* Gaunt factor, but one can obtain a generally much better accuracy by introducing an *energy dependence* with a suitable approximation formula.

Fitting formulae for excitation data. Some time ago Mewe (1972) introduced a parametrized interpolation formula for Ω (or g) that represents the correct

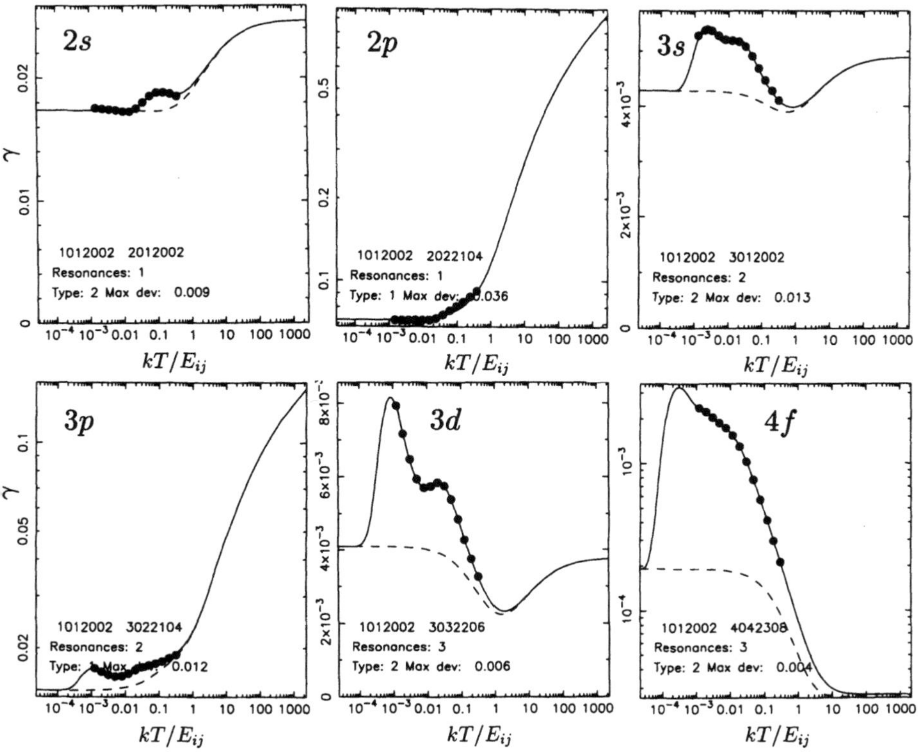

Fig. 11. Fitted effective collision strengths $\gamma = \bar{\Omega}(y)$ (cf. Eq. 67) as a function of $y^{-1} = kT/E_{ex}$ for the transitions from the ground state $1s\ ^2S$ in the H-like ion C VI to the $n = 2$ states $2s\ ^2S$ and $2p\ ^2P$, the $n = 3$ states $3s\ ^2S$, $3p\ ^2P$, and $3d\ ^2D$, and to $4f\ ^2F$ (data from Aggarwal & Kingston 1991). The effect of the resonances is taken into account [see the corresponding Fig. 10 for the non-integrated collision strength $\Omega(U)$]. The solid line indicates the collision strength including resonances, while the dashed line gives that without resonances. Various types of fitting formulae (cf. Table 3) are used.

behaviour both near threshold and asymptotically at high energies (cf. also Mewe & Schrijver 1978a):

$$\Omega(U) = A + \frac{B}{U} + \frac{C}{U^2} + \frac{2D}{U^3} + F \ln U \ . \tag{69}$$

This expression can be integrated over a Maxwellian electron energy distribution. The coefficients can be adjusted to fit both calculated collision strengths as well as measured excitation rates. Insertion of Eq. (69) into Eq. (67) gives the collision strength averaged over the electron Maxwell distribution:

$$\bar{\Omega}(y) = A + (By - Cy^2 + Dy^3 + F)\, E^y E_1(y) + (C + D)\, y - Dy^2 \ , \tag{70}$$

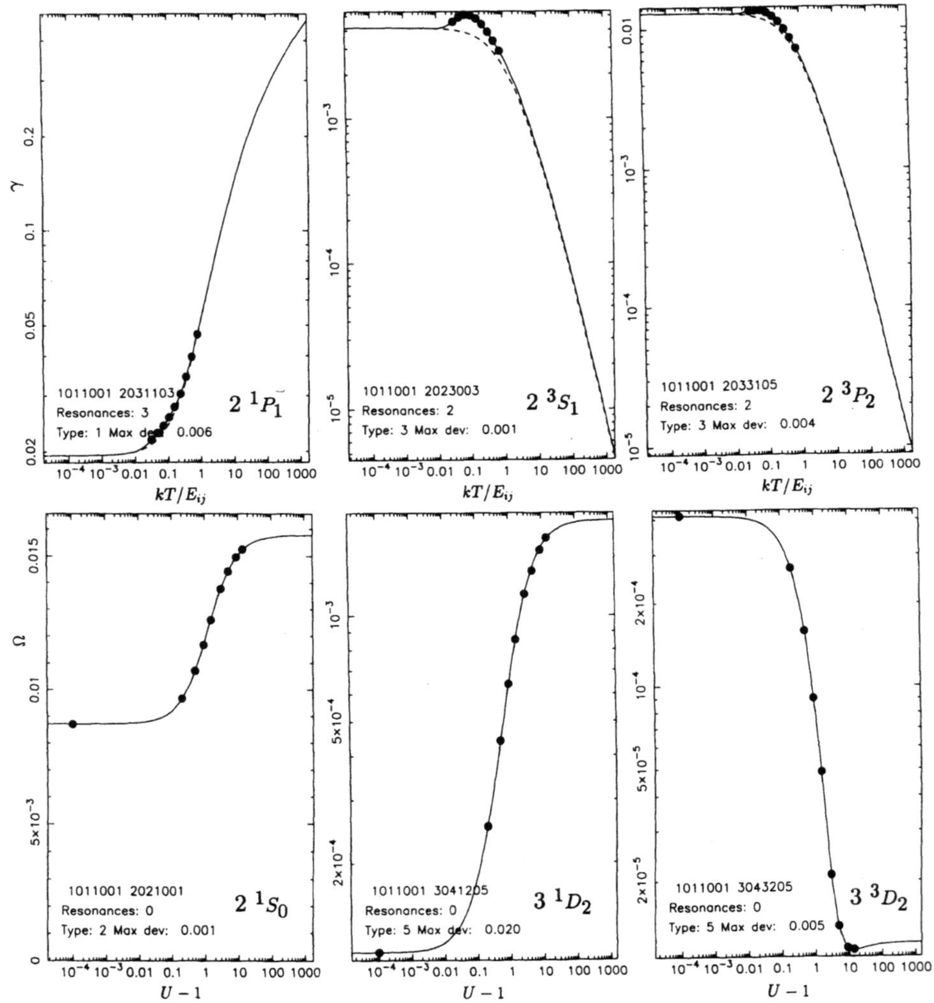

Fig. 12. Fitted effective collision strengths $\gamma = \bar{\Omega}(y)$ as a function of $kT/E_{\rm ij}$ for the transitions from the ground state $1s^2\ ^1S_0$ in the He-like ion O VII to the $n = 2$ states $1s2p\ ^1P_1$, $1s2s\ ^3S_1$, and $1s2p\ ^3P_2$, and collision strengths $\Omega(U)$ as a function of $U - 1$ for transitions to the $n = 2$ state $1s2s\ ^1S_0$ and the $n = 3$ states $1s3d\ ^1D_2$ and $1s3d\ ^3D_2$ (data from Sampson et al. 1983, and Zhang & Sampson 1987). In the top three plots the solid line indicates the collision strength including resonances, while the dashed line gives that without resonances.

Table 3. Parameters for fitting formula (75)

Type[a]	c_0	c_1	c_2	c_3	c_4	c_5	c_6	c_7	s
1	+	+	+	+	0	0	0	+	0
2	+	+	+	+	0	0	0	0	0
3	0	0	+	+	+	0	0	0	+
4	+	+	+	+	0	0	0	+	+
5	+	+	+	+	0	0	0	0	+
6	0	0	+	+	+	+	0	0	+
7	0	0	0	+	+	+	+	0	+

a

1: allowed electric dipole transitions
2: forbidden transitions
3: spin-forbidden transitions
4: type 1 + scaling
5: type 2 + scaling
6: forbidden transitions
7: forbidden transitions

where

$$E_1(y) \equiv \int_1^\infty z^{-1} e^{-yz} dz \ , \tag{71}$$

is the exponential integral (cf. Abramowitz & Stegun 1970, Eqs. 5.1.53 and 5.1.54; for $y < 1$, the absolute accuracy is better than 2×10^{-7}; for $y > 1$, the relative accuracy is better than 5×10^{-5}). Mewe & Schrijver (1978a) have introduced a useful approximation (accurate within 0.2% for $y \geq 0.001$):

$$e^y E_1(y) \simeq \ln\left(\frac{y+1}{y}\right) - \frac{0.36 + 0.03(y+0.01)^\alpha}{(y+1)^2} \approx \frac{1}{y+1} \ (y \gg 1) \ , \tag{72}$$

where $\alpha = 0.5$ ($y \geq 1.5$) or $\alpha = -0.5$ ($y < 1.5$).

From the approximations for (71) we obtain for the collision strength (and the rate S_{ij} from Eq. 68) near threshold or at asymptotically high energies:

$$\bar{\Omega}(y \gg 1) \simeq \Omega(U=1) = A + B + C + 2D, \quad S_{ij} \propto T^{-1/2} e^{-y}, \tag{73}$$

and

$$\bar{\Omega}(y \ll 1) \simeq F/y, \quad S_{ij} \propto T^{1/2}. \tag{74}$$

In fitting collision strengths, Eq. (69) can be used in many cases also for forbidden transitions with the logarithmic term put to zero, but sometimes it is necessary to introduce higher-order terms $\propto 1/U^n$ ($n \geq 4$) and a *scale factor* s by substituting $U \to U + s$. Kaastra & Mewe have recently developed a fitting routine which incorporates a more complicated fitting formula for Ω (or S, Q, g):

$$\Omega(U) = \sum_{i=0}^{i=6} \frac{c_i}{(U+s)^i} + c_7 \ln U \ . \tag{75}$$

In Table 3 I indicate when the coefficients c_i and the scale parameter s are non-zero (indicated by a +) or zero for various types of transitions.

Note that the fitting procedure can also take into account the contribution from possible *resonances* near the threshold of excitation which can be important for the calculation of excitation rates for cool plasmas.

Figs. 11 and 12 give examples of fitting results for the H-like ion C VI and the He-like ion O VII, respectively. Clearly the effect of resonances shows up in various cases. Different types of fitting formulae are used (cf. Eq. 75 and Table 3). In the case of the first three transitions in O VII (fit to γ in Fig. 12) we have applied a combined fitting to the data of Sampson et al. (1983) (without resonances) and those of Zhang & Sampson (1987) (including resonances). The effect of resonances is clearly seen, e.g., in the 3s, 3d, and 4f excitation in C VI and the 2^3S excitation in O VII.

More complex expressions for the line formation in an optically thin plasma (cf. Eq. 51), including possible cascade effects on the excitation rate and also contributions from recombination ($\propto \eta_{z+1}$), inner-shell ionization ($\propto \eta_{z-1}$) and unresolved dielectronic recombination satellite lines ($\propto \eta_z$) have been given elsewhere (Mewe & Gronenschild 1981; Mewe et al. 1985a).

8.1.2 De-excitation rates. The rate coefficient S'_{ji} of de-excitation $j \to i$ is related to S_{ij}, the rate coefficient of the inverse excitation process $i \to j$ by

$$S'_{ji} = \frac{w_i}{w_j} e^{E_{ij}/kT} S_{ij} \ , \tag{76}$$

where w_i and w_j are the statistical weights of the lower level i and the upper level j, respectively. This relation immediately follows by applying the principle of detailed balance (Sect. 4) and the Boltzmann equation for the ratio of the fictitious population densities of levels i and j.

8.1.3 Accuracy of excitation rates. Though in a number of cases the Gaunt factor approximation method gives reasonably accurate results, more accurate values are needed, especially for the weaker lines and optically forbidden transitions. Raymond (1988) gives a brief discussion of the accuracy of several computational methods such as Coulomb-Born (CB) and Distorted Wave (DW) approximations, and the more accurate Close Coupling (CC) method, which properly takes into account the dominant resonances near the threshold of the excitation mentioned in connection with DR. [For a theoretical review see Seaton (1975), for a review on crossed-beam experiments see Dolder & Peart (1986), and for a review on plasma measurements of atomic rates see Griem (1988)].[10] CB tends to overestimate the collision strength near threshold by ~20–50%, while DW gives better results, especially for high-Z ions. Few accurate CC results are available for He-like ions (Kingston & Tayal 1983). For many of the strongest X-ray lines, e.g., from H- and He-like ions and Li- to Ne-like ions, the collision strengths are known with better than 20% accuracy. Near threshold, the strong narrow resonances may spoil the accuracy, but for applications of rate coefficients in plasma diagnostics, these are smoothed out to a large extent in averaging the cross section over the electron energy distribution, and may contribute to the excitation rate only in the case of relatively cool plasmas.

8.2 Radiative transitions

A spectral line corresponds to the electromagnetic radiation of an ion which makes a transition from the excited state (j) to the lower one (i) with a given transition probability A_{ji}. I give here a brief description of several spectroscopic concepts and notations and refer for further details to various textbooks such as Herzberg (1944) and Condon & Shortley (1970).

8.2.1 Notations for energy levels and electron configurations. The *energy state* of an atom or ion is specified by a set of quantum numbers, the most familiar set being described by the Russell-Saunders *LS coupling*, i.e.

$$n_1\ell_1^{q_1}n_2\ell_2^{q_2}\cdots n_m\ell_m^{q_m}\pi\ [^{2S+1}L_J] \equiv \gamma\pi\ [^{2S+1}L_J] \quad ,$$

where γ denotes the electron configuration, $\pi = (-1)^{\sum q_i\ell_i}$ is its "parity", and the symbol between square brackets denotes the energy level (see later). An electron (i) in the configuration γ is described by the *principal quantum number* n_i (characteristic for a *certain shell*), ℓ_i the quantum number of its orbital momentum. Furthermore, s_i denotes the quantum number of the electron spin ($= \pm\frac{1}{2}$) and j_i the resultant total angular momentum of l_i and s_i; $j = \ell \pm \frac{1}{2}(\ell \geq 1)$, $j = \frac{1}{2}(\ell = 0)$. All angular momenta are expressed in units of $h/2\pi$ (where h is the Planck constant). For $\ell = 0, 1, 2, 3, 4, 5, \cdots$ we speak of $s, p, d, f, g, h, \cdots$ electrons according to the sharp, principal, diffuse, and fundamental series in alkali spectra, and then alphabetically.

The building up of the electron configurations (and the associated periodic system of elements) is determined by *Pauli's exclusion principle* which states that in one atom no two identical electrons can occur, i.e., no two electrons can have the same set of quantum numbers [e.g., n, ℓ, j, and m_j (= component of j in a direction of preference)]. This principle prevents the filling of shells

[10] Compilations of excitation data have been reported by Aggarwal et al. (1986), Gallagher & Pradhan (1985), and Lang (1994), while Itikawa et al. (1984) and Itikawa (1991) have published comprehensive bibliographies.

with an arbitrary number of electrons. It follows that only a limited number of electrons can have the same n and ℓ (so-called *equivalent* electrons). With $\ell = -n+1, \cdots, 0, \cdots, n-1$, and $s = \pm\frac{1}{2}$ the building up occurs for the first three closed shells ($n = 1, 2, 3$) as: $1s^2 2s^2 2p^6 3s^2 3p^6 3d^{10}$ (cf. Herzberg 1944).[11] A *closed shell* gives no net contribution to the angular momentum: $\sum_i \boldsymbol{\ell}_i = \sum_i \mathbf{s}_i = \sum_i \mathbf{j}_i = 0, \sum_i \ell_i =$ even, $\sum_i s_i =$ an integer number. For example, we have $1s^2 2s^2 2p^6 3s^2 3p^2\ {}^3P_0$ for the configuration of the ground state of Fe XIII (for term notation see below).

An atomic energy *level* (or summed over all J called: *term*) is denoted by ${}^r(L\text{-symbol})_J$, where L and S are the total orbital and spin momenta and J is the total angular momentum (cf. Sect. 8.2.2). The notation for the levels (terms) is $S, P, D, F, G, H, \cdots$ for $L = 0, 1, 2, 3, 4, 5, \cdots$. For a given value of $L \geq S$ there are $r = 2S+1$ levels that are distinguishable only by a different magnetic interaction of $\mathbf{L}$ and $\mathbf{S}$ and that have sometimes the same energy ("degenerate" levels). One speaks of a term with "*multiplicity*" $r = 2S+1$. Also in the case $L < S$ where the actual multiplicity $= 2L+1$ one still calls formally r the multiplicity (e.g., ${}^3P_{2,1,0}$ is a triplet term but also ${}^3S_1$ is called a triplet term). The notation for the multiplicity is for $S = 0, \frac{1}{2}, 1, 1\frac{1}{2}, 2, 2\frac{1}{2}, \cdots$: singlet, doublet, triplet, quartet, quintet, sextet, etc. terms. If the algebraic sum $\sum_i \ell_i$ is even or odd the term is called *even* or *odd* denoted by the index o ("odd"), e.g., even doublet term ${}^2D$ or odd triplet term ${}^3P^o$.

Since the vector $\mathbf{J}$ has $2J+1$ orientations with respect to a certain direction of preference, e.g., given by a magnetic field, one calls $g_J = 2J+1$ the ***statistical weight*** of the level with quantum number J. In the case of a magnetic field the component M_J (or m_j for a single electron) of $\mathbf{J}$ in the field direction, the so-called magnetic quantum number (Zeeman effect) can take the values: $M_J = J, J-1, \cdots, 0, \cdots, -J+1, -J$. For example, for $L = 2, S = 1, J = 2$ we have a ${}^3D_2$ level with statistical weight $g_2 = 5$.

8.2.2 Coupling schemes. LS-coupling occurs when the individual electron orbital angular momenta $\boldsymbol{\ell}_i$ couple to a total orbital angular momentum $\mathbf{L} = \sum_i \boldsymbol{\ell}_i$, the electron spins $\mathbf{s}_i$ to a total spin angular momentum $\mathbf{S} = \sum_i \mathbf{s}_i$, while $\mathbf{L}$ and $\mathbf{S}$ combine with a smaller coupling to a resultant $\mathbf{J}$ (Herzberg 1944). Symbolically:

$$(\ell_1, \ell_2, ...)(s_1, s_2, ...) = (L, S) = J.$$

LS-coupling holds when the electrostatic interaction $V_{\rm el}$ between the electrons (which causes the distances between the multiplets) is much larger than the relativistic spin(S)-orbit(L) interaction $V_{\rm rel}$ (producing the multiplet splitting), i.e. $V_{\rm el} \gg V_{\rm rel}$; the multiplet splitting is small compared to the energy separation between levels from the same electron configuration but different L. LS-coupling holds for a large number of elements, at least at low nuclear

[11] The building-up order is according to increasing $n+\ell$ and at equal $n+\ell$ according to increasing n, i.e.: $1s2s2p3s3p4s3d4p5s4d5p6s4f5d6p\cdots$ ($4f$ is an inner orbit which is filled up much later for the rare earth elements in the periodic system).

Table 4. Parameters for radiative decay of $n = 2$ levels in He- and H-like ions

Seq.	i	Transition[1]	Type	α_i^2	β_i^2	$A_i(6)^3$	$A_i(12)^3$	$A_i(26)^3$
He	1	$2\,^1S_0 \to 1\,^1S_0$	$2E1$	1.3 10^7	6	3.3 10^5	3.2 10^7	4.1 10^9
	2	$2\,^1P_1 \to 1\,^1S_0(w)$	$E1$	9 10^{12}	4	8.9 10^{11}	2.0 10^{13}	4.6 10^{14}
	3	$2\,^3S_1 \to 1\,^1S_0(z)$	$M1$	1.3 10^4	10	50	7.3 10^4	2.1 10^8
	4	$2\,^3P_2 \to 1\,^1S_0(x)$	$M2$	3 10^6	8	2.7 10^4	1.1 10^7	6.6 10^9
	5	$2\,^3P_1 \to 1\,^1S_0(y)$	$E1$	6 10^9	9	2.9 10^7	3.4 10^{10}	4.3 10^{13}
	6	$2\,^1P_1 \to 2\,^1S_0$	$E1$	4 10^7	3	1.8 10^7	4.9 10^7	5.0 10^8
	7	$2\,^1P_1 \to 2\,^3S_1$	$E1$	4 10^5	7	7.7 10^3	1.2 10^6	3.5 10^8
	8	$2\,^3P_2 \to 2\,^3S_1$	$E1$	1.2 10^8	3	5.9 10^7	1.5 10^8	1.5 10^9
	9	$2\,^3P_1 \to 2\,^3S_1$	$E1$	1.5 10^8	1	5.8 10^7	1.4 10^8	4.8 10^8
	10	$2\,^3P_0 \to 2\,^3S_1$	$E1$	1.5 10^8	1	5.8 10^7	1.3 10^8	3.8 10^8
	11	$2\,^1S_0 \to 2\,^3S_1$	$M1$	0.13	7	0.0020	0.35	100
	12	$2\,^1S_0 \to 2\,^3P_1$	$E1$	350	3.5	< 1	650	350
		$2\,^3S_1 \to 1\,^1S_0$	$2E1$	weak				
		$2\,^3P_2 \to 2\,^3S_1$	$M2$	weak				
		$2\,^3P_0 \to 1\,^1S_0$	$E1M1$	weak				
		$2\,^3P_0 \to 1\,^1S_0$	$E1^4$	weak				
H	13	$2\,^2P_{3/2} \to 1\,^2S_{1/2}(L\alpha_1)$	$E1$	6.27 10^{12}	4	8.1 10^{11}	1.3 10^{13}	2.9 10^{14}
	14	$2\,^2P_{1/2} \to 1\,^2S_{1/2}(L\alpha_2)$	$E1$	6.27 10^{12}	4	8.1 10^{11}	1.3 10^{13}	2.9 10^{14}
	15	$2\,^2S_{1/2} \to 1\,^2S_{1/2}$	$2E1$	8.1 10^6	6	3.8 10^5	2.4 10^7	2.5 10^9
	16	$2\,^2S_{1/2} \to 1\,^2S_{1/2}$	$M1$	2.6 10^4	10	1.5 10^2	1.6 10^5	3.7 10^8
		$2\,^2P_{3/2} \to 1\,^2P_{1/2}$	$E2$	weak				

[1] The resonance line (w), intercombination line (y) (blended by the line x), and the forbidden line (z) form the He-like triplet; $L\alpha 1,2$ is the H-like resonance doublet line.
[2] Formula: $A_i \approx \alpha_i(Z/10)^{\beta_i}$ s^{-1} (He: $Z \gtrsim 10$, and for case i =12 only for $Z = 20$; H: accurate for all Z).
[3] Accurate value of A_i (s^{-1}) for $Z = 6$, 12, and 26.
[4] Induced by hyperfine structure (HFS) mixing.

charge ($Z \lesssim 30$), but with increasing Z, V_{rel} increases more rapidly than V_{el} and the opposite situation occurs ($V_{\text{rel}} \gg V_{\text{el}}$): *jj-coupling*. In this case the individual $\boldsymbol{\ell}_i$ and $\mathbf{s}_i$ combine to give a $\mathbf{j}_i$, the total angular momentum of an individual electron. The $\mathbf{j}_i$'s combine with a smaller coupling to the total $\mathbf{J}$ of the atom or ion. Symbolically:

$$(\ell_1, s_1)(\ell_2, s_2)(\ell_3, s_3)... = (j_1, j_2, j_3...) = J.$$

This holds for larger Z especially in complicated level configurations with many levels close to each other. For highly excited levels (with large principal

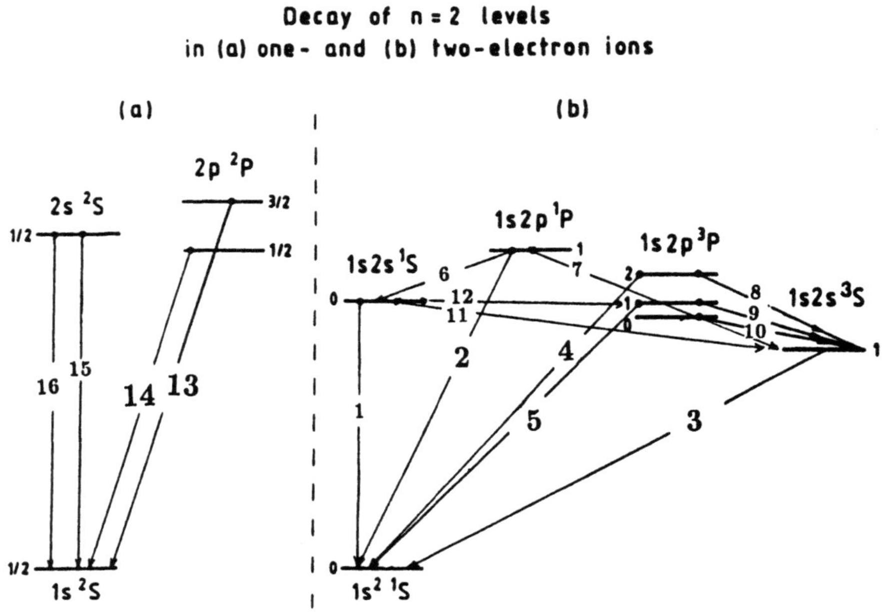

Fig. 13. Various radiative decay modes of the $n = 2$ levels in H-like ions (a) and in He-like ions (b). Lines 13 & 14 are the components of the H-like Lyman α resonance line; the well-known He-like triplet is formed by the resonance line (2), intercombination line (5) (blended by line 4), and forbidden line (3).

quantum number n) the electrostatic interaction decreases, hence jj-coupling occurs already at lower Z than for the lower levels (cf. Herzberg 1944).

An example of the transition from LS- to jj-coupling is provided by the $2\,^1P_1$ and $2\,^3P_{2,1,0}$ terms in He-like ions (cf. Fig. 13). For low Z ($Z \lesssim 30$) the three substates of $2\,^3P$ stay close (form an LS multiplet and are well below the $2\,^1P$ state, but for large Z the $2\,^3P_2$ state moves up close to the $2\,^1P_1$ state forming with this level by $(\frac{3}{2}, \frac{1}{2})$ jj-coupling a J of 2 or 1, while the two $2\,^3P_{1,0}$ states still stay close and form by $(\frac{1}{2}, \frac{1}{2})$ a J of 1 or 0. The two $^3P_1$ and $^3P_0$ levels cross each other several times as Z increases: between Z = 7–44, level 1 is above 0, outside this range on both sides the situation is reversed. Also the $2\,^1S_0$ level crosses the two ^3P levels several times (cf. Lin et al. 1977).

8.2.3 Types of transitions. Quantum selection rules characterize the type of transition that an atom or ion makes under the interaction with electromagnetic radiation. We distinguish the following radiative transitions (cf. Pal'chikov & Shevelko 1995): electric ($E\kappa$) and magnetic ($M\kappa$) 2κ-pole transitions. For example, we have $E1$ or $E2$ = electric dipole or quadrupole radiation, respectively, $M1$ or $M2$ = magnetic dipole or quadrupole radiation, etc., or more involved combinations such as two-photon electric dipole ($2E1$) or two-photon electro-magnetic ($E1M1$) radiation (e.g., Pal'chikov & Shevelko 1995, and references therein). Further, for isotopes with nuclear spin $I \neq 0$ and at high Z ($\gtrsim 30-40$) $E1$ radiation induced by hyperfine structure mixing (HFS) by the magnetic moment of the nucleus can occur (e.g., Dunford et al. 1990; Munger & Gould 1986). As an illustrative example I have shown in Fig. 13 and Table 4 all kinds of possible different radiative decay modes i of the $n = 2$ states in He-like and H-like ions.

8.2.4 Selection rules. We can distinguish between "exact" and "approximate" selection rules. The exact ones follow from the properties of the angular parts of the operators for electric and magnetic transitions and are independent of the coupling scheme. They are formulated for the "exact" quantum numbers, the total momentum J and its projection M_J, and the parity π (cf. Pal'chikov & Shevelko 1995):

$$\Delta J = 0, \pm 1, \pm 2, \cdots, \pm \kappa, \quad J + J' \geq \kappa \ , \tag{77}$$

$$\Delta M_J = 0, \pm 1, \pm 2, \cdots, \pm \kappa, \ , \tag{78}$$

$$\Delta \pi = (-1)^\kappa \ (\text{for } E\kappa), \ \text{and} \ - (-1)^\kappa \ (\text{for } M\kappa) \ . \tag{79}$$

There is no restriction for the principal quantum number n. Finally, the rule of Laporte for a one-electron jump (more-electron jumps are less important) gives:

$$\Delta(\sum_i \ell_i) = \pm 1, \pm 3, \cdots \ , \tag{80}$$

that is even terms combine only with odd, and odd only with even. The approximate selection rules are dependent on the type of coupling scheme. We consider here LS-coupling for which one has, for electric E_κ transitions:

$$\Delta L = 0, \pm 1, \pm 2, \cdots, \pm \kappa \ \ (L + L' \geq \kappa), \Delta S = 0 \ , \tag{81}$$

and for magnetic M_κ transitions:

$$\begin{aligned} \Delta L &= 0, \pm 1, \pm 2, \cdots, \pm(\kappa - 1) \ \ (L + L' \geq \kappa - 1) \ , \\ \Delta S &= 0, \pm 1, \pm 2, \cdots, \pm(\kappa - 1) \ \ (S + S' \geq \kappa - 1) \ . \end{aligned} \tag{82}$$

For "allowed" electric dipole ($E1$) radiation which usually dominates the line emission one has thus: $\Delta L = 0, \pm 1$, $\Delta S = 0, \Delta J = 0, \pm 1, J = 0 \not\leftrightarrow J = 0$, $M_J = 0 \not\leftrightarrow M_J = 0$ (for $\Delta J = 0$), $\Delta \ell = \pm 1$, and $\pi' = -\pi$. Hence S-S transitions are ruled out.

For magnetic monopole ($M1$) radiation one has the same restrictions for L, S, and J as for E_1 and further: $\Delta \ell = 0$, and $\pi' = \pi$.

Finally, for electric quadrupole ($E2$) transitions: $\Delta L = 0, \pm 1, \pm 2$, $\Delta S = 0$, $\Delta J = 0, \pm 1, \pm 2$, $J = 0 \not\leftrightarrow J = 0$, $J = \frac{1}{2} \not\leftrightarrow J = \frac{1}{2}$, $J = 0 \not\leftrightarrow J = 1$, $\Delta \ell = 0, \pm 2$, and $\pi' = \pi$. Here transitions between S terms and between S and P terms are ruled out.

For the most important electric dipole transitions $j \to i$ the transition probability A_{ji} is expressed as (e.g., Wiese et al. 1996):

$$A_{ji} = 6.6703\ 10^{15} (w_i/w_j) f_{ij} \lambda^{-2} \tag{83}$$

where A_{ji} is in units of s^{-1}, w_i and w_j are the statistical weights of the lower and upper level, respectively, f_{ij} is the absorption oscillator strength and λ the wavelength in Å. The general systematic trends in the behaviour of A for the allowed transitions is as follows. For transitions where a change in the principal quantum number n occurs the oscillator strength is nearly comstant and $\lambda \propto Z^{-2}$ so that A approximately scales $\propto Z^4$. For $\Delta n = 0$ transitions approximately $f \propto 1/Z$ and $\lambda \propto 1/Z$, hence approximately $A \propto Z$.

If for a transition one of the above-mentioned selection rules is violated it is called "forbidden". A violation of the selection rules is connected with magnetic interactions (mostly spin-orbital ones) and the forbidden lines can arise from any of the higher multipoles or by a multi-photon mode.

Allowed electric dipole transitions are in most cases the dominant ones and other types of transitions are usually orders of magnitude weaker. As a rule of thumb Herzberg (1944) gives intensity ratios typically $E1 : M1 : E2 \sim 1 : 10^{-5} : 10^{-8}$, and Summers (1988) gives: $\sim 5\ 10^7\ Z^4 : 3\ 10^3 Z^6 : 3\ Z^6$. However, the actual situation is more complicated. For hot plasmas non-electric-dipole transitions can play an important role (e.g., in the density diagnostics, see Sect. 13) because the probabilities of forbidden transitions rapidly increase with increasing Z with powers between $\sim$ 6–10. To give an idea of the relative importance and behaviour of various types of radiation, I have given in Table 4 some typical values for the transition probabilities for all kinds of possible radiative decay modes i of the $n = 2$ states in He-like and H-like ions (cf. also Fig. 13 in which weak transitions are not indicated but mentioned in Table 4 except for the very weak electric quadrupole $E2$ transition $2^2P_{3/2} \to 2^2P_{1/2}$ in H-like ions). The A values are mainly based on data given by Lin et al. (1977) and Wiese et al. (1996). Table 4 gives also the parameters for an approximation of the transition probability A_i of transition i which is generally coarse for the He-like case and accurate for the H-like case, and the accurate values of A_i for $Z = 6$, 12, and 26.

9 Continuum Radiation

In astrophysics there are essentially three physical processes that produce a continuum spectrum in X rays: bremsstrahlung ("braking radiation"), synchrotron radiation, and Compton scattering. These processes are discussed in detail in the Chapters by Longair and Liedahl; here I briefly consider the first process (plus free-bound and two-photon emission) because this is important for an optically thin plasma.

Bremsstrahlung is radiation due to the acceleration of a charged particle in the Coulomb field of another charge and is sometimes also referred to as *free-free emission* (ff) because it originates from a transition of the impacting particle between two "free" or unbound states of the target ion. If the two charges are identical, there is no net acceleration of the electric dipole moment during this binary encounter, so electron-electron and ion-ion free-free radiation power will vanish in the electric-dipole limit and can usually be neglected. Electron-ion collisions that produce a much larger electron acceleration are mainly responsible for the bremsstrahlung in a hot cosmic plasma.

An exact treatment of this process requires a quantum mechanical calculation but it turns out that a classical treatment gives already the correct functional dependence for most of the physical parameters (cf. the Chapter by Longair). The quantum correction on the classical result is then introduced as the socalled *Gaunt factor*, which in a large interesting parameter range is of order unity. In the coronal model we consider only *thermal* bremsstrahlung for a Maxwellian electron velocity distribution, characterized by some temperature T. Integrating over the Maxwell distribution one obtains for the thermal free-free continuum emissivity (power per unit volume per unit frequency range) (cf. Rybicki & Lightman 1979):

$$P_{\mathrm{ff}}(\nu,T) = C_{\mathrm{ff}} Z_i^2\, n_e n_i\, \bar{g}_{\mathrm{ff},i}(\nu,T)\, T^{-1/2} \mathrm{e}^{-h\nu/kT} \quad , \tag{84}$$

where $\bar{g}_{\mathrm{ff},i}$ is the velocity-averaged Gaunt factor for the given target ion (i), Z_i and n_i are the effective charge and density of the target ion, n_e the electron density, and the factor C_{ff} worked out numerically in c.g.s. units (also T in K, densities in cm^{-3}):

$$C_{\mathrm{ff}} = \frac{2^5 \pi e^6}{3 m_e c^3} \left(\frac{2\pi}{3 k m_e}\right)^{1/2} = 6.842 \times 10^{-38} \mathrm{\ erg\ cm^{-3}\ s^{-1} Hz^{-1}} \quad , \tag{85}$$

where e and m_e are the electron charge and mass, and k the Boltzmann constant. Because the Gaunt factor only slowly varies the ff radiation has a rather flat spectrum with a sharp exponential cut-off near $h\nu \approx kT$ which provides a straightforward measurement of the temperature from the spectral shape near the cut-off when the bremsstrahlung is the dominant process creating the continuum. The frequency-integrated ff emission (in erg cm^{-3} s^{-1}) is given by:

$$P_{\mathrm{ff}}^{\mathrm{tot}}(T) = 1.426 \times 10^{-27} Z_i^2\, n_e n_i\, \bar{g}_{\mathrm{ff}}(T)\, T^{1/2} \quad , \tag{86}$$

where $\bar{g}_{\mathrm{ff}}(T)$ is the frequency-average of $\bar{g}_{\mathrm{ff},i}$, which is in the range 1.1–1.5 and can be approximated within about 20% by 1.2 (cf. Rybicki & Lightman 1979).

To calculate the total amount of ff radiation one must sum over all ions (i) present in the plasma. For a plasma with cosmic abundances the main contributions come from hydrogen and helium but other species cannot always be neglected. Gronenschild & Mewe (1978) and Mewe et al. (1986) have introduced a total averaged Gaunt factor $G_{\rm ff}$ by replacing Eq. (84) by

$$P_{\rm ff}(\nu,T) = C_{\rm ff} Z_i^2 \, n_{\rm e}^2 \, G_{\rm ff} \, T^{-1/2} {\rm e}^{-h\nu/kT} \ , \tag{87}$$

where,

$$G_{\rm ff}(\nu,T) = \sum_i \left(\frac{n_i}{n_{\rm e}}\right) \bar{g}_{{\rm ff},i}(\nu,T) Z_i^2 \ , \tag{88}$$

and the effective charge of ion (i) is defined by

$$Z_i = \left(\frac{n_{i-1,0}^2 E_{i-1}}{E_{\rm H}}\right)^{1/2} , \tag{89}$$

where E_{i-1} is the ionization energy from the ground state with principal quantum number $n_{0,i-1}$ in the *recombined* ion (i-1) and $E_{\rm H}$ the ionization energy of hydrogen.

The Gaunt factors have been calculated as a function of ν, T, and Z_i by Karzas & Latter (1961) who presented their results only graphically but since then, Carson (1988) has presented extensive tables reproducing the results of Karzas & Latter. Gronenschild & Mewe (1978) and Mewe et al. (1986) have derived approximation formulae for the free-free Gaunt factors given by Karzas & Latter (1961), based upon the high-energy Born approximation and a correction factor (accuracy better than 15%). Kaastra (1992) has later improved these calculations by interpolating Carson's table. Moreover he has introduced a relativistic correction factor based on the relativistic calculations in the Born limit by Kylafis & Lamb (1982).

In addition to free-free radiation, continuum emission in optically thin plasmas is also produced by the interaction of a free electron with an ion by a *free-bound* (fb) transition or a *two-photon* process (2γ). Free-bound emission occurs when a free electron with kinetic energy $E_{\rm el}$ is captured into a bound state (m) with ionization energy E_m of the ion (radiative recombination). The energy of the emitted photon is $h\nu = E_{\rm el} + E_m$. This process is often important and dominates the continuum radiation in a CIE plasma with temperature $T \ll 1$ MK, whereas for $T \gg 10$ MK free-free radiation is dominant. At $T \sim 1$ (or 10) MK free-bound radiation is dominant for photon energy $h\nu \gtrsim 0.1$ (or 3) keV. In a photo-ionized plasma free-bound emission is even more dominant because recombination is here much more important (see the Chapter by Liedahl). For relatively hot plasmas (e.g., $T \gtrsim$ a few MK) dielectronic recombination may also give a contribution to the continuum radiation but this process was not considered.

Two-photon emission can be important for H- and He-like ions and it occurs when the metastable $2s$ level ($2\,^2S_{1/2}$ for H, $2\,^1S_0$ for He) is excited. From the selection rules (cf. Sect. 8.2.4) it follows that the $2s$ electron cannot decay by a single-photon transition. At high electron density the metastable $2s$ level will be depopulated by collisions with electrons or protons (induced transition $2s \rightarrow 2p$) but if the density is sufficiently low ($n_e \lesssim n_c$) the metastable level can decay by the simultaneous emission of two photons ($2E1$, cf. Table 4) which have together a total energy equal to the excitation energy E_0 of the metastable S level. The spectrum is symmetric around $E = \frac{1}{2}E_0$ and extends from $E = 0$ to $E = E_0$ [see Spitzer & Greenstein (1951) for H-like ions and Dalgarno & Drake (1969) for He-like ions]. Mewe et al. (1986) derive for the critical density $n_c \approx 7\ 10^3 Z^{9.5}$ (H) or $2\ 10^5 (Z-1)^{9.5}$, e.g., a few times 10^{11} cm^{-3} for carbon. Thus for many astrophysical applications collisional depopulation will be unimportant.

For a comparison of the various contributions of ff, fb, and 2γ processes to the total continuum emission I refer to the figures given by Mewe et al. (1986). The fb and 2γ processes were incorporated by Gronenschild & Mewe (1978) and Mewe et al. (1986) in their calculations as a total continuum Gaunt factor G_c given by

$$G_c = G_{ff} + G_{fb} + G_{2\gamma}. \tag{90}$$

For the fb calculations the above-mentioned authors have used a constant approximation for the fb Gaunt factor given by Karzas & Latter (1961), and used a hydrogenic approximation for recombinations to excited levels (cf. Sect. 7.1). Later Kaastra (1992) has taken into account the correct energy dependence of the Gaunt factor and improved the calculations for the recombinations to the ground state by using the photo-ionization cross sections given by Verner et al. (1993) in applying the Milne relation (Eq. 34). Kaastra (1992) has also improved Mewe's approximation for the 2-photon Gaunt factor by applying a more accurate interpolation of the results of Spitzer & Greenstein (1951) and Dalgarno & Drake (1969). All these improvements were implemented in our code SPEX (cf. Sect. 3.2).

Finally, I give here a few alternative expressions for the total continuum emissivity $P_c(\lambda, T)$ written as the energy emissivity at wavelength λ per unit wavelength interval in units of erg cm^{-3} s^{-1} Å^{-1}:

$$P_c(\lambda, T) = 2.461\ 10^{-22}\ n_e N_H\, G_c\, \lambda^{-2} T^{-1/2} \exp[-143.88/(\lambda T)] \ , \tag{91}$$

or, alternatively, as the photon number emissivity $P_c(E, T)$ at photon energy E per unit energy interval in units of phot cm^{-3} s^{-1} keV^{-1}:

$$P_c(E, T) = 3.637\ 10^{-15}\ n_e N_H\, G_c\, E^{-1} T^{-1/2} \exp[-E/kT] \ , \tag{92}$$

where in Eq. (91) T is in MK, λ in Å, and in Eq. (92) E and kT are in keV and where the ratio n_e/N_H is taken equal to 1.2 for a plasma with cosmic abundances (Anders & Grevesse 1989). For an easy estimate we can make use of the following Gaunt factor approximation (λ in Å, T in MK) (Mewe et al. 1985a):

$$G_c \simeq 27.83(T+0.65)^{-1.33} + 0.15\lambda^{0.34}T^{0.422} , \tag{93}$$

which gives a reasonable approximation (within ~10–20%) at $T \geq 3$ MK and a crude estimate (~30–50%) for 0.2 MK $\leq T < 3$ MK for a plasma with cosmic abundances from Allen (1973). However, for wavelengths below the O VIII edge at 16.8 Å, better results can be obtained by properly taking into account the contributions from fb and 2γ emission.

IV. Diagnostics of Plasma Parameters

High-resolution spectroscopy permits the diagnostics of a variety of plasma parameters such as coronal temperature, ionization balance, emission measure, abundances, densities, and velocities (Mewe 1991; for many references, see Keenan 1992, 1995).

10 Electron Temperature

The soft X-ray spectrum depends sensitively on the ionization structure throughout the temperature range 0.1–100 MK. The mere detection of a particular spectral line in the spectrum implies the existence of plasma near the temperature of maximum formation (see Sect. 8) for that line, hence provides an accurate "thermometer" for the source emission region.

The electron temperature T can be diagnosed from the intensity ratio of two collisionally excited lines from different ions of the same element using Eq. (53). Though this line ratio is independent of the elemental abundance, it does depend on the ionization equilibrium. In situations in which deviations from ionization equilibrium occur, one can avoid this constraint by using the ratio of two lines from the same ionization stage, but with a different temperature dependence, e.g., two collisionally excited lines with different excitation energies such as the Lyman α and β lines. Alternatively, one can measure the ratio of lines at nearly the same wavelength for which the excitation functions have a different dependence on T, e.g., in the helium-like $2 \rightarrow 1$ triplet. This consists of a resonance (r) $2^1P \rightarrow 1^1S$ line, a forbidden (f) $2^3S \rightarrow 1^1S$ line and an intercombination (i) $2^3P \rightarrow 1^1S$ line (summed over two close components). The intensity ratio $r/(i+f)$ varies with electron temperature.

Another possibility is to use the intensity ratio of the collisionally excited resonance line and a nearby dielectronic recombination satellite line (see Sect. 15). The helium-like triplets with their satellites provide a very valuable diagnostics for electron temperature, emission measure, abundances, and ionization balance.

11 Elemental Abundances

In both coronal equilibrium and non-equilibrium situations, ions of different elements which have similar ionization potentials usually coexist in close proximity. Thus the ratios of the prominent emission lines from ions of different elements yield in a relatively model-independent way elemental abundances. By using a suitable set of lines from different ionization stages of one given element, e.g., iron, we can determine the temperature distribution by fitting the line intensities. Then by selecting suitable lines from different elements, relative elemental abundances can be determined. For a given line the emissivity $P_\ell \propto n_e N_H A_Z$ (Sect. 8) and the continuum emissivity $P_c \propto n_e N_H G_c$ (Sect. 9). Here the effective Gaunt factor G_c is a complicated function of the element abundances (cf. Gronenschild & Mewe 1978; Mewe et al. 1986), but for relatively hot ($\gtrsim 5$ MK) plasmas it is mainly determined by contributions from H and He. Then the line-to-continuum ratio P_ℓ / P_c is approximately proportional to the element abundance so that from the intensity ratio of a line to the neighbouring unblended continuum, in principle, the absolute abundance with respect to H can be derived.

12 Ionization Balance in NEI

In a transient plasma the line spectrum is sensitive to the deviations from the ionization equilibrium, whereas the bremsstrahlung continuum immediately follows the changing plasma conditions (temperature), so that the line/continuum ratio can provide an indication of the deviation from the ionization equilibrium (e.g., Shapiro & Moore 1977). To study such effects in solar or stellar flares, high-resolution spectroscopy must be applied (e.g., Mewe et al. 1985b). Also resonance/inner-shell excitation satelline ratios can be used for this purpose (cf. Sect. 15.4.2).

13 Electron Density

From spectral fits to the line and/or continuum emission one can determine the emission measure $\int n_e^2 dV$ (cf. Sect. 14), but this alone is not sufficient to model the optically thin source. One needs to know the electron density, n_e from density-sensitive lines to determine the size of the emitting volume V which is a critical parameter in any theoretical modeling of plasma heating and formation.

Electron densities can be measured using density sensitive spectral lines originating from metastable levels or using inner-shell excitation satellites to resonance lines (for reviews, e.g., Feldman 1981; Mason & Monsignori Fossi 1994; Mewe et al. 1991; Masai 1994b). In the first case the helium-like $2 \rightarrow 1$ triplet system lines (cf. Sect. 10) are particularly important (Gabriel & Jordan 1972; Pradhan 1985; Mewe et al. 1985a). The helium-like intensity ratio f/i varies with electron density above a certain critical density n_{ec} due to the collisional coupling between the *metastable* 2^3S upper level of the forbidden line f and the 2^3P upper level of the intercombination line. It does not strongly depend on the model (CIE or XPN) because its density dependence is determined only by the collisional coupling between the two upper line levels (see below), but the singlet/triplet ratio $r/(i+f)$ (which can be used for temperature diagnostics) does and can also serve as an indication of the validity of the coronal model (see Mewe 1990). The f/i line intensity ratio of helium-like ions from carbon through magnesium in the wavelength region 9–42 Å can be used to diagnose coronal plasmas in the density range $n_e = 10^8$–10^{13} cm^{-3} and corresponding temperature range $T \sim$ 1–6 MK.

However, for density diagnostics in active late-type stars one must observe lines formed at temperatures around $T \sim 10$ MK. For this purpose Mewe et al. (1985a, 1991) have considered many density-sensitive lines in the Fe L-shell complex, i.e. 2ℓ–$\geq 3\ell'$ transitions from ions Fe XVII–XXIV (and some corresponding nickel lines from Ni XXI–XXIV) covering the wavelength region 7–13 Å, lines between 90–140 Å from 2ℓ–$2\ell'$ transitions in Fe XVIII–XXIII, and lines between 170–275 Å from 3ℓ–$3\ell'$ transitions in Fe IX–XIV ions. The density dependence arises because the upper line level can be excited from various sub-levels within the ground state which become collisionally coupled at increasing densities (the same holds for the satellite lines). The Fe and Ni lines can be used as tools for diagnosing plasmas in the density range 10^{10}–10^{15} cm^{-3} and temperature range $\sim$ 0.5–15 MK. A selection of density-sensitive lines is given by Mewe et al. (1991).

The forbidden/intercombination line ratio in He-like ions

For the special case of the helium-like forbidden f ($m \rightarrow g$) and intercombination i ($k \rightarrow g$) lines I illustrate the steady-state solution of Eq. (51) for the upper-level populations of these lines, using a simplified 3-level term scheme with levels 1^1S (denoted by g, the ground level), 2^3S (m, the "forbidden" level), and 2^3P (k, the "intercombination" level). Taking into account only the most important atomic processes and neglecting for the moment all recombination processes, the population densities are given by solving the following two coupled equations:

$$N_m(A_{mg} + n_e S_{mk}) = n_e N_g S_{gm} + N_k A_{km},$$
$$N_k(A_{kg} + A_{km}) = n_e N_g S_{gk} + n_e N_m S_{mk} \ . \quad (94)$$

Assuming first a very low population for level k, i.e. $N_k \ll N_g$, hence neglecting in the r.h.s. of the first equation of Eq. (94) the last term, then we obtain for the ratio R of the forbidden and intercombination line intensities $P_f = N_m A_{mg}$ and $P_i = N_k A_{kg}$:

$$R \equiv \frac{P_f}{P_i} = \frac{S_{gm}(T)}{S_{gk}(T)} \frac{1 + (A_{km}/A_{kg})}{1 + (n_e/n_{ec})} , \tag{95}$$

where

$$n_{ec} = \frac{A_{mg}}{S_{mk}} \frac{S_{gk}}{S_{gk} + S_{gm}} , \tag{96}$$

is the *critical density* n_{ec} above which collisional excitation 2^3S to 2^3P begins to dominate the radiative decay from the metastable 2^3S level to the ground state. If $n_e S_{mk} \ll A_{mg}$ then R does not depend on n_e, but when $n_e \gtrsim n_{ec}$ then R becomes sensitive to variations in n_e. As $A_{mg} \sim Z^{10}$ (cf. Table 4) and $S_{mk} \sim Z^{-3}$, n_{ec} scales as Z^{13}, i.e. $n_{ec} \approx 0.3(T/T_m)^{0.3} Z^{13}$ cm^{-3}, where T_m is the temperature of maximum line formation (cf. Mewe & Schrijver 1978a; Mewe et al. 1991). The ratio $S_{gm}(T)/S_{gk}(T)$ in Eq. (95) introduces a temperature dependence of R which is, however, rather weak because the difference $E_k - E_m$ in excitation energies is small. Though Eq. (95) gives the essential behaviour of R, the solution is actually more complicated because the effect of the last term in the first equation of Eq. (94) cannot be neglected completely. The full solution of Eq. (94) reads (cf. Mewe & Gronenschild 1981):

$$R = \frac{[S_{gm} + (1 - BR_i)S_{gk}]BR_f}{S_{gk}BR_i + [S_{gm} + (1 - BR_i)S_{gm}](1 - BR_f)} , \tag{97}$$

where the collision rates S_{gm} and S_{gk} include the effects of cascades, and S_{gk} also contains a contribution from proton excitation (cf. Mewe & Schrijver 1978a,c and also Sect. 8.1), and where

$$BR_f = \frac{A_{mg}}{A_{mg} + n_e S_{mk} BR_i}, \text{ and } BR_i = \frac{A_{kg}}{A_{kg} + A_{km}} \tag{98}$$

are the branching ratios of the forbidden and intercombination line, respectively. Eq. (97) would reduce to Eq. (95) only if $BR_f \simeq 1$ and $BR_i \simeq 1$ which is obviously not the case, but the density behaviour of R is similar.

We have neglected here recombination processes which are usually less important in collision-dominated (CIE) plasmas but are just dominant in X-ray photo-ionized (XPN) plasmas. If we make the substitutions $S_{gm} \rightarrow (N_i/N_g)\alpha_{cm}$ and $S_{gk} \rightarrow (N_i/N_g)\alpha_{ck}$ (where N_i is the density of the next higher H-like ion and α_{cm} and α_{ck} are the recombination rate coefficients to the levels m and k) we obtain the general case. Since it turns out that the ratios S_{gm}/S_{gk} and α_{cm}/α_{ck} are quite similar (e.g., Mewe & Schrijver 1978a) the critical density has the same behaviour in dependence of Z and is approximately the same (within about 20–40%) for the CIE and XPN case (see also the Chapter by Liedahl).

The magnitude of A_{mg} (i.e. the type of transition) determines the density range over which R is n_e–sensitive. Keenan (1992) distinguishes the following three types of transitions:

- Intercombination transitions in low-Z species (e.g., $2s2p\ ^3\mathrm{P}_1 \rightarrow 2s^2\ ^1\mathrm{S}$ in O V) or forbidden transitions in high-Z species (e.g., $2s^22p^2\ ^3\mathrm{P}_1 \rightarrow 2s^22p^2\ ^3\mathrm{P}_0$ in Ca XV) have $A_{mg} \simeq 10^2 - 10^4\ \mathrm{s}^{-1} \Rightarrow n_e$ range $\simeq 10^8 - 10^{13}\ \mathrm{cm}^{-3}$ (applicable, for example, to the solar transition region/corona and late-type stellar atmospheres).
- Forbidden transitions in low-Z species (e.g., $3s3p\ ^3\mathrm{P}_2 \rightarrow 3s^2\ ^1\mathrm{S}$ in Si III) have $A_{mg} \simeq 10^{-3} - 1\ \mathrm{s}^{-1} \Rightarrow n_e$ range $\simeq 10^2 - 10^7\ \mathrm{cm}^{-3}$ (applicable to gaseous nebulae such as H II regions and planetary nebulae).
- Forbidden transitions within the ground term of low-Z species (e.g., $2s^22p\ ^2\mathrm{P}_{3/2} \rightarrow 2s^22p\ ^2\mathrm{P}_{1/2}$ in C II) have $A_{mg} \simeq 10^{-6} - 10^{-3}\ \mathrm{s}^{-1} \Rightarrow n_e$ range $\leq 100\ \mathrm{cm}^{-3}$ (applicable to the interstellar medium and supernova remnants).

14 Differential Emission Measure

From Eq. (53) we find for the line power P integrated over the whole emitting source volume V (cm^3)

$$P \propto EM f(T) \ , \tag{99}$$

where $f(T)$ is the emission function that contains contributions from electron impact excitation, ion balance, and abundances, and

$$EM = \int n_e N_\mathrm{H}\, \mathrm{d}V \tag{100}$$

is the total *emission measure* of the source (in units cm^{-3}). In this case the ratio of two lines from the same element but with different temperature dependence gives T and EM, while the ratio of two lines with the same temperature dependence but from different elements gives the abundance ratio. The line/continuum ratio gives the absolute abundance.

However, in general the situation is more complicated. It is to be expected that the source is not well represented by one single temperature, but that instead plasma at *many different temperatures* will contribute to the X-ray spectrum. Then we write

$$P \propto \int f(T) D(T)\ \mathrm{d}\log T \ , \tag{101}$$

where we introduce the *differential emission measure* (DEM) distribution function $D(T)$ defined by

$$D(T) = n_e(T) N_\mathrm{H}(T) \frac{\mathrm{d}V}{\mathrm{d}\log T} \ . \tag{102}$$

This is the weighting function that measures how strongly any particular temperature contributes to the observed spectrum.

To derive physical parameters from the spectra various techniques can be applied such as multi-temperature fitting or fitting with continuous models to derive $D(T)$ (for references e.g., Kaastra, Mewe, Liedahl et al. 1996; Mewe et al. 1995b, 1996; see also the Chapter by Kaastra). These methods use a pre-calculated set of model spectra $f(\lambda_i, T_j)$. Here $f(\lambda_i, T_j)$ denotes the plasma line plus continuum emission function (phot cm^3 s^{-1}) per unit $n_e N_H$, per wavelength bin λ_i, and for an electron temperature grid T_j. These model spectra are folded through the instrument response, include interstellar absorption and take into account the appropriate abundances. The total spectrum integrated over the temperature distribution is written as

$$F(\lambda_i) = \int f(\lambda_i, T) n_e(T) N_H(T) \, dV(T) \simeq \sum_{j=1}^{M} f(\lambda_i, T_j) D(T_j) \Delta \log(T_j) \ . \quad (103)$$

The DEM distribution is derived by matching the observed line and continuum (wavelength) spectrum $\widetilde{F}(\lambda_i)$ as well as possible to the synthesized model spectrum $F(\lambda_i)$ for a grid of temperatures ranging from T_1 up to T_M and with a step size of typically $\Delta \log(T_j) = 0.1$ (sufficiently small since most lines are formed within $\delta \log T \simeq 0.3$). It is important to include also the continuum spectrum in the analysis since the line–to–continuum ratio contains important information on the abundances (in particular the [Fe/H] abundance ratio) and line scattering effects. I note that the determination of $D(T)$ by the inversion of Eq. (103) is not trivial since in practice the measured spectrum contains noise and a unique solution is never possible (see the Chapter by Kaastra for a discussion). In the analysis with SPEX (cf. Sect. 3.2) five DEM methods can be applied:

- (i) multi–temperature Gauss fitting,
- (ii) regularization method,
- (iii) polynomial method,
- (iv) clean algorithm,
- (v) genetic algorithm.

The first, fourth, and fifth method are especially suited to resolve discrete narrow temperature structures, whereas the second and third method are more suited for smooth temperature distributions (cf. Kaastra et al. 1996; Mewe et al. 1996). A nice example of a comparison between all DEM methods, as applied to the analysis of combined *ASCA* and *EUVE* data of the active solar-type star EK Draconis, is given in the paper by Güdel et al. (1997; their Fig. 4).

15 Diagnostics of Satellite Lines

Satellite lines are systems of lines appearing close to or blended with the resonance lines of highly ionized atoms formed in hot astrophysical and laboratory plasmas. For H- and He-like ions these types of lines have been well known for almost 60 years. Edlén & Tyrén (1939) discovered such lines in the spectra from a vacuum spark and interpreted them as resonance transitions in the presence of an additional bound perturbing ("spectator") electron $n\ell$, for example, the $1s^2n\ell$–$1s2pn\ell$ Li-like satellites to the He-like resonance line $1s^2$–$1s2p$. These lines correspond to the stabilizing transitions in the process of *dielectronic recombination*, which was first recognized by Massey & Bates (1942) as a *two-step* recombination process. The *first* step, radiationless (or "dielectronic") capture of a free electron by a recombining ion, gives rise to a doubly excited autoionizing quasi-bound state. The *second* step (stabilization) is the emission of radiation (*satellite* line) to give a stable, singly excited state in the recombined ion.

The stronger satellites arise from states with $n = 2$ for the perturbing spectator electron and appear on the long-wavelength side of the parent resonance line. The group of $n = 3$ satellites appears closer with some members on the short-wavelength side and/or blended with the resonance line, while those arising from $n \geq 4$ are practically indistinguishable from it.

The *dielectronic recombination* (DR) process has been studied in detail since Burgess (1964, 1965) was able to show that it can provide an important additional recombination mechanism in hot dilute plasmas like those observed in the solar corona, affecting noticeably the ionization equilibrium, and thus the interpretation of the ionization temperature of the corona. Most applications of DR were therefore first concentrated on this particular aspect.

Since the early seventies with the advent of many observations of high-resolution X-ray spectra from hot astrophysical and laboratory plasmas, its importance as a line formation mechanism has been emphasized. In particular, Gabriel (1972) and co-workers have demonstrated that satellite lines provide very powerful tools to diagnose such hot plasmas since these lines are sensitive to variations of *electron temperature* and *density*, and *deviations from a Maxwellian electron distribution*. Moreover, in transient ionizing plasmas and for heavier ions, these lines can also be produced by direct electron impact excitation of a ground-state *inner-shell* electron, which also allows the *(transient) ionization state* of the plasma to be determined. These authors made the first attempt to interpret the satellite intensities and have developed a detailed theory of the He-like ion satellites (for a review see Dubau & Volonté 1980). For references to the numerous observations of satellite lines in high-resolution spectra of hot astrophysical plasmas (essentially the solar corona) and laboratory sources, see Mewe (1988). Note that Mewe & Kaastra have recently written several internal SRON notes on data compilations for K-shell X-ray spectra of ions with $Z = 6$–28 which are available on request.

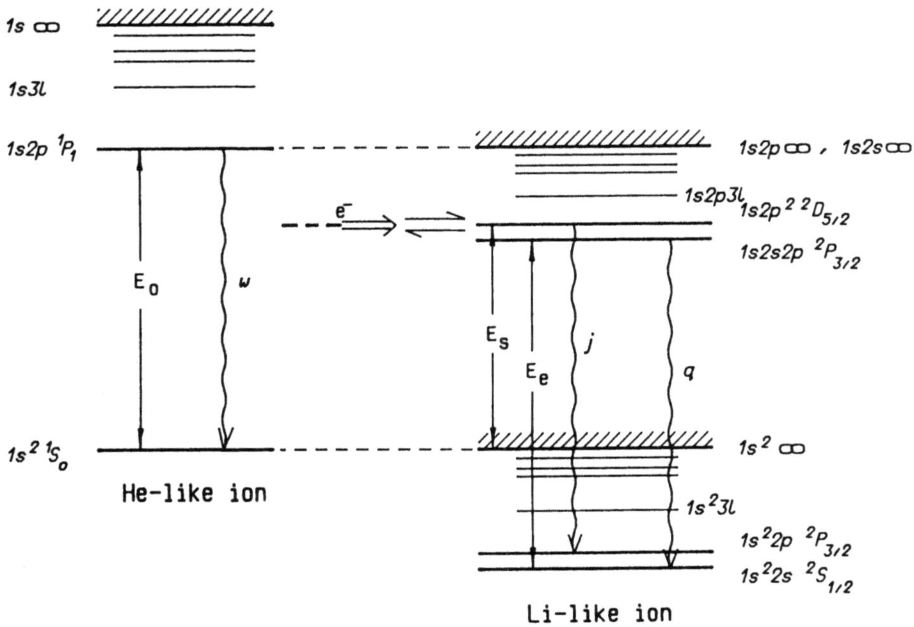

Fig. 14. Energy level diagram showing the parent resonance line w in the He–like ion together with the strong DR satellite j and the IE satellite q in the Li–like ion (from Mewe 1988).

15.1 Dielectronic recombination (DR) satellite intensity

The contribution from DR is derived as follows. In the usual way the emission rate I_{sk}^{DR} (photons cm^{-3} s^{-1}) by dielectronic excitation of the satellite line $s \to k$ is given by

$$I_{sk}^{\mathrm{DR}} = N_{z,s} A_{sk}^{r} \ , \tag{104}$$

where the population density $N_{z,s}$ (cm^{-3}) of state s is determined by the balance between dielectronic capture (rate coefficient C_s^{d}) and autoionization plus radiative decay, i.e.,

$$N_{z+1,1}\, n_{\mathrm{e}}\, C_s^{\mathrm{d}} = N_{z,s}\, A_s(\mathrm{tot}) \ , \tag{105}$$

where n_{e} is the electron density, and $A_s(\mathrm{tot}) = A_s^{\mathrm{a}} + \sum A_s^{\mathrm{r}}$, the total decay rate of satellite level s by autoionization and by radiative decay. Thus:

$$I_{sk}^{\mathrm{DR}} = N_{z+1,1} n_{\mathrm{e}} C_s^{\mathrm{d}} BR_s \ , \tag{106}$$

with C_s^{d} given by Eq. (45) and the branching ratio BR_s given by Eq. (47), so that we obtain for the DR line intensity (photons cm^{-3} s^{-1}):

$$I_{sk}^{\mathrm{DR}} = 2.071 \times 10^{-16}\ N_{z+1} n_{\mathrm{e}} T_{[\mathrm{K}]}^{-3/2}\ B_s\ \exp(-E_s/kT) \ , \tag{107}$$

where the satellite intensity factor B_s is given by Eq. (49).

15.1.1 Density effects. Up to now we have considered the special case of a low-density plasma in which all ions $Z^{+(z+1)}$ and Z^{+z} are in their ground states. At increasing electron density (say $n_e \gtrsim 10^{13}$ cm^{-3}) the excited levels $p > 1$ in the ground-state configuration of ion $Z^{+(z+1)}$ (cf. Fig. 8) will start to become populated significantly, so as to give sometimes also contributions to the DR satellites. The total contribution of all levels p to the emission rate $s \rightarrow k$ is then given by a summation over all p (e.g., Lemen et al. 1986). The *density dependence* of the satellite emission then arises through the density-dependent populations in the fine-structure levels of the ground configuration of ion $Z^{+(z+1)}$.

15.2 Inner-shell excitation (IE)

Sometimes there is a contribution to the satellite intensity from electron *impact excitation* of an inner-shell electron in the recombined ion Z^{+z} from the *ground* state (because we assume the conditions for the low-density case) from which the satellite upper line level s has an excitation energy E_e. This process can become efficient in atomic systems containing many electrons in the inner shell or for satellite lines arising from non-autoionizing states. It is not effective in He-like ions since there it requires simultaneous excitation of two electrons which is much less probable than single-electron excitation. In *transient* plasmas (cf. Sect. 4.1.3) this process can be very important (cf. Mewe & Schrijver 1980).

For the photon emission rate by *inner-shell excitation* (photons cm^{-3} s^{-1}) of the satellite line $s \rightarrow k$ we write

$$I_{sk}^{\mathrm{IE}} = N_z n_e B R_s S_e \quad , \tag{108}$$

where N_z is the density of ion Z^{+z} in the ground state, BR_s is the branching ratio given by Eq. (47) and S_e is the electron impact excitation rate coefficient given by the formulae of Sect. 8.1.1.

This process is most effective under circumstances where simultaneously a *high temperature* [needed to overcome the excitation energy, e.g., E_e in Eq. (111)] and a *low ion stage* [e.g. Li-like in Eq. (111)] occur. Because the (Li-like) ionization potential is much lower than the excitation energy of the satellite (cf. Fig. 14) this is possible only in a *transient* plasma in which the ionization lags behind (see Sect. 4.1.3).

15.2.1 Density effects. At higher electron densities, also other (excited) levels in the ground configuration of ion Z^{+z} can obtain a significant population, thus introducing a *density dependence* in the IE part of the satellite line, similarly as in the DR case (cf. Lemen et al. 1986).

15.3 Inner-shell ionization (II)

Mewe & Gronenschild (1981) and Mewe et al. (1980) have considered electron impact ionization from the inner $1s^2$ shell of ion $Z^{(z-1)}$ which contributes to the production in an optically thin plasma to the formation of a certain line in the next higher ion Z^{+z} (e.g., the satellites to He-like resonance lines from lower Li-, Be-, etc. like ions).

For the same reason as given for IE the II process is especially important in a transient plasma.

It turns out that for inner-shell ionization of Li-like ions the ionization formula of Lotz (1968) (corrected by the *fluorescence yield* $C_{\rm II}$) agrees within 5% with the calculations of Sampson & Zhang (1988). For the photon emission rate we write analogously to Eq. (108)

$$I_{sk}^{\rm II} = N_{z-1} n_e C_{\rm II} S_{\rm i} \ , \tag{109}$$

where the rate coefficient for inner-shell ionization $S_{\rm i}$ is given by Eq. (24) with $\xi = 2$, the total number of electrons in the $1s^2$ shell and χ the ionization energy of the $1s^2$ shell in ion $Z^{+(z-1)}$ and where $C_{\rm II}$ is the fluorescence yield (= probability of emission of a photon per ionization).

Mewe & Schrijver (1975, 1978a,b) have argued that inner-shell ionization of a Li-like ion is an additional mechanism in populating the metastable state of a He-like ion by the process

$$Z^{+(z-1)}(1s^2 2s\ ^2S) + e^- \to Z^{+z}(1s2s\ ^3S) + 2e^- \ , \tag{110}$$

which can give a significant contribution to the formation of the forbidden He-like (z) $2^3S \to 1^1S$ line in a transient plasma (Mewe & Schrijver 1975, 1978b, 1980). In this case $C_{\rm II} = 0.75$ for a low-density plasma.

15.4 Diagnostics

The diagnostic techniques based on the measurement of satellite to resonance line intensity ratios in high-resolution X-ray spectra from hot plasmas provide powerful tools for the determination of electron *temperature*, electron *density* (see above), *departures from steady-state ionization* balance (in transient plasmas), or *departures from Maxwellian energy* distributions (in non-thermal plasmas). As an example, I briefly consider the illustrative case of the He-like Fe XXV resonance line (w) at 1.85 Å with the satellites in the Li-like ion (e.g., the prominent DR satellite j and the IE satellite q, see Fig. 14) (for notations and details, e.g., Gabriel 1972, and Mewe 1988).

From the formulae given above and in Sect. 8, the relative intensities of the resonance (w), DR satellite (j), and IE satellite (q) lines can be approximately written as:

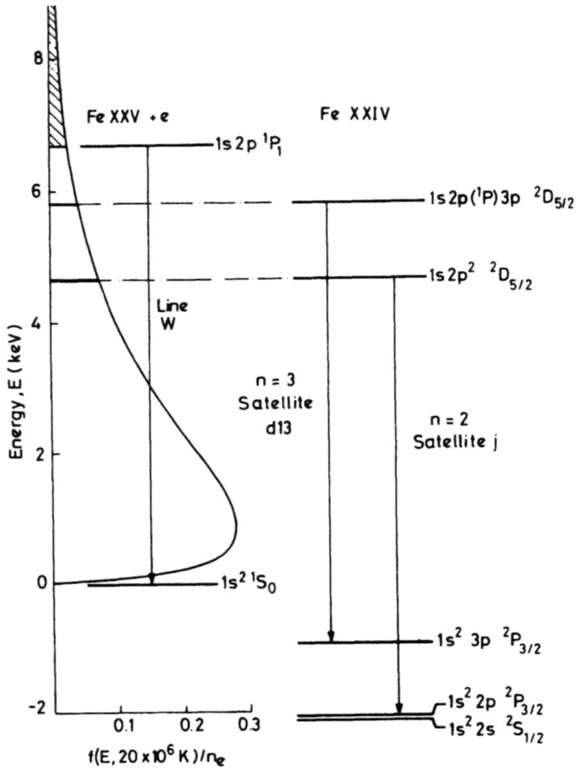

Fig. 15. Energy level diagrams for Fe XXIV and Fe XXV + e^- with line transitions w, j and $d13$ indicated. The normalized free-electron Maxwellian energy distribution, $f(E)/n_e$, is shown for a typical flare temperature of 20×10^6 K (after Gabriel & Phillips 1979).

$$I_w \propto n_e \, N_{\rm He} \, T^{-1/2} \, e^{-E_0/kT} ,$$
$$I_j \propto n_e \, N_{\rm He} \, T^{-3/2} \, e^{-E_s/kT} ,$$
$$I_q \propto n_e \, N_{\rm Li} \, T^{-1/2} \, e^{-E_e/kT} , \qquad (111)$$

where $N_{\rm He}$ and $N_{\rm Li}$ are the densities of He- and Li-like ions determined by the ionization balance, and the excitation energies are E_0, E_s, and E_e (note that $E_0 \simeq 1.4 E_s \simeq E_e$).

15.4.1 Temperature diagnostics. The DR satellite/resonance line ratio

$$j/w \propto T^{-1} \, e^{0.3 E_w/kT} \qquad (112)$$

determines the electron *temperature*, independently of the electron density and the ionization balance.

15.4.2 Ionization balance diagnostics. On the other hand, the IE satellite/resonance line ratio

$$q/w \propto N_{\rm Li}/N_{\rm He} \tag{113}$$

determines the *ion abundance ratio* Li/He, independently of temperature.

It has been common after Gabriel (1972) to describe the derived abundance ratio by the so-called "ionization" temperature, T_z, i.e. the value of the electron temperature for which in ionization equilibrium, the actual ratio $N_{\rm Li}/N_{\rm He}$ is equal to that determined from the measured line ratio ($T = T_z$). With T and T_z derived from such line ratios, departures from ionization equilibrium in a transient plasma can be deduced, $T > T_z$ indicating an ionizing plasma, and $T < T_z$ a recombining plasma.

15.4.3 Non-Maxwellian electron velocity distributions. Departures from a Maxwellian velocity distribution can occur when energy is deposited into (or lost from) the high-energy tail of the distribution at a rate much faster than the Coulomb electron-electron collision rate $t_{\rm ee}^{-1} \sim 100 n_e T^{-3/2}$ s^{-1}, resulting in a high-energy excess (or deficit, respectively). Even when the Coulomb collision rate is too slow, plasma instabilities may cause a relaxation towards a Maxwellian distribution in a much shorter time as is probably the case in the shocks in supernova remnants (SNRs). Non-Maxwellian distributions are most likely to occur in low-density plasmas and at high velocities where collisional relaxation is slow, or in plasmas confined by strong magnetic fields that inhibit the dissipation process. As the electrons in the high-energy tail are just the ones which excite and ionize ions, the excitation and ionization rates are most strongly affected, whereas recombination rates which rely mostly on the electrons in the bulk of the distribution are not significantly affected. However, Itoh (1984b) showed that in the early stages of ionization in the shocks of SNRs the electron population can consist of primary shock-heated electrons plus cooler secondary electrons ejected during the ionization process that cause an enhanced recombination rate and slowing down of the ionization process. This in turn can lead to an underestimate of the age of the SNR (cf. Kaastra 1998).

A deficit of high-energy electrons could arise if the fast electrons lose their energy by exciting or ionizing ions more rapidly than Coulomb collisions shuffle electrons from the bulk of the energy distribution (energy $\sim kT$) to the tail (e.g., Dreicer 1960). The reverse, i.e., the production of a high-energy excess in the tail, may occur in plasmas with a steep temperature gradient where fast electrons can penetrate into cooler regions of the plasma, e.g., in the solar wind (Owocki & Scudder 1983) or in solar flares (Seely et al. 1987).

Various diagnostics can be applied to obtain constraints on the shape of the electron energy distribution: (1) measurements of spectral lines from a number of ion charge stages with a range in ionization thresholds; (2) measurement of the polarization of lines or continuum which is quite sensitive to the presence of electron beams in plasmas; (3) measurements of spectral lines that are excited by different parts of the electron distribution.

For the second method I refer to Pal'chikov & Shevelko (1995) who discuss the theory of polarization of X-ray lines and give many references to crossed-beam experiments and solar spectral measurements. Here I consider the technique of measuring the ratio of two spectral lines that are excited by different portions of the electron energy distribution.

It is generally admitted that energetic non-thermal electrons are responsible for the hard X-ray bursts occurring in the impulsive phase of flares, which may enhance the high-energy tail (e.g. > 10 keV) of the Maxwellian distribution of the emitting background plasma, and as a consequence this may result in an enhancement of the resonance line intensity relative to the DR satellite intensities. This is connected with the resonant character of the DR process as is illustrated in Fig. 15 showing the energy level diagrams for the He-like Fe XXV resonance line w and its DR Li-like satellites j and $d13$. A DR satellite is produced only by free electrons having an energy exactly equal to [within the small autoionization width δE_ζ, see Eq. (62)] the excitation energy E_s of the satellite line level, whereas the resonance line (w) can be excited by all electrons having any energy above the threshold value E_0 [$\simeq 1.4E_s$ (or $\simeq 1.16E_s$) for $n = 2$ (or 3) satellites]. Therefore the appearance of an enhanced high-energy tail in the distribution will result in an increase of the resonance line but will have no effect on the satellite lines. As a consequence, the satellite/resonance ratio decreases giving larger values of the apparent electron temperature. On the other hand, the ratio of two $n = 2$ and $n = 3$ satellites remains unaffected, but gives a measure of the actual temperature referring to the Maxwellian part of the electron energy distribution. Gabriel & Phillips (1979) have developed such a detection method using the Fe XXV resonance line and two strong $n = 2$ (j) and $n = 3$ ($d13$) satellites (Fig. 15). Seely et al. (1987) have applied this method to high-resolution spectra measured with the SOLFLEX spectrometer onboard the *P78-1* spacecraft from several solar limb flares. They have found that non-thermal electron energy distributions occur during the first few minutes of the impulsive phase, near the time when hard X-ray bursts are observed at the onset of the flare. For the discrete electron energies (4.7 and 5.8 keV) at which the j and $d13$ satellites are formed, the electron energy distributions were observed to have a bump or to be flat at this time.

15.4.4 Line and continuum emission for non-thermal distributions.

Line emission

As an illustration I compare the intensities of dielectronic recombination (DR) and inner-shell excitation (IE) lines for various normalized[12] electron energy distributions $f(E)$: a Maxwell (M), a power law (PL), and a mono-energetic (ME) distribution. The Maxwell distribution is given by Eq. (20) and the power law distribution is given by:

$$f_{\mathrm{PL}}(E) = (\delta - 1)E_1^{\delta-1}E^{-\delta} \ (E \geq E_1), \text{or } f_{\mathrm{PL}}(E) = 0 \ (E < E_1) \ , \tag{114}$$

where we assume $\delta > 1$. The mono-energetic distribution is:

$$f_{\mathrm{ME}}(E) = (\Delta E)^{-1} \tag{115}$$

in a small energy interval ΔE around electron energy $E = E_m$.

For an arbitrary electron energy distribution the intensity of a DR satellite line with excitation energy E_s is given by

$$I^{\mathrm{DR}} = \left(\frac{2E_s}{m_e}\right)^{1/2} \frac{h^3}{16\pi m_e} \frac{f(E_s)}{E_s} \beta \ , \tag{116}$$

where $\beta = N_{z+1}n_eB_s$ [cf. Eq. (107)]. Inserting the numerical constants, expressing now energies and temperature T in keV we obtain for the line intensity in phot cm^{-3} s^{-1} for the Maxwellian distribution:

$$I_{\mathrm{M}}^{\mathrm{DR}} = 5.24 \ 10^{-27} \ T^{-3/2}e^{-E_s/T}\beta \ . \tag{117}$$

For the power law distribution:

$$I_{\mathrm{PL}}^{\mathrm{DR}} = 4.64 \ 10^{-27} \ (\delta - 1)E_1^{\delta-1}E_s^{-\delta-\frac{1}{2}}\beta \ , \tag{118}$$

where $E_s > E_1$.

For a mono-energetic distribution:

$$I_{\mathrm{ME}}^{\mathrm{DR}} = 4.64 \ 10^{-27} \ (\Delta E)^{-1}E_s^{-1/2}\beta \ , \tag{119}$$

where $E_m = E_s$, and the autoionization width $\delta E_\zeta \ll \Delta E$.

For an IE line (or similarly for the resonance line) with excitation energy E_0 we obtain from Eqs (108) and (68) with the approximation for the Gaunt factor $g(E) \approx 0.1 + 0.276 \ln U$ $(U = E/E_0)$ for the Maxwell distribution:

$$I_{\mathrm{M}}^{\mathrm{IE}} = 5.0 \ 10^{-10} \ T^{-1/2}E_0^{-1}f \ \bar{g}_{\mathrm{M}} \ e^y \ \gamma \ , \tag{120}$$

where $\bar{g}_{\mathrm{M}} = 0.1 + 0.276 e^y E_1(y)$ with $y = E_0/T$, $E_1(y)$ the exponential integral (Eq. 71), f the oscillator strength, and $\gamma = N_z n_e B R_s$. Again energies and T are in keV.

[12] $\int_{E_1}^{\infty} f(E)\,\mathrm{d}E = 1$; $E_1 = 0$ for the Maxwell distribution.

We can integrate the excitation cross section over a power law distribution using Eqs (19) and (114) and derive:

$$I_{\mathrm{PL}}^{\mathrm{IE}} = 4.43\ 10^{-10} \left(\frac{\delta - 1}{\delta - \frac{1}{2}}\right) E_1^{-1/2} E_0^{-1} f\ \bar{g}_{PL}\ \gamma\ , \tag{121}$$

where $\bar{g}_{\mathrm{PL}} = 0.1 + 0.276[\ln(E_1/E_0) + (\delta - \frac{1}{2})^{-1}]$, and $E_1 > E_0$.

For the mono-energetic distribution:

$$I_{\mathrm{ME}}^{\mathrm{IE}} = 4.43\ 10^{-10}\ E^{-1/2} E_0^{-1} f\ g_{\mathrm{ME}}\ \gamma\ , \tag{122}$$

where $g_{\mathrm{ME}} = g(E)$ and $E > E_0$.

Continuum emission

As an example I consider only ff emission and follow essentially the treatment of Brown (1971) who calculated the electron-proton bremsstrahlung spectrum for a power law distribution of non-thermal electrons in a solar flare. The ff emissivity at photon energy ε (in photons/sec per unit volume and per unit ε) for electron-ion (i) encounters can be written for an arbitrary electron distribution as:

$$P^{\mathrm{ff}}(\varepsilon) = n_{\mathrm{e}} n_{\mathrm{i}} Z_{\mathrm{i}}^2 \int_{\varepsilon}^{\infty} \sqrt{\frac{2E}{m_{\mathrm{e}}}} Q(\varepsilon, E, Z_{\mathrm{i}}) f(E)\,\mathrm{d}E\ , \tag{123}$$

where the cross section (area per unit ε) is written as:

$$Q(\varepsilon, E, Z_{\mathrm{i}}) = \left(\frac{C}{\varepsilon E}\right) g(\varepsilon, E, Z_{\mathrm{i}})\ , \tag{124}$$

where Z_{i} is the ion charge and

$$C = \frac{8\pi}{3\sqrt{3}} \alpha r_0^2 m_{\mathrm{e}} c^2 \tag{125}$$

and where $r_0 = 2.8179\ 10^{-13}$ cm is the classical electron radius, $\alpha = 1/137.04$ the fine-structure constant, and $g(\varepsilon, E, Z_{\mathrm{i}})$ is the Gaunt factor. Brown (1971) uses the Born-Heitler non-relativistic high-energy approximation

$$g(\varepsilon, E, Z_{\mathrm{i}}) = \frac{\sqrt{3}}{\pi} \ln\left[\frac{1 + \sqrt{1 - \varepsilon/E}}{1 - \sqrt{1 - \varepsilon/E}}\right]\ . \tag{126}$$

The Born Gaunt factor becomes zero at threshold $\varepsilon{=}E$, whereas it should assume here a finite value (cf. Karzas & Latter 1961). On the basis of the results of Karzas & Latter I have corrected for this effect and averaged g for each ion over a power-law electron distribution (Eq. 114) assuming $\varepsilon > E_1$:

$$\bar{g}(\varepsilon, \delta, Z_{\mathrm{i}}) = \left(\delta - \frac{1}{2}\right) \varepsilon^{\delta - \frac{1}{2}} \int_{\varepsilon}^{\infty} E^{-\delta - \frac{1}{2}} g(\varepsilon, E, Z_{\mathrm{i}})\,\mathrm{d}E \tag{127}$$

Moreover, I calculated a total Gaunt factor G_{PL} by summing over all ions (i) in the plasma in a similar way as we did before (cf. Eq. 88) (Brown considered only protons and took $Z_{\mathrm{i}} = 1$). We express all energies in keV, inserting all physical constants into C, take $m_{\mathrm{e}}c^2 = 511.0$ keV then we obtain $C = 1.4321\ 10^{-24}$, and take $\sqrt{2E/m_{\mathrm{e}}} = 1.8755\ 10^9\sqrt{E}$. Integration of Eq. (123) yields finally for the free-free emissivity in phot cm^{-3} s^{-1} keV^{-1}:

$$P_{\mathrm{PL}}^{\mathrm{ff}}(\varepsilon) = 2.686\ 10^{-15} G_{\mathrm{PL}} \left(\frac{\delta - 1}{\delta - \frac{1}{2}}\right) \left(\frac{E_1}{\varepsilon}\right)^{\delta - 1} \varepsilon^{-3/2} n_{\mathrm{e}} N_{\mathrm{H}}\ . \tag{128}$$

Comparing this formula with the corresponding one for a Maxwell distribution (Eq. 92) we find comparable results only when $\varepsilon \sim T$ and $\varepsilon \sim E_1$.

For a mono-energetic distribution ($E = \varepsilon$) we obtain:

$$P_{\mathrm{ME}}^{\mathrm{ff}}(\varepsilon) = 2.686\ 10^{-15} G_{\mathrm{PL}}\ \varepsilon^{-3/2}\ n_{\mathrm{e}} N_{\mathrm{H}}\ . \tag{129}$$

The total Gaunt factors are of order unity and our calculations approximately give $G_{\mathrm{PL}} \sim 2\varepsilon^{-0.13}\delta^{-0.33}$ and $G_{\mathrm{ME}} \sim \varepsilon^{-0.3}$ for $\varepsilon \sim$ 1–20 keV and $\delta \sim$ 2–5. Finally, I have estimated that relativistic effects are less than about $15(\varepsilon/50)$%, e.g., $\lesssim 15\%$ for $\varepsilon \lesssim 50$ keV.

16 Comparison of Calculated Spectra and Accuracy

Raymond (1988) has compared various spectral calculations and discussed the differences resulting from different ionization balance calculations and from different treatments of the line excitation. He comes to the conclusion that for the strongest X-ray lines from astrophysical plasmas, those of the H- and He-like ions, the agreement generally approaches about 20% (which is important because the He-like lines can be used for density diagnostics), whereas for other cases (e.g., iron lines around 10–12 Å and silicon and sulphur lines around 40–50 Å) discrepancies of a factor of two may exist.

The ionization balance strongly determines the overall appearance of the X-ray spectra. Mewe (1990) has made a few spectral calculations using different ionization balance calculations for a spectral resolution of 0.05 Å, typical for the high-resolution spectrometers to be flown on the *AXAF* and *XMM* missions (see the Chapter by Paerels). The use of different ionization balances may lead to appreciable differences in the derived plasma parameters as has been recently shown by Masai (1997) in an analysis of simulated *ASCA* spectra.

Raymond (1988, 1990) has made an attempt to test the existing models by comparing with high-resolution solar X-ray observations. From his comparison of a model calculation with the composite X-ray spectrum of a solar flare he concludes that uncertainties in atomic rates or the breakdown of simplifying model assumptions may lead to errors on the order of ~50 % in the predicted line strengths. A comparison of coronal models with *EXOSAT*

(Lemen et al. 1989) and *EUVE* (Mewe et al. 1995b; Schrijver et al. 1995) spectral X-ray observations with moderate resolution (~3 Å) of a few late-type stars by using differential emission measure distributions show a satisfactory agreement, but also indicate that for the interpretation of the future *AXAF* and *XMM* X-ray spectra better spectral model calculations will be needed.

Kaastra & Mewe have started a major undertaking to revise and extend all atomic data used in the current MEKAL code (cf. Sect. 3.2) to meet the demands of the analysis of the future data. This project includes a systematic search through the literature and a critical evaluation of the data used. A comparison of new and old data for the H-sequence (e.g., the $n \rightarrow 1$ transitions) shows that for the most important temperature ranges the newly calculated collision strengths are accurate to ~10% for the Lyman α (2–1) transitions, but for Lyman β (3–1) and Lyman γ (4–1) the differences can be up to 40–70%, especially near threshold (low temperature) where resonance effects are important (cf. Kaastra 1998).

17 Summary

High-resolution X-ray spectroscopy has applications to a wide range of optically thin hot astrophysical and laboratory plasmas. Its significance as a tool in understanding the physics of these sources depends on the reliability of the theoretical models used to interpret the spectra. We have considered the coronal model in describing the ionization and excitation of hot plasmas. In particular, we have discussed the processes of ionization, recombination and excitation, including the formation of dielectronic recombination satellite lines. The accuracy with which the emergent X-ray spectrum can be predicted has been briefly considered. Various effects leading to deviations from the coronal model such as high density, optical depth, transient ionization, and non-Maxwellian electron distributions, have been discussed. It is obvious that the complexity of plasma physics and the atomic parameters involved are such that a sound verification of plasma theories and atomic physics which are applied will be required for the interpretation of future high-resolution spectra such as can be obtained with the future space missions *AXAF*, *XMM*, and *ASTRO-E*.

Acknowledgements. The Space Research Organization of the Netherlands (SRON) is supported financially by NWO, the Netherlands Organization for Scientific Research. I thank J.M. Braun for assistance in preparing the figures.

References

Abramowitz, M. & Stegun, I.A. (1970): *Handbook of Math. Functions*, (9th Ed., Dover)

Acton, L.W. & Brown, W.A. (1978): ApJ 225, 1065

Aggarwal, K.M., Berrington, K., Eissner, W. & Kingston, A.E. (1986): Report on Recommended Data, Atomic Data Workshop, Daresbury Lab.
Aggarwal, K.M. & Kingston, A.E. (1991): J. Phys. B 24, 4583
Aldrovandi, S.M.V. & Péquignot, D. (1973): A&A 25, 137; erratum 47, 321 (1976)
Allen, C.W. (1973): *Astrophys. Quantities* (3rd ed., The Athlone Press, London)
Anders, E. & Grevesse, N. (1989): Geochim. Acta 53, 197
Arnaud, M. & Raymond, J.C. (1992): ApJ 398, 394
Arnaud, M. & Rothenflug, R. (1985): A&AS 60, 425
Bates, D.R. & Dalgarno (1962): in *Atomic and Molecular Processes*, Ed. D.R. Bates, Academic Press Inc., New York, p. 245
Bates, D.R., Kingston, A.E. & McWhirter, R.W.P. (1962a): Proc. Roy. Soc. (London) A 267, 297
Bates, D.R., Kingston, A.E. & McWhirter, R.W.P. (1962b): Proc. Roy. Soc. (London) A 270, 155
Bell, R.H. & Seaton, M.J. (1985): J. Phys. B.: At. Mol. Phys. 18, 1589
Bely, O. & van Regemorter, H. (1970): ARA&A 8, 329
Bely-Dubau, F., Gabriel, A.H. & Volonté, S. (1979): MNRAS 189, 801
Blaha, M. (1971): Bull. Am. Astron. Soc. 3, 246
Brown, J.C. (1971): Solar Phys. 18, 489
Burgess, A. (1964): ApJ 139, 776
Burgess, A. (1965): ApJ 141, 1588
Burgess, A. et al. (1977): MNRAS 179, 275
Burgess, A. & Chidichimo, M.C. (1983): MNRAS 203, 1269
Burgess, A. & Summers, H.P. (1969): ApJ 157, 1007
Burgess, A. & Tworkowski, A.S. (1976): ApJ 205, L105
Butler, S.E. & Dalgarno, A. (1980): ApJ 241, 838
Carson, T.R. (1988): A&A 189, 319
Condon, E.U. & Shortley, G.H. (1970): *The Theory of Atomic Spectra*, Cambridge University Press, U.K., 9th ed.
Cooper, J. (1966): Rep. Progr. Phys. 29, 35
Dalgarno, A. & Drake, G.W.F. (1969): in *Les Transitions Interdites dans les Spectres des Astres*, Les Congrès et Colloques de l'Université de Liège, Belgium Vol. 54, p. 69
Dolder, K. & Peart, B. (1986): Adv. At. Mol. Phys. 22, 197
Drawin, H.-W. (1961): Z. Physik 164, 513; 168, 238
Dreicer, H. (1960): Phys. Rev. 117, 343
Dubau, J. & Volonté, S. (1980): Rep. Prog. Phys. 43, 199
Düchs, D. & Griem, H.R. (1966): Phys. Fluids 9, 1099
Dunford, R.W., Church, D.A. & Liu, C.J. et al. (1990): Phys. Rev. A 41, 4109
Edlén, B. & Tyrén, F. (1939): Nature 143, 940
Elwert, G. (1952): Z. Naturf. 7a, 432; 703
Fabian, A.C. & Ross, R.R. (1981): MNRAS 195, 29P
Feldman, U. (1981): Physica Scripta 24, 681
Gabriel, A.H. & Jordan, C. (1972): *Case Studies in Atomic and Collisional Physics* 2, Eds E.W. McDaniel & M.R.C. McDowell, N.-H. Publ. Co., Amsterdam, p. 209
Gabriel, A.H. (1972): MNRAS 160, 99
Gabriel, A.H. & Phillips, K.J.H. (1979): MNRAS 189, 319

Gallagher, J.H. & Pradhan, A.K. (1985): JILA Data Center Report No. 30 (JILA, Univ. of Colorado, Boulder)
Gaunt, J.A. (1930): Proc. Roy. Soc. (London), Ser. A 126, 654
Gregory, D.C. et al. (1987): Phys. Rev. A 34, 3657; A 35, 3526
Griem, H.R. (1964): *Plasma Spectroscopy*, McGraw-Hill, New York; Univ. Microfilms Internatl. 212 00000 7559, Ann Arbor, Mich.
Griem, H.R. (1988): *Principles of Plasma Spectroscopy*, Cambridge Monographs on Plasma Physics 2, Cambridge Univ. Press, Cambridge, UK
Griem, H.R. (1997): J. Quant. Spectr. Rad. Transf. 40, 403
Gronenschild, E.H.B.M. & Mewe, R. (1978): A&AS 32, 283
Gronenschild, E.H.B.M. & Mewe, R. (1982): A&AS 48, 305
Güdel, M., Guinan, E.F. & Mewe, R. et al. (1997): ApJ 479, 416
Hahn, Y. (1985) : Adv. At. Mol. Phys. 21, 123
Hahn, Y. et al. (1980): J. Quant. Spectr. Rad. Transf. 23, 65
Hahn, Y. & LaGattuta, K.J. (1988): Phys. Rep. 166, 195
Hamilton, A.J.S., Sarazin, C.L.S. & Chevalier, R.A. (1983): ApJS 51, 115
Hearn, A.G. (1966): Proc. Phys. Soc. 88, 171
Herzberg, G. (1944): *Atomic Spectra & Atomic Structure*, Dover Publications, New York, 2nd Ed.
Holt, S. & McCray, R. (1982): ARA&A 20, 323
Huaguo Teng, Baifei Sheng, Wengqi Zhang & Zhizhan Xu (1994a): Phys. Scr. 49, 463
Huaguo Teng, Baifei Sheng, Wengqi Zhang & Zhizhan Xu (1994b): Phys. Scr. 49, 468
Huaguo Teng, Baifei Sheng, Wengqi Zhang & Zhizhan Xu (1994c): Phys. Scr. 49, 696
Huaguo Teng, Baifei Sheng, Wengqi Zhang & Zhizhan Xu (1994d): Phys. Scr. 50, 55
Huaguo Teng & Zhizhan Xu (1996): J. Quant. Spectr. Rad. Transf. 56, 443
Itikawa, Y. (1991): At. Data Nucl. Data Tables 49, 209
Itikawa, Y., Kato, T. & Sakimoto, K. (1995): The Inst. of Space and Astronautical Sc. Report No. 657 (Kanagawa, Japan)
Itikawa, Y., Takayanagi, K. & Iwai, T. (1984): At. Data Nucl. Data Tables 31, 215
Itoh, H. (1984a): Physica Scripta T7, 19
Itoh, H. (1984b): ApJ 285, 601
Jacobs, V.L. et al. (1977): ApJ 211, 605; 215, 690
Kaastra, J.S. (1992): *An X-ray spectral code for optically thin plasmas*, Internal SRON-Leiden report, version 2.0.
Kaastra, J.S. (1998): in *The Hot Universe*, Eds M. Itoh, S. Kitamoto & K. Koyama, Proc. IAU Symp. No. 188, Kluwer Academic Publishers, p. 43
Kaastra, J.S. & Jansen, F.A. (1993): A&AS 97, 873
Kaastra, J.S. & Mewe, R. (1993): Legacy 3 (Journal of HEASARC, NASA/GSFC (Greenbelt), p. 16
Kaastra, J.S. & Mewe, R. (1998): in preparation
Kaastra, J.S., Mewe, R. & Liedahl, D.A. et al. (1996): A&A 314, 547
Kaastra, J.S., Mewe, R. & Nieuwenhuijzen, H. (1996): in *UV and X-ray Spectroscopy of Astrophysical and Laboratory Plasmas*, (Eds. K. Yamashita & T. Watanabe, Universal Academy Press, Inc., Tokyo, p. 411

Kahn, S.M. & Liedahl, D.A. (1995): in *Physics with Multiply Charged Ions*, Ed. D. Liesen, NATO Advanced Study Institute Series, Plenum Press, New York, p. 169
Kallman, T.R. & McCray, R. (1982): ApJS 50, 263
Karim, K.R. & Bhalla, C.P. (1988): Phys. Rev. A 37, 2599
Karim, K.R. & Bhalla, C.P. (1989): Phys. Rev. A 39, 3548
Karzas, W.J. & Latter, R. (1961): ApJS 6, 167
Kato, T., Masai, K. & Arnaud, M. (1991): Research Report Nucl. Inst. Fusion Science-Data Ser. No. 14
Keenan, F.P. (1992): in *UV and X-ray Spectroscopy of Astrophysical and Laboratory Plasmas*, Eds E.H. Silver & S.M. Kahn, Cambridge University Press, Cambridge, p. 44
Keenan, F.P. (1995): Space Sc. Rev. 75, 537
Kingston, A.E. & Tayal, S.S. (1983): J. Phys. B. 16, 3465
Kramers, H.A. (1923): Philos. Mag. J. Sci. 46, 836
Kunze, H.-J. (1972): Space Sci. Rev. 13, 565
Kylafis, N.D. & Lamb, D.Q. (1982): ApJS 48, 239
Landini, M. & Monsignori Fossi, B.C. (1990): A&AS 82, 229
Lang, J. (ed.) (1994): At. Data Nucl. Data Tables 57, 1-332
Lemen, J.R., Mewe, R., Schrijver, C.J. & Fludra, A. (1989): ApJ 341, 474
Lemen, J.R. et al. (1986): J. Appl. Phys. 60 (6), 1960
Liedahl, D.A., Kahn, S.M., Osterheld, A.L. & Goldstein, W.H. (1991): ApJ 350, L37
Liedahl, D.A., Osterheld, A.L. & Goldstein, W.H. (1995): ApJ 438, L115
Lightman, A.P., Lamb, D.Q. & Rybicki, G.B. (1981): ApJ 248, 738
Lin, C.D., Johnson, W.R. & Dalgarno, A. (1977): Phys. Rev. A 15, 154
Lotz, W. (1968): Z. Physik 216, 441
Masai, K. (1984): Ap&SS 98, 367
Masai, K. (1994a): ApJ 437, 770
Masai, K. (1994b): J. Quant. Spectr. Rad. Tranf. 51, 211
Masai, K. (1997): A&A 324, 410
Mason, H.E. & Monsignori Fossi, B.C. (1994): A&AR 6, 123
Massey, H.S.W. & Bates, D.R. (1942): Rep. Progr. Phys. 9, 62
McCray, R. (1984): Physica Scripta T7, 73
McWhirter, R.W.P. (1965): in *Plasma Diagnostic Techniques*, Eds R.H. Huddlestone & S.L. Leonard, Acad. Press, New York, p. 201
McWhirter, R.W.P. & Hearn, A.G. (1963): Proc. Phys. Soc. 82, 641
Merts, A.L., Cowan, R.D. & Magee, N.H. (1976): Los Alamos Sci. Lab. Rep. LA-6220-MS
Mewe, R. (1967): Brit. J. Appl. Phys. 18, 107
Mewe, R. (1970): Z. Naturf. 25a, 1798
Mewe, R. (1972): A&A 20, 215
Mewe, R. (1984): Physica Scripta T7, 5
Mewe, R. (1988): in *Astrophysical and Laboratory Spectroscopy*, Eds R. Brown & J. Lang, Scottish Univ. Summer School in Phys. Publ. 33, p. 129
Mewe, R. (1990): in *Physical Processes in Hot Cosmic Plasmas*, Eds W. Brinkmann, A.C. Fabian & F. Giovanelli, Kluwer Acad. Publ., Dordrecht-Holland, p. 39
Mewe, R. (1991): A&AR 3, 127

Mewe, R. (1992): in *The Physics of Chromospheres, Coronae and Winds*, Eds C.S. Jeffery & R.E.M. Griffin, Cambridge University Printing Service, p. 33

Mewe, R. & Gronenschild, E.H.B.M. (1981): A&AS 45, 11

Mewe, R., Gronenschild, E.H.B.M. & van den Oord, G.H.J. (1985a): A&AS 62, 197

Mewe, R. & Kaastra, J.S. (1994): European Astron. Soc. Newsletter 8, p. 3

Mewe, R., Kaastra, J.S. & Liedahl, D.A. (1995a): Legacy 6, 16

Mewe, R., Kaastra, J.S., Schrijver, C.J., van den Oord, G.H.J. & Alkemade, F.J.M. (1995b): A&A 296, 477

Mewe, R., Kaastra, J.S., White, S.M. & Pallavicini, R. (1996): A&A 315, 170

Mewe, R., Lemen, J.R. & van den Oord, G.H.J. (1986): A&AS 65, 511

Mewe, R., Lemen, J.R., Peres, G. et al. (1985b): A&AS 152, 229

Mewe, R., Lemen, J.R. & Schrijver, C.J. (1991): Ap&SS 182, 35

Mewe, R. & Schrijver, J. (1975): Astrophys Space Sci. 38, 345

Mewe, R. & Schrijver, J. (1978a): A&A 65, 99

Mewe, R. & Schrijver, J. (1978b): A&A 65, 115

Mewe, R. & Schrijver, J. (1978c): A&AS 33, 311

Mewe, R. & Schrijver, J. (1980): A&A 87, 261

Mewe, R., Schrijver, J. & Sylwester, J. (1980): A&AS 40, 323

Milne, E.A. (1924): Philos. Mag. J. Sci. 47, 209

Moores, D.L. (1988): in *Astrophysical and Laboratory Spectroscopy*, Eds R. Brown & J. Lang, Scottish Univ. Summer School in Phys. Publ. 33, p. 75

Müller, T. et al. (1987): Phys. Rev. A 36, 599

Munger, C.T. & Gould, H. (1986): Phys. Rev. Lett. 57, 2927

Nussbaumer, H. & Storey, P.J. (1983): A&A 126, 75

Owocki, S.P. & Scudder, J.D. (1983): ApJ 270, 758

Pal'chikov, V.G. & Shevelko, V.P. (1995): *Reference Data on Multicharged Ions:*, Springer-Verlag, Berlin

Peacock, N.J. (1996): Ap&SS 237, 341

Pradhan, A.K. (1985): ApJ 288, 824

Pradhan, A.K. (1987): Physica Scripta 35, 840

Raymond, J.C. (1988): in *Hot Thin Plasmas in Astrophysics*, Ed. R. Pallavicini, Kluwer Acad. Publ., Dordrecht, p. 3

Raymond, J.C. (1990): in *High Resolution X-ray Spectroscopy of Cosmic Plasmas*, Eds P. Gorenstein & M.V. Zombeck, Proc. IAU Coll. 115, Cambridge, U.S.A., Reidel Publ. Co.

Raymond, J.C. & Brickhouse, N.S. (1996): Ap&SS 237, 321

Raymond, J.C. & Smith, B.W. (1977): ApJS 35, 419

Reilman, R.F. & Manson, S.T. (1979): ApJS 40, 815

Romanik, C.J. (1988): ApJ 330, 1022

Ross, R.R. (1979): ApJ 233, 334

Rybicki, G.B. & Lightman, A.P. (1979): *Radiative Processes in Astrophysics:*, Wiley InterSc. Publ., New York

Sampson, D.H. (1982): J. Phys. B 15, 2087

Sampson, D.H., Goett, S.J. & Clark, R.E.H. (1983): At. Data Nucl. Data Tables 29, 467

Sampson, D.H. & Zhang, H.L. (1988): Phys. Rev. A 37, 3765

Schrijver, C.J., Mewe, R., van den Oord, G.H.J. & Kaastra, J.S. (1995): A&A 302, 438

Seaton, M.J. (1959): MNRAS 119, 81
Seaton, M.J. (1962): in *Atomic and Molecular Processes*, Ed. D.R. Bates, Academic Press Inc., New York, p. 414
Seaton, M.J. (1975): Adv. At. Mol. Phys. 11, 83
Seely, J.F., Feldman, U. & Doschek, G.A. (1987): ApJ 319, 541
Shapiro, P.R. & Moore, R.T. (1977): ApJ 217, 621
Smith, B.W. et al. (1985): ApJ 298, 898
Spitzer, L., Jr. (1962): *Physics of Fully Ionized Gases*, 2nd Ed., Intersc. Publ., New York
Spitzer, L., Jr. & Greenstein, J.L. (1951): ApJ 114, 407
Summers, H.P. (1974): MNRAS 169, 663; Appleton Lab. Rep. AL-R-5
Summers, H.P. (1988): in *Astrophysical and Laboratory Spectroscopy*, Eds R. Brown & J. Lang, Scottish Univ. Summer School in Phys. Publ. 33, p. 15
Sylwester, B. et al. (1986): Solar Phys. 103, 67
Unsöld, A. (1955): *Physik der Sternatmosphären:*, 2nd Ed., Springer-Verlag, Berlin
van Regemorter, H. (1962): ApJ 136, 906
Vernazza, J.E. & Raymond, J.C. (1979): ApJ 228, L29
Verner, D.A., &Ferland, G.J. (1996): ApJS 103, 467
Verner, D.A. & Yakovlev, D.G. (1995): A&AS 109, 125
Verner, D.A., Yakovlev, D.G., Band, I.M. & Trzhaskovskaya, M.B. (1993): At. Data Nucl. Data Tables 55, 233
Vrtilek, S.D., McClintock, J.E., Seward, F.D. et al. (1991): ApJS 76, 1127
Wang, J.-S. et al. (1986): Phys. Rev. A 33, 4293
Wang, J.-S. et al. (1987): Phys. Rev. A 36, 951
Wang, J.-S. et al. (1988): Phys. rev. A 38, 4761
Wiese, W.L., Fuhr, J.R. & Deters, T.M. (1996): J. Phys. Chem. Ref. Data, Monograph No. 7
Wilson, R. (1962): J. Quant. Spectr. Rad. Transf. 2, 477
Younger, S. (1981): J. Quant. Spectr. Rad. Transf. 26, 329
Zhang, H.L. & Sampson, D.H. (1987): ApJS 63, 487
Zygelman, B. & Dalgarno, A. (1987): Phys. Rev. A 35, 4085

The X-Ray Spectral Properties of Photoionized Plasmas and Transient Plasmas

Duane A. Liedahl

Department of Physics and Space Technology
Lawrence Livermore National Laboratory
P.O. Box 808, L-41, Livermore, CA 94550, U.S.A.

Abstract. I present the fundamental concepts of X-ray spectral formation in X-ray photoionized plasmas and transient plasmas, emphasizing the role of atomic kinetics in determining the attendant spectral characteristics.

Objects which harbor compact sites of hard X-ray production — active galactic nuclei, X-ray binaries, and cataclysmic variables — are the domains of photoionized plasmas. With a focus on the basic elements of discrete spectroscopy, I discuss the unique properties of X-ray spectra in such environments. As a prelude, an introduction to the theory of Comptonization is provided, including a full derivation of the Kompaneets Equation, and its application to heating and cooling in X-ray nebulae. In the discussions on line spectroscopy, the mechanisms behind various plasma diagnostics are described, including recombination continua, $\Delta n = 0$ dielectronic recombination, density diagnostics, and Kα fluorescence.

The second topic of this chapter, the effect on X-ray spectra of time-dependent ionization conditions, is restricted to ionizing plasmas dominated by electron-ion impact processes. This scenario is motivated by impulsively heated plasmas, such as occur in supernova remnants and solar flares. Treated in this way, transient ionization is an extension of collisional ionization equilibrium. I present the fundamentals of ionization dynamics, followed by a numerical solution of the ionization equations for oxygen, a case study intended to illustrate the spectroscopic consequences of rapid ionization.

1 Introduction

We begin by considering a highly ionized atom, in ionization stage i, immersed in an ion-electron plasma, where the velocities of the electrons are described by a Maxwellian distribution. A variety of mechanisms can produce discrete X rays, with a wavelength distribution that depends on the energy level spacings in ion i. For example, electron impact excitation from low-lying energy levels of ion i to higher levels will produce X-ray line emission. Recombination from the next-higher charge state, $i+1$, can produce X-ray lines resulting from a radiative cascade following capture of a free electron. Furthermore, ionization of non-valence electrons (inner-shell ionization) of the next-lower charge state, $i-1$, leaves the ion in an excited state that may decay by X-ray emission. Thus one may think of three sources of line production – the three adjacent ions $i-1$, i, and $i+1$ – with different line-forming mechanisms associated with each source. It follows that the emissivity of an arbitrary line of ion i can be considered as the sum of three "component emissivities",

$$j_i = n_{i-1}R(i-1 \rightarrow i) + n_i R(i \rightarrow i) + n_{i+1}R(i+1 \rightarrow i). \qquad (1)$$

The notation $R(k \rightarrow i)$ represents the sum of temperature-dependent atomic processes that couple ion k to ion i and which lead to production of the line in question. For example, $R(i \rightarrow i)$ accounts for processes internal to ion i that lead to emission of the line under consideration. Summing over all lines gives the discrete *local* spectrum of ion i, which is thus seen to be the superposition of three "component spectra". In the optically thin limit, the spectrum observed at infinity is proportional to the volume integral of the local emissivity, assuming that allowance for attenuation by the intervening medium is given. In the more general case, line transfer must be considered.

If one could isolate and examine each of the three component spectra, it would be discovered that they are often easily differentiated. This might be expected, since the individual atomic processes that produce them are quite different. In measuring a spectrum from a real plasma, we would like to use the relative weightings of each component to infer something about the physical state of the plasma. Clearly, the charge state distribution (CSD), the temperature, and the ambient radiation field together determine the weightings of the component spectra. We now consider the relation of the CSD and the temperature.

X-ray emitting plasmas in *most* cosmic X-ray sources are characterized either by collisional ionization equilibrium or by photoionization equilibrium. Transient plasmas, which are evolving toward one of these two equilibria, can be thought of as constituting a third class. In collisional ionization equilibrium (CIE), the CSD, hence, the weightings of the component spectra, is determined primarily by the temperature, T.[1] Denote by $(T_{\rm peak})_{\rm CIE}$ the temperature at which the ionic fraction of a given charge state attains its maximum value in CIE. It turns out that, for T near $(T_{\rm peak})_{\rm CIE}$, charge state i itself is the dominant source of X-ray line emission, because of the high efficiency of collisional excitation by the same electron population responsible for collisional ionization. In other words, the second term on the right-hand side of (1) dominates, and the overall spectrum most closely resembles that produced by collisional excitation. alone. The atomic processes that couple adjacent ions, though dictating the ionization balance, produce relatively subtle effects (for example, satellite lines) in the spectra.

In photoionization equilibrium (PIE), as the name implies, the physical conditions in the plasma are controlled, in part, by the radiation field. Obviously, recombination rates depend on T, but, as we will see, T can be parametrized by the ratio of the ionizing flux to the electron density (denoted

[1] Neither the effects on the ionization balance of high densities, nor optical depth effects are considered here. Also, see the Chapter by Rolf Mewe in this Volume.

by ξ). For optically thin irradiated plasmas, the CSD and temperature structure are both functions of ξ. Still, it is often useful to think of line emissivities in terms of T, rather than ξ. In PIE, the T-CSD relationship is such that $(T_{\rm peak})_{\rm PIE} \ll (T_{\rm peak})_{\rm CIE}$. To understand the difference between CIE spectra and PIE spectra, imagine fixing the CSD of a highly ionized plasma while lowering the temperature from $(T_{\rm peak})_{\rm CIE}$ to $(T_{\rm peak})_{\rm PIE}$. Generally speaking, recombination increases as T is lowered, while collisional excitation decreases. Therefore, for highly ionized PIE plasmas, the dominant weighting in (1) shifts from charge state i to charge state $i+1$, and the spectrum produced by such a region most closely resembles that produced by pure recombination.

X-ray spectra from plasmas in photoionization equilibrium actually have a dual character. They may consist of either recombination emission or fluorescent emission. Fluorescence is the radiative decay of an ion from an initial state with an energy above the first ionization potential (§3.7). These highly excited electronic configurations are formed by inner-shell photoionization, so that the production rate of fluorescence emission in charge state i depends on the population density of charge state $i-1$.[2] Note the distinction that has been drawn here: emission from CIE plasmas is dominated by charge state i, while that from PIE plasmas is dominated by those charge states that "surround" charge state i. The dominance of fluorescence over recombination, or *vice versa*, is related to the CSD; highly ionized plasmas are recombination dominated, while colder, less ionized plasmas are fluorescence-dominated. There is nothing deep or mysterious about this. It simply reflects the fact that wavelengths of transitions to the valence shell, which result from recombination cascades, move out of the X-ray band to lower energies as a given element becomes more neutral. By contrast, wavelengths of iron K-shell fluorescence lines, to take an example, change only slightly as iron becomes more neutral, since the atomic potential at radii characteristic of the inner shells is dominated by the nuclear charge, and changes in the screening of the nuclear charge by outer-shell electrons are of small consequence.

The possibility of observing line emission from every charge state of an element — from neutral up to hydrogenic — in the same object, and with the same instrument, is a unique feature of X-ray spectroscopy. In such cases the received X-ray spectrum then appears as a composite of fluorescent and recombination emission. For historical reasons, in the context of compact X-ray sources, the fluorescent component has received more attention than the recombination component, and its presence in X-ray spectra is well established. Recombination lines are not as well studied, and, while there is no controversy over their existence, the observational evidence is sparse. This is easily attributable to two observational factors, [1] the difficulty of extracting lines of low equivalent width from the bright continuum that characterizes

[2] Multiple Auger emission can couple emission in ion i to the ionic fractions of charge states $i-1$, $i-2$, $i-3$, and lower, a complication we ignore here.

sources expected to exhibit recombination spectra, and [2] the low resolving powers of the instruments used to observe these objects to date.

X-ray spectroscopy of photoionized plasmas is the subject of §3. As a prelude, I provide in §2 a brief treatment of Comptonization, the process through which an electron population exchanges energy with a radiation field. I emphasize those aspects of Comptonization theory which pertain to X-ray spectroscopy of compact X-ray sources.[3]

For the more general case of a time-dependent plasma, the T-CSD relationship is case-specific. In a transient plasma, T_{peak} depends upon the initial conditions and the time profile of the temperature. Charge state i may have a relatively short lifetime as the plasma evolves toward equilibrium. During its lifetime, its radiative properties may be characterized by an increased weighting of charge state $i-1$ (transient ionization), of charge state $i+1$ (transient recombination), or of ion i itself. The most commonly treated case involves transient ionization, but, in any case, it is impossible to make more than qualitative generalizations. The topic of transient ionization is taken up in §4.

There is apparently no agreement as to the proper designation for the ionization state of a transient plasma. The most commonly used term is *non-equilibrium ionization*, with its acronym NEI – emphasis on "non-equilibrium". The term *non-ionization equilibrium* is also used. One may raise an objection to the use of this latter term on the grounds that, if the emphasis is applied in the same manner ("non-ionization"), it suggests, confusingly, an equilibrium in which ionization plays no role. Unfortunately, taken literally, the term *non-equilibrium ionization* is synonymous with *transient ionization*, which is a more specific physical scenario than what is implied by "NEI", and appears to exclude transient recombination, which it should not. Consider the term *ionization equilibrium* (IE). Plasmas that deviate from IE are then characterized simply by *non*–IE, i.e., NIE, emphasis on "ionization equilibrium", rather than on "non-ionization". Thus I propose a more systematic nomenclature: CIE, PIE, and NIE, the C, P, and N serving on behalf of the prefixes *collisional*, *photo-*, and *non-*, respectively.

In this Chapter, the focus is on PIE and NIE. As treated here, NIE is a generalization (or an extension) of CIE, in that we will emphasize only collision dominated population kinetics, but retain the explicit time dependence of the CSD. A more thorough treatment might consider time-dependent photoionization and recombination, such as may occur in, for example, an X-ray pulsar. As becomes evident after reading the Chapter by Rolf Mewe, the study of atomic processes relevant to low-density collisionally ionized plasmas is quite advanced, although there is no dearth of problems to attack. The impetus to develop the basic coronal equilibrium theory has, of course, derived from the availability of high-quality solar X-ray data, and from ter-

[3] Compton scattering is treated at a more fundamental level in the Chapter by Malcolm Longair in this Volume.

restrial experiments in which high-temperature plasmas are created (for example, tokamak plasmas). By contrast, note that there are no large-scale X-ray photoionized plasmas in the solar system, and no intensive laboratory research programs involving high-energy radiation dominated plasmas. Consequently, there are currently many more uncertainties associated with X-ray spectroscopy of photoionized plasmas. However, in the spirit of offering a primer, most of this Chapter involves "safe" topics, i.e., elements of the theory of X-ray spectroscopy that are unlikely to be threatened by new observations. As the writing of this Chapter predates the launch of *AXAF* by only a few months, this concern is not entirely groundless. As we begin to look in more detail at X-ray spectra from extrasolar X-ray sources, we may find numerous occasions for emendations not only of our spectral models, but perhaps also of our preconceptions concerning the use and overall aim of X-ray spectroscopy.

2 Comptonization

At photon energies $\hbar\omega$ above $\sim$ 10 keV, the total (summed over all ions) photoelectric cross section of an astrophysical plasma can be smaller than the Compton cross section. Therefore, the opacity can be dominated by Compton scattering. When the electron temperature becomes sufficiently high, or the ionizing radiation field sufficiently intense, such that ions are fully or mostly stripped of their electrons, Compton scattering is the dominant photon-matter interaction, even for lower-energy photons.

Since electrons and photons exchange energy through Compton scattering, the process can modify the spectrum, and it can affect, even control, the electron temperature. Two effects come into play, the Doppler effect and electron recoil. When a photon is scattered by an electron[4] it will undergo a frequency shift $\Delta\omega/\omega \sim v/c \sim (kT/m_e c^2)^{1/2}$ as a result of the Doppler effect. The recoil effect, which imparts energy to the electron in its rest frame, always leads to a "downscatter" of the photon: $\Delta\omega/\omega \sim -\hbar\omega/m_e c^2$. While these individual shifts are, by themselves, negligible, the accumulated effect of repeated scatterings can give rise to substantial spectral deformation. The process by which this occurs is called *Comptonization*. The literature devoted to the various aspects of Compton scattering and Comptonization is quite extensive (see, for example, Holt & McCray 1979, and references therein). There have been numerous efforts to derive analytical solutions that describe the accompanying spectral modifications. However, approximate analytic solutions are usually compared to Monte Carlo simulations, the ultimate theoretical arbiter in this complex situation (see Pozdnyakov, Sobol & Sunyaev 1983).

With some exceptions, such as the Sunyaev-Zeldovich effect, Compton scattering is most often discussed in the context of compact X-ray sources.

[4] Except where otherwise noted, only the non-relativistic regime of Compton scattering will be considered here ($\hbar\omega \ll m_e c^2$ and $kT \ll m_e c^2$).

Comptonization can act as a "thermostat" in the highly ionized, X-ray irradiated gas in the circumsource media of accreting objects. Comptonization also provides a means by which to produce power law continua (the traditional non-thermal spectrum) from purely thermal processes. To begin, let us evaluate the Compton optical depth for an idealized case:

Example: Compton depth through a spherically symmetric accretion flow

Assume steady-state collisionless spherical free fall onto a star of radius R at a total mass accretion rate $\dot{M}$. The optical depth along a radial vector from the stellar surface, of radius R, to a point r is given by

$$\tau_{\rm c}(r) = \sigma_{\rm c} \int_R^r {\rm d}r' \; n_{\rm e}(r'), \tag{2}$$

where $\sigma_{\rm c}$ is the Compton cross section.[5] For simplicity, assume that the plasma is composed only of ionized hydrogen, so that $n_{\rm p} = n_{\rm e}$, where the subscript p refers to protons. Then the electron density can be found from the equation of continuity,

$$\dot{M} = 4\pi r^2 m_{\rm p} v(r) n_{\rm e}(r), \tag{3}$$

where $v(r) = (2GM/r)^{1/2}$. The accretion luminosity L can be defined by an accretion efficiency $\eta_{\rm acc}$, such that $L = \eta_{\rm acc}\dot{M}c^2$. For stars (i.e., stars with surfaces), $\eta_{\rm acc} \sim GM/Rc^2$, which is ~ 0.1 for neutron stars and $\sim 10^{-4}$ for white dwarfs. Substituting for $\dot{M}$ and $v(R)$ in Eq. (3), the density of the accreting gas can be written as

$$n_{\rm e}(r) = \frac{L}{4\pi\eta_{\rm acc} m_{\rm p} c^2 (2GMr^3)^{1/2}}. \tag{4}$$

Upon integration, we find for the Compton scattering depth

$$\tau_{\rm c} = \frac{L\sigma_{\rm c}}{4\pi\eta_{\rm acc} m_{\rm p} c^2 \, (2GMR)^{1/2}} \left[1 - \left(\frac{R}{r}\right)^{1/2}\right]. \tag{5}$$

For most cases of interest $r \gg R$. Inserting typical numbers for the case of accretion onto a neutron star in a luminous X-ray binary, we have, numerically

$$\tau_{\rm c} = 2.1 \; L_{38} \; (R_6)^{-1} \left(\frac{\eta_{\rm acc}}{0.1}\right)^{-1} \left(\frac{M}{{\rm M}_\odot}\right)^{-1/2}, \tag{6}$$

where L_{38} is the luminosity in units of 10^{38} erg s^{-1}, and R_6 is the radius of the neutron star in units of 10^6 cm. The assumption of spherically symmetric

[5] Throughout, we will find it sufficient to use the non-relativistic form of the Compton cross section, i.e., the energy independent Thomson cross section. The subscript c is retained for the sake of formality. For the relativistically correct Klein-Nishina form, see Jauch & Rohrlich (1955).

free fall results in a conservative estimate of the optical depth. For example, if mass accretion proceeds through a disk, asymmetry in the matter distribution, combined with the fact that the inflow velocity is less than the free-fall velocity, results in a much enhanced scattering optical depth near the disk plane. Thus we find that Compton scattering optical depths can often be substantial.

The expression (5) for the Compton depth can be expressed somewhat more elegantly. In terms of the Eddington luminosity ($L_{\rm Edd} = 4\pi GMm_{\rm p}c\sigma_{\rm c}^{-1}$; for a derivation, see Frank, King & Raine 1992), and for $r \gg R$,

$$\tau_{\rm c} = \left(\frac{2}{\eta_{\rm acc}}\right)^{1/2} \frac{L}{L_{\rm Edd}}. \tag{7}$$

This form is often used for accretion onto black holes, where $\eta_{\rm acc}$ cannot be expressed in terms of a stellar radius.

2.1 Energy transfer in a single Compton scatter

In order to establish some notational conventions, and to obtain a formula that we will need for the derivation of the Kompaneets Equation in §2.3, we work through the four-vector derivation of the energy exchanged by a photon and an electron in a scatter observed in the "lab frame". In what follows, boldfaced type denotes an ordinary three-vector, and unit three-vectors are indicated with a caret. Momentum four-vectors will be designated by Π. Let the subscript ω refer to photons, and the subscript e refer to electrons. Primed variables denote post-scattering quantities. The quantities $\beta = v/c$ and $\gamma = (1-\beta^2)^{-1/2}$ are used to represent characteristics of the electron kinematics. We define the four-momentum by $\Pi = (E/c, \mathbf{p})$ [(time component, space component)]. The electron and photon four-momenta before and after the collision are

$$\Pi_{\rm e} = (E/c,\ \mathbf{p}) \qquad \Pi_{\rm e}' = (E'/c,\ \mathbf{p}') \tag{8}$$

$$\Pi_\omega = \frac{\hbar\omega}{c}\,(1,\hat{n}) \qquad \Pi_\omega' = \frac{\hbar\omega'}{c}\,(1,\hat{n}') \tag{9}$$

Conservation of four-momentum requires

$$\Pi_\omega' + \Pi_{\rm e}' = \Pi_\omega + \Pi_{\rm e}, \tag{10}$$

which can be rearranged to isolate the final electron four-momentum,

$$\Pi_{\rm e}' = \Pi_\omega - \Pi_\omega' + \Pi_{\rm e}. \tag{11}$$

Recalling the invariance properties of four-momenta (Jackson 1975), $\Pi_{\rm e}\cdot\Pi_{\rm e} = (mc)^2$ and $\Pi_\omega \cdot \Pi_\omega = 0$, square both sides of Eq. (11) to give

$$\Pi_\omega \cdot \Pi_{\rm e} - \Pi_\omega \cdot \Pi_\omega' - \Pi_{\rm e} \cdot \Pi_\omega' = 0. \tag{12}$$

Substituting (8) and (9) into (12) gives for the energy of the scattered photon

$$\hbar\omega' = \hbar\omega \, \frac{E - c\mathbf{p}\cdot\hat{n}}{E - c\mathbf{p}\cdot\hat{n}' + \hbar\omega(1 - \hat{n}\cdot\hat{n}')}. \tag{13}$$

It will be useful to have an expression for the energy exchange in a collision. Using expression (13), this is

$$\hbar\omega' - \hbar\omega = \hbar\omega \, \frac{c\mathbf{p}\cdot(\hat{n}' - \hat{n}) - \hbar\omega\,(1 - \hat{n}'\cdot\hat{n})}{E - c\mathbf{p}\cdot\hat{n}' + \hbar\omega(1 - \hat{n}'\cdot\hat{n})}. \tag{14}$$

If the collision is viewed in the rest frame of the pre-collision electron ($\mathbf{p} = 0$, $E = mc^2$), then Eq. (13) reduces to the more familiar result for the energy of the photon after losing energy through the effect of electron recoil:

$$\hbar\omega' = \hbar\omega \, \frac{1}{1 + (\hbar\omega/mc^2)(1 - \hat{n}\cdot\hat{n}')}. \tag{15}$$

It is of interest to calculate the highest energy to which a photon may be scattered. The photon gains the most energy when it is scattered by 180° after a head-on collision with the electron. Rewriting (13) for the case of a head-on collision yields

$$\hbar\omega' = \hbar\omega \, \frac{E + cp}{E - c\mathbf{p}\cdot\hat{n}' + \hbar\omega(1 - \hat{n}\cdot\hat{n}')}. \tag{16}$$

The maximum energy exchange corresponds to $\mathbf{p}\cdot\hat{n}' = p$ and $\hat{n}\cdot\hat{n}' = -1$, so the scattered photon has an energy

$$\hbar\omega' = \hbar\omega \, \frac{E + cp}{E - cp + 2\hbar\omega}. \tag{17}$$

Since $E \pm cp = \gamma mc^2(1 \pm \beta)$,

$$\hbar\omega' = \hbar\omega \, \frac{\gamma(1 + \beta)}{\gamma(1 - \beta) + 2(\hbar\omega/mc^2)}. \tag{18}$$

For $\gamma \gg 1$, we can use the approximation $\beta \approx 1 - (2\gamma^2)^{-1}$, so that

$$\hbar\omega' \;=\; \hbar\omega \, \frac{4\gamma^2}{1 + 4\gamma(\hbar\omega/mc^2)}. \tag{19}$$

For low-energy photons, $\hbar\omega' \approx 4\gamma^2\hbar\omega$. In other words, amplification factors of $\sim 4\gamma^2$ are attainable. For example, an IR photon, with $\hbar\omega = 0.1$ eV, can be scattered by a $\gamma = 10^3$ electron to an energy as high as 400 keV. For higher-energy photons scattering off of relativistic electrons, where the second term in the denominator of Eq. (19) exceeds unity, the maximum attainable energy in a single scatter is $\sim \gamma mc^2$.

The effects of Compton scattering on line profiles can get quite complicated (Ross, Weaver & McCray 1978; Sunyaev & Titarchuk 1980). Here is the simplest example, involving a single scatter.

Example: line profile after single scatter in cold plasma (see Pozdnyakov, Sobol & Sunyaev 1979)

Suppose that a monochromatic photon distribution is modified by single Compton scatters, and that the scattering electrons are initially stationary in the observer's frame. What is the resulting line shape? The distribution of scattering angles is given by the Thomson differential cross section, normalized over the interval $[-1, 1]$ in $\cos\theta$,

$$P(\theta)\ \mathrm{d}(\cos\theta) = \frac{3}{8}\ (1+\cos^2\theta)\ \mathrm{d}(\cos\theta) \tag{20}$$

This is related to the frequency distribution of scattered photons $P(\omega')$ according to

$$P(\theta)\ \mathrm{d}(\cos\theta) = P(\omega')\ \mathrm{d}\omega' \tag{21}$$

where, in the limits that $\hbar\omega \ll mc^2$ and $kT \to 0$, the kinematic relation between ω' and θ, according to Eq. (13), can be approximated by

$$\hbar\omega' = \hbar\omega\left[1 - \frac{\hbar\omega}{mc^2}\ (1-\cos\theta)\right]. \tag{22}$$

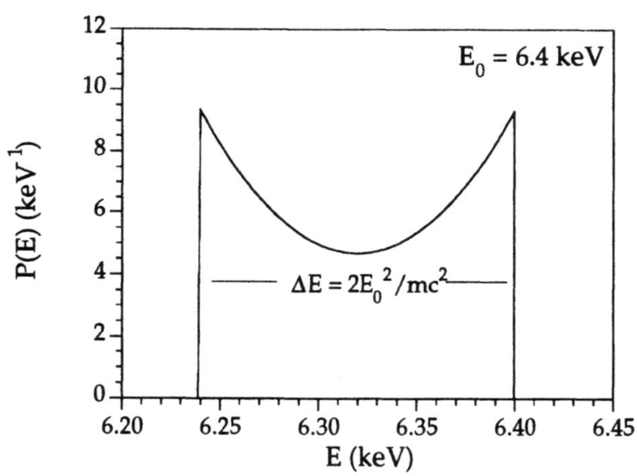

Fig. 1. Normalized line profile profile after single Compton scatter in a cold plasma. Initial line energy is 6.4 keV, comparable to an Fe Kα line from a near-neutral species.

Inverting for $\cos\theta$ yields

$$\cos\theta = 1 - \frac{mc^2}{\hbar\omega}\left(1 - \frac{\omega'}{\omega}\right) \qquad \mathrm{d}(\cos\theta) = \frac{mc^2}{\hbar\omega^2}\ \mathrm{d}\omega' \tag{23}$$

The range of $\cos\theta$ restricts the range of scattered energies to

$$\left(1 - 2\,\frac{\hbar\omega}{mc^2}\right)\omega \le \omega' \le \omega, \tag{24}$$

which shows that the line width is $\Delta\omega' = 2\hbar\omega^2/mc^2$. Therefore, for this range of energies,

$$P(\omega') = \frac{3}{8}\,\frac{mc^2}{\hbar\omega^2}\left[1 + \left(1 - \frac{mc^2}{\hbar\omega}\left(1 - \frac{\omega'}{\omega}\right)\right)^2\right] \tag{25}$$

or, after some algebra

$$P(\omega') = \frac{3}{8}\,\frac{mc^2}{\hbar\omega^2}\left[1 + \left(\frac{mc^2}{\hbar\omega^2}\right)^2 (\omega' - \omega_{\text{min}})^2\right] \tag{26}$$

where $\omega_{\text{min}} = \omega(1 - \hbar\omega/mc^2)$. The line shape is "double-horned", as shown in Figure 1.

2.2 The Compton y parameter

To determine whether or not Comptonization is likely to play an important role in modifying the spectrum of a source, there is a simple attribute of the scattering medium to which we may refer, the Compton y parameter. This parameter is the product of two factors, the number of scatters multiplied by the average fractional energy exchanged per scatter.

First, we want to derive an expression for the average energy transfer in a collision for an isotropic radiation field and a Maxwellian electron distribution at temperature T. From expression (13) for the scattered photon energy in the observer's frame, we have, for $\hbar\omega \ll mc^2$,

$$\frac{\hbar\omega'}{\hbar\omega} \approx \frac{E - c\mathbf{p}\cdot\hat{n}}{E - c\mathbf{p}\cdot\hat{n}'}. \tag{27}$$

For definiteness, assume that $\mathbf{p} = p\hat{z}$. Then

$$\frac{\Delta(\hbar\omega)}{\hbar\omega} = \frac{\cos\theta' - \cos\theta}{(E/cp) - \cos\theta'}, \tag{28}$$

where $\Delta(\hbar\omega) = \hbar\omega' - \hbar\omega$. A proper averaging of this quantity is somewhat tedious, and we will not try it here (see Pozdnyakov, Sobol & Sunyaev 1983). Instead, we simply take advantage of the fact that quite elegant methods of deriving the energy loss rate of an electron undergoing Compton scattering are given in Blumenthal & Gould (1970), and in the textbook by Longair (1992), where it is shown that

$$\frac{\mathrm{d}E}{\mathrm{d}t} = -\,\frac{4}{3}\,\gamma^2\beta^2 c\sigma_{\text{c}} U, \tag{29}$$

where U is the frequency-integrated radiation energy density. We turn this around, and think of the negative of $\mathrm{d}E/\mathrm{d}t$ as the energy gained by the photon population. Since the number of scatters per unit time is $c\sigma_{\mathrm{c}}U/\hbar\omega$, the average energy gained by photons per scatter, in the non-relativistic limit $(\gamma \to 1)$, is

$$\left\langle \frac{\Delta(\hbar\omega)}{\hbar\omega} \right\rangle_{+} = \frac{4}{3}\,\beta^2, \tag{30}$$

where the angled brackets indicate angle-averaging, and the "+" denotes "gain". Note that the terms linear in β have cancelled when averaged over all angles. We are interested in scattering of photons by a Maxwellian distribution of electrons, so it is convenient to write the previous expression as

$$\left\langle \frac{\Delta(\hbar\omega)}{\hbar\omega} \right\rangle_{+} = \frac{4kT}{mc^2}, \tag{31}$$

since β^2 averaged over a Maxwellian is just $3kT/mc^2$. From this gain term we need to subtract an energy loss term that accounts for the recoil effect. From Eq. (15), to lowest order in $\hbar\omega/mc^2$ this is

$$\left[\frac{\Delta(\hbar\omega)}{\hbar\omega} \right]_{-} = -\,\frac{\hbar\omega}{mc^2}\,(1 - \hat{n}\cdot\hat{n}'). \tag{32}$$

The subscript "−" denotes that this is the loss term. An average of this term over angles gives $-\hbar\omega/mc^2$. Thus the net energy exchanged in a collision can be approximated by the expression

$$\left\langle \frac{\Delta\hbar\omega}{\hbar\omega} \right\rangle = \frac{4kT - \hbar\omega}{mc^2}. \tag{33}$$

This suggests that, on average, photons gain energy whenever the electron temperature exceeds $\langle\hbar\omega\rangle/4$, and lose energy otherwise.

To find the overall change in the radiation energy density resulting from Compton scattering, we start with the previous expression, (33). Let $N_\omega\, d\omega$ be the number of photons per unit volume with frequencies in the interval $[\omega,\ \omega + d\omega]$. Then a change in the radiation energy density in that frequency interval is given by

$$\Delta(N_\omega \hbar\omega\ \mathrm{d}\omega) = N_\omega\,\frac{\hbar\omega}{mc^2}\,(4kT - \hbar\omega)\ \mathrm{d}\omega. \tag{34}$$

The time rate of change in the energy density, $\partial U/\partial t$, is found by multiplying the right-hand side of this equation by the Compton scattering rate, $n_{\mathrm{e}}c\sigma_{\mathrm{c}}$, and integrating over frequency, which gives

$$\frac{\partial U}{\partial t} = \frac{n_{\mathrm{e}}\sigma_{\mathrm{c}}}{mc}\int_0^\infty \mathrm{d}\omega\ U_\omega(\omega)\ (4kT - \hbar\omega). \tag{35}$$

We return to this result in §2.4, where it is derived somewhat more rigorously.

We now need an expression for the number of scatters. This is given simply by $\max(\tau_c, \tau_c^2)$, which is derived from a random-walk argument (Rybicki & Lightman 1979). Note that the second term in Eq. (33) corresponds to the recoil energy loss. However, the Compton y parameter is most commonly used for situations in which $4kT \gg \hbar\omega$, so that the Doppler effect is far more important than the recoil effect. In this case, y is given by

$$y = \frac{4kT}{mc^2} \max(\tau_c, \tau_c^2). \tag{36}$$

The role of y in the Comptonization process can be made clearer by the following example (see Rybicki & Lightman 1979). Imagine an isothermal medium of temperature T with electron scattering optical depth $\tau_c > 1$. Assume that a photon of low initial energy ($\hbar\omega_0 \ll kT$) is injected into the medium, and suffers multiple scatters. How does the energy of the emergent photon depend upon τ_c? Equation (33) can be approximated as

$$\frac{\mathrm{d}(\hbar\omega)}{\mathrm{d}N} = \left(\frac{4kT}{mc^2} - \frac{\hbar\omega}{mc^2}\right) \hbar\omega, \tag{37}$$

where N is the number of scatterings, and which is treated here as a continuous variable. We define a dimensionless variable $\epsilon \equiv \hbar\omega/mc^2$, and a dimensionless constant $A \equiv 4kT/mc^2$. Then the previous equation can be written as

$$\frac{\mathrm{d}\epsilon}{\mathrm{d}N} = A\epsilon - \epsilon^2. \tag{38}$$

This integrates to

$$\frac{\epsilon}{A - \epsilon} = \kappa e^{AN}, \tag{39}$$

where the constant of integration $\kappa = \epsilon_0/(A - \epsilon_0)$, which, with the assumptions used here, is well approximated by $\kappa \approx \epsilon_0/A$. Using $AN = y$, and inverting the above expression for $\hbar\omega$, we find

$$\hbar\omega = \hbar\omega_0 \frac{e^y}{1 + (\hbar\omega_0/4kT)\, e^y}. \tag{40}$$

Therefore, for small y, the photon emerges from the scattering medium with an energy that has exponentiated in accordance with y. For large y, we see that, consistent with Eq. (33), $\hbar\omega \to 4kT$, and no further energy exchange occurs. At this point, the Comptonization process is said to be *saturated*. It is worthwhile noting that for large y, the photon energy is independent of $\hbar\omega_0$ — the photon distribution "forgets" its initial condition.

To quantify the saturation of Comptonization, we define a critical y parameter, y_{crit}, such that a photon has attained 1/2 of its energy at saturation ($= 2kT$) when $y = y_{\mathrm{crit}}$. Setting $\hbar\omega = 2kT$ in Eq. (40) gives

$$y_{\mathrm{crit}} = \ln\left(\frac{4kT}{\hbar\omega_0}\right). \tag{41}$$

In terms of a critical electron scattering optical depth, $(\tau_c)_{\rm crit}$, where we assume that $N \sim \tau_c^2$, Eq. (41) shows that

$$(\tau_c)_{\rm crit} = \left(\frac{mc^2}{4kT} \ln \frac{4kT}{\hbar\omega_0}\right)^{1/2}, \tag{42}$$

where Eq. (36) has been used to express $y_{\rm crit}$ in terms of $(\tau_c)_{\rm crit}$. For a given temperature and initial photon energy, this gives an estimate of the scattering optical depth at which saturation begins to occur. For example, if we inject an optical photon ($\hbar\omega_0 \sim 4$ eV) into a gas with temperature $kT \sim 10$ keV, then $(\tau_c)_{\rm crit} \sim 10$. In other words, after about 100 scatters the photon energy begins to approach $4kT$.

This example contains the essence of Comptonization. However, ultimately, we are interested in the broadband spectral properties induced by Comptonization. For that, we need the more sophisticated approach involving the Kompaneets Equation.

2.3 The Kompaneets Equation

The Kompaneets Equation[6] describes the time evolution of the distribution of photon occupancies in the case where photons and electrons are interacting through Compton scattering. In this case photon number is conserved, and the photons will evolve toward a Bose-Einstein distribution with a chemical potential μ which is non-zero. For future reference, the Bose-Einstein monochromatic energy density is

$$(U_\omega)_{\rm B-E} = \frac{\hbar\omega^3}{\pi^2c^3} \frac{1}{e^{(\hbar\omega-\mu)/kT} - 1}. \tag{43}$$

The occupancy $n(\omega)$ in this case is $[\exp(\hbar\omega - \mu)/kT) - 1]^{-1}$. In most cases of interest here, μ will be negative with a large absolute value.

First, we need the Boltzmann Equation that describes the evolution of the photon occupation number. Application of such an equation is valid as long as fractional energy changes are small. The relevant equation is

$$\frac{\partial n(\omega)}{\partial t} = c \int d^3p \, d\Omega \, \frac{d\sigma}{d\Omega} \, [f(\mathbf{p}')n(\omega')(1+n(\omega)) - f(\mathbf{p})n(\omega)(1+n(\omega'))]. \tag{44}$$

The first term describes the rate at which the photon occupancy at frequency ω is increased by scattering photons from frequency ω'. A quantum mechanical effect comes into play here. The scattering rate *into* ω is the sum of two terms: *spontaneous Compton scattering* and *stimulated Compton scattering*. The factor $1+n(\omega)$ accounts for this, the "1" for the former, and the $n(\omega)$ for

[6] The original paper by Kompaneets is somewhat terse, and although derivations can be found in a number of other places (for example, Rybicki & Lightman 1979; Katz 1987; Peebles 1993), a "fleshed out" version is, nevertheless, presented here.

the latter. In other words, since photons are bosons, the stimulated scattering rate is proportional to the occupancy of the state. The second term on the right-hand side of Eq. (44) describes the rate at which the occupation of states with frequency ω are destroyed, and, again, contains a factor to accommodate stimulated scattering. Obviously, n is also a function of time; $n = n(\omega, t)$. For simplicity of notation, the t-dependence has been suppressed.

After we relate $\mathbf{p}'$ to $\mathbf{p}$, below, we will use the basic kinematic relations to relate ω, ω', and $\mathbf{p}$. The electron distribution function is

$$f(\mathbf{p}) = n_e c^3 (2\pi mc^2 kT)^{-3/2}\ e^{-p^2/2mkT}. \tag{45}$$

The first step is to expand $n(\omega')$ around ω to second order,

$$n(\omega') = n(\omega) + (\omega' - \omega)\ \frac{\partial n}{\partial \omega} + \frac{1}{2}\ (\omega' - \omega)^2\ \frac{\partial^2 n}{\partial \omega^2} + \cdots \tag{46}$$

Introduce two dimensionless variables, one to serve as the independent frequency variable, and one to describe the energy exchange following electron-photon interactions:

$$x = \frac{\hbar\omega}{kT} \qquad \delta \equiv \frac{\hbar(\omega' - \omega)}{kT} \tag{47}$$

In terms of x and δ, the expansion of $n(\omega')$ becomes

$$n(x') = n(x) +\ \delta\ \frac{\partial n}{\partial x} + \frac{1}{2}\ \delta^2\ \frac{\partial^2 n}{\partial x^2} + \cdots \tag{48}$$

We can rewrite Eq. (45) as

$$f(E') = n_e c^3 (2\pi mc^2 kT)^{-3/2}\ e^{-E'/kT}, \tag{49}$$

which we expand around E to second order. Since $E' - E = -kT\delta$, and

$$\frac{\partial^k f(E)}{\partial E^k} = (-1)^k (kT)^{-k}\ f(E), \tag{50}$$

the expansion of f takes the simple form

$$f(E') = f(E)\ \left(1 + \delta + \frac{1}{2}\ \delta^2 + \cdots\right). \tag{51}$$

Substituting Eqs (46) and (51) into Eq. (44), we find that the time rate of change of the photon occupancy is the sum of a term proportional to δ and a term proportional to δ^2;

$$\frac{1}{c}\ \frac{\partial n}{\partial t} = \left[\frac{\partial n}{\partial x} + n(1+n)\right]\ I_1 + \left[\frac{1}{2}\ \frac{\partial^2 n}{\partial x^2} + (1+n)\frac{\partial n}{\partial x}\ +\ \frac{1}{2}\ n(1+n)\right]\ I_2, \tag{52}$$

where I_1 and I_2 remain to be evaluated, and are given by

$$I_1 = \int \mathrm{d}^3p \int \mathrm{d}\Omega \, \frac{\mathrm{d}\sigma}{\mathrm{d}\Omega} \, f(E) \, \delta \tag{53}$$

and

$$I_2 = \int \mathrm{d}^3p \int \mathrm{d}\Omega \, \frac{\mathrm{d}\sigma}{\mathrm{d}\Omega} \, f(E) \, \delta^2. \tag{54}$$

An explicit expression for δ in terms of the incident electron momentum is derived easily from the expression for the energy exchange in a collision, which, from (14), is

$$\delta = \frac{xc\mathbf{p} \cdot (\hat{n}' - \hat{n}) - x^2kT \, (1 - \hat{n}' \cdot \hat{n})}{E - c\mathbf{p} \cdot \hat{n}' + xkT(1 - \hat{n}' \cdot \hat{n})} \approx \frac{x\mathbf{p} \cdot (\hat{n}' - \hat{n})}{mc}, \tag{55}$$

where the approximation holds in the non-relativistic limit.

In completing the derivation, we follow Kompaneets in evaluating I_2 explicitly, then deducing I_1 through some sleight-of-hand (see also Rybicki & Lightman 1979). For the integration over d^3p, the coordinate axes can be conveniently oriented so that the polar angle, denoted by ψ, corresponds to the angle between $\mathbf{p}$ and $\hat{n}' - \hat{n}$. Then $\mathrm{d}^3p = 2\pi p^2 \mathrm{d}p \, \sin\psi \, \mathrm{d}\psi$, and

$$I_2 = \int \mathrm{d}^3p \int \mathrm{d}\Omega \, \frac{\mathrm{d}\sigma}{\mathrm{d}\Omega} \left(\frac{x}{mc}\right)^2 f(\mathbf{p}) \, p^2 \cos^2\psi \, |\hat{n}' - \hat{n}|^2. \tag{56}$$

Since the angle between the incident and scattered electrons is independent of $\mathbf{p}$ in the non-relativistic limit, the (five-dimensional) integral can be decomposed into a simpler product of two integrals:

$$I_2 = \left(\frac{x}{mc}\right)^2 \left(\int \mathrm{d}^3p \, f(p) \, p^2 \cos^2\psi\right) \left(\int \mathrm{d}\Omega \, \frac{\mathrm{d}\sigma}{\mathrm{d}\Omega} \, |\hat{n}' - \hat{n}|^2\right). \tag{57}$$

The integral over momentum space becomes

$$\int \mathrm{d}^3p \, \cdots = \frac{2\pi n_{\mathrm{e}}}{(2\pi mkT)^{3/2}} \int_0^\infty \mathrm{d}p \, p^4 e^{-p^2/2mkT} \int_0^\pi \mathrm{d}\psi \, \sin\psi \, \cos^2\psi, \tag{58}$$

which is easily evaluated to give $n_{\mathrm{e}}mkT$. Thus an intermediate expression for I_2 is

$$I_2 = x^2 n_{\mathrm{e}} \frac{kT}{mc^2} \int \mathrm{d}\Omega \, \frac{\mathrm{d}\sigma}{\mathrm{d}\Omega} \, |\hat{n}' - \hat{n}|^2. \tag{59}$$

Now, we orient the coordinate system such that the polar angle θ corresponds to the scattering angle, i.e., $|\hat{n}' - \hat{n}|^2 = 2 \, (1 - \cos\theta)$, and substitute the explicit expression for the Thomson differential cross section:

$$\frac{\mathrm{d}\sigma}{\mathrm{d}\Omega} = \frac{3\sigma_{\mathrm{c}}}{16\pi} \, (1 + \cos^2\theta). \tag{60}$$

The integral in expression (59) becomes

$$\int \mathrm{d}\Omega \, \cdots = \frac{3\sigma_{\mathrm{c}}}{4} \int_0^\pi \mathrm{d}\theta \, \sin\theta \, (1 + \cos^2\theta)(1 - \cos\theta). \tag{61}$$

We are left with

$$I_2 = 2x^2 n_e \sigma_c \, \frac{kT}{mc^2}. \tag{62}$$

We evaluate I_1 by invoking the fact that Compton scattering in a homogeneous infinite medium conserves photon number density. The number density is given by the frequency integral of the density of photon states ($\propto \omega^2$) times the occupancy. Therefore, conservation requires

$$\frac{\mathrm{d}}{\mathrm{dt}} \int_0^\infty \mathrm{d}\omega \, \omega^2 \, n(\omega) = 0. \tag{63}$$

This is equivalent to

$$\frac{\mathrm{d}}{\mathrm{d}t} \int_0^\infty \mathrm{d}x \, x^2 \, n(x) = 0 \quad \Rightarrow \quad \int_0^\infty \mathrm{d}x \, x^2 \, \frac{\partial n}{\partial t} = 0. \tag{64}$$

Since n obeys a conservation law, we can write a continuity equation by defining a "current", $\mathbf{J}$, as follows:

$$\frac{\partial n}{\partial t} + \nabla \cdot \mathbf{J} = 0, \tag{65}$$

or, using the spherical form of the divergence,

$$\frac{\partial n}{\partial t} = -\frac{1}{x^2} \frac{\partial}{\partial x} \left[x^2 \, J(x)\right] = -\frac{2}{x} \, J - \frac{\partial J}{\partial x}. \tag{66}$$

We can think of $J(x)$ as a "radial" flow of quanta in **k**-space.

Equation (66) gives us the functional form of the right-hand side of the equation we seek, the Kompaneets Equation. All we need to do is determine J. But we already have an expression for $\partial n/\partial t$ from Eq. (52), which can be rewritten as

$$\frac{\partial n}{\partial t} = C_1(x) \, \frac{\partial^2 n}{\partial x^2} \; + \; C_2(n,x) \, \frac{\partial n}{\partial x} \; + \; C_3(n,x), \tag{67}$$

and which must have the same functional form as Eq. (66). Evidently, J can have no frequency derivatives higher than first order. If we assume that J contains a term of the form $A(n,x)\ (\partial n/\partial x)$, upon differentiation we would have factors of $(\partial n/\partial x)^2$, which cannot appear. If we consider a term of the form $A(x)\ (\partial n/\partial x)$, then

$$\frac{\partial n}{\partial t} = -A(x) \, \frac{\partial^2 n}{\partial x^2} - \left[\frac{2}{x} \, A(x) + \frac{\partial A(x)}{\partial x}\right] \frac{\partial n}{\partial x}. \tag{68}$$

We still need a term to play the role of $C_3(n,x)$, and the coefficient of $\partial n/\partial x$ needs to be a function of n, as well as of x. The simplest form to try is $J(x) = A(x)\ (\partial n/\partial x) + B(n,x)$. This gives

$$\frac{\partial n}{\partial t} = -A(x) \, \frac{\partial^2 n}{\partial x^2} - \left[\frac{2}{x} \, A(x) + \frac{\partial A(x)}{\partial x} + \frac{\partial B(n,x)}{\partial n}\right] \frac{\partial n}{\partial x} - \frac{\partial B(n,x)}{\partial x}, \tag{69}$$

which is acceptable. Therefore,

$$J(x) = g(x) \left[\frac{\partial n}{\partial x} + h(n,x)\right], \tag{70}$$

where g and h are to be determined.

If $J = 0$, then $\partial n/\partial t = 0$ and n is given by the Bose-Einstein occupancy. From Eq. (43), in terms of x, $n(x) = (e^{x-\alpha} - 1)^{-1}$, where $\alpha \equiv \mu/kT$, and where α is negative with a large absolute value. When $J = 0$, from Eq. (70), $\partial n/\partial x + h(n,x) = 0$, from which it follows that

$$h(n,x) = n(1+n). \tag{71}$$

Updating, we now have for J,

$$J(x) = g(x) \left[\frac{\partial n}{\partial x} + n(1+n)\right], \tag{72}$$

where g is still to be determined.

Rewriting Eq. (52) by substituting the explicit form of I_2 gives

$$\frac{1}{c}\frac{\partial n}{\partial t} = \left[\frac{\partial n}{\partial x} + n(1+n)\right] I_1 + x^2 n_e \sigma_c \frac{kT}{mc^2}\left[\frac{\partial^2 n}{\partial x^2} + 2(1+n)\frac{\partial n}{\partial x} + n(1+n)\right]. \tag{73}$$

This equation can be compared with that obtained by substituting our latest version of J into Eq. (66), yielding

$$\frac{\partial n}{\partial t} = -\left[\frac{\partial^2 n}{\partial x^2} + \left(1 + 2n + \frac{2}{x}\right)\frac{\partial n}{\partial x} + \frac{2}{x}\, n(1+n)\right] g - \left[\frac{\partial n}{\partial x} + n(1+n)\right]\frac{\partial g}{\partial x}. \tag{74}$$

By comparing the coefficients of $\partial^2 n/\partial x^2$, we see that

$$g(x) = -cx^2 n_e \sigma_c \frac{kT}{mc^2}. \tag{75}$$

With an explicit form for g, we now have an exact form for J, obviating the need to explicitly evaluate I_1.[7]

Now we substitute J into Eq. (66) to give the Kompaneets Equation:

$$\frac{\partial n}{\partial t} = n_e c \sigma_c \frac{kT}{mc^2}\frac{1}{x^2}\frac{\partial}{\partial x}\left[x^4\left(\frac{\partial n}{\partial x} + n + n^2\right)\right]. \tag{76}$$

Typically, one defines a dimensionless time variable,

$$t_c = n_e c \sigma_c \int^t \mathrm{d}t'. \tag{77}$$

[7] If you are curious, the value of I_1 turns out to be $(kT/mc^2) n_e \sigma_c x(4-x)$, which can be obtained by substituting g and $\partial g/\partial x$ back into Eq. (74), then comparing with Eq. (73).

The quantity t_c can be thought of as a time given in units of a "Compton scattering time" $(n_e c\sigma_c)^{-1}$. Finally, in dimensionless form, the Kompaneets Equation is

$$\frac{\partial n}{\partial t_c} = \frac{kT}{mc^2}\frac{1}{x^2}\frac{\partial}{\partial x}\left[x^4\left(\frac{\partial n}{\partial x} + n + n^2\right)\right]. \tag{78}$$

By design, the Bose-Einstein distribution (including the special case of a Planck distribution) is a steady-state solution to the Kompaneets Equation, as can be easily verified.

Blandford (1990) shows how the three terms in the Kompaneets Equation can be interpreted in terms of the diffusion of photons in frequency space, which was described above by $J(x)$. For example, consider the recoil of an electron in a Compton scatter. From the expression for the fractional energy exchange per collision, (Eq. 33), the energy lost by photons through recoil is $\Delta x/x = -(kT/mc^2)x$. Recalling also the Compton scattering frequency, $n_e c\sigma_c$, we have

$$\frac{\mathrm{d}x}{\mathrm{d}t} = -n_e c\sigma_c \frac{kT}{mc^2}x^2. \tag{79}$$

The current associated with the recoil is just $n(x)(\mathrm{d}x/\mathrm{d}t)$, so we can write

$$J_{\mathrm{recoil}}(x) = -n_e c\sigma_c \frac{kT}{mc^2}x^2 n(x). \tag{80}$$

Substituting this into the continuity equation for J, (Eq. 66), gives

$$\frac{\partial n(x)}{\partial t} = n_e\sigma_c c \frac{kT}{mc^2}\frac{1}{x^2}\frac{\partial}{\partial x}\left[x^4 n(x)\right]. \tag{81}$$

In terms of the Compton scattering time, this is

$$\frac{\partial n(x)}{\partial t_c} = \frac{kT}{mc^2}\frac{1}{x^2}\frac{\partial}{\partial x}\left[x^4 n(x)\right], \tag{82}$$

which shows that the second term in the Kompaneets Equation accounts for the electron recoil effect.

Example: Unsaturated Comptonization

More realistic astrophysical settings require us to find steady-state solutions to a modified version of the Kompaneets Equation, in which photons, after being produced, are upscattered and allowed to escape from a finite medium. To represent the production of soft photons, one introduces a source term $Q(x)$, which describes the number of source photons produced per Compton scattering time (t_c) per state. Let $Q(x) = Q_0(x)$ for $x < x_s < 1$, and $Q(x) = 0$ for $x > x_s$. In other words, assume a soft photon input, with negligible production by the source above the frequency x_s. Assume that the escape probability per Compton scatter is just the inverse of the average number of scatters. We neglect the stimulated scattering term ($n \ll 1$). Then, from

Eq. (78), the modified Kompaneets Equation (Shapiro, Lightman & Eardley 1976) for this case is

$$\frac{\partial n}{\partial t_c} = \frac{kT}{mc^2}\frac{1}{x^2}\frac{\partial}{\partial x}\left[x^4\left(\frac{\partial n}{\partial x}+n\right)\right] + Q(x) - \frac{n}{\max(\tau_{es},\ \tau_{es}^2)}. \tag{83}$$

We want to find a steady-state solution, so set $\partial n/\partial t_c = 0$. We now consider the spectrum for frequencies far above the cutoff x_s. Therefore, although we have taken the trouble to define the source term, we now disregard it. Let $x \gg x_s$. Then

$$\max(\tau_{es},\ \tau_{es}^2)\ \frac{kT}{mc^2}\frac{\partial}{\partial x}\left[x^4\left(\frac{\partial n}{\partial x}+n\right)\right] - x^2 n = 0, \tag{84}$$

or, parametrizing by the Compton y parameter,

$$\frac{1}{4}\,y\,\frac{kT}{mc^2}\frac{\partial}{\partial x}\left[x^4\left(\frac{\partial n}{\partial x}+n\right)\right] - x^2 n = 0. \tag{85}$$

For $x \gg 1$, the second term can also be neglected, and we have

$$\frac{\partial n}{\partial x} + n = \frac{constant}{x^4}. \tag{86}$$

The solution for n approaches e^{-x} asymptotically with x. Therefore, I_ω, which is proportional to $\omega^3 n$) has the Wien shape in this spectral regime. Next, consider frequencies such that $x \gg x_s$ but $x \ll 1$. In that case, we may neglect n compared to $\partial n/\partial x$. The modified Kompaneets Equation is

$$\frac{1}{4}\,y\,\frac{kT}{mc^2}\frac{\partial}{\partial x}\,x^4\left(\frac{\partial n}{\partial x}\right) - x^2 n = 0 \tag{87}$$

Trying a power law form $n \propto x^m$, we find

$$m = -\frac{3}{2} \pm \left(\frac{9}{4}+\frac{4}{y}\right)^{1/2} \tag{88}$$

Thus we find the important result that a power law continuum can arise from thermal processes (discussed by Katz 1976). For small y, take the negative root. [For large y, Comptonization becomes saturated, and $m = 0$ corresponds to the low-frequency limit of the Wien spectrum (Rybicki & Lightman 1979)]. If we join these two solutions graphically, the resulting spectral shape resembles the curve shown in Figure 2. If one obtains the slope and the cutoff from observed spectra, and assumes that unsaturated Comptonization accounts for the spectral shape, then the electron temperature and the Compton depth can be simultaneously determined. Sunyaev & Titarchuk (1980), for example, use this technique to interpret the high-energy continuum spectrum of the black-hole candidate Cygnus X-1.

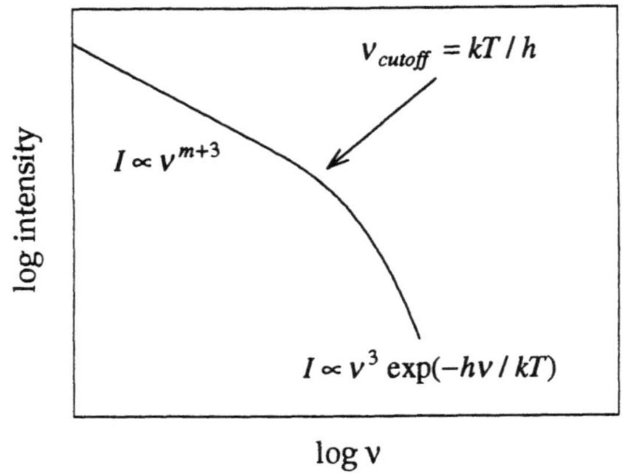

Fig. 2. Schematic representation of the high-energy portion of a spectrum modified by unsaturated Comptonization. A power law is joined smoothly to a Wien "tail", the high-frequency limit.

2.4 Compton heating and cooling

This section and the one following concern the energy balance in plasmas that are ionized to the extent that atomic processes (bound-bound, free-bound, or bound-free transitions) are unimportant. We have shown that the photon field can lose or gain energy from the electron population (§2.2). We rederive Eq. (35) as another example of the power of the Kompaneets Equation, and as a way to provide a more rigorous determination of the relationship between the photon spectrum and the electron temperature.

From the expression for the monochromatic radiation energy density, $U_\omega = (\hbar\omega^3/\pi^2 c^3)\, n(\omega)$, the total radiation energy density is

$$U = \int_0^\infty d\omega\, \frac{\hbar\omega^3}{\pi^2 c^3}\, n(\omega). \tag{89}$$

Expressing U in terms of $x \equiv \hbar\omega/kT$,

$$U = \frac{(kT)^4}{\pi^2(\hbar c)^3} \int_0^\infty dx\, x^3\, n(x). \tag{90}$$

First, we want to compare with an earlier result for the exponential increase of photon energy after injection into a hot plasma [see Eq. (40)]. Here we work out the time development of the radiation energy density. By differentiating U with respect to time,

$$\frac{\partial U}{\partial t} \propto \int_0^\infty dx\, x^3 \frac{\partial n}{\partial t}, \tag{91}$$

we see that the Kompaneets Equation can be used to replace $\partial n/\partial t$ in the integrand with terms involving frequency derivatives. Assume that $n \ll 1$ initially, so that only the first term on the right-hand side of the Kompaneets Equation is needed. Then

$$\frac{\partial U}{\partial t} = \frac{(kT)^4}{\pi^2(\hbar c)^3}\, n_{\rm e} c \sigma_{\rm c}\, \frac{kT}{mc^2} \int_0^\infty {\rm d}x\; x\, \frac{\partial}{\partial x}\left(x^4\, \frac{\partial n}{\partial x}\right). \tag{92}$$

Integrating by parts gives for the integral over x

$$\int {\rm d}x \cdots = -\int_0^\infty {\rm d}x\; x^4\, \frac{\partial n}{\partial x}. \tag{93}$$

A second integration by parts gives

$$\int {\rm d}x\; \cdots = 4\int_0^\infty {\rm d}x\; x^3\, n(x) = 4\, \frac{\pi^2(\hbar c)^3}{(kT)^4}\, U, \tag{94}$$

referring back to Eq. (90). We are left with the trivial differential equation,

$$\frac{\partial U}{\partial t} = n_{\rm e} c \sigma_{\rm c}\, \frac{4kT}{mc^2}\, U. \tag{95}$$

Recalling the definition of the Compton scattering time, this becomes

$$U = U_0\, \exp\left(\frac{4kT}{mc^2}\, t_{\rm c}\right). \tag{96}$$

This is reminiscent of our earlier result for the rate at which the energy of a photon increases as it traverses a medium of Compton depth $\tau_{\rm c}$. In fact, the results are essentially identical if we make the assignment $n_{\rm e}\sigma_{\rm c}ct = N$, where N is the number of scatterings ($\sim \tau_{\rm c}^2$) suffered in time t. The e-folding time for the increase of the radiation energy density is

$$t_{\rm rad} = \frac{mc^2}{4kT}\, t_{\rm c} = \frac{mc}{4\sigma_{\rm c} n_{\rm e} kT}. \tag{97}$$

Now, consider the rate at which the plasma is heated by Compton scattering. Define a Compton heating function $\Gamma_{\rm c}$ (c.g.s. dimensions erg s^{-1}) according to

$$n_{\rm e}\Gamma_{\rm c} = -\frac{\partial U}{\partial t}. \tag{98}$$

Thus we are reversing the sense of the Kompaneets Equation to reflect the different "point of view". Equivalently, this is no more than a statement of energy conservation. Modifying Eq. (92) in order to account for the heating effect of Compton recoil, we find for the volumetric heating rate

$$n_{\rm e}\Gamma_{\rm c} = -\frac{(kT)^4}{\pi^2(\hbar c)^3}\, n_{\rm e} c \sigma_{\rm c}\, \frac{kT}{mc^2} \int_0^\infty {\rm d}x\; x\, \frac{\partial}{\partial x}\left[x^4\left(\frac{\partial n}{\partial x} + n\right)\right]. \tag{99}$$

Two integrations by parts, as before, gives for the integral of the first term in the integrand

$$\int_1 \mathrm{d}x \ \cdots = 4\int_0^\infty \mathrm{d}x\ x^3 n(x) = 4\,\frac{\pi^2(\hbar c)^3}{(kT)^4}\int_0^\infty \mathrm{d}\omega\ U_\omega . \qquad (100)$$

For the integral of the second term, a single integration by parts gives

$$\int_2 \mathrm{d}x \ \cdots = -\int_0^\infty \mathrm{d}x\ x^4 n(x) = -\,\frac{\pi^2(\hbar c)^3}{(kT)^5}\int_0^\infty \mathrm{d}\omega\ \hbar\omega\ U_\omega . \qquad (101)$$

Therefore, we find, in agreement with our result from §2.2, the simple expression

$$n_\mathrm{e}\Gamma_\mathrm{c} = \frac{n_\mathrm{e}\sigma_\mathrm{c}}{mc}\int_0^\infty \mathrm{d}\omega\ U_\omega\ (\hbar\omega - 4kT). \qquad (102)$$

Thus the Compton heating term, when used in the equation of energy balance, actually expresses a *net* heating rate — the excess of Compton heating over Compton cooling. The result (Eq. 102) is, of course, not exact. For example, when considering the effect of scattering of very-high-energy photons, the Klein-Nishina cross section must be used. For the purposes at hand, we work with the original equation derived by Kompaneets, in which case, Eq. (102) follows, *if* stimulated scattering is neglected.

2.5 The Compton temperature

Equation (102) shows that for sufficiently low temperatures, the gas is heated by the electron recoil effect, while at high temperatures, the gas cools through inverse Compton scattering. Therefore, there is a temperature at which Compton heating and Compton cooling balance, the *Compton temperature*, given by

$$4kT_\mathrm{c} = \frac{\int \mathrm{d}\omega\ \hbar\omega\ U_\omega}{\int \mathrm{d}\omega\ U_\omega}. \qquad (103)$$

It is important to realize that the Compton temperature does not depend on the total flux, just the spectral shape. Let us calculate a few examples, since the Compton temperature arises in many applications.

Example: Planck spectrum

First, consider a Planck spectrum with a blackbody temperature kT_BB, which may be diluted (the dilution factor cancels). We will need to evaluate two integrals of the form

$$I(m) = \int_0^\infty \mathrm{d}x\ \frac{x^m}{e^x - 1}. \qquad (104)$$

This can be evaluated by rewriting in the form

$$I(m) = \int_0^\infty \mathrm{d}x \, \frac{e^{-x} x^m}{1 - e^{-x}}, \tag{105}$$

then Taylor expanding $(1 - e^{-x})^{-1}$, which yields

$$I(m) = m! \left(\sum_{p=1}^{\infty} p^{-(m+1)} \right), \tag{106}$$

where the factor in parentheses is the Riemann zeta function (denoted by ζ; see Abramowitz & Stegun 1972; hereafter, AS72). Therefore,

$$I(m) = m! \, \zeta(m+1). \tag{107}$$

Referring back to Eq. (103), substituting the Planck distribution gives

$$kT_{\mathrm{c}} = \frac{kT_{\mathrm{BB}}}{4} \frac{\int_0^\infty \mathrm{d}x \, (1 - e^{-x})^{-1} \, x^4 \, e^{-x}}{\int_0^\infty \mathrm{d}x \, (1 - e^{-x})^{-1} \, x^3 \, e^{-x}}, \tag{108}$$

which, using Eqs (105) and (107) can be simplified to

$$kT_{\mathrm{c}} = \frac{kT_{\mathrm{BB}}}{4} \frac{I(4)}{I(3)} = \frac{\zeta(5)}{\zeta(4)} \, kT_{\mathrm{BB}} \approx 0.97 \, kT_{\mathrm{BB}}. \tag{109}$$

At this point it is worth making the distinction between Eq. (103) and the commonly quoted $4kT_{\mathrm{c}} = \langle \hbar\omega \rangle$. If by $\langle \hbar\omega \rangle$, we intend the mean photon energy, then it is given by

$$\frac{\int \mathrm{d}\omega \, U_\omega}{\int \mathrm{d}\omega \, (\hbar\omega)^{-1} \, U_\omega}, \tag{110}$$

not the ratio in Eq. (103). For example, if we used the formula $4kT_{\mathrm{c}} = \langle \hbar\omega \rangle$ to find the Compton temperature for a Planck distribution, we would find $kT_{\mathrm{c}} \approx 0.68 \, kT_{\mathrm{BB}}$, a substantial error. As long as we agree in this context to equate $\langle \hbar\omega \rangle$ to the ratio in Eq. (103), there will be no problem.

Example: Power law with exponential cutoff

The "cutoff power law", $U_\omega \propto \omega^\alpha \, e^{-\hbar\omega/\epsilon_0}$, with $\alpha \geq 0$, is typical of continuum spectra for X-ray pulsars (White, Swank & Holt 1983). The "cutoff energy" is given by ϵ_0.

$$kT_{\mathrm{c}} = \frac{1}{4} \frac{\int_0^\infty \mathrm{d}\omega \, \hbar\omega \, \omega^\alpha \, e^{-\hbar\omega/\epsilon_0}}{\int_0^\infty \mathrm{d}\omega \, \omega^\alpha \, e^{-\hbar\omega/\epsilon_0}} = \frac{\epsilon_0}{4} \frac{\int_0^\infty \mathrm{d}x \, x^{\alpha+1} \, e^{-x}}{\int_0^\infty \mathrm{d}x \, x^\alpha \, e^{-x}}. \tag{111}$$

In terms of the gamma function,

$$kT_{\mathrm{c}} = \frac{\epsilon_0}{4} \frac{\Gamma(\alpha+2)}{\Gamma(\alpha+1)}. \tag{112}$$

Using the recurrence relation $\Gamma(a+1) = a\Gamma(a)$ (AS72), we have simply

$$kT_c = (\alpha + 1)\, \frac{\epsilon_0}{4}. \tag{113}$$

For example, a pure exponential ($\alpha = 0$) gives $kT_c = \epsilon_0/4$. A Wien shape ($\alpha = 3$) gives $kT_c = \epsilon_0$, which is obviously quite close to the Planck value found above. Of course, this result is consistent with our earlier discussion of the evolution of the spectrum to the Bose-Einstein distribution in the limit where stimulated scattering is unimportant. Recall that this latter stipulation, equivalent to ignoring the n^2 term in the Kompaneets Equation, was used in deriving Γ_c above. The neglect of stimulated processes is equivalent to using the Wien distribution in place of the Bose-Einstein distribution.

3 Spectroscopy of X-ray Photoionized Plasmas

Accretion powered X-ray sources constitute the most important applications of the theory of X-ray photoionized plasmas. The power contained in the radiation continuum originates in a compact region, as evidenced, for example, by rapid, high-amplitude time variability. In accretion powered objects, the radiation energy is produced through the conversion of gravitational potential energy near the surface of a collapsed object.[8] Circumsource material, some fraction of which is destined to accrete onto the compact star, intercepts the central X-ray continuum, and *reprocesses* it into line emission, recombination continua, two-photon continua, and bremsstrahlung emission. Interactions of the primary continuum with the accreting gas thus "deform" the continuum spectrum in a way which allows us, in principle, to infer the physical state and geometrical distribution of the circumsource material.

The enormous variety of ways in which the continuum can be deformed, or imprinted, by its interaction with ionized gas makes the study of X-ray photoionized plasmas especially rich. Consequently, we are forced to narrow somewhat the scope of this Chapter, and neglect some major topics (e.g., Compton reflection, line transfer, absorption spectroscopy). Emphasis is placed instead upon selected aspects of atomic emission line spectroscopy.

As discussed in the Introduction, the character of X-ray line spectra in X-ray photoionized plasmas is determined partly by recombination processes, and partly by fluorescence. In either case, the summed line power of a given ion can be related to the local flux in the ionizing continuum of that ion. This is not to say that X-ray lines in accretion powered objects are powered exclusively by photoionization. Admittedly, it would be difficult to contrive an origin *other* than fluorescence following photoionization for the near-neutral iron K lines observed near 6.4 keV in, for example, AGN (Nandra et al. 1989) and X-ray pulsars (White, Swank & Holt 1983). On the other hand,

[8] An additional source of power is the spin energy of a Kerr black hole. See the review article by Blandford (1990).

there is much less evidence that would allow us to support or to refute the assertion that X-ray *resonance* lines in compact X-ray sources are produced by recombination following photoionization.

In the 1980s, the failure of plasma emission models to explain line spectra from X-ray binaries was taken as evidence that the emission line regions are not in CIE (Kahn, Seward & Chlebowski 1984; Vrtilek et al. 1986), but attempts to interpret the spectra in terms of recombination were also unsuccessful. More recently, with the higher statistical quality afforded by the *ASCA* observatory,[9] convincing evidence for recombination dominated line emission in X-ray binaries has begun to accumulate (Angelini et al. 1995; Liedahl & Paerels 1996; Sako et al. 1998a). The situation for Seyfert galaxies is less clear. Observations of "warm absorbers" in Seyfert 1 galaxies, and their response to variations in the X-ray continuum, provide compelling evidence that the gas is photoionization dominated (e.g., Otani et al. 1996). The existence of soft X-ray recombination emission is likely, but has yet to be convincingly identified (George et al. 1998, and references therein). In Seyfert 2 galaxies, while there is no question that observable soft X-ray line emission exists, its nature is uncertain. The best example is the *ASCA* spectrum of NGC 1068. Ueno et al. (1994) fit the soft component to a two-component CIE plasma, while Netzer & Turner (1997) argue that the spectrum is consistent with photoionization. The fact that either model is statistically acceptable should warn against overinterpreting the data. Sako et al. (1998b) argue that neither pure PIE nor pure CIE can account for the soft emission. Iwasawa, Fabian & Matt (1997) suggest that Fe XXV and Fe XXVI K lines result from resonant scattering of the central continuum. This confusing state of affairs is likely to be ameliorated when high-resolution spectra become available in the near future.

3.1 X-ray nebulae

The theory of X-ray photoionization has been investigated extensively. Following the suggestion that line emission in the X-ray binary Scorpius X-1 is likely to be affected by the presence of a hard X-ray source (Shklovsky 1967), theoretical efforts to determine the response of a gas to hard X rays were begun in earnest. These models are called *X-ray nebular models*. They represent the extreme case of pure photoionization heating, that is, no account is taken of additional heating sources, such as shocks or magnetic-field annihilation.

Models of static gas clouds irradiated by hard X rays have been developed to address the basic issues of heating, cooling, ionization, recombination, and the global structure of X-ray photoionized gases (Tarter, Tucker, & Salpeter 1969; Halpern & Grindlay 1980; Kallman & McCray 1982; Netzer 1990, and references therein). Such a model begins with a cloud of specified size or

[9] Results from the *ASCA* mission are discussed by Jelle Kaastra, this volume.

column density, shape, and elemental composition placed in the vicinity of a point source of continuum X rays. Usually, either the particle density or the gas pressure is fixed. Local heating is dominated by the thermalization of photoelectrons and Auger electrons produced through photoionization by the continuum source. Photoelectrons and Auger electrons are assumed to deposit their energy at the site of photoionization, hence, suprathermal electrons are not treated explicitly. Energy flow throughout the gas is in the form of radiation. The charge state distribution is determined by a balance between photoionization and recombination. Level populations may be determined by the explicit solution of the rate equations. Typically, the model plasma is free of gravitational[10] and magnetic fields. The primary X-ray spectrum is modified in its passage through the nebula by geometrical dilution and absorption. The spectrum observed at infinity is, therefore, a superposition of the modified X-ray point source spectrum and the "diffuse" spectrum of the irradiated gas. These various aspects constitute the *nebular approximation*. The physical manifestation of a model of this type is referred to as an *X-ray nebula*, in analogy with, for example, a planetary nebula, in which a quasi-spherical distribution of gas is dominated by the influence of a central point source of UV continuum emission.

In calculating the CSD, the temperature, which determines the magnitude of the recombination rate coefficients, must be known. But, since the local heating and cooling rates depend upon the CSD, the energy equation is coupled to the equations of ionization balance. Among the aims of nebular model calculations is to determine a self-consistent solution to these equations, which provides, among other things, the CSD as a function of temperature. Several instructive examples are provided in Kallman & McCray (1982). As mentioned in the Introduction, it is found that $(T_{\rm peak})_{\rm PIE} \ll (T_{\rm peak})_{\rm CIE}$. Another way of stating this is to say that the gas is overionized relative to CIE. We now investigate this quantitatively, and introduce the concept of the ionization parameter.

3.2 The ionization parameter: overionization in the nebula

A measure of the relative importance of collisional processes and photoionization is provided by the *ionization parameters*, $\xi = 4\pi F/n$ (Tarter, Tucker & Salpeter 1969), or $\Xi = F/nkT_{\rm e}c$ (Krolik, McKee & Tarter 1981). In these expressions, F is the ionizing energy flux, n is the particle number density, $T_{\rm e}$ is the electron temperature, and the other symbols have their conventional meanings. X-ray emission in photoionized plasmas occurs for values of ξ in the range $10 - 10^4$ and Ξ in the approximate range $1 - 10$. Since Ξ is roughly the ratio of radiation energy density to thermal energy density, X-ray emission line regions in photoionized nebulae are said to be *radiation dominated*.

[10] Models of accretion disk coronae, in which the vertical structure of an irradiated disk is calculated, also include the effects of a gravitational field (Ko & Kallman 1991; Raymond 1993; Murray et al. 1994).

To see how the ionization parameter arises, first write the equation of ionization equilibrium, including the photoionization rate β, the collisional ionization rate C_i, and the recombination, α_{i+1}, where we represent the sum of dielectronic recombination and radiative recombination by a single term:

$$n_e C_i n_i + \beta_i n_i = n_e \alpha_{i+1} n_{i+1}. \tag{114}$$

The monochromatic radiation energy density at energy ϵ, and at a distance r from a point source, is given by

$$U_\epsilon(\epsilon) = \frac{L_\epsilon(\epsilon)}{4\pi r^2 c} = \frac{L f_\epsilon(\epsilon)}{4\pi r^2 c}, \tag{115}$$

where $f_\epsilon(\epsilon)$ is a normalized spectral shape function,

$$\int \mathrm{d}\epsilon \; f_\epsilon(\epsilon) = 1. \tag{116}$$

Denote the photoionization cross section for charge state i by $\sigma_i(\epsilon)$, and the photoionization threshold energy by χ_i. Then, the rate at which charge state i is photoionized into charge state $i+1$ is given by

$$\beta_i = \int_{\chi_i}^{\infty} \mathrm{d}\epsilon \; c\sigma_i(\epsilon)\epsilon^{-1} \; U_\epsilon(\epsilon) = \frac{L}{r^2}\left[\frac{1}{4\pi}\int_{\chi_i}^{\infty} \mathrm{d}\epsilon \; \epsilon^{-1}\sigma_i(\epsilon) f_\epsilon(\epsilon)\right], \tag{117}$$

which can be abbreviated as

$$\beta_i = \frac{L}{r^2}\,\Phi_i. \tag{118}$$

The equation of ionization balance becomes

$$n_e C_i n_i + \frac{L}{r^2}\,\Phi_i \; n_i = n_e \alpha_{i+1} n_{i+1}. \tag{119}$$

At this point, we introduce the *ionization parameter* (Tarter, Tucker & Salpeter 1969),

$$\xi = \frac{L}{n_e r^2}. \tag{120}$$

The c.g.s. dimensions of ξ are erg cm s^{-1}. Substituting into Eq. (119) gives

$$(C_i + \xi\Phi_i)\; n_i = \alpha_{i+1} n_{i+1}. \tag{121}$$

Therefore, we see that formulating the problem in this way allows us to "collapse" three parameters — the ionizing luminosity, the distance from the source, and the electron density — into one. After some rearrangement of Eq. (121), the equations of ionization balance become

$$\frac{n_{i+1}}{n_i} = \frac{C_i}{\alpha_{i+1}}\left(1 + \frac{\xi\Phi_i}{C_i}\right). \tag{122}$$

Recognizing the factor C_i/α_{i+1} as the CIE value of n_{i+1}/n_i, we have

$$\left(\frac{n_{i+1}}{n_i}\right)_{\rm PIE} = \left(\frac{n_{i+1}}{n_i}\right)_{\rm CIE} \left(1 + \frac{\xi \Phi_i}{C_i}\right). \tag{123}$$

Therefore, we may take the magnitude of $\xi\Phi_i/C_i$ compared to unity as a measure of the degree of overionization.

Let's work this out numerically for a simple case: hydrogenic iron (χ = 9.3 keV) irradiated by a 1 keV diluted blackbody flux ($kT_{\rm BB}$=1 keV). The spectral shape function, which satisfies Eq. (116), is

$$f_\epsilon = \frac{15}{(\pi k T_{\rm BB})^4} \frac{\epsilon^3}{e^{\epsilon/kT_{\rm BB}} - 1}. \tag{124}$$

Assume that the photoionization cross section can be adequately represented by $\sigma(\epsilon) = \sigma(\chi)(\epsilon/\chi)^{-3}$. Combining these expressions into the integrand for Φ_i [refer to Eqs (117) and (118)] gives

$$\int_\chi^\infty {\rm d}\epsilon \, \frac{1}{\epsilon} \frac{1}{e^{\epsilon/kT_{\rm BB}} - 1}. \tag{125}$$

The integral can be evaluated as a sum of first exponential integrals:

$$\Phi_i = \frac{15\sigma(\chi)\chi^3}{4\pi^5 (kT_{\rm BB})^4} \sum_{n=0}^\infty E_1\left[\frac{(n+1)\chi}{kT_{\rm BB}}\right]. \tag{126}$$

Since $\chi/kT_{\rm BB}$ is significantly larger than unity, we can take advantage of the asymptotic expansion of E_1 (AS72) to find $E_1(z) \approx z^{-1}e^{-z}$, which allows us to neglect all but the first term in the summation. Therefore,

$$\Phi_i = \frac{15}{4\pi^5} \frac{\sigma(\chi)\chi^2}{(kT_{\rm BB})^3} e^{-\chi/kT_{\rm BB}}. \tag{127}$$

Taking 9×10^{-21} cm^2 for the threshold cross section (Saloman, Hubble & Scofield 1988),

$$\Phi_i = 5.4 \times 10^{-16} \ {\rm cm^2 \ erg^{-1}}. \tag{128}$$

The collisional ionization rate coefficient is approximately (Shull & van Steenberg 1982)

$$C(T_{\rm e}) = 2.6 \times 10^{-16} \ T_{\rm e}^{1/2} \ e^{-\chi/kT_{\rm e}} \ {\rm cm^3 \ s^{-1}}, \tag{129}$$

where the electron temperature $T_{\rm e}$ is in Kelvins. The magnitude of $\xi\Phi_i/C_i$ is plotted against the electron temperature for two values of ξ in Figure 3. Clearly, collisional ionization is entirely negligible under the conditions found in X-ray nebulae.

In §2, we discussed the processes of Compton heating and cooling. It was shown that, if these processes are dominant, the Compton temperature is independent of the local photon flux, hence, independent of position in the nebula. Suppose, however, that additional cooling processes contribute to the

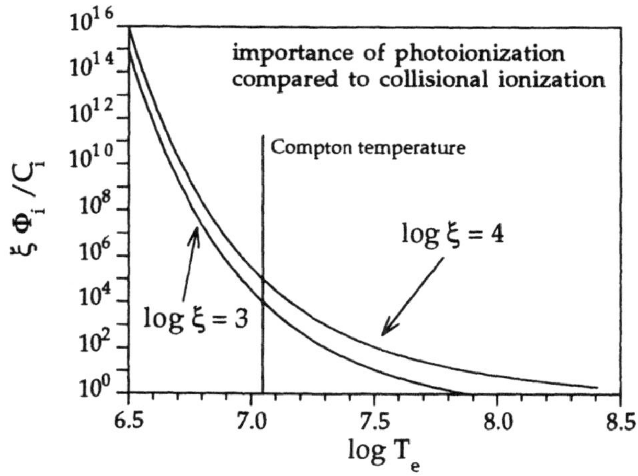

Fig. 3. The quantity $\xi\Phi_i/C_i$, which determines the degree of overionization relative to CIE, plotted against the electron temperature, assuming an irradiating spectrum with a Planck form and a radiation temperature of 1 keV. The electron temperature must be less than the Compton temperature, indicated by a vertical line. Curves for two values of ξ are shown.

local energy balance. The precision to which electrons match the Compton temperature depends on the degree to which other cooling processes come into play. Let us consider the simple (though, relevant) example of adding bremsstrahlung cooling to the energy equation.

Example: the effect of bremsstrahlung cooling in a Compton heated gas.

Denote the bremsstrahlung (free-free) radiative power by $\Lambda_{\rm ff}$ (c.g.s. dimensions erg cm^3 s^{-1}). Then, equating the net volumetric Compton heating rate to the volumetric bremsstrahlung cooling rate gives us an equation of local energy balance,

$$n_{\rm e}\Gamma_{\rm c} = n_{\rm e}^2\Lambda_{\rm ff}, \tag{130}$$

where the Compton heating function, $\Gamma_{\rm c}$, is defined in Eq. (102). By retaining appropriate powers of $n_{\rm e}$ in this equation, we illustrate the simple, but important, facts that Compton heating/cooling is a one-body/one-photon process, whereas bremsstrahlung is a two-body process. Also, since the left-hand-side is proportional to the radiative flux, we suspect that ξ will play a role in determining the relative contributions of Compton heating/cooling and bremsstrahlung cooling. To simplify the calculations, we ignore the free-free Gaunt factor, which leaves us with the following explicit form of Eq. (130):

$$\frac{\sigma_c}{mc}\int_0^\infty d\omega\; U_\omega(\hbar\omega - 4kT) = \left(\frac{8}{3\pi}\right)^{1/2} \alpha_f n_e \sigma_c\, mc^3 \left(\frac{kT}{mc^2}\right)^{1/2}, \qquad (131)$$

where α_f is the fine-structure constant. Using Eq. (103), we find

$$4U(kT_c - kT) = \left(\frac{8}{3\pi}\right)^{1/2} \alpha_f n_e (mc^2)^2 \left(\frac{kT}{mc^2}\right)^{1/2}. \qquad (132)$$

The magnitude of the term on the right-hand-side of this equation is proportional to the magnitude of the deviation of the electron temperature and the Compton temperature. If we consider gas surrounding a point source of ionizing radiation, we can substitute for the radiation energy density $U = L/4\pi r^2 c$. Then, after dividing both sides by n_e, and rearranging,

$$\xi\ (kT_c - kT) = \left(\frac{8\pi}{3}\right)^{1/2} (mc^2)^{3/2}\, \alpha_f c\ (kT)^{1/2}. \qquad (133)$$

Let $\Theta = T/T_c$. Then, the energy equation (130) can be written

$$\frac{\xi}{\xi_c}\ (1 - \Theta) = \Theta^{1/2}, \qquad (134)$$

where ξ_c can be defined as the "Compton ionization parameter", and provides an approximate value of ξ above which the temperature is close to the Compton temperature (as $\xi \to \infty$, $\Theta \to 1$). ξ_c is given by

$$\xi_c = \alpha_f mc^3 \left(\frac{8\pi}{3}\frac{mc^2}{kT_c}\right)^{1/2}. \qquad (135)$$

Numerically, for a typical value of T_c,

$$\xi_c = 3.7 \times 10^3 \left(\frac{kT_c}{10\ \text{keV}}\right)^{-1/2}. \qquad (136)$$

The solution to Eq. (134), which yields Θ as a function of ξ, is

$$\Theta = 1 - \frac{1}{2}\left(\frac{\xi_c}{\xi}\right)^2 \left[\sqrt{1 + 4(\xi/\xi_c)^2} - 1\right]. \qquad (137)$$

We should be careful not to extend this equation beyond its range of validity. As ξ decreases, recombination will begin to play a role in the cooling. As electrons become bound to nuclei, photoelectron thermalization begins to contribute to the heating. Thus the situation becomes considerably more complicated. For $\xi > \xi_c$, Eq. (137) provides a good approximation.

3.3 Differential emission measure distributions

The emissivity of a given line varies with the ambient plasma conditions. If there is a sufficient "amount" of gas in a source with the appropriate conditions to produce the line, then the line luminosity will be such that it is observable. We need to be careful with what we mean by "amount". For recombination lines, the emissivity depends upon the square of the density, since recombination (radiative and dielectronic) is a two-body process. Therefore, the source-integrated line luminosity depends upon a volume weighted by the square of the density, the *emission measure*, $EM \propto n_e^2 V$. Note that this is quite different from what we might think of as the "amount" of gas, which is proportional to $n_e V$.

Therefore, we could characterize an X-ray source, in part, by reporting the emission measure inferred from a given line flux. If we knew from theory the value of ξ at which the line forms, we could report the emission measure of gas existing at that ξ. Corollary information, such as a direct measurement of the density using density-sensitive line ratios, or knowledge of the size of the emitting region from photometric data, may allow us to decompose the emission measure into a density and a volume. A little thought, however, convinces us that we have not properly defined this concept. For example, suppose that we measure fluxes from the O VII lines near 22 Å, and also from the Ne IX lines near 13.5 Å. Do we infer an emission measure at a single value of ξ that simultaneously accounts for these lines, or should we try to infer emission measures for two values of ξ, corresponding to each of the two line-forming zones? There is no easy answer to this. There are a number of plausible scenarios that could be consistent with the observed line emission. Obviously, the more information that is available, the more secure the interpretation.

Undoubtedly, spectral analysis would be greatly simplified if, for a given source, all lines formed at a single value of ξ, but this is not likely. In a realistic situation, it is more likely that conditions vary continuously over a range of plasma conditions.[11] Therefore, the source-integrated spectrum depends upon the distribution of conditions throughout the emitting volume. This leads to the concept of a differential emission measure distribution, which we now develop (see also the Chapters by Mewe and Kaastra).

First, we express the line luminosity for a given line of charge state i, with energy E_l, as a volume integral of its emissivity:

$$L_l = \int_V \mathrm{d}V \, j_l. \tag{138}$$

where the emissivity (c.g.s. dimensions erg cm^{-3} s^{-1}) is given by

[11] At present, the extent to which the plausibly accessible parameter space (temperature, density, velocity, ionization parameter) is occupied by X-ray emission line regions in compact X-ray sources is not known.

$$j_l = n_e n_{i+1} \eta \alpha(T) E_l . \tag{139}$$

Note that the emissivity is proportional to the population density of charge state $i+1$, since recombination is the dominant mechanism by which excited states are populated. The factor η is the fraction of recombinations that leads to emission in the line under consideration. Thus η embodies the cascade kinetics.

Now, expressing the emissivity in terms of the charge state fraction of the recombining ion, f_{i+1}, and the elemental abundance relative to hydrogen, A_Z,

$$L_l = \int_V \mathrm{d}V \; n_e^2 \; \frac{n_H}{n_e} \; A_Z \; f_{i+1}(T) \; \eta \; \alpha(T) E_l . \tag{140}$$

The hydrogen particle density, n_H, irrespective of its charge state is denoted by n_H. Of course, in the ionization regime of interest, hydrogen is almost entirely in the form H^+. Moreover, the factor n_H/n_e is virtually constant, since it depends almost exclusively upon the ionization state of hydrogen and helium, and helium is almost fully stripped, as well.

In the usual formulation, the volume integral (140) is converted to a temperature integral. Here, however, we make ξ the independent variable and convert to an integral over ξ:

$$L_l = \int_\xi \mathrm{d}\ln\xi \left(n_e^2 \, \frac{dV}{d\ln\xi} \right) \left(\eta A_Z \; \frac{n_H}{n_e} \; f_{i+1}(\xi) \; \alpha(\xi) E_l \right) . \tag{141}$$

The second factor in the integrand is the *line power* (c.g.s. dimensions erg $\mathrm{cm}^3\ \mathrm{s}^{-1}$), while the first factor is the *differential emission measure (DEM) distribution*. In abbreviated form, the previous expression is written

$$L_l = \int_\xi \mathrm{d}\ln\xi \left(\frac{\mathrm{d}EM}{\mathrm{d}\ln\xi} \right) P_l(\xi) . \tag{142}$$

This formalism effects a decomposition of the line luminosity into a macrophysical part and a microphysical part.

The DEM distribution derived here is analogous to that which is often encountered in the context of CIE plasmas ($\mathrm{d}EM/\mathrm{d}\ln T$). Also by analogy to spectral fitting methods applied to CIE plasmas (Kaastra et al. 1996), one may, in practice, convert the integral (142) into a sum over "ξ-bins", thereby representing the spectrum as a linear superposition of "iso-ξ" spectra. In fitting spectroscopic data, each ξ-component has a weight that acts as a free fitting parameter. The ensemble of weights constitutes the DEM distribution.

Example: gas with uniform density illuminated by a point source

Start with $\mathrm{d}EM = n^2 \; dV$. Assume that a point source of X rays is embedded in a uniform medium. Then

$$\mathrm{d}EM = 4\pi n(\xi)^2 r(\xi)^2 \; \frac{\mathrm{d}r(\xi)}{\mathrm{d}\xi} \; \xi \; \mathrm{d}\ln\xi \tag{143}$$

For n a constant over space, $n = n_0$, substitute for r from $r = (L/n_0\xi)^{1/2}$. This gives

$$\frac{\mathrm{d}EM}{\mathrm{d}\ln\xi} = 2\pi L\ (Ln_0)^{1/2}\ \xi^{-3/2}. \tag{144}$$

One can foresee the usefulness of developing this technique, since the DEM distribution can be derived theoretically, based upon a global model of a given X-ray source, and can be inferred experimentally, based upon fits to spectroscopic data. The DEM distribution then serves as a "watering hole" for theorists and observers.

3.4 Radiative recombination continua

In this section, we discuss the basics of radiative recombination in order to derive a simple plasma diagnostic, the shape of the radiative recombination continuum (RRC; used for "radiative recombination continua", also). Throughout this section, let ϵ denote a photon energy.

When a plasma is overionized, by definition $kT \ll \chi$. For the conditions that prevail in an X-ray photoionized gas, $kT/\chi \sim 10^{-2}$–10^{-1} (Kallman & McCray 1982). The energy of a typical RRC photon is $\epsilon = \chi + kT$ and the width of the RRC is approximately $\Delta\epsilon = kT$. Therefore, $\Delta\epsilon/\epsilon \approx kT/\chi$, so that the RRC are narrow, allowing them to be contrasted against the bright continuum spectrum produced by the compact X-ray source. Equally important, under recombination/cascade kinetics,[12] the RRC should be observable whenever lines are observable. For example, for hydrogenic ions in a purely recombining plasma, the RRC intensity should be roughly comparable to the Lyα analog line ($2p \rightarrow 1s$). This feature of RRC makes them simple-to-use signatures of photoionization dominance (Hatchett, Buff & McCray 1976), as has been demonstrated in the analysis of the *ASCA* spectrum of the X-ray binary Cygnus X-3 (Liedahl & Paerels 1996).

In terms of the RR cross section, σ_{rr}, the monochromatic RRC emissivity (c.g.s. dimensions erg cm^{-3} s^{-1} erg^{-1}) is given by

$$j_\epsilon = n_e n_{i+1} v f(v)\ \sigma_{\mathrm{rr}}(v)\ \epsilon\ \frac{\mathrm{d}v}{\mathrm{d}\epsilon}, \tag{145}$$

where $f(v)$ is the free electron velocity distribution, which we will assume to be the Maxwellian:

$$f(v) = \left(\frac{2}{\pi}\right)^{1/2} \left(\frac{mc^2}{kT}\right)^{3/2} \frac{v^2}{c^3}\ e^{-mv^2/2kT}. \tag{146}$$

The population density of the initial energy level of charge state $i+1$ is denoted n_{i+1}, and does not necessarily coincide with the level-summed population of the recombining ion. The energy of an outgoing photon is equal to

[12] We restrict the discussion to densities such that three-body recombination does not contribute substantially to the line flux (cf., Bautista et al. 1998).

the sum of the kinetic energy of the recombining electron plus the ionization potential of the recombined level,

$$\epsilon = \chi + \frac{1}{2}\, mv^2 . \tag{147}$$

The Milne relation (consult the Chapter by Mewe, and Osterbrock 1989) is a detailed-balance relationship that expresses the RR cross section in terms of the corresponding photoionization cross section $\sigma(\epsilon)$;

$$\sigma_{\rm rr}(v) = \frac{g_i}{g_{i+1}}\, \frac{\epsilon^2}{(mvc)^2}\, \sigma(\epsilon), \tag{148}$$

where, g_i and g_{i+1} are statistical weights associated with the recombined and recombining levels, respectively. In terms of the total angular momentum J, $g = 2J + 1$.

Combining the four previous equations gives us the monochromatic RRC emissivity as a function of electron temperature,

$$j_\epsilon = \left(\frac{2}{\pi}\right)^{1/2} n_e n_{i+1}\, \frac{g_i}{g_{i+1}}\, c\sigma(\epsilon) \left(\frac{\epsilon}{\chi}\right)^3 \left(\frac{\chi^2}{mc^2kT}\right)^{3/2} e^{-(\epsilon-\chi)/kT} . \tag{149}$$

We now show how the shape of the RRC can be related to the electron temperature. The following discussion is illustrative. In practice, of course, the process of determining the temperature from the RRC should be automated, and should use the exact photoionization cross section. The monochromatic RRC emissivity falls with increasing energy from its maximum at the RRC "edge" ($\epsilon = \chi$). The rate at which it falls depends on the functional form of the photoionization cross section and the electron temperature. Define an energy ϵ' such that $j_\epsilon(\epsilon') = e^{-1} j_\epsilon(\chi)$, and a width $\Delta\epsilon$ such that $\Delta\epsilon = \epsilon' - \chi$. Using Eq. (149), this gives

$$\epsilon'^3\, \sigma(\epsilon')\, e^{-(\epsilon'-\chi)/kT} = e^{-1}\, \chi^3\, \sigma(\chi) \tag{150}$$

which can be inverted for the temperature as a function of the observed RRC width:

$$kT = \Delta\epsilon \left[1 + \ln\, \frac{\sigma(\epsilon')\epsilon'^3}{\sigma(\chi)\chi^3}\right]^{-1} \tag{151}$$

If we parametrize the photoionization cross section as a power law, $\sigma(\epsilon) = \sigma(\chi)(\epsilon/\chi)^{-\gamma}$, then

$$kT = \Delta\epsilon \left[1 + (3-\gamma)\, \ln\left(1 + \frac{\Delta\epsilon}{\chi}\right)\right]^{-1} \tag{152}$$

For hydrogenic ions, the approximation $\gamma = 3$ is adequate, and, to sufficient accuracy, $kT = \Delta\epsilon$. For steeper cross sections, the width will be less than kT, while for shallower cross sections, the width is greater than kT.

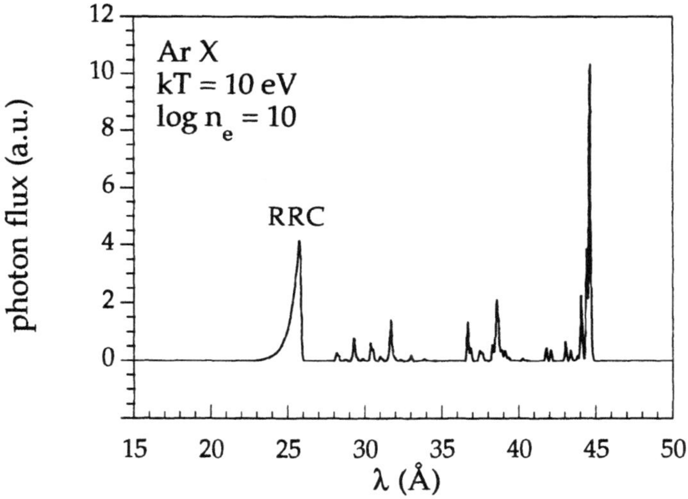

Fig. 4. Model recombination dominated spectrum of F-like Ar X with conditions as indicated. Spectra are convolved with Gaussian resolution kernel with FWHM of 0.05 Å.

An example to illustrate the appearance of RRC compared to emission lines is provided in Figure 4, which shows a model spectrum of F-like Ar X. Spectral features include $n \to 2$ emission, the brightest being the $3s-2p$ lines near 45 Å and the RRC near 26 Å. At this low density ($n_e = 10^{10}$ cm^{-3}, nearly all of the population of the recombining ion, O-like Ar, is in the $2s^2 2p_{1/2}^2 2p_{3/2}^2$ ground state. Since the only allowed recombination from the continuum to the L shell is to the $2p_{3/2}$ subshell, the RRC is made up of a single feature. At higher densities, where the O-like ion can have a substantial population with three electrons in the $2p_{3/2}$ subshell, thereby, leaving a vacancy in $2p_{1/2}$, the RRC can become more complex, with a number of individual edges visible.

Recalling our discussion of the differential emission measure distribution, clearly we have thus far taken a simplistic view of RRC. In cases where a non-zero temperature gradient ($\mathrm{d}T/\mathrm{d}\xi$) exists, the source-integrated RRC shape will be determined by the n_e^2 weighting of each temperature component. Modifying Eq. (142) to accommodate the monochromatic RRC power, the monochromatic source-integrated luminosity is

$$L_\epsilon(\epsilon) = \int_\xi \mathrm{d}\ln\xi \left(\frac{\mathrm{d}EM}{\mathrm{d}\ln\xi}\right) P_\epsilon(\epsilon,\xi). \qquad (153)$$

Ignoring factors associated with the overall RRC normalization, the RRC shape is found by substituting from Eq. (149) into Eq. (153), yielding

$$L_\epsilon(\epsilon) \propto \int_\xi \mathrm{d}\ln\xi \left(\frac{\mathrm{d}EM}{\mathrm{d}\ln\xi}\right) f_{i+1}(\xi)\, T(\xi)^{-3/2} e^{-\epsilon/kT(\xi)}. \qquad (154)$$

If the emitting medium is not isothermal, the RRC shape will deviate from an exponential. If the temperature does not change drastically over the zone of formation of the feature, then the profile should not change drastically from an exponential, and, in any case, should provide a reasonable estimate of the electron temperature. Although this complicating effect is not discussed in Liedahl & Paerels (1996), the *ASCA* spectrum of Cygnus X-3 shows strong evidence for a rising temperature as ξ increases. This is expected, of course, according to the results of photoionization models. An in-depth study of this problem has not been performed. We will not attempt one here, as it would take us too far afield.

3.5 Spectral signatures of recombination kinetics

As noted in the previous section, RRC are obvious signatures of recombination in an overionized plasma. RRC can be detected even in spectroscopic data acquired with instruments of moderate resolving power, such as the *ASCA* Solid-State Imaging Spectrometer (SIS). Other, more subtle, effects are predicted to be observable when higher spectral resolution data become available. We discuss two classes of *known* (but not necessarily observed) spectral signatures of recombination dominance. Undoubtedly, others will arise (e.g., Savin et al. 1997).

The key to understanding the first, essentially qualitative, class of diagnostics, is to identify the dominant process leading to excitation of the upper levels of strong X-ray lines. Consider two energy levels of charge state i, u and ℓ, where u (upper level) has a higher energy than ℓ (lower level), and a downward radiative transition $u \rightarrow \ell$ connecting them (see Figure 5). In CIE, most bright lines are driven by *direct* collisional excitation from the ground state. The production of such a line in CIE can be represented by the two-step process $1(i) \rightarrow u \rightarrow \ell$, where $1(i)$ denotes the ground state of charge state i. By contrast, some bright lines observed in CIE plasmas, say $u' \rightarrow \ell'$, are produced *indirectly*. In this case, while the ultimate source of population influx may be the ground state, another level, w, with energy greater than that of level u', intervenes in the population flux chain, as follows: $1(i) \rightarrow w \rightarrow u' \rightarrow \ell'$. For lines driven by this indirect process, u' is usually the lower level of a large number of transitions, i.e., w represents a large number of levels, so it is appropriate to express this chain as follows: $1(i) \rightarrow \Sigma w_k \rightarrow u' \rightarrow \ell'$. We are assuming that lines can be distinguished according to whether they are driven directly or indirectly, and that, in the context of this section, intermediate cases are not relevant.

As has been discussed, $(T_{\rm peak})_{\rm PIE} \ll (T_{\rm peak})_{\rm CIE}$, with consequent reductions in the rate coefficients for the electron impact excitation steps $1(i) \rightarrow u$

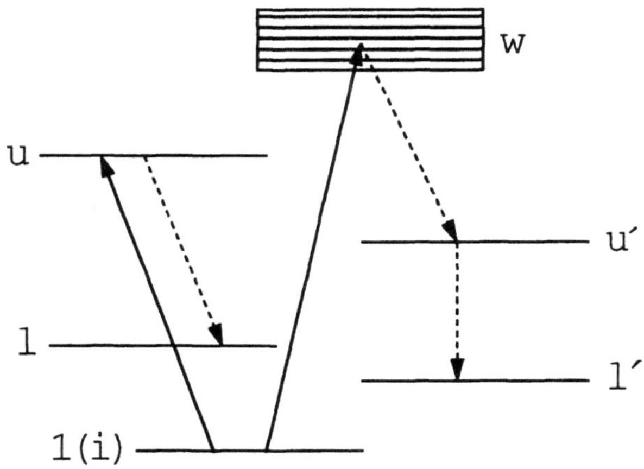

Fig. 5. Schematic diagram of population kinetics illustrating the distinction between direct and indirect level feeding. The transition $u \to l$ is driven directly, while the transition $u' \to l'$ is driven indirectly. An intermediate set of levels w intervenes in channeling population flux into the upper state u' of the indirectly driven line.

and $1(i) \to \Sigma w_k$. This is accompanied, however, by an increase of the recombination rates. The dominant population flux chains for each of these lines is then modified by simply supplanting $1(i)$ by $1(i+1)$. Instead of collisional excitation, the first step is recombination: $1(i+1) \to u \to \ell$ for the directly driven line, and $1(i+1) \to \Sigma w_k \to u' \to \ell'$ for the indirectly driven line. The crucial factor is that the emissivity of $u \to \ell$ can be drastically reduced compared to $u' \to \ell'$ since the several sources for influx into u', the levels w_k, now benefit from recombination from charge state $i+1$, whereas, the same factors that conspire to render indirect processes unimportant for the production of $u \to \ell$ in CIE work toward the same end in PIE. Therefore, the signature of recombination kinetics in this case is a greatly reduced $I(u \to \ell)/I(u' \to \ell')$ intensity ratio relative to CIE.

One of the best examples of this effect is the Ne-like Fe XVII $n = 3 \to n = 2$ (or, simply, 3-2) spectrum (Liedahl et al. 1990). In order to justify the following detailed discussion, it suffices to point out that the basic mechanism to be described applies to other L-shell ions, as well, including those of other elements. Therefore, the following discussion is actually of a substantially broader scope than might be guessed.

The situation is depicted in Figure 6 for the case of recombination kinetics. The ground levels of Fe XVII[13] ($2s^2 2p_{1/2}^2 2p_{3/2}^4$) and F-like Fe XVIII ($2s^2 2p_{1/2}^2 2p_{3/2}^4$) are shown.[14] Lying above the Fe XVII ground level are the four levels belonging to $2s^2 2p^5 3s$ (abbreviate $3s$). The total angular momentum J of each of these four levels is indicated ($J = 2$, 1, 0, and 1). At higher energies are the *manifolds* $2s^2 2p^5 3p$ (ten levels; abbreviate $3p$) and $2s^2 2p^5 3d$ (twelve levels; abbreviate $3d$). We ignore configurations of the form $2s 2p^6 3l$, since they do not play an important role in the population kinetics.

There are three electric dipole (E1; $\Delta J = 1$, initial and final states have opposite parity) transitions connecting $3d$ to the $J = 0$ ground level. In CIE, these lines are bright (McKenzie et al. 1980), and driven by direct excitation from ground. They are exemplary of the line $u \rightarrow \ell$ described above. There are two E1 transitions from $3s$ to ground. In CIE, these lines are bright also, but driven by indirect processes, in this case, collisional excitation to higher levels, followed by cascade down to the $3s$ levels. The line from the $J = 2$ level at 17.10 Å, though produced by way of a magnetic quadrupole (M2) transition, is also bright at densities below $\sim 10^{15}$ cm^{-3}, since a transition to ground is the only available single-photon decay channel. The $3s$ lines provide examples of the line $u' \rightarrow \ell'$ described above. Lines from $3p$ to ground are relatively weak, since they are parity-forbidden.

We look first at the $3d$ levels. In CIE, collisional excitation from the ground state dominates the transfer of flux into $3d$. By contrast, in PIE, the importance of collisional excitation is drastically diminished, owing to the $T^{-1/2} \exp(-E/kT)$ dependence of this process, where E is the energy of the excited state relative to ground. Increases in the RR rates into Fe XVII partially offset the reduction of collisional excitation. However, the rate at which a level is fed by RR is in proportion to its statistical weight, and the levels responsible for the bright lines observed in CIE are of lower statistical weight ($2J + 1 = 3$) than most levels in $3d$. Since only $3d$ levels with angular momenta $J = 1$ decay to ground, and $3d \rightarrow 3s$ is parity-forbidden, nine of the twelve $3d$ levels decay preferentially to $3p$. This is in contrast to the behavior of the $3d$ levels in CIE, where the dominant sink, in terms of population flux, is the ground level. Note that RR directly to $3p$ [indicated in Figure 6], as well as discrete transitions from higher energy levels, adds to the population influx. Since $3p \rightarrow 2p$ transitions are parity-forbidden, nearly all of the flux through $3p$ populates $3s$ levels. Again, account must be taken of direct RR

[13] In all references to specific configurations, the presence of the inactive closed K shell $1s2$ is implied.

[14] Subscript labels to non-relativistic subshells give the coupling of the azimuthal quantum number l to the electron spin ($j = l + 1/2$, $l - 1/2$). Superscripts indicate the occupancy of the relativistic subshell. An implied $j = 1/2$ holds for all s subshells. An unsubscripted p or d subshell implies any allowed permutation of relativistic subshells consistent with the occupancy, for example, $p^5 \rightarrow p_{1/2}^2 p_{3/2}^3$, $p_{1/2} p_{3/2}^4$, or $p_{3/2}^5$.

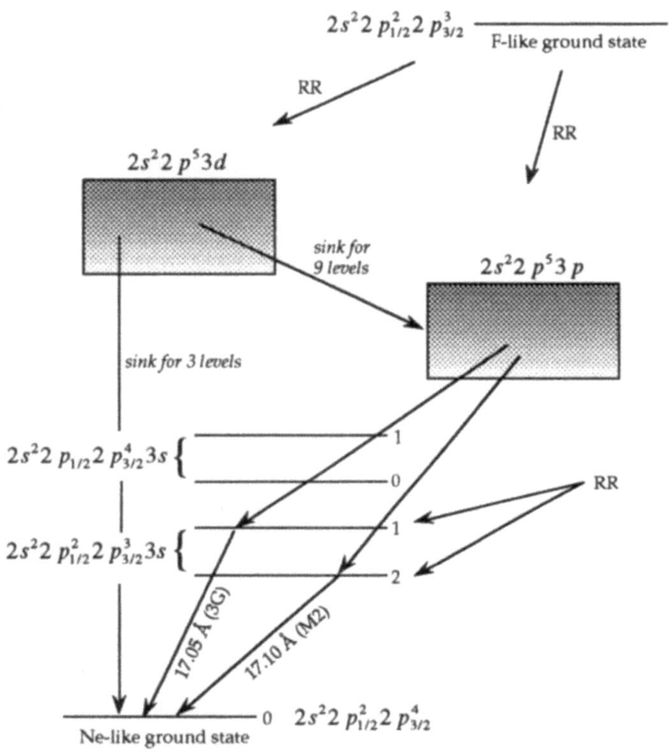

Fig. 6. Level population mechanisms in Ne-like Fe XVII. Ground states of Fe XVII and Fe XVIII, and 26 excited states of Fe XVII are represented. The X-ray lines at 17.05 Å and 17.10 Å dominate the Fe XVII spectrum in overionized plasmas.

plus the cascade contribution into $3s$. These various effects result in $3s \to 2p$ transitions that produce bright lines compared to $3d \to 2p$ lines.

An additional consequence of recombination kinetics involves the distribution of line power among the $3s - 2p$ lines. Note that, in Figure 6, only two $3s - 2p$ lines are indicated. This is because the $[2s^22p_{1/2}2p_{3/2}^43s\ (J=1)]$ upper level of the third E1 line, λ16.77, which is bright in CIE, has an excited core. That is, the $2s^22p_{1/2}2p_{3/2}^4$ core is excited relative to the core of the F-like ground state, $2s^22p_{1/2}^22p_{3/2}^3$. At most astrophysical densities, nearly all of the F-like ion will be found in its ground state, which, after recombination, leads to excited Ne-like states of the form $2s^22p_{1/2}^22p_{3/2}^3\ nlj$. The ensuing cascade usually proceeds with a "frozen" core — the configuration changes owing only to changes in the quantum numbers of the outer electron. This effect strongly favors the two lines $\lambda 17.05$ and $\lambda 17.10$. In CIE, by contrast, configurations with the excited core arise from direct collisional excitation out of the Ne-like $2p_{1/2}$ subshell. Obviously, in this case, the level population distribution in the F-like ion has no bearing on the creation of the $2s^2 2p_{1/2} 2p_{3/2}^4$ core.

In PIE, a substantial population in F-like $2s^2 2p_{1/2} 2p_{3/2}^4$ can be built up at sufficiently high electron densities (Liedahl et al. 1992), in which case $\lambda 16.77$ can be bright (the density dependence of emission lines is the subject of §3.6). Moreover, the population of the excited F-like core can be enhanced by dielectronic recombination (Savin et al. 1997), which, if the temperature falls below $\sim 10^5$ K, somewhat complicates the description given above. The spectroscopic consequences of this latter effect have not yet been explored in detail.[15]

Therefore, for L-shell ions, it is found that the rules of cascade kinetics favor the longer wavelength members of a given $n \to n'$ series. This means simply that population flux down through the cascade chain favors excited levels with the lowest energies, as shown in the previous example. A similar situation occurs in He-like ions. Without the contribution from collisional excitation, the resonance line ($1s^2\ ^1S_0 - 1s2p\ ^1P_1$) is weak compared to the intercombination ($1s^2\ ^1S_0 - 1s2p\ ^3P_1$) and forbidden ($1s^2\ ^1S_0 - 1s2s\ ^3S_1$) lines (Pradhan 1985).

There is one more spectral signature that we discuss in this section. It is well known that iron L-shell ions dominate the discrete X-ray spectrum of high-temperature plasmas, if the elemental abundance distribution resembles that of the solar photosphere (see, for example, the solar flare spectra in Phillips et al. 1982). It has been found that this is not the case in X-ray photoionized gas (Kallman et al. 1996). This can be explained by the n-scaling of collisional excitation and RR, where n refers to a principal quantum number. For collisional excitations $n \to n'$ we have, roughly, $\gamma \propto n^4$ (Sobelman, Vainshtein & Yukov 1981), where γ denotes a collisional excitation rate coefficient. Therefore, in CIE, the efficiency of iron L-shell emission compared to, say, oxygen K-shell emission results, in part, from the relative ease of collisional excitations out of $n = 2$ compared to $n = 1$. When we look at the case of recombination kinetics in PIE, we note that, for low n, $\alpha_{rr} \propto n^{-1}$ (Sobelman, Vainshtein & Yukov 1981). Therefore, again comparing oxygen K-shell emission to iron L-shell emission, we see that not only has iron lost its efficient collisional excitation mechanism, but that recombination favors the K-shell ions, since their valence shells correspond to a smaller n. The existence of this phenomenon is supported by the analysis of Sako et al. (1998a) of the eclipse spectrum of the X-ray pulsar Vela X-1. It is found that the soft emission is dominated by K-shell ions of elements lighter than iron (e.g., neon

---

[15] Dielectronic recombination of L-shell ions can create a vacancy in the $2s$ subshell (§3.8), and, although it may lead to an overall modification of the line spectrum, operates virtually independently of the mechanism discussed in this section.

and magnesium). The presence of bright iron K emission in the spectrum would appear to be inconsistent with a possible alternative explanation for the apparent weakness of iron L-shell emission, namely, that iron is simply underabundant relative to the lighter elements. An unequivocal determination requires higher spectral resolution, so that soft X-ray iron lines can be isolated.

3.6 Density diagnostics in X-ray photoionized plasmas

We here examine the mechanisms underlying two types of density diagnostics. They both depend on collisional depopulation of metastable levels (refer to Rolf Mewe's Chapter in this Volume). One may be inclined to doubt this, since the point has been made that electron-ion impact processes are of reduced importance in photoionized plasmas. An examination of the mechanisms involved, however, shows that the relevant collisional transitions have energies of order 10 to 100 eV, energies which are commensurate with the electron temperatures in X-ray photoionized gases. Therefore, in PIE, He-like density diagnostics behave kinetically and spectroscopically in a manner which is virtually identical to their behavior in CIE. As we show below, however, there is a difference in the spectroscopic behavior of L-shell ions, although the basic mechanism is similar to that which operates in CIE.

A special circumstance pertains to densities of photoionized plasmas. Going back to our discussion of ξ in §3.2, we recall that the volumetric recombination rate (a two-body process) depends on the square of the density, but that the volumetric ionization rate depends on a single power of density. This should be contrasted to the situation in CIE plasmas, where the volumetric rates for both ionization and recombination scale as the square of the density. In a photoionized plasma, for a fixed radiation field, increasing the density decreases the average effective charge of a given element – the plasma is more recombined. Looking at this another way, for a fixed value of ξ, increasing the density moves the ionization zone of a given ion closer to the source of ionizing X rays, according to $R_{\rm ion} = (L/n\xi)^{1/2}$. Therefore, not only can density measurements tell us something about the size of the emission line region, but they can help us to fix its position.

He-like ions constitute the best-known class of density diagnostics in X-ray spectroscopy. The mechanism as it operates in CIE is described in Gabriel & Jordan (1969). A schematic of the mechanism as it functions in an X-ray photoionized gas is presented here (see Figure 7). For ease of notation, let F denote the upper level ($1s2s\ ^3S_1$) of the forbidden line, and let I denote the upper level ($1s2p\ ^3P_1$) of the intercombination line. The diagnostic ratio is denoted R, and is given by the emissivity ratio of the forbidden and intercombination lines:

$$R = \frac{j_F}{j_I} = \frac{n_F}{n_I}\frac{A_F}{A_I} \qquad (155)$$

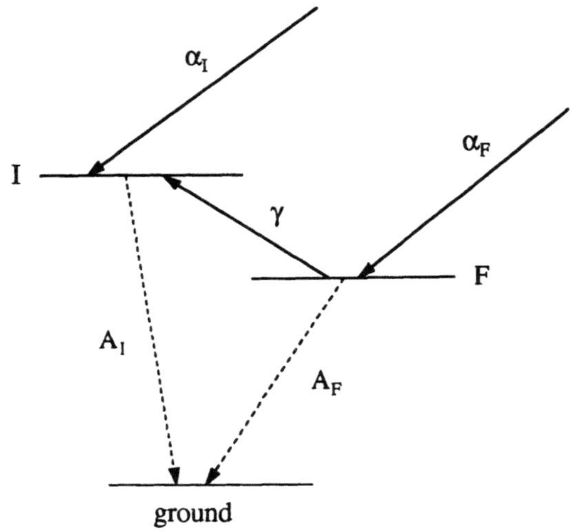

Fig. 7. Schematic of population mechanisms responsible for behavior of He-like R ratio in overionized plasma. Three levels are represented, the $1s^2$ ground state, the $1s2s\ ^3S_1$ "forbidden level" (F), and the $1s2p\ ^3P_1$ "intercombination level" (I). Population influx from recombination and cascades are denoted by α_I and α_F, collisional transitions from F to I by γ, and radiative rates are indicated by dotted lines.

where the emissivities (j) are expressed in terms of level populations (n) and Einstein coefficients (A). A single collisional excitation rate coefficient, γ (c.g.s. dimensions $\mathrm{cm^3\ s^{-1}}$), connects the closely spaced F and I levels. We do not need to include similar collisional coupling between the ground state and the levels F and I, since the energy spacings are several hundred eV, rendering collisional excitation from ground ineffective in populating these levels in overionized plasmas.

Letting n_h represent the population density of the hydrogenic ground state, we see from Figure 7 that

$$n_F = \frac{n_\mathrm{e} n_\mathrm{h} \alpha_F}{A_F + n_\mathrm{e}\gamma} \qquad n_I = \frac{n_\mathrm{e} n_\mathrm{h} \alpha_I + n_\mathrm{e} n_F \gamma}{A_I} . \tag{156}$$

Substitute the expression for n_F into that for n_I:

$$n_I = n_\mathrm{e} n_\mathrm{h} \frac{\alpha_I A_F + n_\mathrm{e}\gamma(\alpha_I + \alpha_F)}{A_I(A_F + n_\mathrm{e}\gamma)} . \tag{157}$$

Combining everything, we have

$$R = \frac{\alpha_F A_F}{\alpha_I A_F + n_\mathrm{e}\gamma(\alpha_I + \alpha_F)} . \tag{158}$$

In terms of a critical density, which is defined here to be

$$n_{\rm crit} = \frac{A_F}{\gamma} \frac{\alpha_I}{\alpha_I + \alpha_F}, \tag{159}$$

this becomes

$$R = \frac{\alpha_F}{\alpha_I} \frac{1}{1 + (n_{\rm e}/n_{\rm crit})}. \tag{160}$$

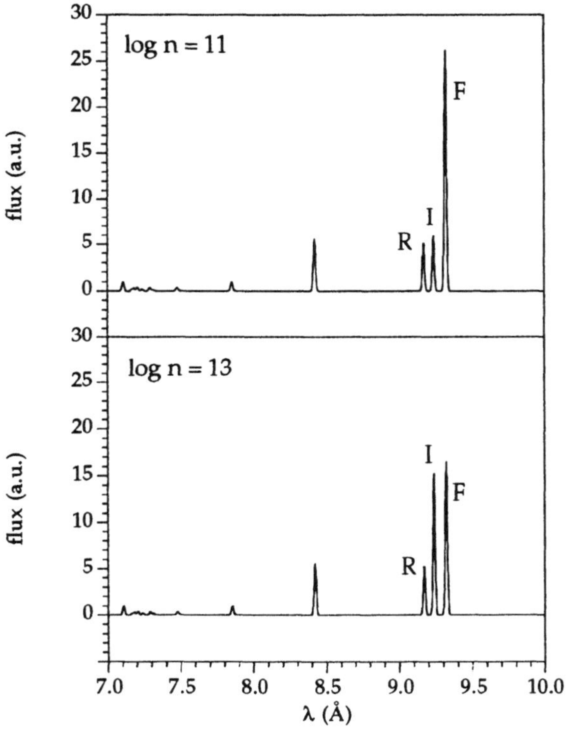

Fig. 8. Mg XI at two densities, as indicated, in recombination dominated plasma, with $kT = 20$ eV. The resonance (R), intercombination (I), and forbidden (F) lines are labeled. In the lower panel, where a density near $n_{\rm crit}$ has been chosen, the ratio F/I has begun to invert from its low-$n_{\rm e}$ value.

This basic functional form is illustrated in Figure 10. The curve suggests that the ratio shows rapid variation over roughly two orders of magnitude in density. Outside this range, R does not vary appreciably, and one can infer upper or lower limits only. At low densities, $R = \alpha_F/\alpha_I$, while at densities higher than $n_{\rm crit}$, $R \to 0$ as the forbidden line disappears. When $n_{\rm e} = n_{\rm crit}$, then $R = (1/2)\alpha_F/\alpha_I$, i.e., one-half of the maximum value of R, which motivates the definition given in Eq. (159). To take a specific example, Figure 8 shows Mg XI spectra in an overionized plasma for two densities, one below and one near $n_{\rm crit}$. The inversion of R occurs over a range of two orders of magnitude in density. Plots of R in CIE for a number of He-like ions as a function of electron density can be found in Pradhan (1982). There it can be seen that values of $n_{\rm crit}$ range from $\sim 10^9\ {\rm cm}^{-3}$ (for C V) to $\sim 10^{17}\ {\rm cm}^{-3}$ (for Fe XXV).

While the simple schematic of the kinetics presented earlier is sufficient to describe the basic mechanism, a detailed model of a He-like ion shows that n_F and n_I are dependent upon a number of additional levels. Therefore, although we have described the qualitative behavior of these ions, an exact quantitative treatment of the behavior of R requires extensive calculations.

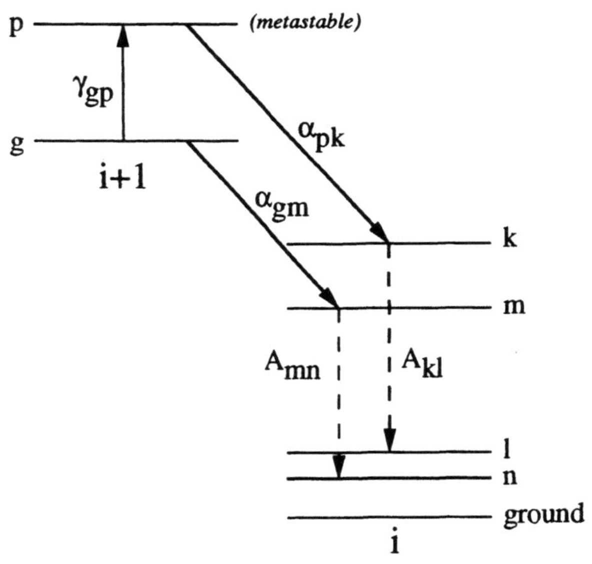

Fig. 9. Schematic of population mechanisms responsible for density-sensitive behavior of L-shell spectra. Two charge states, i and i+1 are coupled by recombination (α_{gm} and α_{pk}). Level p is fed by collisional excitation (γ_{gp}) from the ground state g of $i+1$. X-ray transitions are indicated by dotted lines, and labeled according to their Einstein coefficients (A_{mn} and A_{kl}).

The other class of density diagnostics involves L-shell ions. The mechanism as it operates in CIE is described in Mason et al. (1979). Modifications to the basic mechanism needed for PIE are described in Liedahl et al. (1992).

We develop the underlying mechanism schematically, as before. Referring to Figure 9, assume that two neighboring charge states, i and $i+1$, are coupled by the RR rate coefficient α_{gm}, which leads to emission in the line $m \to n$. Assume that charge state $i+1$ has a low-lying metastable state p. As opposed to the metastable forbidden level (F) in He-like ions, L-shell ions (with the exception of Ne-like Fe XVII) have a number of energy levels lying just a few tens of eV above ground. Therefore, they can be fed through direct collisional excitation from ground. As the density increases, the population of p increases, and recombination from level p becomes more important. Radiative cascades initiated by α_{pk} will begin to result in X-ray lines ($k \to l$) that are faint or non-existent at lower densities.

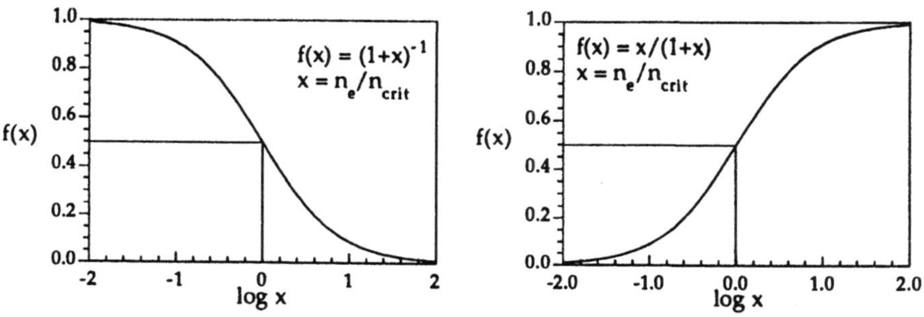

Fig. 10. Functional forms of density-dependent line ratios for two classes of density diagnostics. (*Left panel*) He-like R ratio. (*Right panel*) L-shell ratios. Abscissa is scaled in multiples of the critical density, and the ordinate is normalized to unity. At the critical density, each ratio has a value equal to 1/2 of its maximum value, as indicated. Both types of ratio have maximal sensitivity to changes in density over a range of two orders of magnitude centered on n_{crit}.

We want to derive an expression for the ratio of the emissivity of the line $k \to l$ to that of $m \to n$, which we call S. In terms of the level populations of the upper levels, n_m and n_k, of the radiative transitions, S is given by

$$S = \frac{j_{kl}}{j_{mn}} = \frac{n_k}{n_m} \frac{A_{kl}}{A_{mn}}. \tag{161}$$

The population densities of levels k and m are

$$n_k = \frac{n_e n_p \alpha_{pk}}{\sum_q A_{kq}} \qquad n_m = \frac{n_e n_g \alpha_{gm}}{\sum_r A_{mr}}. \tag{162}$$

If we introduce the radiative branching ratios

$$B_{kl} = \frac{A_{kl}}{\sum_q A_{kq}} \qquad B_{mn} = \frac{A_{mn}}{\sum_r A_{mr}} \tag{163}$$

we can write

$$S = \frac{n_p}{n_g} \frac{\alpha_{pk}}{\alpha_{gm}} \frac{B_{kl}}{B_{mn}}. \tag{164}$$

We need an expression for the population density of level p. Since p is metastable, we include collisional de-excitation in the denominator. Then

$$n_p = \frac{n_e n_g \gamma_{gp}}{A_{pg} + n_e \gamma_{pg}} \tag{165}$$

Equation (165) gives us the ratio n_p/n_g. Finally, introducing a critical density, $n_{\rm crit} = A_{pg}/\gamma_{pg}$, we have

$$S = \left(\frac{\gamma_{gp}}{\gamma_{pg}} \frac{\alpha_{pk}}{\alpha_{gm}} \frac{B_{kl}}{B_{mn}} \right) \frac{n_e/n_{\rm crit}}{1 + n_e/n_{\rm crit}}. \tag{166}$$

For iron L-shell ions, critical densities are $\sim 10^{13}$ cm^{-3}.

Equation (166) shows that, in the "zero-density limit", $S \to 0$. This means that in the absence of additional excitation mechanisms, the line $k \to l$ is unobservable. At high densities, S "saturates" to a value given by the parenthesized factor above. The functional dependence is shown in Figure 10. Qualitatively, we expect an enrichment of the discrete spectral structure of L-shell ions as the density increases, as lines that are unobservable at low densities begin to appear. Obviously, there has to be a tradeoff. Since the total recombination rate is more or less constant, irrespective of density, lines that are bright at low densities weaken with increasing density. To illustrate this aspect, in Figure 11 model spectra of F-like Ar X are plotted for two densities, 10^{11} cm^{-3} and 10^{15} cm^{-3} (refer also to Figure 4). The lower of these two densities is below $n_{\rm crit}$, while the higher of the two is above $n_{\rm crit}$. Although a simple case, it is clear that, at the higher density, the spectral structure is more complex. The total line power is approximately the same in either case. This is attributable to the redistribution of population among the low-lying states of the recombining ion, O-like Ar XI, to which level p in our example is analogous.

3.7 Fluorescent K-shell emission

Iron K-shell transition energies fall into the band covered by proportional counters (Makishima 1986), and iron K lines have been observed in virtually all classes of cosmic X-ray source. An entire astrophysics conference[16] devoted to the subject is testament to the wide-ranging interest in iron K emission inspired by the last two decades of X-ray observations since its discovery

[16] See the conference proceedings, *Iron Line Diagnostics in X-Ray Sources*, ed., A. Treves, G.C. Perola & L. Stella (Springer-Verlag), 1991.

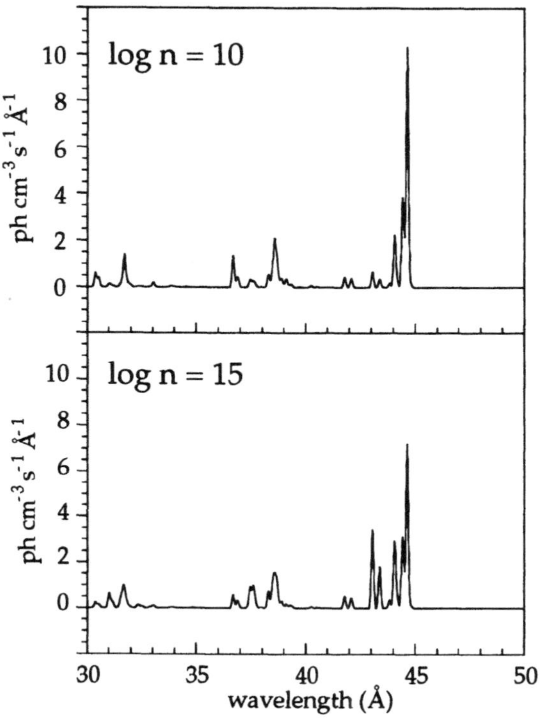

Fig. 11. Ar X line power spectra at two electron densities assuming recombination kinetics with $kT = 10$ eV. Ordinates are arbitrarily, but identically, scaled in order to show the enrichment of a typical L-shell spectrum at densities exceeding $n_{\rm crit}$. Bright features result from $3s - 2p$ transitions. Spectra are convolved with a Gaussian resolution kernel with FWHM of 0.05 Å.

in clusters of galaxies (Mitchell et al. 1976; Serlemitsos et al. 1977). In the context of this Chapter, we are primarily interested in fluorescence lines from low-charge states, which are powered by a hard X-ray continuum. This is to be contrasted with iron K line emission from clusters, which is a tracer of very-high-temperature gas that is in or near CIE.

Results from the *ASCA* mission have provided examples of neutral (or near-neutral) fluorescence from elements besides iron. For example, the presence of K fluorescence lines from neon, magnesium, silicon, sulfur, argon, and calcium is claimed in the spiral galaxy NGC 6552 (Fukazawa et al. 1994). Ebisawa et al. (1996) identify silicon and magnesium fluorescence in the X-ray pulsar Cen X-3. Sako et al. (1998a) identify partially ionized magnesium, silicon, sulfur, argon, and calcium in the eclipse spectrum of the X-ray pulsar Vela X-1 (cf., Nagase et al. 1994; Nagase 1996). Unfortunately, the resolving power of the *ASCA* SIS is not high enough to permit identifications of individual charge states among near-neutral ions. For example, the Kα lines of the ions Si I – Si IV lie within a $\sim$ 40 eV band near 1.75 keV (Kaastra & Mewe 1993). A resolving power ($E/\Delta E$) near 175 would be needed to resolve these lines, whereas the *ASCA* SIS has a resolving power of $\sim$ 30 at this energy. In the future, when instruments of high resolving powers and large collecting areas are flown, it may be possible to perform *ion fluorescence spectroscopy*, where fluorescent lines from individual ions are used to study, in detail, cold material in compact X-ray sources.

Since iron K line emission is so firmly ensconced in the "lore" of X-ray astronomy, it is worth our time to look at fluorescent emission in some detail. In the case of a slab that is optically thick to the ionizing continuum and the K lines themselves, radiative transfer cannot be disregarded, and numerical techniques must be used (Kallman 1991). In the optically thick case, one must also make a distinction between transmitted spectra and reflected spectra. In the case of transmission spectra, the continuum source backlights the slab, and is viewed through the slab. The spectrum is a superposition of the attenuated backlighter spectrum and line features. In the case of reflection, scattered continuum and slab emission, which may be superimposed on the irradiating continuum, make up the spectrum. An important application of reflection spectra occurs in the case of irradiated accretion disks. The so-called *Compton reflection hump*, a broad peak in the spectrum at a few tens of keV, is a signature of reflection.[17] Hard X-ray photons "burrow" into the cold material, where they are absorbed through K-shell photoionization,[18] resulting in fluorescent emission. Overlying layers are maintained at high levels of ionization. Therefore, fluorescent lines, as they propagate toward the observer through regions of increasingly higher ionization levels, are subject to a reduced opacity compared to K line photons that propagate in the other direction, to greater depths in the accretion disk. Introductions to this topic in the context of accretion disks in AGN may be found in Fabian & George (1991) and in George & Fabian (1991), where it is shown that, from simple considerations, an equivalent width in iron K of $\sim$ 150 eV should be typical. In the latter paper, this estimate is verified by detailed calculations, although the actual *observed* value of the equivalent width is shown to depend upon a number of factors, such as the inclination of the irradiated disk with respect to the observer. Here, we will consider only the simpler case of transmission in the optically thin case, since it can be worked out reasonably accurately

[17] The hump results from the deformation of the spectrum of the incident flux (usually a power law form in the case of AGN) through Compton downscattering of hard X rays on cold material, coupled with photoelectric absorption of soft X rays (Lightman & White 1988).

[18] Because of the rapid fall-off in energy of photoionization cross sections, the opacity at energies above the K edge of an ion is dominated by the total K-shell cross section, even when L-shell or M-shell electrons are present.

without the need for numerical methods. We point out that if the slab is optically thin, there is no difference in the K line luminosity when comparing reflection and transmission (Kallman 1991).

At the atomic level, the production of inner-shell K photons involves two or more steps. In the first step, a photon with an energy $\hbar\omega$, which is above the K-shell photoionization threshold energy of an atom A in charge state i, ionizes a $1s$ electron:

$$A_i + \hbar\omega \rightarrow A_{i+1}^{**} + e \quad (1s \text{ photoionization}). \tag{167}$$

The double asterisk denotes a level with an energy above the first ionization potential, an autoionizing level. The excited state can decay through photon emission. This is known as *fluorescence*.

$$A_{i+1}^{**} \rightarrow A_{i+1}^{*} + \hbar\omega' \text{ (fluorescence)} \tag{168}$$

In the radiative decay, a bound electron fills the $1s$ vacancy created in the photoionization step, which may leave a hole in another shell. Therefore, after the first radiative decay, the ion may still be in an excited state (denoted by a single asterisk). It may also happen that the atom, after the first autoionization, is still in an autoionizing configuration.

More often, instead of decaying by a radiative transition, states with holes in the K shell will autoionize (Cowan 1981), ejecting a second electron (the *Auger effect*[19]), which leaves the atom in the next highest stage of ionization, possibly in an excited state:

$$A_{i+1}^{**} \rightarrow A_{i+2}^{*} + e' \quad \text{(autoionization)}. \tag{169}$$

A detail that should be noted involves the excitation state of the atom (A_{i+2}) after the first Auger decay. In our example, although the ion resulting from the autoionization is excited, its energy lies below the first ionization limit — it is *stabilized*, and can autoionize no further. However, it often happens that the first Auger decay does not lead to stabilization, i.e., the atom is left in an autoionizing state of ion $i+2$. It may undergo further Auger decays until it is stabilized. It is common for a series of Auger decays (an *Auger cascade*) to occur in this way. In setting up the equations of ionization balance, we see that Auger cascades can couple several charge states (Weisheit 1974). Moreover, at any step, the atom may stabilize radiatively by a variety of routes. Therefore, the calculation of the overall process can become quite involved, since one may need to calculate millions of transition probabilities to follow the Auger cascade initiated by a single ionization (Kaastra & Mewe 1993).

The most probable radiative decay following the creation of a $1s$–hole state involves a $n = 2 \rightarrow n = 1$ transition — a Kα transition (Kβ for

[19] The terms "Auger effect" and "Auger decay" are reserved for autoionizations for which the initial state configuration has at least one vacancy in an inner shell.

$n = 3 \to n = 1$, etc.). The K *fluorescence yield* is $Y_K = I_K/n_K$, where I_K is the total number of K photons ($n \to 1$) produced from a large sample of atoms, and n_K is the number of K-shell vacancies produced in the sample (Bambynek et al. 1972). The fluorescence yield is a rapidly increasing function of Z. A typical value of Y_K for neutral and near-neutral iron is 0.34, while, for example, Y_K for neutral neon Y_K is only 0.018 (McGuire 1969).

Lacking spectrometers of high resolution, the near-neutral iron K lines are blended in all data from extrasolar X-ray sources. One often speaks of the "iron K line", when it is actually made up of hundreds of lines. Although the excitation conditions in solar flares and tokamaks differ from those considered here, the spectra produced by photoionization should be comparably rich. Thus it is fair to say that some appreciation of the complexity of the iron K spectrum produced in an X-ray photoionized plasma can be gained upon examination of the tokamak spectra shown in Beiersdorfer et al. (1993), and the solar flare spectra shown in Seely, Feldman & Safranova (1986).

An additional, and potentially important, complication involves the excitation state of the ion immediately preceding photoionization. For example, in an iron ion with valence electrons in the M shell, say, Fe IX (Ar-like), at densities exceeding $\sim 10^{10}$ cm^{-3} most of the population is in levels with configurations of the form [Ne] $3s^2 3p^5 3d$. Compared to ionization out of the ground state, creation of a K-shell vacancy from this excited configuration results in a much larger number of autoionizing states, which may lead to changes in the spectral distribution of Kα and Kβ, and may affect the fluorescence yields, as well. A preliminary investigation of this problem in the context of the high-temperature solar corona is presented in Jacobs et al. (1989) for the case of iron K fluorescence in iron L-shell ions. There it is shown that the Kα spectrum is sensitive to changes in the electron density. Needless to say, fully accounting for the level population distribution in a manner consistent with what is known about the ambient plasma conditions and radiation field, and then calculating the resulting CSD and spectrum represents a formidable challenge.

We turn now to the global production of Kα in the optically thin case. We invoke the "cartoon" picture of a Seyfert (Sy) galaxy, although most of the discussion carries over naturally to other sources. The geometrical model derives from the purported connection between Sy 1 and Sy 2 galaxies, which posits that there is no intrinsic difference between the two classes, and that *apparent* differences are simply the consequence of viewing the nuclear regions from different angles (Antonucci & Miller 1985). The orientation enters owing to the presence of a molecular torus which surrounds the nucleus. In the Sy 1 case, we are viewing the nuclear region more or less along the toroidal axis, and obscuration by the torus is not important. In the Sy 2 case, we are viewing the nucleus at larger inclinations, and part of the nuclear region is obscured.

A somewhat abbreviated version of the following derivation appears in Krolik & Kallman (1987), where further discussion may be found. In terms of the geometric picture described above, it is important to note that the space interior to the toroidal boundary is filled with ionized gas, which may extend to high altitudes and be directly visible above the top of the torus, even in the extreme case of a 90° inclination (Antonucci & Miller 1985; Krolik & Begelman 1986).

Let Y_i denote the fluorescence yield for production of Kα. The source-integrated K line luminosity is

$$L_K = \sum_i \int \mathrm{d}V \; n_i(r)\beta_i(r)Y_i\epsilon_{K_{i+1}}, \tag{170}$$

where $\epsilon_{K_{i+1}}$ is the average K line energy of charge state $i+1$, and β_i is the photoionization rate of charge state i at a distance r from the point X-ray source.[20] The photoionization rate is given by

$$\beta_i(r) = \frac{1}{4\pi r^2} \int_{\chi_i}^{\infty} \mathrm{d}\epsilon \; \epsilon^{-1} L_\epsilon \; \sigma_i(\epsilon). \tag{171}$$

where L_ϵ is the monochromatic luminosity of the central source, and χ_i is the K-shell ionization potential of charge state i. This form is consistent with the assumption that the source spectrum is geometrically diluted but not attenuated by absorption.

Adopting a spherical coordinate system, we substitute Eq. (171) into Eq. (170) which gives

$$L_K = \int_\Omega \frac{\mathrm{d}\Omega}{4\pi} \sum_i Y_i\epsilon_{K_{i+1}} \int_{\chi_i}^{\infty} \mathrm{d}\epsilon \; \epsilon^{-1} L_\epsilon \int_{r_0}^{r} \mathrm{d}r \; n_i(r) \; \sigma_i(\epsilon). \tag{172}$$

Distances are measured from the central point source. It should be understood that $\mathrm{d}\Omega$ is a unit of solid angle subtended at the X-ray continuum source by a parcel of gas in the reprocessing medium. The lower limit on the radial integral is set to a nonzero value in order to account for the possibility that part of the fluorescing medium is hidden from our view by the molecular torus.

The optical depth from the nucleus to a "field point" at distance r involves the following sum over all ion stages:

$$\tau(\epsilon, r) = \sum_j \sigma_j(\epsilon) \int_0^r \mathrm{d}s \; n_j(s), \tag{173}$$

[20] Note that we adopt a convention in which fluorescence yields and photoionization rates are labeled such that they refer to the initial (target) ion, but the line energy is labeled according to the ion in which the transition actually occurs. Thus we may speak, for example, of an Fe I fluorescence yield, but clearly, the line observed subsequent to production of a 1s hole through photoionization will correspond to Fe II.

where n_j is the number density of charge state j, and from which

$$d\tau(\epsilon, r) = \sum_j \sigma_j(\epsilon) n_j(r)\, dr. \tag{174}$$

If we make the approximation that $Y_i \epsilon_{K_{i+1}}$ is independent of i, then these two factors can be removed from the summation in Eq. (172), and we let $Y_i \epsilon_{K_{i+1}} = Y \epsilon_K$ for all i. Furthermore, let us assume that χ_i does not vary appreciably from ion to ion, so that the sum can be taken inside the energy integral. Then, using Eq. (174), the radial integral can be expressed in terms of τ, which leads to

$$L_K = Y \epsilon_K \int_\Omega \frac{d\Omega}{4\pi} \int_\chi^\infty d\epsilon\, \epsilon^{-1} L_\epsilon\, [\tau(\epsilon, r) - \tau(\epsilon, r_0)]. \tag{175}$$

Assume that the photoionization cross sections (embedded in the definition of τ) do not vary appreciably from ion to ion. Then dropping the subscript from $\sigma_i(\epsilon)$ gives for the optical depth

$$\tau(\epsilon, r) - \tau(\epsilon, r_0) = \sigma(\epsilon) \sum_i \int_{r_0}^r ds\, n_i(s) = \sigma(\epsilon) N_Z, \tag{176}$$

thus leaving a simple expression for the optical depth in terms of the total column density for the element, N_Z. Updating the expression for the K line luminosity,

$$L_K = Y \epsilon_K \int_\Omega \frac{d\Omega}{4\pi} N_Z \int_\chi^\infty d\epsilon\, \epsilon^{-1} L_\epsilon\, \sigma(\epsilon). \tag{177}$$

Turning now to the energy integral, we assume an inverse cube form for the photoionization cross section: $\sigma(\epsilon) = \sigma(\chi)\, (\epsilon/\chi)^{-3}$. For the ionizing continuum, assume a power law shape $L_\epsilon \propto \epsilon^{-\nu}$, and normalize to a luminosity L on the energy interval $[\epsilon_0,\ \epsilon_1]$. With the normalization requirement, the ionizing spectrum can be written as

$$L_\epsilon = (\nu - 1)\, C_\nu \frac{L}{\epsilon_0} \left(\frac{\epsilon}{\epsilon_0}\right)^{-\nu} \tag{178}$$

where

$$C_\nu = \left[1 - \left(\frac{\epsilon_0}{\epsilon_1}\right)^{\nu - 1}\right]^{-1}. \tag{179}$$

(Note that this form is valid for either $\nu > 1$ or $\nu < 1$.) Substituting Eq. (178) into Eq. (177) yields, after evaluation of the energy integral,

$$L_K = LY\, C_\nu \frac{\nu - 1}{\nu + 3} \left(\frac{\epsilon_0}{\chi}\right)^{\nu - 1} \frac{\epsilon_K}{\chi} \int_\Omega \frac{d\Omega}{4\pi} N_Z\, \sigma(\chi). \tag{180}$$

The remaining integral over solid angle is just the spherical average of the photoionization optical depth at threshold, so that

$$L_K = LY\, C_\nu \,\frac{\nu-1}{\nu+3} \left(\frac{\epsilon_0}{\chi}\right)^{\nu-1} \frac{\epsilon_K}{\chi} \,\langle\tau(\chi)\rangle_\Omega . \tag{181}$$

The subscript Ω signifies that the average is taken over 4π steradians. It is more useful to perform the angle integration only over that portion of solid angle $\Delta\Omega$ with a line of sight to the nucleus which is not obstructed by the molecular torus. In that case

$$L_K = LY\, C_\nu \,\frac{\nu-1}{\nu+3} \left(\frac{\epsilon_0}{\chi}\right)^{\nu-1} \frac{\epsilon_K}{\chi}\,\frac{\Delta\Omega}{4\pi} \,\langle\tau(\chi)\rangle_{\Delta\Omega} . \tag{182}$$

We need to make a distinction between the intrinsic line luminosity and the line luminosity that would be inferred by an observer, where the difference arises from geometrical effects associated with the molecular torus. Likewise for the ionizing continuum. This can be accomplished by studying the observed K line equivalent widths for two scenarios, and relates to our earlier discussion concerning the connection between the Sy1/Sy2 classification scheme and the angle at which we view the nuclear region. Two extreme cases are treated: (1) our vantage point allows full access to the nuclear continuum source, and K lines are produced in an overlying layer of gas, and (2) the central X-ray source is completely blocked, so that we observe only those continuum X rays that are scattered into our line of sight by gas which lies above the torus. This gas also reprocesses the continuum into lines, including fluorescent K lines. The observable difference, in terms of K emission, is that the line equivalent width should be much larger in the Sy2 case, since only a fraction of the continuum is observed. Of course, this presupposes that there is a sufficient quantity of gas with electrons bound to nuclei above the top of the torus. Intermediate cases also exist. For example, the torus need not absorb all of the photons near 6–7 keV (Mulchaey, Mushotzky & Weaver 1992). In that case, the observed continuum is a superposition of a continuum component that is heavily absorbed, and a continuum component with relatively little absorption (the scattered component).

For the Sy1 case, where we view the continuum X-ray source directly, the equivalent width of the summed K features can be written as

$$EW\Big|_{\mathrm{Sy\ 1}} = \frac{L_K}{L_\epsilon(\epsilon_K)} = \frac{\Delta\Omega}{4\pi}\,\frac{1}{\nu+3}\, Y\epsilon_K \left(\frac{\epsilon_K}{\chi}\right)^{\nu} \langle\tau(\chi)\rangle_{\Delta\Omega} \tag{183}$$

For the Sy2 case, we need to determine the fraction of the intrinsic continuum flux that is scattered into our line of sight. The luminosity of scattered continuum radiation is

$$(L_{\mathrm{sc}})_\epsilon = \int_V \mathrm{d}V\, [j_{\mathrm{sc}}(r)]_\epsilon . \tag{184}$$

We assume coherent scattering (Thomson scattering) for the local scattered monochromatic emissivity, so that

$$[j_{\rm sc}(r)]_\epsilon = \frac{1}{4\pi r^2}\, n_{\rm e}\sigma_{\rm c} L_\epsilon, \tag{185}$$

which, when substituted back into Eq. (184) gives

$$(L_{\rm sc})_\epsilon = L_\epsilon \int_{\Delta\Omega} \frac{{\rm d}\Omega}{4\pi} \int_{r_0}^{r} {\rm d}s\; n_{\rm e}(s)\sigma_{\rm c} = L_\epsilon \int_{\Delta\Omega} \frac{{\rm d}\Omega}{4\pi} \int_{\tau_{\rm c}(r_0)}^{\tau_{\rm c}(r)} {\rm d}\tau_{\rm c}' . \tag{186}$$

Denote by $\Delta\tau_{\rm c}$ the difference $\tau_{\rm c}(r) - \tau_{\rm c}(r_0)$. Then

$$(L_{\rm sc})_\epsilon = L_\epsilon \left(\frac{\epsilon}{\chi}\right)^{-\nu} \frac{\Delta\Omega}{4\pi} \langle\Delta\tau_{\rm c}\rangle_{\Delta\Omega} \tag{187}$$

Therefore, the equivalent width for this scenario is

$$EW\Big|_{\rm Sy\ 2} = \frac{1}{\nu+3}\, Y\epsilon_K \left(\frac{\epsilon_K}{\chi}\right)^{\nu} \frac{\langle\Delta\tau(\chi)\rangle_{\Delta\Omega}}{\langle\Delta\tau_{\rm c}\rangle_{\Delta\Omega}} \tag{188}$$

It is straightforward to show that, with the approximations we have made here, the ratio of angle-averaged optical depths in Eq. (188) is equal to the angle-averaged ratio of the optical depths, where this ratio can be expressed in terms of the elemental abundance A_Z as

$$\frac{\Delta\tau(\chi)}{\Delta\tau_{\rm c}} = A_Z\, \frac{\sigma(\chi)}{\sigma_{\rm c}}\, \frac{n_{\rm H}}{n_{\rm e}} . \tag{189}$$

Finally, we have for the equivalent width for the extreme Sy 2 case,

$$EW\Big|_{\rm Sy\ 2} = \frac{1}{\nu+3}\, A_Z\, Y\epsilon_K \left(\frac{\epsilon_K}{\chi}\right)^{\nu} \frac{\sigma(\chi)}{\sigma_{\rm c}}\, \frac{n_{\rm H}}{n_{\rm e}} . \tag{190}$$

Again, we will use iron as an example. If we take the iron abundance to be 4×10^{-5} (Anders & Grevesse 1989), and a typical K photoionization cross section to be 2×10^{-20} cm^2, then the ratio in Eq. (189) is close to unity. Therefore, we obtain a very simple result:

$$EW\Big|_{\rm Sy\ 2,\ Fe} = \frac{1}{\nu+3}\, Y\epsilon_K \left(\frac{\epsilon_K}{\chi}\right)^{\nu} \frac{A_{\rm Fe}}{4\times 10^{-5}} \tag{191}$$

As discussed in Krolik & Kallman (1987), the predicted EW for this ideal case is nearly free of geometrical uncertainties, as well as the details of atomic physics and radiative transfer. Numerically, we expect the typical iron K equivalent width for a Sy2 to be ~ 500 eV.

3.8 Dielectronic recombination in X-ray photoionized plasmas

In §3.7 we showed that discrete quantum states with energies above the ionization limit — autoionizing levels — can be created by inner-shell photoionization. Another way to create autoionizing levels is known as *radiationless capture*. Let us introduce some notation to aid in the description of the process. Radiationless capture occurs when charge state z, in quantum state i (initial), captures a free electron, forming the autoionizing level d (doubly excited) in the adjacent charge state $z-1$ (see Figure 12),

$$A_{z,i} + e \rightarrow A^{**}_{z-1,d}. \tag{192}$$

This "reaction" must, of course, conserve energy. The energy lost by the free electron, instead of being carried away by a photon, as occurs in RR, is expended in exciting the core of the ion. The term "doubly-excited" is intended to imply that two electrons occupy subshells above those which they would occupy if the ion were in its ground configuration. The level d is a short-lived "compound" atomic state, which, though discrete, is coupled to the continuum. A perturbing Hamiltonian composed of electron-electron interactions can break the compound state into its component parts (autoionization), a free electron and an ion. Therefore, radiationless capture is the inverse of autoionization, and capture rates can be found by invoking a detailed balance argument, in analogy with the use of the Milne relation to relate RR cross sections to photoionization cross sections (see Rolf Mewe's Chapter in this Volume).

If, instead of autoionizing, the doubly excited ion emits a photon, then it may stabilize. Denoting by f the stabilized level of charge state $z-1$, stabilization can be represented by

$$A^{**}_{z-1,d} \rightarrow A^{*}_{z-1,f} + \hbar\omega. \tag{193}$$

The two-step process ($i \rightarrow d \rightarrow f$) described by Eqs (192) and (193) is called *dielectronic recombination* (DR) and is, along with RR, one of the most important recombination processes in X-ray photoionized plasmas.

The radiationless capture cross section is highly structured, which can be understood by considering the simple example of capture into high-n levels. The contribution by the captured electron to the total energy of the doubly-excited level can be approximated by $-(Z^2_{\rm eff}/n^2)$ Ry, where $Z_{\rm eff}$ is the "effective" nuclear charge. If we think of the captured electron as a test particle in the field of a screened Coulomb potential, then $Z_{\rm eff} = Z - N + 1$ for N bound electrons, including the captured electron. The Rydberg energy ($\approx$ 13.6 eV) is denoted Ry. The energy lost by the captured electron excites a core electron, with the core excitation energy given by ΔE. Energy conservation requires

$$\frac{1}{2}mv^2 = \Delta E - \frac{Z^2_{\rm eff}}{n^2}\,\text{Ry} \tag{194}$$

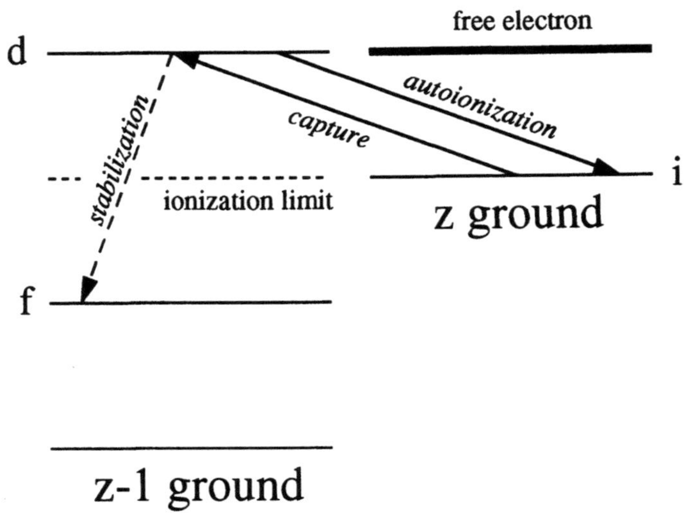

Fig. 12. Schematic of the DR process. The ground states of two adjacent charge states, z and $z-1$, are shown. Levels f and d are a bound level and an autoionizing level, respectively, of $z-1$ (refer to text). Thick line in upper right of diagram indicates a free electron with a kinetic energy such that the resonance condition is met, thus enabling radiationless capture into d.

This defines a resonance condition which must be met by free electrons for radiationless capture to occur. (Actual calculations must, of course, account for the detailed energy level structure.) Equation (194) also implies that there is a minimum value of n for which energy conservation can be satisfied: $n > n_{\rm min} = Z_{\rm eff}(\Delta E/{\rm Ry})^{-1/2}$. This is a simple statement of the requirement that the level formed by radiationless capture must lie above the ionization limit. For small values of the core excitation energy ΔE, $n_{\rm min}$ is relatively large, compared to the opposite case of large ΔE. Equation (194) also shows us that when ΔE is relatively small (large), only electrons with comparably small (large) kinetic energies can meet the resonance condition.

Looking at this last point in another way, we can say that the lower the plasma temperature, the more that radiationless capture favors low core excitation energies and, therefore, capture into high-n levels. This is of great importance for DR in PIE, and can be quantified. Let $A^{\rm a}$ and $A^{\rm r}$ represent an autoionization rate and a radiative transition rate, respectively. Bates & Dalgarno (1962) give the rate coefficient (c.g.s. dimensions $\rm cm^3\ s^{-1}$) for DR through the channel $i \to d \to f$,

$$\alpha_{idf} = 4\pi^{3/2} a_0^3 \, \frac{g_d}{g_i} \, e^{-E_{di}/kT} \left(\frac{kT}{\rm Ry}\right)^{-3/2} \frac{A^{\rm a}_{di} A^{\rm r}_{df}}{\sum_m A^{\rm a}_{dm} + \sum_n A^{\rm r}_{dn}}, \qquad (195)$$

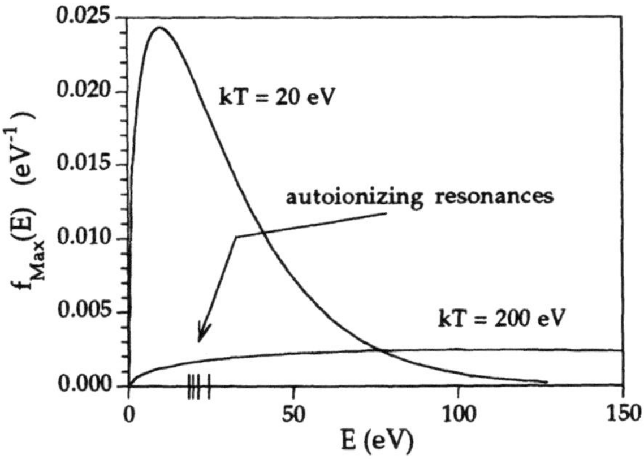

Fig. 13. Maxwellian electron energy distributions for two temperatures. Positions of capture resonance energies corresponding to $\Delta n = 0$ captures are superimposed, indicated by arrow. Peaking of Maxwellian in the low-temperature case shows that, in photoionized plasmas, $\Delta n > 0$ DR cannot compete with $\Delta n = 0$ DR, since resonances corresponding to $\Delta n > 0$ DR have energies of several hundred eV.

where a_0 is the hydrogen Bohr radius, and g refers to a statistical weight. The *resonance energy*, $E_{di} \equiv E_d - E_i$. [Note that the right-hand-side of Eq. (194) is an approximation of E_{di}.] Subscripts on A^{a} and A^{r} represent the initial state and final state, respectively, of the transition. The levels indicated by subscripts n are levels in charge state $z-1$ which can be reached through downward radiative transitions from d. The levels indicated by subscripts m are bound states in charge state z to which d may autoionize. When energetically possible, d may autoionize to a level in charge state z other than i, i.e., $m \neq i$. If the energy of m exceeds that of i, then the overall process resembles a collisional excitation — an excited ion and an outgoing electron with an energy less than it had before the excitation. This is known as *resonance excitation*, since only electrons meeting the resonance criterion are eligible to contribute to the excitation.

A triple summation over i, d, and f in Eq. (195) gives the total DR rate coefficient for charge state z recombining to $z-1$. For most cases of astrophysical interest, the sum over autoionization rates is far larger than the sum over radiative rates in the denominator of Eq. (195). Moreover, it is often found that the term dominating the sum over autoionization rates corresponds to $m = i$. Then α_{idf} depends primarily upon the Einstein coefficient A^{r}_{df}. Since electric dipole transitions scale as Z^4_{eff}, DR has a much greater relative importance for high-Z elements than for low-Z elements. Here, however, we are not so much interested in the total rates as in the spectroscopic consequences of DR in overionized plasmas.

Equation (195) shows that, for a given temperature, DR through a particular autoionizing resonance d is a maximum when $E_{di} = (3/2)kT$. This reinforces our earlier conclusion that low-lying resonances dominate the DR rate in PIE, whereas higher-lying resonances dominate in CIE. This point is illustrated in Figure 13.

A specific example helps to clarify the difference between DR in CIE and DR in PIE. Consider the case of F-like iron recombining to a Ne-like iron through DR (inactive $1s^2$ implied),

$$2s^2 2p^5 + e \rightarrow 2s^2 2p^4\ nl\ n'l' \qquad (n,\ n' \geq 3) \tag{196}$$

When the core excitation involves a transition to a higher shell, as it does here, then the capture is called a $\Delta n > 0$ capture. The core excitation energy corresponding to $2p \rightarrow 3l$, for example, is ~ 800 eV. The binding energy of the captured electron is ~ 400 eV for $n' = 3$, with smaller binding energies for higher n'. Therefore, E_{di} is at least ~ 400 eV. For CIE plasmas, this is just the right range of temperatures for $\Delta n > 0$ DR to be effective, since F-like iron has a high fractional abundance at the temperature (~ 800 eV) for which there is a large population of electrons with energies near 400 eV (Arnaud & Raymond 1992). For comparison, the photoionization equilibrium models of Kallman & McCray (1982) show that electron temperatures in the Fe^{17+} zone are 10-20 eV, so that typical electrons have energies far below what is needed to populate doubly-excited levels via $\Delta n > 0$ captures.

Suppose that the core excitation involves a transition to a higher-energy subshell within the same shell. The capture is called a $\Delta n = 0$ capture. For example,[21] again considering the capture into Ne-like iron starting from F-like iron,

$$2s^2 2p^5 + e \rightarrow 2s 2p^6 nl \qquad (n \geq 6). \tag{197}$$

Note that $n_{\rm min}$ is greater in this case than for a $\Delta n > 0$ capture. For smaller n, even electrons hovering just above the ionization limit lose too much energy to be compensated by the core excitation $2s \rightarrow 2p$. The core excitation energy is 136 eV in F-like iron, so, according to Eq. (194), only electrons with energies below 136 eV can be captured.

Based upon the previous example, our conclusion concerning the prevalence of captures involving small core excitation energies for overionized plasmas can be refined as follows: *under conditions of PIE, radiationless capture, hence DR, proceeds primarily through $\Delta n = 0$ core excitations*. Since H-like and He-like ions have no available $\Delta n = 0$ excitations, DR in these species occurs only through $\Delta n > 0$ channels. Thus it follows that DR is not as

[21] An additional core excitation, $2p_{1/2} \rightarrow 2p_{3/2}$, should be included in calculations of the $\Delta n = 0$ DR rate at low temperatures (Savin et al. 1997). For the recombination $Fe^{17+} \rightarrow Fe^{16+}$, $\Delta E = 12.7$ eV (Shirai et al. 1990), and $n_{\rm min} = 18$.

important for K-shell ions as for L-shell ions. In fact, for K-shell ions in PIE, RR completely dominates the total recombination. rate. By contrast, DR can dominate the total recombination rate in L-shell ions, depending on the charge state and the temperature.

Compared to plasmas in CIE, DR in X-ray photoionized plasmas takes on a somewhat different significance, for the same reason that RR is more important in PIE than in CIE, namely, that recombination controls not just the charge state distribution, but also the X-ray line spectrum. For the reasons given above, we may restrict the discussion to recombinations in which the initial charge state has at least one L-shell electron.

The first step in $\Delta n = 0$ DR of L-shell ions has the general form

$$2s^q 2p^r \rightarrow 2s^{q-1} 2p^{r+1}\ nl \qquad (n \geq n_{\mathrm{min}}). \tag{198}$$

In addition to allowed autoionization channels, there are two types of radiative transition that may follow the capture step, the relaxation of the core

$$2s^{q-1} 2p^{r+1}\ nl \rightarrow 2s^q 2p^r\ nl + \hbar\omega, \tag{199}$$

or a jump of the outer electron to a lower energy subshell,

$$2s^{q-1} 2p^{r+1}\ nl \rightarrow 2s^{q-1} 2p^{r+1}\ n'l' + \hbar\omega, \tag{200}$$

which stabilizes the ion, if $n' < n_{\mathrm{min}}$.

The line emitted when the core decays, as in Eq. (199), is what is usually referred to as a *satellite line.* The presence of the outer electron, the *spectator electron*, in both the initial and final configurations, causes the wavelength of the satellite line to be shifted, usually to a longer wavelength, with respect to the so-called *parent line*, produced by the transition $2s^{q-1}2p^{r+1} \rightarrow 2s^q 2p^r$ in the adjacent charge state. As n increases, the effect of the spectator on the atomic potential gradually disappears, and the wavelengths of corresponding satellite lines converge to that of the parent line. Satellite lines produced by $2p \rightarrow 2s$ transitions lie in the EUV band for most ions relevant to X-ray spectroscopy. Since the core decay stabilizes the ion, the outer electron will decay radiatively in a series of steps, and the ion eventually reaches the ground configuration. Transitions to the L-shell, the terminal step in the cascade, simply add flux to the same X-ray lines produced by cascade following RR.

The alternate radiative decay given by Eq. (200) can, in many cases, be the first step in a radiative cascade in which the $2s$ hole remains open — the core is "frozen". The ion may reach its ground configuration through emission of a $3p \rightarrow 2s$ X-ray line, which terminates the cascade. The efficient opening of the $2s$ subshell by $\Delta n = 0$ radiationless capture leads to the unique signature of DR, namely, bright $3p \rightarrow 2s$ lines, in an X-ray photoionized plasma. Since the temperature dependences of DR and RR differ, the intensities of lines driven selectively by DR compared to RR-driven lines forms the basis of a new class of temperature diagnostic in X-ray photoionized plasmas. Some preliminary work to develop these diagnostics for iron L-shell ions is discussed briefly in Kahn & Liedahl (1995), and at greater length in Liedahl (1992).

4 Transient Phases of Ionization Disequilibrium

A plasma can be characterized by several time scales, for example, its cooling time scale or its dynamical time scale. When a given property of a plasma is forced to change, the remaining properties that describe the plasma state will evolve until a new equilibrium state is attained. Thus a common scenario in astrophysics involves a time-dependent CSD, which is responding to a variable temperature. The time scale for a change in the temperature may be considerably shorter than that for changes in the CSD. The radiative properties of an ion in such a plasma can differ substantially from those found in CIE (Shapiro & Moore 1977; Mewe & Schrijver 1978).

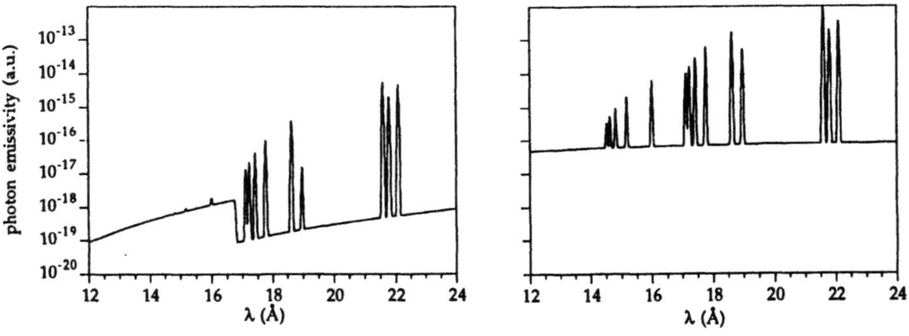

Fig. 14. (*Left panel*) Oxygen spectrum in CIE, $T = 10^6$ K. (*Right panel*) The result of instantaneously stepping the temperature to $T = 10^{6.8}$ K. Emission lines are from O VII and O VIII. Model spectra have been convolved with a Gaussian resolution kernel with FWHM of 0.05 Å. The ordinate is arbitrarily scaled and has dimensions photons $\mathrm{cm}^{-3}\ \mathrm{s}^{-1}\ \text{Å}^{-1}$.

This picture is motivated by the theory of supernova remnants, wherein gas is impulsively heated by a shock wave propagating more or less radially with respect to the explosion site. The rate at which the temperature rises is not matched by the rate at which the ionization level rises — the plasma is initially underionized. Eventually, the time profile of the local temperature will level off, and the CSD approaches CIE asymptotically. Deviations from

CIE can lead to observable spectroscopic effects (Shull 1982), and have been identified in X-ray data from supernova remnants (Winkler et al. 1981).

Crucial to the formation of the X-ray spectrum is that the approach to equilibrium is characterized by the condition $(T_{\rm peak})_{\rm NIE} > (T_{\rm peak})_{\rm CIE}$. The ions that dominate the CSD at any given time interact with a population of electrons with a distribution weighted towards higher kinetic energies compared to the electron distribution corresponding to a comparable level of ionization in CIE.

The mismatch (relative to CIE) of electron temperature and ionization level has a simple spectroscopic signature, which is illustrated in Figure 14. The left panel shows the spectrum of a hydrogen-helium-oxygen plasma in CIE with $T = 10^6$ K. The oxygen CSD is dominated by O VII and O VIII,[22] as evidenced by bright emission lines from these ions. Certain spectral features are worth pointing out here, since we return to a discussion of oxygen spectra in §4.5. The emission lines are primarily from O VII, including the resonance $(1s^2\ ^1S_0 - 1s2p\ ^1P_1)$, intercombination $(1s^2\ ^1S_0 - 1s2p\ ^3P_1)$, and forbidden $(1s^2\ ^1S_0 - 1s2s\ ^3S_1)$ lines near 22 Å. The O VIII Lyα analog is at 19 Å, and the Lyβ analog is at 16 Å. The continuum consists of hydrogen and helium bremsstrahlung and RRC from O VII and O VIII. The O VII RRC can be seen at 16.7 Å. In order to obtain realistic line equivalent widths, the oxygen and helium abundances relative to hydrogen have been set to their solar photospheric values.

The spectrum immediately after a step-function temperature increase to $10^{6.8}$ K is shown in the right panel of Figure 14. Accompanying the temperature rise is a large increase in both the line and bremsstrahlung emissivities. The CSD, however, is identical to its initial configuration. The CSD cannot change arbitrarily fast, since it depends upon the ionization and recombination rate coefficients (see §4.1). If the plasma were in CIE at this higher temperature, then oxygen would be fully stripped, and the oxygen line emission would be negligible. Therefore, this continuum, accompanied by bright line emission, is inconsistent with CIE. One (faulty) interpretation of such a spectrum might be that emission from two plasmas, each with a different temperature, are superposed. The presence of the lower temperature component would be invoked to account for O VII and O VIII emission. This "two-temperature" appearance is a characteristic of an isothermal plasma in which ionization equilibrium has not been reached.

Before further studying the spectroscopic consequences of NIE, we look at the basics of ionization dynamics.

---

[22] An ion, for example, He-like oxygen, should be designated by O^{6+}, rather than O VII. The Roman numeral form is reserved for line labeling. For convenience, we dispense with this formality, and use the latter form for both.

4.1 Equilibration time and ionization time

Let n_i represent the number density of charge state i in a volume element of arbitrary size. Formally, the time-dependent ionization equations arise from a generic rate equation, which, for unspecified sources and sinks, can be written as

$$\frac{\partial n_i}{\partial t} + \nabla \cdot (n_i \mathbf{v}_i) = \text{sources} - \text{sinks}. \tag{201}$$

The divergence term accounts for net flow across the surface bounding the volume element. Let α_k denote the total recombination rate coefficient out of charge state k ($k \to k-1$), and let C_k denote the collisional ionization rate coefficient out of charge state k ($k \to k+1$). Both α_k and C_k have c.g.s. dimensions $\mathrm{cm^3\,s^{-1}}$. If we assume a static medium, or enter a comoving frame of the volume element, and assume that only two-body processes are important, Eq. (201) simplifies to

$$\frac{\mathrm{d}n_i}{\mathrm{d}t} = n_\mathrm{e} C_{i-1} n_{i-1} - n_\mathrm{e}(C_i + \alpha_i) n_i + n_\mathrm{e} \alpha_{i+1} n_{i+1} \tag{202}$$

The sink terms are ionization and recombination out of i, with rates proportional to n_i. The source terms are recombination from $i+1$ to i, with a rate proportional to n_{i+1}, and ionization from $i-1$ to i, with a rate proportional to n_{i-1}. In cases where photoionization is more important than collisional ionization, Eq. (202) becomes

$$\frac{\mathrm{d}n_i}{\mathrm{d}t} = \beta_{i-1} n_{i-1} - (\beta_i + n_\mathrm{e}\alpha_i) n_i + n_\mathrm{e} \alpha_{i+1} n_{i+1} \tag{203}$$

where β denotes a photoionization rate. Hereafter, we focus on NIE plasmas for which electron impact ionization dominates over photoionization.

A dimensional analysis of Eq. (202) shows that the relevant time scale in the approach to ionization equilibrium is roughly $t_\mathrm{eq} \sim [n_\mathrm{e}(C_i + \alpha_i)]^{-1}$, where we have defined the *ionization equilibration time scale*. In other words, we expect solutions to contain terms with decaying exponentials of the form $\exp(-t/t_\mathrm{eq})$. Since CIE is, for most cases of astrophysical interest, fully characterized by the electron temperature, changes in the temperature on time scales shorter than t_eq will initiate a phase of NIE conditions. For rate coefficients $\sim 10^{-11}\ \mathrm{cm^3\,s^{-1}}$, t_eq is hundreds of years in a supernova remnant, but only a few seconds for stellar coronae.

In cases where the rate coefficients are functions of temperature but not of density, one usually defines the *ionization time*,[23] whose differential is given by $\mathrm{d}\tau = n_\mathrm{e}\,\mathrm{d}t$. The equations of ionization balance are then given by

$$\frac{\mathrm{d}n_i}{\mathrm{d}\tau} = C_{i-1} n_{i-1} - (C_i + \alpha_i) n_i + \alpha_{i+1} n_{i+1} \tag{204}$$

[23] The term *ionization parameter*, which is securely entrenched in the lexicon of X-ray nebulae, should not be used here.

The ionization time has dimensions cm^{-3} s, the inverse of rate coefficients. To take an example, assume that at $t = 0$ a rapid rise in the local electron temperature occurs in a gas with a density of 1 cm^{-3}. If the density remains constant, then after 10^3 years, the gas has an ionization time $\sim 3 \times 10^{10}$ cm^{-3} s. In more realistic cases, where the electron density changes in time, the ionization time can be expressed as a time integral of the electron density,

$$\tau = \int_0^t \mathrm{d}t' \, n_{\mathrm{e}}(t'). \tag{205}$$

4.2 A two-stage system

Let us first work out the simplest possible example: a two-stage system described by the level population densities n_1 and n_2. From Eq. (204), the system of equations to be solved is

$$\frac{\mathrm{d}n_1}{\mathrm{d}\tau} = -Cn_1 + \alpha n_2 \tag{206}$$

$$\frac{\mathrm{d}n_2}{\mathrm{d}\tau} = Cn_1 - \alpha n_2. \tag{207}$$

The subscripts on the rate coefficients have been dropped, since there is no risk of ambiguity. Adding these equations shows that the total population $n_1 + n_2$ is constant in time, which we denote by n. Suppose that, at $t = 0$ (i.e., $\tau = 0$), $n_1 = n$ and $n_2 = 0$. Then the solutions are given by

$$n_1(\tau) = n \, \frac{\alpha}{C + \alpha} \left(1 + \frac{C}{\alpha} \, e^{-(C+\alpha)\tau} \right) \tag{208}$$

$$n_2(\tau) = n \, \frac{C}{C + \alpha} \left(1 - e^{-(C+\alpha)\tau} \right) \tag{209}$$

The transient behavior is "damped out" after a time $[n_{\mathrm{e}}(C + \alpha)]^{-1}$, corresponding to t_{eq} discussed above. The populations n_1 and n_2 evolve to their CIE values — $n_1 \to n\alpha/(C + \alpha)$, and $n_2 \to nC/(C + \alpha)$ — which can be obtained by setting the time derivatives in Eqs (206) and (207) to zero, and invoking the constraint $n_1 + n_2 = n$.

4.3 A three-stage system

While the two-stage system is trivial, the tedium involved in obtaining closed-form solutions of the time-dependent equations for systems with more stages begins to test the limits of human patience. Here we will work out the solutions for a three-stage system, but make a few simplifying assumptions. The purpose of this exercise is principally to provide a "toy" atom, wherein the

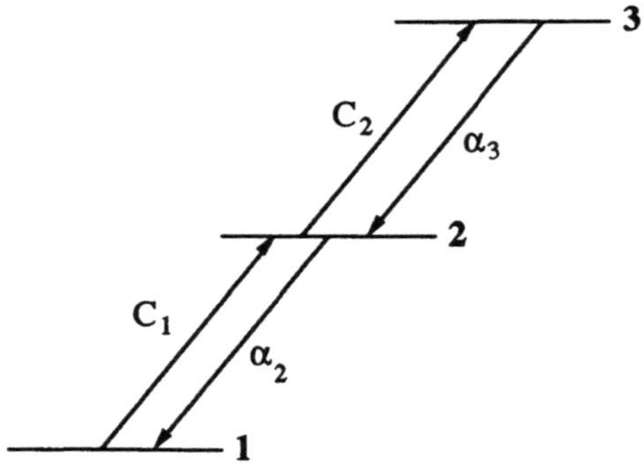

Fig. 15. Schematic of the three-level system used in §4.3. Charge states, 1, 2, and 3 are indicated. Collisional ionization rate coefficients are labeled C, while recombination rate coefficients are labeled α.

relative rate coefficients can be easily adjusted and the consequent time evolution studied. More complex models are best studied numerically, an example of which is provided in §4.5.

A schematic of a three-level system is shown in Figure 15. From this figure, we can write down the system of equations to be solved.

$$\frac{dn_1}{d\tau} = -C_1 n_1 + \alpha_2 n_2 \tag{210}$$

$$\frac{dn_2}{d\tau} = C_1 n_1 - (C_2 + \alpha_2) n_2 + \alpha_3 n_3 \tag{211}$$

$$n_3 = n - n_1 - n_2 \tag{212}$$

These equations can be combined to give

$$\frac{d^2 n_2}{d\tau^2} + \epsilon \frac{dn_2}{d\tau} + \phi n_2 = C_1 \alpha_3 n \tag{213}$$

where, for the sake of notational simplicity, we let

$$\epsilon \equiv C_1 + C_2 + \alpha_2 + \alpha_3 \qquad \phi \equiv C_1 C_2 + C_1 \alpha_3 + \alpha_2 \alpha_3 \tag{214}$$

There is a similar equation for n_1. After solving for n_1 and n_2, Eq. (212) then gives n_3. To simplify the ensuing algebra, assume $C_1 = C_2 \equiv C$ and $\alpha_2 = \alpha_3 \equiv \alpha$. Also, define a dimensionless ratio of the two rate coefficients $\phi \equiv C/\alpha$.

Now, we need initial conditions. Since the differential equation is second order, we need initial conditions for n_1, n_2, and their first derivatives. Let $n_1(0) = n$; $n_2(0) = 0$; $n_3(0) = 0$. We determine $\dot{n}_1(0)$ and $\dot{n}_2(0)$ by referring back to Eqs (210) and (211): $\dot{n}_1(0) = -Cn; \dot{n}_2(0) = Cn$. Thus we find

$$\frac{n_1}{n} = p(\phi)\left[1 + \left(\frac{\phi^2 + \phi^{3/2} + \phi}{2}\right) e^{-F_-\tau} + \left(\frac{\phi^2 - \phi^{3/2} + \phi}{2}\right) e^{-F_+\tau}\right] \quad (215)$$

$$\frac{n_2}{n} = \phi p(\phi)\left[1 + \left(\frac{\phi^{3/2} - 1}{2}\right) e^{-F_-\tau} - \left(\frac{\phi^{3/2} + 1}{2}\right) e^{-F_+\tau}\right] \quad (216)$$

$$\frac{n_3}{n} = \phi^2 p(\phi)\left[1 - \left(\frac{\phi + \phi^{1/2} + 1}{2\phi^{1/2}}\right) e^{-F_-\tau} + \left(\frac{\phi - \phi^{1/2} + 1}{2\phi^{1/2}}\right) e^{-F_+\tau}\right] \quad (217)$$

where $p(\phi) = (\phi^2+\phi+1)^{-1}$, $F_+ = \alpha(\phi+\phi^{1/2}+1)$, and $F_- = \alpha(\phi-\phi^{1/2}+1)$. The form of the equilibration time is somewhat more complex than that found for the two-level system. In Figure 16, where an example of the behavior of this system is shown for the case $\phi = 2$, the charge state fractions are plotted as a function of $n_e\alpha t$.

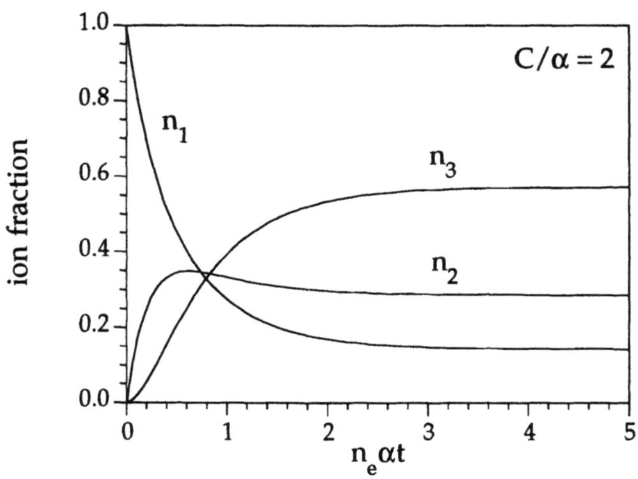

Fig. 16. Evolution of a three-stage system subjected to a step function temperature impulse for the case $\phi \equiv C/\alpha = 2$. Time axis is given as a multiple of the recombination time scale.

Note that the choice of ϕ is determined implicitly by the choice of temperature, and is constant in this simple case, since the temperature is constant. A more general case involves a varying temperature, in which case the rate coefficients themselves become functions of time. Such cases must be handled numerically.

The behavior of this system in the limit $\phi \gg 1$ represents the time-dependent ionization of a system subjected to a severe temperature impulse. A reduction of Eqs (215)–(217) by repeated application of L'Hopital's Rule can be avoided by simply rewriting Eqs (210)–(212) for this special case. The condition $\phi \gg 1$ can be approximated by setting the recombination rate coefficients to zero. The solutions are then easily obtained:

$$n_1 = ne^{-C\tau} \tag{218}$$

$$n_2 = nC\tau e^{-C\tau} \tag{219}$$

$$n_3 = n\left[1 - (1 + C\tau)e^{-C\tau}\right] \tag{220}$$

The solutions are similar to those in the previous example, except that, instead of approaching finite steady-state charge state fractions, n_1 and n_2 vanish as $t \to \infty$, and all of the population ends up in level 3.

4.4 Metastable energy levels in rapidly ionizing plasmas

In this example we investigate the possibility of detecting transient ionization by measuring line ratios involving metastable levels. It also allows us to introduce the important concept of a quasi-steady-state plasma. As illustrated in Figure 17, our model system consists of three charge states, $i-1$, i, and $i+1$, and four energy levels, labeled 1–4. Unlike the previous examples in §4, we include processes that are internal to an ion, in this case, two rates internal to charge state i: a radiative rate $A \equiv A_{21}$ and a collisional rate coefficient $\gamma \equiv \gamma_{12}$. Level 2 is assumed to be metastable against radiative decay to level 1. For initial conditions, assume that at $t = 0$, all of the population is in charge state $i - 1$, i.e., $n_3(0) = n$. The scenario to be modeled is that of a rapidly ionizing plasma, so that we neglect all recombinations. This allows us to treat level 4 as a sink for population flux, and to neglect its effect as a source. Our neglect of recombinations approximates a situation in which the temperature jumps to a value sufficiently high so that the charge state fractions of our three ions approach zero as the system evolves to steady state. In other words, we assume the presence of several charge states more ionized than $i + 1$. We make a further simplification and assume that all collisional ionization rate coefficients are equal. Finally, we can write down the ionization equations.

$$\frac{dn_1}{dt} = -n_e(\gamma + C)\, n_1 + An_2 + n_e C n_3 \tag{221}$$

$$\frac{dn_2}{dt} = n_e \gamma n_1 - (A + n_e C)\, n_2 \tag{222}$$

$$\frac{dn_3}{dt} = -n_e C n_3 \tag{223}$$

Note that level 4 is being treated as a "phantom" level, and is not explicitly included in the system of equations.

The detailed set of initial conditions are as follows:

$$n_1(0) = 0 \qquad n_2(0) = 0 \quad n_3(0) = n \tag{224}$$

$$\frac{dn_1}{dt}(0) = n_e C n \quad \frac{dn_2}{dt}(0) = 0 \quad \frac{dn_3}{dt}(0) = -n_e C n \tag{225}$$

Note that we are not using the ionization time in this example, since we must consider a process with a rate that does not scale with the electron density, namely, spontaneous radiative decay.

The three coupled equations can be combined to give a second-order differential equation for n_1:

$$\frac{d^2 n_1}{dt^2} + [A + n_e(\gamma + 2C)]\frac{dn_1}{dt} + n_e C\, [A + n_e(\gamma + C)]\, n_1 = n n_e A C e^{-n_e C t} \tag{226}$$

The solution is

$$n_1 = n\, \frac{n_e C}{A + n_e \gamma}\, e^{-n_e C t} \left[At + \frac{n_e \gamma}{A + n_e \gamma}\, (1 - e^{-(A + n_e \gamma)t}\,) \right] \tag{227}$$

A solution for n_2 is then obtained from a manipulation of Eq. (222),

$$n_2 = n_e \gamma \int_0^t dt'\, n_1(t')\, e^{-n_e (A+C)(t-t')}, \tag{228}$$

which gives

$$n_2 = n\, \frac{n_e \gamma}{A + ne\gamma}\, e^{-n_e C t} \left[n_e C t - \frac{n_e C}{A + n_e \gamma}\, (1 - e^{-(A + n_e \gamma)t}\,) \right] \tag{229}$$

Consistent with our neglect of recombination, both n_1 and n_2 vanish as $t \to \infty$. Note also the presence of factors $A + n_e\gamma$, consistent with our assumption that level 2 is metastable. Otherwise, we would make the approximation $A + n_e \gamma \to A$.

The population of the middle stage of ionization, n_i (see Figure 17), is just the sum $n_1 + n_2$. Adding Eqs (227) and (229) gives $n_i = n n_e C t e^{-n_e C t}$, which

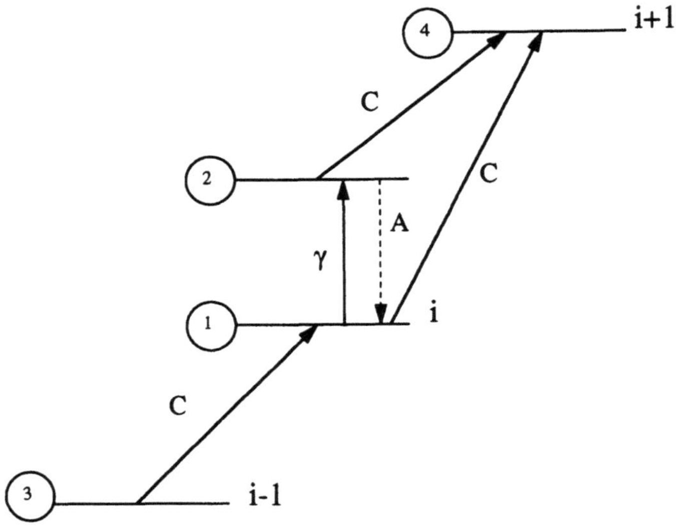

Fig. 17. Schematic of the four-level system used in §4.4. Levels are labeled by encircled numerals. Ground states of three charge states, $i-1$, i, and $i+1$ are indicated. Coupling between levels is indicated by arrows. Collisional ionization rate coefficients are labeled C, while levels 1 and 2 are coupled by a radiative transition A and a collisional excitation rate coefficient γ.

agrees with our earlier result for a three-stage system in a rapidly ionizing plasma [see Eq. (219)]. The more complex time dependences contained in Eqs (227) and (229) account for the internal adjustments to the level populations as they approach equilibration.

For sufficiently large t, the second terms in brackets on the right-hand sides of Eqs (227) and (229) are small compared to the first terms, so that both n_1 and n_2 can be described individually by the time dependence $te^{-n_e Ct}$.

$$n_1 = n \frac{A}{A + n_e\gamma} n_e Cte^{-n_e Ct} = n_i \frac{A}{A + n_e\gamma} \tag{230}$$

$$n_2 = n \frac{n_e\gamma}{A + n_e\gamma} n_e Cte^{-n_e Ct} = n_i \frac{n_e\gamma}{A + n_e\gamma} \tag{231}$$

This shows that after internal equilibration is achieved, the populations of level 1 and level 2 can be found by fixing the total ion population and calculating the steady-state distribution, assuming that the internal populations sum to unity, then multiplying by n_i, which acts as a scaling factor. Each level population then varies in time simply according to the time dependence of n_i. Typically, since radiative decay rates are fast compared to changes in the CSD, internal adjustments occur, for practical purposes, almost instantaneously, and the time dependence of the processes leading to internal equilibration are not considered explicitly. Rather, the level populations are determined by assuming steady-state conditions, as above, while still accounting for the NIE distribution of ionic fractions. This is called the *quasi-steady-state (QSS) approximation* (see, for example, Goldstein et al. 1987).

The QSS approximation is assumed implicitly in almost all treatments of astrophysical X-ray spectroscopy. However, as we continue this example, we will find that metastable level populations may differ from their QSS values to such an extent that they have an observable spectroscopic effect (Feldman 1992; cf., Liedahl, Mewe & Kaastra 1998).

We continue by first rewriting Eq. (229), labeling the rates for collisional excitation and radiative decay,

$$n_2 = n \, \frac{n_e\gamma_2}{A_2 + n_e\gamma_2} \, e^{-n_e C t} \left[n_e C t - \frac{n_e C}{A_2 + n_e\gamma_2} \left(1 - e^{-(A_2 + n_e\gamma_2)t}\right) \right]. \quad (232)$$

Now, suppose that there is another level in charge state i, which we label k, that is only weakly coupled to level 2. Assume that k is coupled to level 1 by a collisional rate coefficient γ_k and a radiative rate A_k, and for which $A_k \gg n_e\gamma_k$. Taking the appropriate limiting values, and relabeling subscripts, n_k can be determined directly from Eq. (229), and is given by

$$n_k = n \, \frac{n_e\gamma_k}{A_k} \, e^{-n_e C t} \left[n_e C t - \frac{n_e C}{A_k} \left(1 - e^{-A_k t}\right) \right]. \quad (233)$$

By stipulating that $A_k \gg n_e\gamma_k$, Eq. (233) shows that $n_k \ll n$, from which it follows that the addition of level k represents only a minor perturbation to the solutions already obtained for this system. Therefore, we simply assume that Eq. (232) for the level population of level 2 is sufficiently accurate.

Our goal is to calculate the observed line ratio N_k/N_2, where N_k and N_2 are the number of photons incident on a spectrometer in the lines produced by the transitions $k \to 1$ and $2 \to 1$, respectively, assuming an arbitrary integration time, and ignoring any possible effects of attenuation by intervening matter. The number of photons emitted in each line during a transient ionization event can be written as a time integral of the level populations multiplied by the radiative decay rates:

$$N_2(t) \propto A_2 \int_0^t \mathrm{d}t' \, n_2(t') \qquad N_k(t) \propto A_k \int_0^t \mathrm{d}t' \, n_k(t') \quad (234)$$

Let $t \to \infty$ (a suitable approximation, assuming that the ensuing recombination phase is not recorded). Then

$$N_2 \propto \frac{\gamma_2}{C} \, \frac{A_2}{n_e C + n_e\gamma_2 + A_2} \qquad \text{(slow decay)} \quad (235)$$

$$N_k \propto \frac{\gamma_k A_k}{C(n_e C + n_e \gamma_k + A_k)} \approx \frac{\gamma_k}{C} \qquad \text{(fast decay)} \tag{236}$$

Therefore, a measurement of the time-integrated line ratio would show

$$\frac{N_2}{N_k} = \frac{\gamma_2}{\gamma_k} \frac{A_2}{A_2 + n_e(\gamma_2 + C)}, \tag{237}$$

illustrative of the fact that a collisional ionization rate which is non-negligible in comparison to the decay rate of a metastable level ($n_e C \sim A_2$) will reduce the observed line ratio to one below that which one would calculate using the QSS approximation. Loosely speaking, we may say that the metastable level does not have time to decay before it is ionized, thereby reducing the number of line photons from level 2 produced in the event.

This mechanism is unlikely to be important for diffuse, low-density, plasmas (e.g., supernova remnants), since we know of no observable X-ray transitions with sufficiently long radiative lifetimes. However, many classes of X-ray source (e.g., stellar coronae, compact X-ray sources) are characterized by densities high enough for the mechanism to affect observed line ratios. It remains to be seen whether or not ionization (or recombination) time scales are rapid enough.

4.5 A worked example: transient ionization of oxygen

Having developed some of the basic concepts relevant to the ionization dynamics of NIE plasmas, we are now in a position to consider some of the associated spectroscopic effects. Rather than attempt to formulate general concepts, in this example we deal with a specific, more realistic, example. We subject a hydrogen-helium-oxygen plasma, initially in a low level of ionization, to an increase in temperature, where the temperature rise time is shorter than the equilibration time. This gives rise to NIE effects in the oxygen X-ray spectrum. We have already taken a preliminary look at this problem in §5.1. Here we will follow the time evolution of the system, and, without attempting an exhaustive account, discuss a few of the spectral signatures of NIE.

Since oxygen consists of nine charge states, this example is far too complex to solve by analytic means, and we resort to a numerical solution. The ionization equations (204) are solved using a fourth-order Runge-Kutta technique (Press et al. 1986). Rate coefficients for collisional ionization, dielectronic recombination, and radiative recombination are taken from Shull & van Steenberg (1982). Only the X-ray spectrum will be treated, which, to first order, means that we need only consider the detailed spectroscopy of O VII and O VIII. The atomic structure of these two ions, as well as collisional excitation rate coefficients and radiative rates, are calculated with the atomic physics package HULLAC (Hebrew University/Lawrence Livermore Atomic

Code; Klapisch et al. 1977; Bar-Shalom et al. 1988). Bremsstrahlung continuum emission from hydrogen and helium is included in the spectral models according to the formula provided in Mewe, Lemen & van den Oord (1986). Radiative recombination continuum emission for transitions of free electrons to the K shells of O VII and O VIII is included by applying the Milne relation to the photoionization cross sections taken from Saloman, Hubble & Scofield (1988). We ignore contributions to the spectrum from two-photon continuum emission, but include the two-photon decay rate for the $2s$ level of O VIII (Parpia & Johnson 1982) in modeling the population kinetics of this ion. Abundances of helium and oxygen are set to their solar photospheric values as given by Anders & Grevesse (1989).

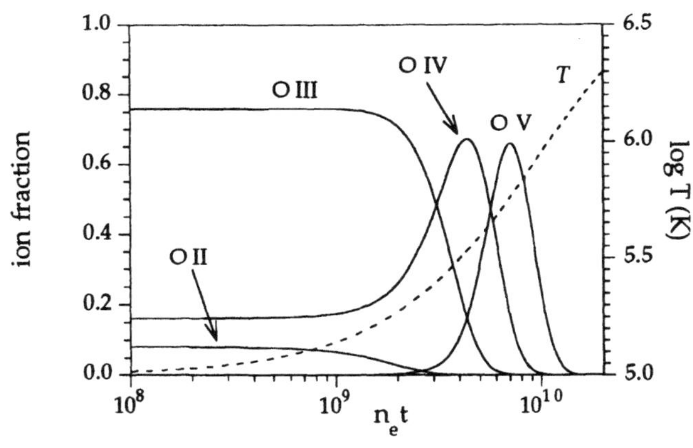

Fig. 18. Early phase of oxygen CSD evolution. Charge state fractions are indicated by solid lines (left ordinate). Temperature profile is shown as the dotted line (right ordinate).

Let the independent variable be the ionization time, and suppose that the temperature rises exponentially according to the expression

$$T = T_0 + (T_{\rm max} - T_0)[1 - \exp(-\tau/\tau_{\rm rise})]. \tag{238}$$

Therefore, $T(\tau = 0) = T_0$ and $T(\tau \to \infty) = T_{\rm max}$. For this example, let the initial temperature be $T_0 = 10^5$ K, and $T_{\rm max} = 10^{6.5}$ K. Let the rise time $\tau_{\rm rise} = 10^{10}$ cm^{-3} s.

Results for the time-dependent CSD are shown in three figures. Each figure represents a different phase of the time evolution, which we refer to as

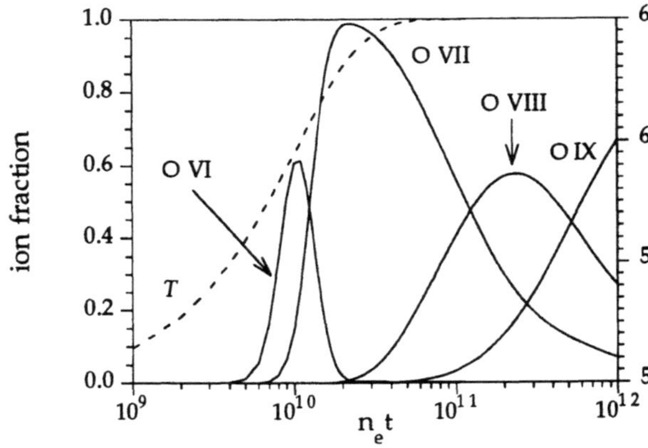

Fig. 19. Continuation of Figure 18 – intermediate phase.

early, *intermediate*, and *late*, respectively. (These designations have no formal significance here, but are chosen simply to display the results more clearly.) The early phase is shown in Figure 18. Beginning from a steady state, at $\tau = 0$ (off scale), O III is the dominate ion, corresponding to CIE for $T = 10^5$ K. The ionization time axis starts at 10^8 cm^{-3} s, which is still early enough to see the initial CSD, i.e., the CSD has had insufficient time to evolve. As T increases, the ionic fractions of all charge states from O III and lower decrease, while all higher charge state fractions begin to increase. If $(T_{\rm peak})_{\rm CIE} < T_{\rm max}$, these higher charge states show a "rise-and-fall" behavior. Each charge state, in turn, dominates the CSD over a range of ionization time.

In the intermediate phase, shown in Figure 19, the temperature continues to rise. The ionic fractions show the same rise-and-fall behavior, but the lifetimes of the ions that dominate this phase are longer than those that dominate the early phase. This is a result of the larger collisional ionization cross sections of the lower charge states that dominate the early phase, and their role in setting the ionization time scale.

We now discuss an important spectroscopic effect that occurs during this intermediate phase. But first, to touch base with the standard nomenclature associated with the diagnostic uses of He-like ions, we introduce the G ratio, which is defined as the emissivity ratio (Gabriel & Jordan 1969)

$$G = \frac{j(1s^2\ {}^1S_0 - 1s2s\ {}^3S_1) + j(1s^2\ {}^1S_0 - 1s2s\ {}^3P_1)}{j(1s^2\ {}^1S_0 - 1s2s\ {}^1P_1)}, \tag{239}$$

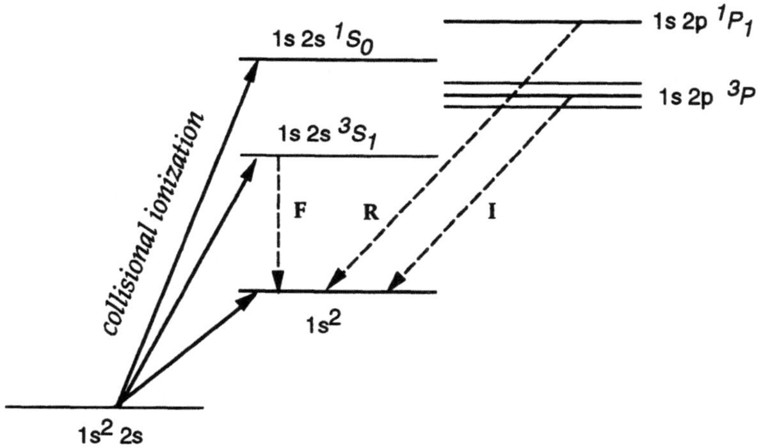

Fig. 20. Schematic of mechanism responsible for enhancing the He-like forbidden line ($1s^2\ ^1S_0 - 1s2s\ ^3S_1$) in NIE plasma. Ground states of O VI ($1s^22s$) and O VII ($1s^2$), and the first six excited levels of O VII are shown. Radiative transitions are indicated by dotted lines, and include the forbidden (F), resonance (R), and intercombination (I) lines. Collisional ionizations are indicated by solid lines.

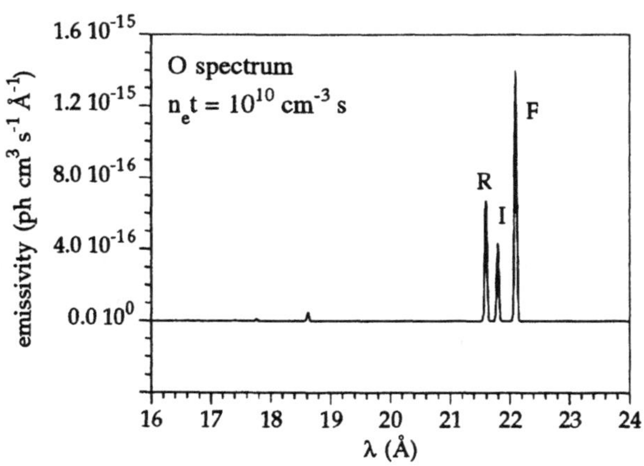

Fig. 21. Oxygen continuum-subtracted spectrum taken from intermediate phase CSD, $\tau = 10^{10}$ cm^{-3} s.

or, more simply, $G = [j(F) + j(I)]/j(R)$, where F, I, and R denote the forbidden, intercombination, and resonance lines, respectively. We will show how the G ratio can be used to detect prevailing NIE conditions.

Among the dominant charge states during the intermediate phase is Li-like O VI. Inner-shell collisional ionization of O VI populates excited states of He-like O VII (see Figure 20). Starting with the ground configuration of O VI, $1s^2\,2s$, ionization to O VII $1s2s\ ^3S_1$, followed by the magnetic transition to the $1s^2$ ground state, yields F line photons. (Note that this is a case of inner-shell ionization that does *not* produce an autoionizing configuration.) By contrast, ionization into O VII $1s2s\ ^1S_0$ adds to the emissivity of the two-photon continuum, produced in decays to the ground state, but does not contribute to X-ray line emission. At most astrophysical densities, the population of O VI $1s^2 2p$ levels is negligible, so that collisional ionization does not couple O VI to O VII $1s2p$ levels. In other words, neither the I nor R lines are driven by this process. Obviously, this mechanism works in CIE, as well as in NIE, although, in CIE it is not especially important. There are major differences in the NIE case, however, which render this process much more effective. First, note the relatively high fraction attained by O VI; $f(\mathrm{O\,VI}) \approx 0.6$ near $\tau = 10^{10}$ cm^{-3} s (Figure 19). Comparing with Figure 23, we see that this is significantly higher than the maximum CIE value. Second, again referring to these same figures, note that the temperature at which the fraction peaks is higher in NIE than in CIE. These last two effects conspire to enhance the rate at which O VII $1s2s\ ^3S_1$ is populated by inner-shell collisional ionization. Finally, $f(\mathrm{O\,VI})$ can actually exceed $f(\mathrm{O\,VII})$ over a small range in ionization time. The significance of this fact becomes clear when we realize that both the F and the I line emissivities depend upon $f(\mathrm{O\,VII})$. The net spectroscopic consequence of these various factors is a bright F line compared to the I and R lines, i.e., a large G ratio, as was first shown by Mewe & Schrijver (1978). A "snapshot" spectrum, taken at 10^{10} cm^{-3} s, is shown in Figure 21.

Referring back to the Introduction, He-like ions in an ionizing plasma provide an example of a superposition of two-component spectra, one whose source is the charge state i (the He-like stage), and one whose source is charge state $i-1$ (the Li-like stage). The mechanism discussed above is common to all Li-like/He-like pairs. Moreover, a similar mechanism operates in Na-like/Ne-like pairs in an ionizing plasma. The example of Na-like Fe XVI acting as a source for Ne-like Fe XVII $3s - 2p$ emission after collisional ionization out of the Fe XVI $2p$ subshell is discussed in Kahn & Liedahl (1991).

Although the detection of a large G ratio may imply NIE conditions, as just shown, absence of such does not imply CIE. As shown below, as the plasma evolves away from the Li-like stage, the F line, compared to the R line, actually becomes weaker than in CIE — the R line will appear to be enhanced. It is also worth remembering that the enhanced G ratio in He-like ions is also a signature of PIE, as pointed out in §3.5. Suppose one were to

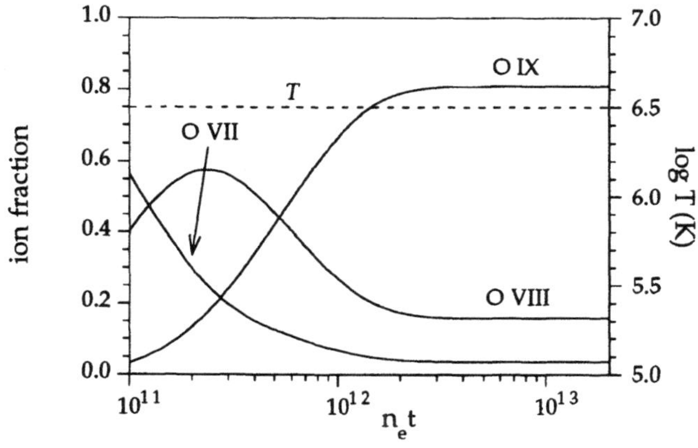

Fig. 22. Continuation of Figure 19 – late phase.

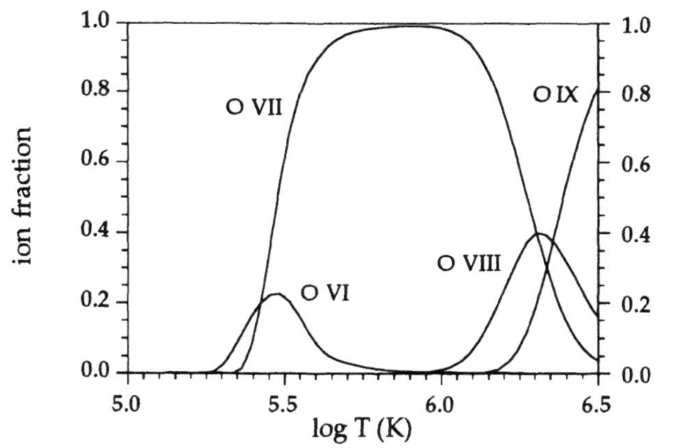

Fig. 23. Oxygen (O VI–O IX) charge state distribution in CIE. Based on Shull & van Steenberg (1982).

record a spectrum of an active galaxy in which the F line was brighter than the R line. Would this imply PIE in the presence of a hard X-ray continuum or NIE in shocked gas? We may not be able to answer that question without access to corollary spectroscopic information from other charge states and other elements. In general, we need to bring to bear as many diagnostics as are available in order to make an internally consistent interpretation of an X-ray spectrum.

In the late phase (see Fig. 22) the temperature reaches its maximum value. The O IX fraction begins to dominate after a few times 10^{11} cm^{-3} s. The O VII and O VIII fractions, having attained their maximum values, decay to the level of trace constituents. Steady-state ionization is, for practical purposes, reached in 3×10^{12} cm^{-3} s. The ionization fractions at later ionization times should converge to their CIE values. This can be checked by comparing to Figure 23 for a temperature $T = T_{\rm max} = 10^{6.5}$ K.

Another example of the spectroscopic complications associated with NIE plasmas can be seen by noting that the O VIII fraction is not one-to-one with respect to the ionization time. Referring to Figure 22, we see that, for example, $f(\mathrm{O\,VIII}) = 0.4$ when $n_e t \approx 10^{11}$ cm^{-3} s, and again when $n_e t \approx 6 \times 10^{11}$ cm^{-3} s. Note that the temperature has already attained its maximum value for each of these ionization times. Therefore, the O VIII spectrum is *nearly* identical at two different intervals of ionization time during the approach to equilibrium; same line equivalent widths, same set of line ratios. (There is a slight difference owing to the recombination contribution from fully-stripped oxygen, whose fraction is obviously not the same at the two ionization times noted.) But $f(\mathrm{O\,VII})$ drops by a factor ~ 5 over the same τ interval. Therefore, the overall oxygen spectrum is drastically different at these two ionization times.

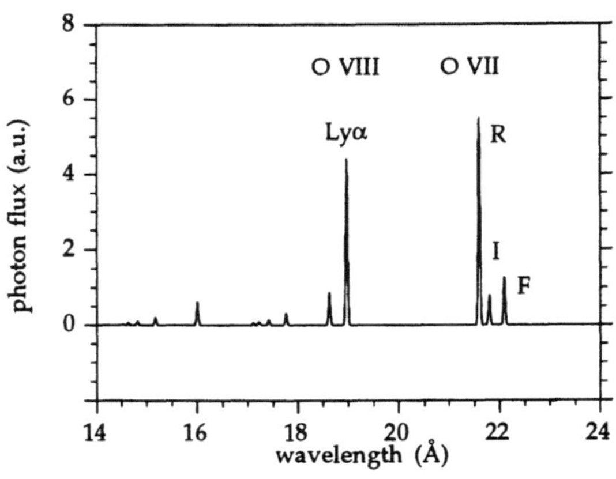

Fig. 24. Oxygen spectrum at $\tau = 10^{11}$ cm^{-3} s, where $f(\mathrm{O\,VII}) \approx f(\mathrm{O\,VIII})$. Note the small G ratio compared with earlier phases (Figure 21).

We end this section with one final example. As mentioned earlier, as the CSD evolves away from the Li-like stage, the G ratio diminishes. Emission from O VII will eventually be rivaled by O VIII emission. In that case, we can use both O VII and O VIII to infer a deviation from CIE. We take a snapshot of the system at $\tau = 10^{11}$ cm^{-3} s, at which time $f(\text{O VII}) \approx f(\text{O VIII})$, and the temperature has reached its steady-state value of $10^{6.5}$ K. The spectrum is shown in Figure 24. We imagine that we observe this spectrum in an astrophysical source. How can we determine whether or not the emission is produced under CIE conditions? In Figure 25, we plot both the theoretical values of the O VII G ratio in CIE, and the ratio of O VIII Lyα to the O VII resonance line R, also for CIE. From Figure 24, $G = 0.37$, which from Figure 25 implies that $T > 500$ eV, *if* the plasma in in CIE. The model NIE spectrum also gives $I(\text{Ly}\alpha)/I(R) = 0.81$, which, from Figure 25 implies $T < 200$ eV for CIE. The gross disparity of the allowed temperature ranges required by the "observations" force us to to conclude that NIE conditions must obtain.

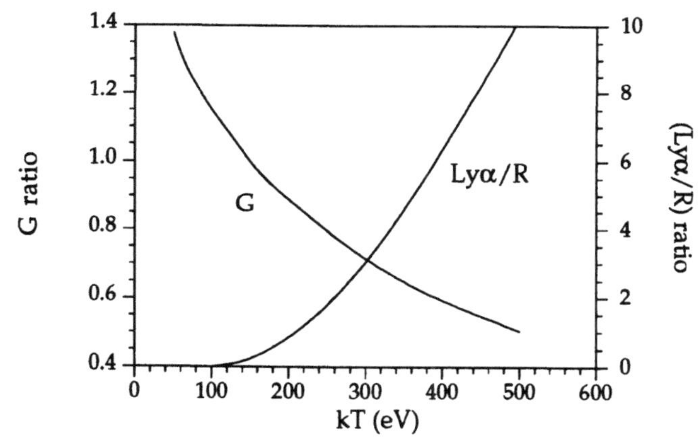

Fig. 25. He-like oxygen G ratio (*left ordinate*) and ratio of O VIII Lyα to O VII resonance line R (*right ordinate*) vs. temperature, assuming CIE.

Acknowledgements

I am grateful for the generous hospitality shown to me by Jan van Paradijs during the EADN School. Thanks also to Jane Ayal for her prompt and cheerful logistical support. The text has been improved owing to a careful reading and helpful comments by Daniel Savin, and from useful conversations with Mau Chen, Rolf Mewe, Frits Paerels, and Masao Sako. Work at Lawrence Livermore National Laboratory was performed under the auspices of the U.S. Department of Energy, Contract No. W-7405-Eng-48.

References

Abramowitz, M. & Stegun, I.A. (1972): *Handbook of Mathematical Functions* (Dover Publications: New York) (AS72)
Anders, E. & Grevesse, N. (1989): Geochimica et Cosmochimica Acta 53, 197
Angelini, L., White, N.E., Nagase, F., Yoshida, A., Takeshima, T., Becker, C., Kallman, T.R. & Paerels, F. (1995): ApJ 449, L41
Antonucci, R.R.J. & Miller, J.S. (1985): ApJ 297, 621
Arnaud, M. & Raymond, J.C. (1992): ApJ 398, 394
Bambynek, W. et al. (1972): Rev. Mod. Phys. 44, 716
Bar-Shalom, A., Klapisch, M. & Oreg, J. (1988): Phys. Rev. A 38, 1773
Bates, D.R. & Dalgarno, A. (1962): in *Atomic and Molecular Processes*, Ed. D.R. Bates (New York: Academic Press), p. 258
Bautista, M., Kallman, T.R., Angelini, L., Liedahl, D.A. & Smits, D.P. (1998): ApJ (in press)
Beiersdorfer, P., Phillips, T., Jacobs, V.L., Hill, K.W., Bitter, M., von Goeler, S. & Kahn, S.M. (1993): ApJ 409, 846
Blandford, R.D. (1990): in *Active Galactic Nuclei: Saas-Fee Advanced Course 20*, Eds T.J.-L. Courvoisier & M. Mayor (Springer-Verlag)
Blumenthal, G.R. & Gould, R.J. (1970): Rev. Mod. Phys. 42, 237
Cowan, R.D. (1981): *The Theory of Atomic Structure and Spectra* (University of California Press, Berkeley)
Ebisawa, K. et al. (1996): PASJ 48, 425
Fabian, A.C. & George, I.M. (1991): in *Iron Line Diagnostics in X-ray Sources*, Eds A. Treves, G. Perola, & L. Stella (Berlin: Springer-Verlag Lecture Notes in Physics), p. 169
Feldman, U. (1985): ApJ 385, 758
Frank, J., King, A. & Raine, D. (1992): *Accretion Power in Astrophysics*, Second Edition, (Cambridge University Press)
Fukuzawa, Y. et al. (1994): PASJ 46, L141
Gabriel, A.H. & Jordan, C. (1969): MNRAS 145, 241
George, I.M. & Fabian, A.C. (1991): MNRAS 249, 352
George, I.M. et al. (1998): ApJS 114, 73
Goldstein, W.H., Whitten, B.L., Hazi, A.U. & Chen, M.H. (1987): Phys. Rev. A 36, 3607
Halpern, J.P. & Grindlay, J.E. (1980): ApJ 242, 141
Hatchett, S., Buff, J. & McCray, R. (1976): ApJ 206, 847
Holt, S.S. & McCray, R. (1982): ARA&A 20, 323
Illarionov, A.F. & Sunyaev, R.A. (1972): Sov. Astron. 16, 45
Iwasawa, K., Fabian, A.C. & Matt, G. (1997): MNRAS 289, 443
Jackson, J.D. (1975): *Classical Electrodynamics*, Second Edition (John Wiley and Sons)
Jacobs, V.L., Doschek, G.A., Seely, J.F. & Cowan, R.D. (1989): Phys. Rev. G 39, 2411
Jauch, J.M. & Rohrlich, F. (1955): *The Theory of Photons and Electrons*, (Addison-Wesley Publ. Co., Inc.)
Kaastra, J.S. & Mewe, R. (1993): A&AS 97, 443
Kaastra, J.S., Mewe, R., Liedahl, D.A., Singh, K.P., White, N.E, & Drake, S.A. (1996): A&A 314, 547

Kahn, S.M. & Liedahl, D.A. (1991): in *Iron Line Diagnostics in X-ray Sources*, Eds A. Treves, G. Perola, & L. Stella (Berlin: Springer-Verlag Lecture Notes in Physics), p. 3

Kahn, S.M. & Liedahl, D.A. (1995): in *Physics with Multiply Charged Ions*, Ed. D. Liesen (Plenum Press: New York), p. 169

Kahn, S.M., Seward, F.D., & Chlebowski, T. (1984): ApJ 283, 286

Kallman, T.R. (1991): in *Iron Line Diagnostics in X-ray Sources*, Eds A. Treves, G. Perola, & L. Stella (Berlin: Springer-Verlag Lecture Notes in Physics), p. 87

Kallman, T.R., Liedahl, D.A., Osterheld, A.L., Goldstein, W.H. & Kahn, S.M. (1996): ApJ 465, 994

Kallman, T.R. & McCray, R. (1982): ApJS 50, 263

Katz, J.I. (1976): ApJ 206, 910

Katz, J.I. (1987): *High Energy Astrophysics* (Addison-Wesley Publishing)

Klapisch, M. et al. (1977): J. Opt. Soc. Am. 67, 148

Ko, Y.-K. & Kallman, T.R. (1991): ApJ 374, 721

Kompaneets, A.S. (1957): Soviet Physics JETP 4, 730

Krolik, J.H. & Begelman, M.C. (1986): ApJ 308, L55

Krolik, J.H. & Kallman, T.R. (1987): ApJ 320, L5

Krolik, J.H., McKee, C.F. & Tarter, C.B. (1981): ApJ 249, 422

Liedahl, D.A. (1992): Ph. D. Thesis, University of California at Berkeley

Liedahl, D.A., Kahn, S.M., Osterheld, A.L, & Goldstein, W.H. (1990): ApJ 350, L37

Liedahl, D.A., Kahn, S.M., Osterheld, A.L, & Goldstein, W.H. (1992): ApJ 391, 306

Liedahl, D.A., Mewe, R. & Kaastra, J.S. (1998): in preparation.

Liedahl, D.A. & Paerels, F.B.S. (1996): ApJ 468, L33

Lightman, A.P. & White, T.R. (1988): ApJ 335, 57

Longair, M.S. (1992): *High Energy Astrophysics: Volume 1* (Second Edition; Cambridge University Press)

Makishima, K. (1986): in *The Physics of Accretion Onto Compact Objects*, Eds K.O. Mason, M.G. Watson & N.E. White (Springer-Verlag), p. 249

Mason, H.E., Doschek, G.A.,Feldman, U. & Bhatia, A.K. (1979): A&A 73, 74

McGuire, E.J. (1969): Phys. Rev. 185, 1

McKenzie, D.L., Landecker, P.B., Broussard, R.M., Rugge, H.R., Young, R.M., Feldman, U. & Doschek, G.A. (1980): ApJ 241, 409

Mewe, R., Lemen, J. & van den Oord, G.H.J. (1986): A&AS 65, 511

Mewe, R. & Schrijver, J. (1978): A&A 65, 115

Mitchell, R.J., Culhane, J.L., Davison, P.J. & Ives, J.C. (1976): MNRAS 175, 29P

Mulchaey, J.S., Mushotzky, R.F.& Weaver, K.A. (1992): ApJ 390, L69

Murray, S.D., Castor, J.I., Klein, R.I. & McKee, C.F. (1994): ApJ 435, 631

Nagase, F. (1996): in *UV and X-Ray Spectroscopy of Astrophysical and Laboratory Plasmas*, Eds K. Yamashita & T. Watanabe (Universal Academy Press: Tokyo), p. 189

Nagase, F. et al. (1994): ApJ 436, L1

Nandra, K., Pounds, K.A, Stewart, G.C., Fabian, A.C. & Rees, M.J. (1989): MNRAS 236, 39P

Netzer, H. (1990): in *Active Galactic Nuclei: Saas-Fee Advanced Course 20*, Eds T.J.-L. Courvoisier & M. Mayor (Springer-Verlag)

Netzer, H. & Turner, T.J. (1997): ApJ 488, 694

Novikov, I.D. & Thorne, K.S. (1973): in *Black Holes*, Eds C. DeWitt & B. DeWitt (Gordon and Breach: New York)

Osterbrock, D.E. (1989): *Astrophysics of Gaseous Nebulae and Active Galactic Nuclei* (University Science Books).

Otani, C., et al. (1996): PASJ 48, 211

Parpia, F.A. & Johnson, W.R. (1982): Phys. Rev. A 26, 1142

Peebles, P.J.E. (1993): *Principles of Physical Cosmology* (Princeton University Press)

Phillips, K.J.H. et al. (1982): ApJ 256, 774

Pozdnyakov, L.A., Sobol, I.M. & Sunyaev, R.A. (1979): A&A 75, 214

Pozdnyakov, L.A., Sobol, I.M. & Sunyaev, R.A. (1983): Astrophys. and Space Phys. Rev. 2, 189

Pradhan, A.K. (1982): ApJ 263, 477

Pradhan, A.K. (1985): ApJ 288, 824

Press, W.H., Flannery, B.P., Teukolsky, S.A. & Vetterling, W.T. (1986): *Numerical Recipes* (Cambridge University Press)

Raymond, J.C. (1993): ApJ 412, 267

Ross, R.R., Weaver, R. & McCray, R. (1978): ApJ 219, 292

Rybicki, G.B. & Lightman, A.P. (1979): *Radiative Processes in Astrophysics* (Wiley Interscience)

Sako, M., Liedahl, D.A., Paerels, F. & Kahn, S.M. (1998a): ApJ (submitted)

Sako, M., Paerels, F., Liedahl, D.A. & Kahn, S.M. (1998b): in preparation.

Saloman, E.B., Hubble, J.H. & Scofield, J.H. (1988): At. Data Nucl. Data Tables 38, 1

Savin, D.W. et al. (1997): ApJ 489, L115

Seely, J.F., Feldman, U. & Safranova, U.I. (1986): ApJ 304, 848

Serlemitsos, P.J. et al. (1977): ApJ 211, L63

Shapiro, P.R. & Moore, R.T. (1977): ApJ 217, 621

Shapiro, S.L., Lightman, A.P. & Eardley, D.M. (1976): ApJ 204, 187

Shirai, T., Funatake, Y., Mori, K., Sugar, J., Wiese, W.L. & Nakai, Y. (1990): J. Phys. Chem. Ref. Data 19, 127

Shklovsky, I.S. (1967): ApJ 148, L1

Shull, J.M. (1982): ApJ 262, 308

Shull, J.M. & van Steenberg, M. (1982): ApJS 48, 95

Sobelman, I.I., Vainshtein, L.A. & Yukov, E.A. (1981): *Excitation of Atoms and Broadening of Spectral Lines* (Springer-Verlag)

Spitzer, L. (1979): *Physical Processes in the Interstellar Medium* (Wiley: New York)

Sunyaev, R.A. & Titarchuk, L.G. (1980): A&A 86, 121

Tarter, C.B., Tucker, W.H. & Salpeter, E.E. (1969): ApJ 156, 943

Ueno, S. et al. (1994): PASJ 46, L71

Vrtilek, S.D., Helfand, D.J., Halpern, J.P., Kahn, S.M. & Seward, F.D. (1986): ApJ 308, 644

Weisheit, J.C. (1974): ApJ 190, 735

White, N.E., Swank, J.H. & Holt, S.S. (1983): ApJ 270, 711

Winkler, P.F., Canizares, C.R., Clark, G.W., Markert, T.H., Kalata, K. & Schnopper, H.W. (1981): ApJ 246, L27

X-ray Spectroscopic Observations with *ASCA* and *BeppoSAX*

Jelle S. Kaastra[1]

SRON, Sorbonnelaan 2, 3584 CA Utrecht, The Netherlands

Abstract. This chapter presents an overview of the X-ray spectroscopic observations with *ASCA* and *BeppoSAX*. After an introduction to both missions a few general notes on spectral data fitting are given. In a systematic overview the spectroscopic achievements of both missions are summarized. Subsequently, cool and hot stars, cataclysmic variables, high- and low-mass X-ray binaries, supernova remnants, normal galaxies (including the galactic center), Seyfert 1 and Seyfert 2 galaxies, quasars and clusters of galaxies are discussed.

1 Introduction

1.1 X-ray spectroscopy

X-ray spectroscopy is a powerful tool for analyzing high-energy phenomena in the universe. Up to 1993 (the launch of *ASCA*), most X-ray missions had low spectral resolution. Exceptional instruments with medium spectral resolution were the SSS detector flown on board of the *Einstein* observatory (which, however, covered the limited energy band of 0.5–4 keV, had no imaging capability, and was considerably less sensitive than the *ASCA* SIS-detector), and the gas scintillation proportional counter (GSPC) flown on board of *EXOSAT* (which likewise had no imaging capabilities and could only observe the strongest cosmic sources). High-resolution X-ray missions have been rare and had low sensitivity; among these were the FPCS detector of *Einstein* as well as the transmission gratings flown on *Einstein* and *EXOSAT*.

The launch of *ASCA* introduced the era of spatially resolved spectroscopy with high sensitivity and medium spectral resolution. The launch of *BeppoSAX* in 1996 with its broad energy band yielded a significant addition to the observational tools available for X-ray astronomers. Both missions are explorers for the coming series of high-throughput, high-resolution missions to be launched between 1998–2000, such as *AXAF*, *XMM* and *ASTRO-E*.

In this series of lectures I focus mainly upon the *spectroscopic* achievements of *ASCA* and *BeppoSAX*. Exciting new results obtained by these missions based upon only their sensitivity, imaging and timing capabilities are not treated here. Among the topics that are not included or only marginally outlined are solar-system observations, planetary nebulae, isolated pulsars, low-mass X-ray binaries, BL Lac objects, low-luminosity AGN, starburst galaxies, normal galaxies, γ-ray bursts and the cosmic background spectrum.

Nevertheless, I hope that these lectures will give a good flavour of what has been achieved with *ASCA* and *BeppoSAX*, and that the readers are encouraged to study these topics in more detail, not in the least place by proposing observations with these instruments and with the next generation of X-ray satellites.

Although most attention is focused upon *ASCA* and *BeppoSAX*, I also discuss some of the new results obtained with *EUVE*, a mission with high spectral resolution but low sensitivity which supplements in several cases the observations at higher energies.

The present review is mostly based upon publications that appeared in the refereed literature. Only in a few cases do I refer to conference proceedings. This implies that not everything is covered completely, but the most important highlights are presented here. Readers are encouraged to search the literature further for a deeper study of specific topics treated here. My literature search ends at September 1997, i.e., the time these lectures were given. Since both *ASCA* and *BeppoSAX* are ongoing missions, many newer results cannot be discussed here.

1.2 The *ASCA* and *BeppoSAX* missions

ASCA (Advanced Satellite for Cosmology and Astrophysics), Japan's fourth X-ray satellite, was launched on 20 February 1993. It carries four identical sets of nested thin-foil, conical focusing mirrors, with two CCD imaging spectrometers (SIS0 and SIS1) and two gas scintillation proportional-counter (GSPC) imaging spectrometers (GIS2 and GIS3). More details can be found in Tanaka et al. (1994). The various components and instruments have been developed by groups in Japan and the USA (see also Serlemitsos et al. 1995; Ohashi et al. 1996; Makishima et al. 1996).

SAX (Satellite italiano per Astronomia X), an Italian/Dutch satellite, was launched on 30 april 1996, and was renamed shortly afterwards *BeppoSAX*, in honour of Giuseppe (Beppo) Occhialini, one of Italy's X-ray pioneers. It consists of four imaging GSPC telescopes, two non-imaging high-energy instruments and a pair of wide-field cameras. Three of the imaging GSPCs (MECS1, MECS2, MECS3, Boella et al. 1997b) are sensitive in the 1–10 keV band, while the other GSPC (LECS, Parmar et al. 1997a) is sensitive in the broader 0.1–10 keV band. Above 10 keV a non-imaging GSPC (HPGSPC, Manzo et al. 1997) as well as a phoswich detector (PDS, Frontera et al. 1997) are operational. More details about the *BeppoSAX* mission can be found in Boella et al. (1997a).

I do not explain the technical details of the satellites here; for those I refer to the above-mentioned papers. Instead I compare some properties of both satellites and indicate the stronger and weaker aspects of each mission.

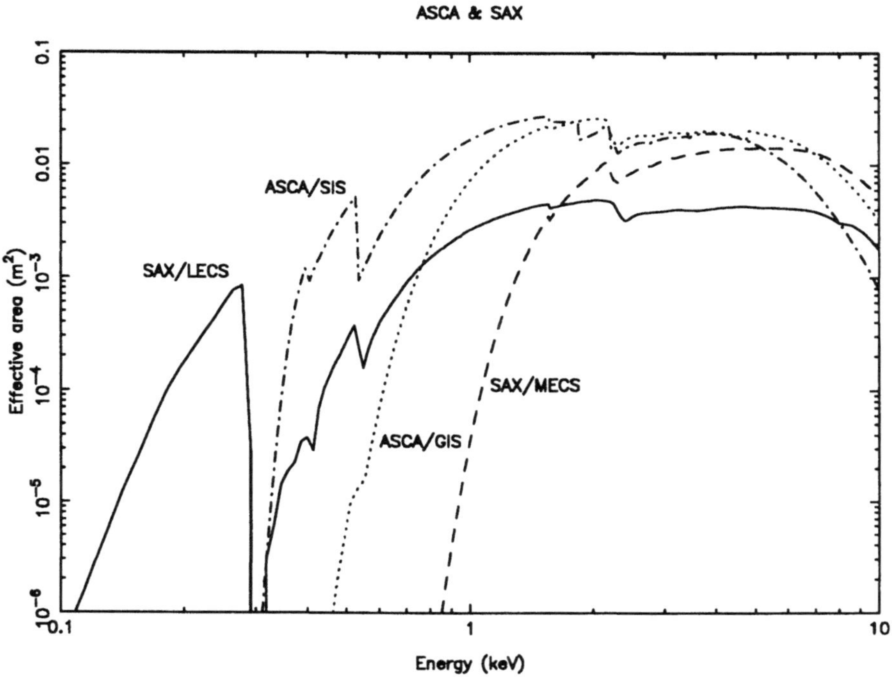

Fig. 1. Effective area of *ASCA* and *BeppoSAX*.

Fig. 1 shows the effective area of the different instruments of both satellites. The effective area of the two SIS detectors and the two GIS detectors of *ASCA* have been combined, as well as the effective area of the three MECS units of *BeppoSAX*. It is seen that in the 2–5 keV band, both *ASCA* detectors have a comparable efficiency; above 5 keV, the GIS detector has better sensitivity. In the same energy bands, the effective area of *BeppoSAX* is comparable to *ASCA*, and at the highest energies above about 7 keV *BeppoSAX* has a slightly larger effective area. In the low-energy band of 0.5–1 keV, the *ASCA* SIS detectors are by far the most sensitive instruments; below 0.5 keV the effective area of the SIS decreases rapidly and its calibration is less well understood, so that effectively the *BeppoSAX* LECS detector is the only medium-resolution instrument available below 0.5 keV.

Fig. 2 shows the spectral resolution of these four instruments. The resolution is defined here as the FWHM of the energy redistribution function. The three GSPC-based instruments all have a similar resolution which increases roughly proportional to the square root of the energy; of course the resolving power ($E/\Delta E$) *increases* with increasing energy. The figure illustrates nicely the advantage of CCD detectors over GSPC detectors: the resolving power of the SIS is 3–4 times better than that of the GSPCs. At the lowest energies,

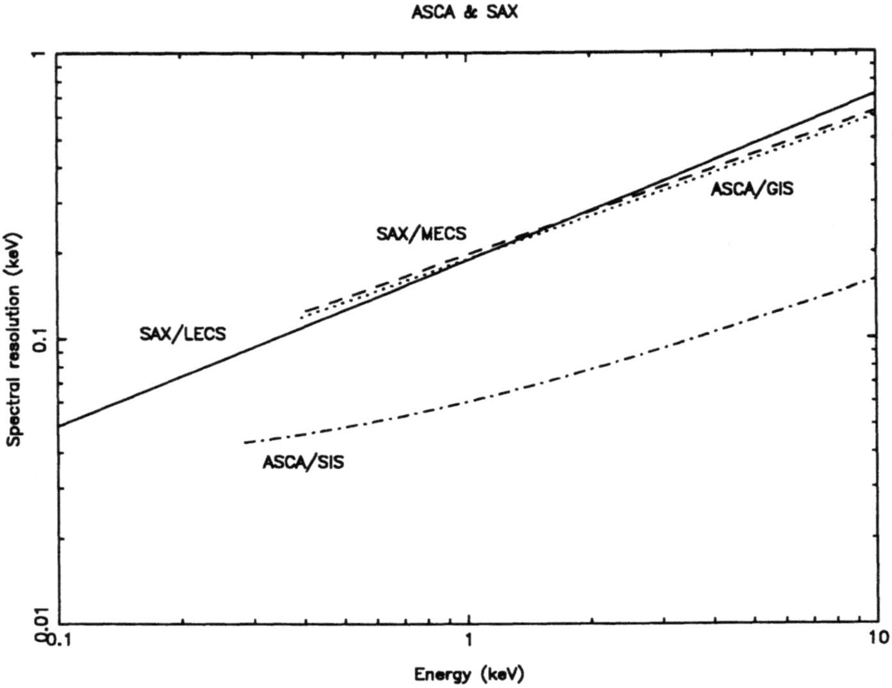

Fig. 2. Spectral resolution (FWHM) of *ASCA* and *BeppoSAX*.

the *BeppoSAX* LECS detector has a considerable better energy resolution than, e.g., the *ROSAT* PSPC detector.

There are substantial differences in the spatial resolution of both satellites (see Fig. 3). Although the *ASCA* telescopes have a point-spread function (PSF) with a rather narrow core, the wings of this PSF are very strong, much stronger than for the *BeppoSAX* telescopes. Note that for the *ASCA*/GIS detector the resolution is worse than for the SIS detectors.

In summary, the strongest merits of *ASCA* are its high spectral resolution and large effective area in the 0.5–2 keV band, and of *BeppoSAX* the broader energy band and the better spatial resolution.

1.3 Most prominent spectral features observable with *ASCA* and *BeppoSAX*

With *ASCA* and *BeppoSAX*, prominent spectral features of the abundant elements from carbon to nickel are visible. In Table 1 I summarize the most important emission lines that can be seen, and give their energies (in keV). For shortness, I label the He-like 2–1 transitions (actually four lines) as Heα, the 3–1 transitions as Heβ, and similarly for the H-like transitions, and finally the Kα transition of the neutral atom as Kα.

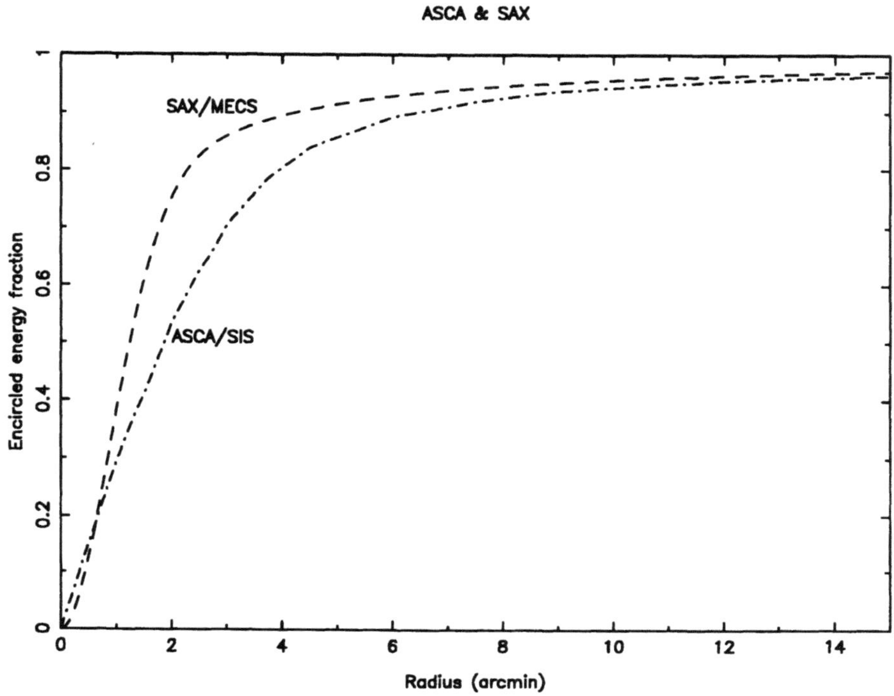

Fig. 3. Encircled energy fraction for *ASCA*/SIS and *BeppoSAX*/MECS at 4.5 keV.

Comparing this table with Fig. 2, it is seen that the *ASCA*/SIS detectors can easily resolve the main H-like and He-like lines of each element. The resolution is insufficient to resolve the He-like lines into the different components. Sometimes care should be taken because the H-like 3–1 transition is very close to the He-like 2–1 transition of another element, like is the case for Mg, Si and S. Also, the SIS detector is able to discriminate between fluorescence from neutral material and He-like emission.

With all GSPC-based detectors it is only marginally possible to resolve the H-like from the He-like components; the known line centroids of these lines help to resolve the individual contributions.

In a similar way, the ionization edges of the elements from C–Ni can be observed. In general the K-edges from oxygen (0.54, 0.57, 0.59, 0.62, 0.65, 0.67, 0.74 and 0.87 keV for O I–O VIII respectively) and from iron (7.12 keV for Fe I to 9.28 keV for Fe XXVI) are the most important absorption edge features.

Table 1. Some important emission lines

element	Kα	Heα	Heβ	Hα	Hβ
C	0.28	0.30–0.31	0.36	0.37	0.44
N	0.39	0.42–0.43	0.50	0.50	0.59
O	0.52	0.56–0.57	0.67	0.65	0.78
Ne	0.85	0.91–0.92	1.07	1.02	1.21
Mg	1.25	1.33–1.35	1.58	1.47	1.75
Si	1.74	1.84–1.86	2.18	2.01	2.38
S	2.31	2.43–2.46	2.87	2.62	3.11
Ar	2.96	3.11–3.14	3.70	3.32	3.94
Ca	3.69	3.86–3.90	4.59	4.11	4.88
Fe	6.40	6.63–6.70	7.90	6.97	8.21
Ni	7.48	7.75–7.80	9.05	8.06	9.46

2 A Few Notes on Spectral Data Fitting

2.1 Introduction

In many X-ray sources the presence of spectral features is evident. For example, often a strong and isolated iron line in the 6.4–6.9 keV band is visible. However, near the Fe-L band around 1 keV there are so many spectral lines that it is impossible to resolve them individually using *ASCA* or *BeppoSAX* data. In these cases one must rely upon *spectral fitting*. Also, many sources show a complicated structure due to the presence of more than one spectral component; the presence and significance of these additional components and the corresponding parameters must be assessed carefully. Unfortunately, this is not always done in the most appropriate way. I outline here a few general guidelines that may be helpful in critically assessing the merits of published papers and analyzing data sets obtained with *ASCA*, *BeppoSAX* or other X-ray missions.

2.2 Data binning

Spectral data for *ASCA* and *BeppoSAX* are collected as a histogram of number of counts per data channel. When a beam of photons with a given energy E enters the instrument, it is distributed over the data channels according to the energy redistribution function. This redistribution function has a typical full width at half maximum (FWHM) of ΔE. From the Shannon information theorem it can be derived that the optimal data binning for a Gaussian redistribution function is about $\Delta E/3$, for spectra with typically 10^3–10^5 counts, as is the case for most *ASCA* and *BeppoSAX* data. No information is lost if this binning is used. The *ASCA* and *BeppoSAX* data as delivered to the guest observers are over-sampled, however, and *rebinning is therefore necessary*.

2.3 Model binning

The response matrices for the *ASCA* SIS detectors as delivered standard to observers and as used in, e.g., the XSPEC package bin the model photon spectrum from 0.2–12 keV into 1180 bins with a uniform bin width of 0.01 keV (the *BeppoSAX*/MECS detectors use the same binning; the *ASCA*/GIS detectors only use 200 bins). Spectral models like, e.g., the *mekal* model that calculate thermal spectra from optically thin plasmas evaluate the model spectrum by binning all flux onto the input model grid as delivered by the response matrix, thereby assuming that all flux is located at the bin center. This gives an error of at most half a model bin width or 0.005 keV for the SIS. However, for a strong line near 1 keV containing N photons, the resolution of the SIS is about 0.05 keV (see Fig. 2), hence the line centroid can be determined[1] with an accuracy of $0.05/2.35\sqrt{N}$ keV, implying that already for a line of 20 photons the statistical accuracy of the line centroid is smaller than the error made by the model binning! Thus it is recommended to *use smaller model bins in case of plasmas with strong line features.*

2.4 Calibration uncertainties

No instrument has ever flown with a perfect calibration. *ASCA* and *BeppoSAX* are no exception to this. Known problems for *ASCA* are the effective-area calibration near the gold edges around 2 keV, and the calibration at low energies, around the oxygen edge at 0.53 keV and lower. In this last case, the instrument efficiency decreases rapidly with decreasing energy, and also background subtraction (geocoronal oxygen) is not always trivial. It is perhaps wise to add a systematic uncertainty of about 2 % at energies above 0.8 keV and 5 % below that energy when analyzing *ASCA* data. By proper data binning and including systematic errors, Kaastra et al. (1996) were able to obtain acceptable fits ($\chi^2 = 239$ for 211 degrees of freedom) to the combined *ASCA* SIS/GIS and *ROSAT* PSPC spectrum of AR Lac. This should be compared to the original fit to only the SIS data by White et al. (1994), which yielded $\chi^2 = 448$ for 300 degrees of freedom. Inclusion of systematic effects has also the advantage that for thermal plasmas the strong Fe-L complex near 1 keV with its larger uncertainty in the plasma models, does get somewhat less weight in spectral fits of high signal-to-noise spectra.

2.5 Spectral deconvolution

Some authors tend to present their spectra as deconvolved photon spectra. This is usually done by dividing the observed spectrum (counts/s/keV) by

[1] For a Gaussian instrument profile with FWHM=2.35σ

the nominal effective area, or by plotting the model spectrum binned to the data resolution, and corrected with a factor that equals the observed count rate divided by the predicted count rate. Although this is illustrative in the sense that it shows what the model spectrum looks like, it is also potentially dangerous, because the details of the model at scales smaller than the resolution of the detector may be artificially and erroneously enhanced. For example, two adjacent lines in the model spectrum that are not resolved by the detector would appear as very significant in such a plot, while perhaps a fit with a single line could give a fit of similar quality.

2.6 Statistics

As was shown before, no information is lost if the spectral data are binned to about one third of the detector FWHM resolution. But is this binning really necessary? The answer is yes, for the following reason. Suppose that we observe the spectrum of an X-ray source that has a given model spectrum M_0. Unfortunately, we do not know this true spectrum, and we test a certain class of models M by calculating the χ^2 statistic, given by

$$\chi^2 = \sum_{i=1}^{n} (S_i - M_i)^2/\sigma_i^2 \ . \tag{1}$$

Here S_i is the number of observed counts in bin i, n is the number of data bins, M_i the expected number of counts for the model M, and σ_i^2 is the expected variance of the data. If M is the true model M_0, then the χ^2 statistic has an expected value of $n - m$ and an expected r.m.s. variation of $\sqrt{2(n-m)}$, where m is the number of free parameters of the model. In general, we reject the model M if χ^2 gets too large, in practice if $\chi^2 > n - m + f\sqrt{2(n-m)}$, with f a factor of order unity corresponding to the precise confidence level chosen. If model M is not equal to the true model M_0, the expected χ^2 value that will be calculated is no more $n - m$ but $n - m + r$ where it is easy to show that r is approximately given by

$$r = N \int [\frac{(f(E) - f_0(E))}{f(E)}]^2 f(E) \mathrm{d}E \ , \tag{2}$$

with f and f_0 the probability distributions for the photons in model M and M_0 respectively, and N the total number of counts in the spectrum. It is seen that r does not depend upon the number of bins in the spectrum. Now the false model M has a large probability of being accepted if r is smaller than the expected r.m.s. variation of χ^2, which equals $\sqrt{2(n-m)}$. In other words, if the spectrum is binned in too many bins (if n is large), the probability of falsely accepting wrong models increases.

A similar situation holds if two different models M_1 and M_2 are to be compared. For example, the observer wants to know whether a fit with non-solar abundances is significantly better than a fit with solar abundances. In this case, the F-test (variance-ratio test) should be used. The test statistic is given by

$$F = \frac{\chi_1^2/(n-m_1)}{\chi_2^2/(n-m_2)} . \qquad (3)$$

For large n and small values of r, the expected value for F is approximately given by $1 + (r_1 - r_2)/n$ and the expected r.m.s. variation is $2/\sqrt{n}$, again showing that for large n the probability of falsely preferring model M_1 over model M_2 is large.

The conclusion must be: *never over-sample your spectra.*

2.7 Low count rates

Apart from binning the data in accordance with the spectral resolution of the instrument, it is sometimes also wise to do some further binning, in particular if the count rates per bin are small (say less than 20 counts). This holds in particular for e.g., the high-energy ends of *ASCA* and *BeppoSAX* spectra above the Fe-K lines where most spectral models have a rather simple shape, determined by, e.g., only two parameters (a normalization and photon index or temperature). Of course, this binning should be done with care in the case that there is an indication for significant line emission in that energy range.

The reason for rebinning low count rate bins further is that in spectral fitting χ^2 (see Eq. 1) is not calculated using the *expected* variance σ_i^2 per bin, but the *observed* variance, which is essentially the number of counts N_i in the bin. This gives problems for small values of N_i. In particular for $N_i = 0$ χ^2 is not defined. Some packages resolve this by arbitrarily replacing N_i by 1 in the last case. Although that procedure formally avoids the singularity, it still yields a bias. This is easy to show. The $1/N_i$ factors in Eq. (1) effectively operate as weights in establishing the r.m.s. differences between observed and model count rates. In those cases where by chance N_i appears to be small the corresponding weight is large, and thus such a bin gets too much weight. Bins with the same expected variance σ_i^2 but that have by chance a higher number of counts get too little weight. This issue is explained in more detail by Wheaton et al. (1995).

This effect can be illustrated by the following example. In a simulation of the high-resolution *XMM*/RGS spectrum of the bright stellar corona of Capella, a spectral fit to the simulated spectrum yielded 10 % overabundance of all metals. This was due to the bias introduced in the low count rate continuum, biassing the fitted continuum towards a 10 % lower value; the metal lines with a higher count rate did not suffer from this bias, and hence the apparent abundances became 10 % too high.

Wheaton et al. show that this effect can play a role as long as there are less than 100 counts in a bin. They also showed a solution to the problem. First a spectrum can be fitted using the observed counts N_i as weights. When a nearly acceptable fit is obtained, that model can be used to predict the expected number of counts σ_i^2, and this expected count rate is used in another fit. After one more repetition of this process, convergence has been reached. This procedure can be followed within the SPEX package (Kaastra et al. 1996).

2.8 Data presentation

In several *ASCA* papers, the observed spectra with both SIS or GIS detectors are plotted in the same graph, without applying any shifts. This prohibits the reader to judge from the plots whether certain fit residuals are caused by, e.g., the SIS0 or SIS1 detector, or by both. If the presence of a spectral feature depends on the detector used, this should be mentioned in the text; otherwise, the differences between the fits could be used to assess the possible systematic uncertainties of the data for both detectors, and these systematic errors should be included in the fit; then the added spectra for both detectors could be fitted and plotted.

2.9 Plasma models

For the analysis of *ASCA* data of X-ray sources with thermal emission components several models have been used. Sometimes one encounters models using a bremsstrahlung continuum with in addition a few Gaussian lines added where necessary. The use of such models should be discouraged however for most cases except for very high temperatures. The reason is that for temperatures below about 2 keV, a considerable, or even dominant, fraction of the 1–10 keV continuum spectrum, in particular at higher energies, is not due to bremsstrahlung but to the free-bound recombination continuum or two-photon emission. Moreover, in this temperature range the number of lines below 2 keV becomes very large, and line emission dominates the spectrum at those energies and temperatures.

Several proper plasma models are currently in use. Among the oldest are the Raymond-Smith (*RS*; Raymond & Smith 1977) and Mewe-Gronenschild models (Mewe et al. 1985, 1986). Minor updates to this last code have resulted in the *meka* model (Kaastra 1992); major updates (most importantly the ionization balance and the treatment of the Fe-L complex) resulted in the *mekal* code (Mewe, Kaastra & Liedahl; see Mewe et al. 1995b). In Japan often the plasma code of Masai (1984), and in the EUV range the code of Landini & Monsignori Fossi (1990) is sometimes used. The *RS*, *meka* and *mekal* codes are included in the XSPEC fitting package (Arnaud 1996), the latest *mekal* code is incorporated in the SPEX package (Kaastra et al. 1996). Both Masai's code and SPEX contain non-equilibrium ionization (NEI) modes. All these codes differ in details, see Brickhouse et al. (1995) for an overview. Most important for the analysis of *ASCA* data is the ionization balance that is used for iron and the treatment of the Fe-L complex.

Masai (1997) has shown the important role of the ionization balance calculations that are used. He simulated an *ASCA* spectrum with a temperature of 1 keV and solar abundances using the Masai code. Then he fitted this spectrum using also the Masai code for the plasma emission but with *(a)* the ionization balance for iron due to Arnaud & Rothenflug (1985) and *(b)* due to Arnaud & Raymond (1992). In case *(a)*, the best-fit temperature and iron abundance are 0.85 keV and 0.7 times solar; in case *(b)* 1.0 keV and 0.9 times solar. Thus, the use of different ionization balances may lead to rather large differences in the derived parameters.

In a similar way, the treatment of the Fe-L complex can give large differences. For example, Matsushita (1997) has fitted the spectrum of the elliptical galaxy NGC 4636 with different plasma codes. Comparing the *RS* and *mekal* codes, she finds for the α-elements (O–S) abundances of 0.42 and 0.73 times solar, respectively, and for iron of 0.44 and 0.70. The temperatures derived from both codes are 0.76 and 0.66 keV, respectively.

In general, at present the *mekal* code is the recommended code to be used, although there are certainly still large uncertainties in several aspects of that code. For more details I refer the reader to Rolf Mewe's contribution to this Volume.

3 Stellar Coronae

3.1 Introduction

Stellar coronae are the hot outer parts of cool stars. They are the sites of a variety of high-energy processes that are fundamental for the understanding of stellar evolution, the interaction of stars with the interstellar medium, the physics of particle acceleration and the physics of hot plasmas. Key questions include the coronal chemical composition, its heating mechanism, its structure and evolution. In many cases, our Sun can be used as a benchmark for obtaining high-resolution spectra and testing models, but in general stellar coronae show a much wider distribution of parameters.

The activity is strongly connected to magnetized coronae. There is a well-defined correlation between the rotation period of the star and the ratio of the X-ray luminosity to the bolometric luminosity. Our Sun emits only 10^{-7} of its luminosity in X rays, while the fastest rotating stars emit 10^{-3} of their luminosity in the X-ray band. At rotation periods of about 1 day or less, the X-ray luminosity does not increase further as the rotation period decreases, maybe due to the fact that in those cases the entire surface area is covered by magnetic loops.

Stellar coronae cannot be modeled by isothermal (single-temperature) spectra. Observations with broad-band instruments show that in many cases two or more temperature components are needed in order to model the corona. For this reason, I show in the next section some tools that can be used to analyze the X-ray spectra of stellar coronae.

3.2 Differential emission measure distribution techniques

The temperature distribution in stellar coronae is rather complex. Neglecting for the moment density effects, the spectrum $S(E)$ of the corona is described by

$$S(E) = \int f(E,T)(\mathrm{d}Y/\mathrm{d}T)\mathrm{d}T \tag{4}$$

where $f(E,T)$ is the emissivity at temperature T, and the emission measure $Y = n_\mathrm{e}n_\mathrm{H}V$, with n_e the electron density, n_H the hydrogen density and V the emitting volume. The quantity $\mathrm{d}Y/\mathrm{d}T$ in Eq. (4) is usually referred to as the differential emission measure. This quantity, or more often simply $(\mathrm{d}Y/\mathrm{d}T)\Delta T$ can be determined from the observed X-ray spectrum by inversion of Eq. (4). This inversion is not a trivial thing to do, since the spectrum contains noise and a unique solution can never be obtained. This non-uniqueness looks worse than it is, since in many cases different solutions have certain similarities: in general, for a given temperature range $\Delta \log T \lesssim 0.3$, the *integrated* emission measure over that interval is rather method-independent (provided the method converged). Only the *details* of the emission measure distribution within $\Delta \log T$ are not unique. For example, with *ASCA* and *BeppoSAX* it is not well possible to distinguish *(a)* two isothermal temperature components at 0.8 and 1.2 keV temperature from *(b)* a broad uniform temperature distribution between 0.8–1.2 keV. However, with good statistics and higher spectral resolution than *ASCA* one might be able to distinguish either model from an isothermal plasma with a temperature of 1 keV. But most methods are able to distinguish between both models if the limiting temperatures are, e.g., 0.6 and 2.4 keV (see Fig. 4). An overview of inversion methods for Eq. (4) is given by Kaastra et al. (1996).

3.3 Temperature structure

The temperature distribution in several stellar coronal sources appears to be bimodal. This was already confirmed by the first *ASCA* spectrum of the eclipsing RS CVn system AR Lac (White et al. 1994), and later also in many other stars. The initial analysis of the *ASCA* data of AR Lac by White et al. has been improved by adding simultaneous *ROSAT* PSPC observations (Singh et al. 1996), thereby extending the energy range of the spectrum downwards and increasing the sensitivity to lower-temperature components. These authors also attempted the first DEM-analysis on the AR Lac data. This analysis was improved and extended by Kaastra et al. (1996). In Fig. 5 the spectrum is shown and in Fig. 6 the temperature distribution derived from this spectrum. There is some uncertainty in the temperature distribution, depending on the precise deconvolution method used (see Section 3.2), but the bimodal distribution is a common feature of all solutions. The cool component near 0.6 keV appears to be rather narrow, while the hotter component

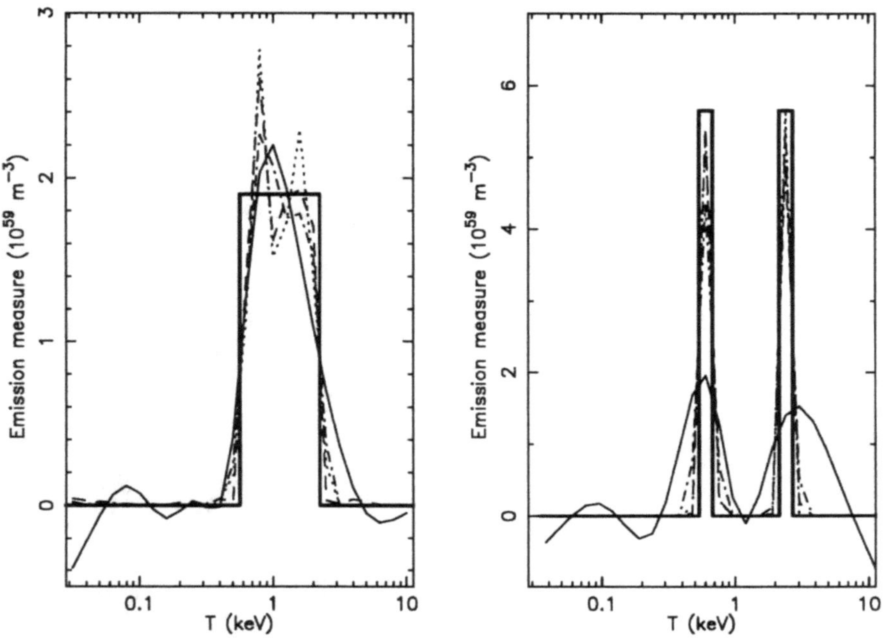

Fig. 4. Reconstructed emission measure distribution for a model consisting of a block profile (left) and two δ functions in temperature (right). Thick solid line: the input model (the sum of the two δ functions are plotted here for clarity as histograms with a width equal to the temperature bin size used); thin solid line: regularization method; dot-dashed line: polynomial method; dashed line: genetic algorithm; dotted line: two broadened Gaussian components (after Kaastra et al. 1996).

is definitely not narrow; it may have some substructure near 1 and 2.4 keV. There is no significant emission below 0.3 keV.

Using a reverberation technique on the phase-dependent *ASCA* light curves of AR Lac, Siarkowski et al. (1996) have attempted to reconstruct the location of the emission in this close binary, which consists of a 1.54 $R_\odot$ G2 IV primary and a 2.81 $R_\odot$ K0 IV secondary, at a mutual distance of 9.22 $R_\odot$. The orbital period is only 1.98 days, and the inclination angle of 87° causes the eclipses. Siarkowski et al. find that both stars are X-ray active. Most (but not all) of the cooler plasma ($T \lesssim 2$ keV) appears to originate from the G-type star, while most of the hot plasma ($T \gtrsim 2$ keV) is due to the K-type star. The emission tends to be concentrated in the region between both stars.

The first *BeppoSAX* observations of stellar coronae (Favata et al. 1997a,b) confirmed the bimodal temperature distribution in VY Ari and Capella. In the case of Capella, *BeppoSAX* sees the two main peaks of the temperature distribution, but is less sensitive to the low temperature tail below 0.3 keV that was observed with *EUVE*.

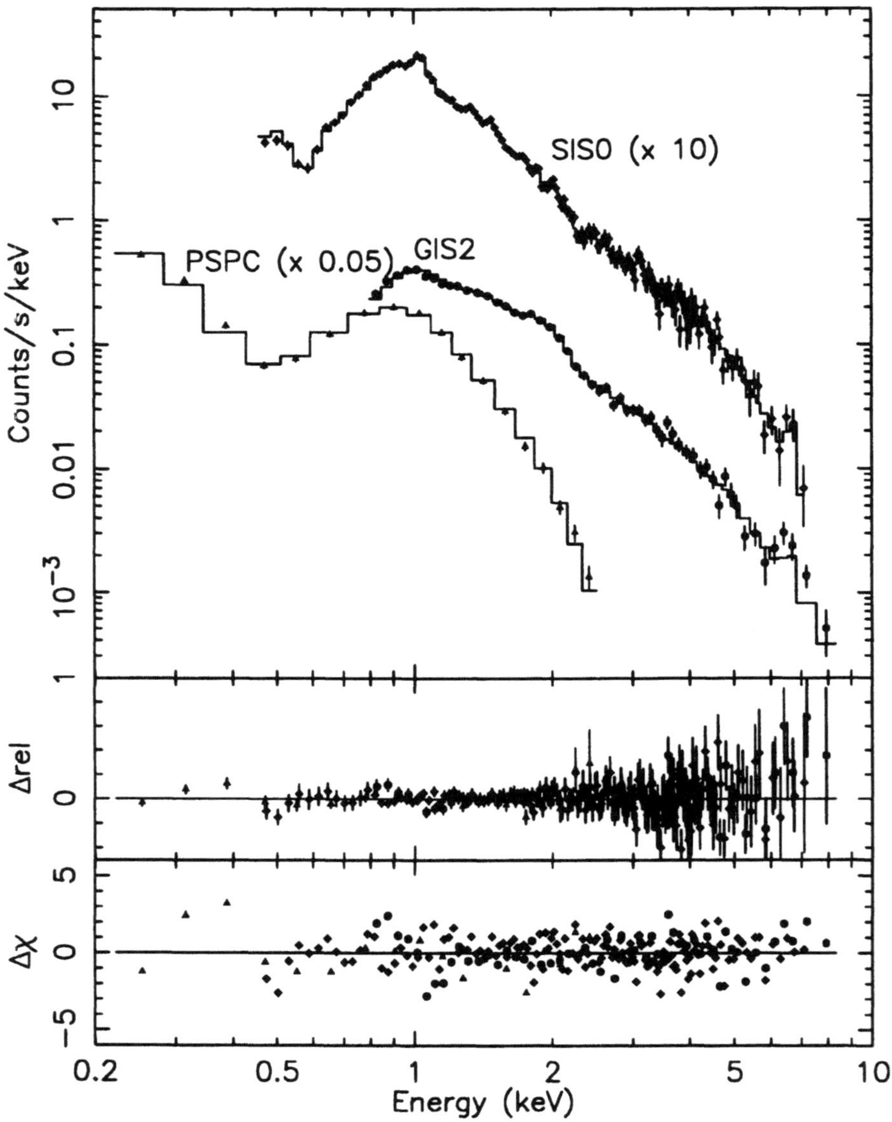

Fig. 5. AR Lac spectrum with best-fit three-temperature model. Triangles: *ROSAT* PSPC data; diamonds: *ASCA* SIS0 data; circles: *ASCA* GIS2 data. Upper panel: observed spectrum with best-fit model (solid histogram). For clarity of representation, the PSPC data have been multiplied by 0.05 and the SIS0 data by 10. Middle panel: relative fit residuals Δrel: (observed - model) / model. Lower panel: fit residuals $\Delta\chi$: (observed - model) / error. (From Kaastra et al. 1996).

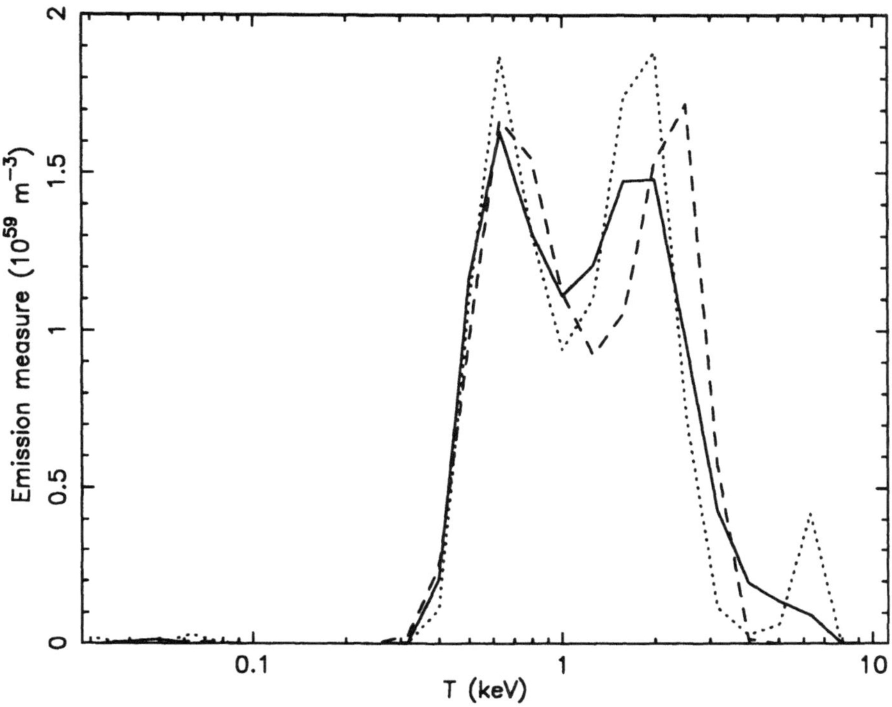

Fig. 6. Emission measure distribution of AR Lac as fitted by polynomials of different order n. Dashed line: $n = 8$. Solid line: $n = 9$. Dotted line: $n = 10$. All fits are equally acceptable in a χ^2 sense (From Kaastra et al. 1996).

3.4 Abundances

One of the big surprises resulting from the *ASCA* observations of stellar coronae is the low metallicity that is found in many systems. The first report of this by White et al. (1994) showed that AR Lac had abundances of O, Mg, Si, S, Ar, Ca and Fe that are 2–4 times lower than the solar photospheric value. Only the Ne and Ni abundances appeared to be higher than solar, but the key line features of these elements are heavily mixed up with the strong Fe-L complex and therefore less reliable. Low abundances were subsequently reported for other systems, like the Algol-type semi-detached binaries Algol (Antunes et al. 1994) and RZ Cas (Singh et al. 1995), the RS CVn system II Peg (Mewe et al. 1997), the active single G-type stars β Cet and π^1 UMa (Drake et al. 1994), the dM1e binary YY Gem (part of the Castor system, Gotthelf et al. 1994), and the young and fast rotating single late F-type star HD 35850 (Tagliaferri et al. 1997).

However, there are also cases showing higher abundances. The young solar-like star EK Dra (Güdel et al. 1997a) shows an iron abundance of 0.83 times solar and a Mg abundance of 1.69 times solar. Also the active binary

Capella (Favata et al. 1997a) as observed by *BeppoSAX* has near-solar abundances, consistent with results derived from *EUVE* observations.

Sub-solar abundances are not only deduced from *ASCA* spectra, but there is also evidence for low abundances from *EUVE* spectra. There have only been a few simultaneous observations of stellar coronae with *ASCA* and *EUVE*. Published up to now is AB Dor (Mewe et al. 1996). The simultaneous data obtained with both instruments yield consistent results and show the typical bimodal temperature distribution, with an under-abundance of O, Mg, Si, S and Fe; only Ne appears to have a solar abundance.

BeppoSAX observations of the active binary VY Ari (Favata et al. 1997b) using a broad energy band (0.1–10 keV) also showed a sub-solar abundance (0.4 times solar) for iron.

The interpretation of the sub-solar abundances is yet unclear, and perhaps it needs not to be the same from source to source. Several effects may play a role simultaneously. For the EUV band, the following effects have been proposed as explanations of the apparently low line-to-continuum ratio: a very hot bremsstrahlung continuum (Mewe et al. 1995a), resonance scattering in the strongest lines (Schrijver et al. 1994), and missing lines in the spectral code (Schmitt et al. 1996). None of these effects appear to be consistent with the *ASCA* spectra of several of the most active systems, so that the low abundances are most likely real.

3.5 Flares

In the most active stellar coronae flares occur frequently. Flares have peak luminosities in the X-ray band up to 10^{24} W and radiate a total energy up to 10^{28} J. In most cases observed with *ASCA*, the flaring activity seems to be associated with the hotter of the two main temperature components, and it is in particular the emission measure and not the temperature that increases most strongly. This is the case in flares observed in Algol (Antunes et al. 1994) and YY Gem (Gotthelf et al. 1994).

A flare observed in the RS CVn star II Peg (Mewe et al. 1997) showed both a doubling of the temperature and the emission measure. There is evidence that during the maximum of the flare the iron abundance increased from 0.1 times solar to 0.4 times solar. A similar increase in iron abundance during a flare was observed in AB Dor (White et al. 1995). A possible explanation for this increase is that the flare causes chromospheric evaporation; since the X-ray emitting corona has lower abundances than the chromosphere, evaporation leads to enhanced abundances in the flaring plasma as compared to the quiescent corona. The decay of the flare in II Peg can be interpreted in terms of cooling-loop models, and yields important parameters such as the loop height (10^9 m) and density (10^{17} m^{-3}).

The wide-field cameras of *BeppoSAX* are able to find the strongest flares that occur in active stars. This instrument detected an exceptional strong flare ($2\,10^{24}$ W) in the flare star BY Dra (Mewe et al. 1997).

3.6 Stellar evolution

Güdel et al. (1997b) used *ASCA* and *ROSAT* in order to study the coronae of nine solar-like single G stars with ages ranging from 0.07–9 10^9 year and X-ray luminosities 1–500 times that of the quiet Sun. Each of these stars shows a bimodal temperature distribution. In the younger stars, a considerable fraction of the emission measure is at very high temperatures, up to ~20–30 MK in EK Dra. At the Sun such temperatures are found only during short flaring episodes. This high-temperature component rapidly decays within the first few 10^8 year, roughly as $T \sim t^{-0.3}$, while also for this component the luminosity decreases with temperature as $L \sim T^4$. Apparently, when the star ages and its rotation slows down, coronal heating is less efficient. Güdel et al. interpret the hot component as an unresolved superposition of flares, and argue that, in addition, the spectrum is quite similar to solar-flare spectra.

Fleming & Tagliaferri (1996) studied the old population II star HD 89499. This G3 VI star is the most metal poor system known to have a corona. The iron abundance deduced from optical data is 0.008 times solar. *ASCA* data yielded an upper limit of 0.03 for the coronal metallicity. The temperature of 2.2 keV is rather high for such an old star. Perhaps the near absence of metals gives a low bound-bound and free-bound cooling rate, thereby keeping the temperature high.

4 Hot Stars

4.1 Introduction

Hot main-sequence stars differ fundamentally from their cooler counterparts in that they do not contain an outer convection zone. Therefore, the usual dynamo mechanisms that generate the magnetic fields and thereby the hot coronal plasma do not work for these objects. Nevertheless, many hot stars do show thermal X-ray emission. The most common reason for this is turbulence and shocks in the strong stellar winds that these systems possess, which cause localized regions of strong heating and X-ray emission. In this chapter I describe a few interesting examples of X-ray emission from hot stars, showing various kinds of X-ray spectra.

4.2 Normal O and B stars

The first *ASCA* observations of normal O-type stars were reported by Corcoran et al. (1994). The spectra of δ Ori and λ Ori could be described by two-temperature models with additional absorption by the warm wind. The hot components have temperatures of 0.6 and 2 keV, respectively. The cooler components (0.3 keV) have the largest emission measure. The hot component does not show significant absorption, while the cooler component is absorbed by ~3 10^{25} m^{-2} of cool gas, corresponding to the total column density of the stellar wind contained within about 2 stellar radii. There is evidence that the iron abundance is 0.2 times solar.

Four bright OB stars in the Cygnus OB association were investigated by Kitamoto & Mukai (1996). The Cygnus OB association is a 3 million years old star forming region at 1.7 kpc distance. Surprisingly, these four stars showed metallicities in their X-ray spectra in the range of 0.1–0.4 times solar, at temperatures between 0.6–2.5 keV. Enhanced He and N abundances (effectively increasing the continuum relative to the lines at X-ray energies) cannot explain these low metallicities; the over-abundances needed are not consistent with optical data.

In the extremely young B0 V star τ Sco, Cohen (1996) found a very hard spectrum (temperature 2.4 keV), with emission line complexes from highly ionized Mg, Si, S and Fe. The temperature is too high to be explained by line-force instability wind shocks. Cohen suggest that in this and some other sources coronal mechanisms might be active.

4.3 Luminous blue variables

Tsuboi et al. (1997) report the *ASCA* spectrum of the LBV η Car. This star of $100\,M_\odot$ had a major outburst in 1843; it was then the second brightest star in the sky. They find a hot (5 keV) point-like source that is highly absorbed ($5\,10^{26}\,m^{-2}$) and is variable on long time scales, and a cooler (0.7 keV) extended component (1 light year radius) without significant absorption above the galactic value. This cool component is interpreted as the shock heated outer shell that was ejected in 1843. Its temperature is consistent with the expansion velocity as derived from optical observations. Tsuboi et al. suggest that for this cooler component NEI effects may play a role: they estimate that $n_e t$ is about $5\,10^{16}\,m^{-3}\,s$. Unfortunately, they did not make a fit to the spectrum using NEI models.

The spectrum of η Car shows a strong line at 0.5 keV, interpreted as the Lyman α line of highly ionized nitrogen. Tsuboi et al. derive abundances for N, O, Si and S of 200, 6, 4 and 3 times solar respectively, while Ne, Mg and Fe have solar abundances. The large N/O ratio is consistent with what is found in optical spectra of this source. A possible explanation is that in the CNO cycle the reaction $^{14}N + p \rightarrow {}^{15}O + \gamma$ is the slowest of all, and hence during the cycle a pile-up of nitrogen might occur. The absolute N abundance with respect to H and He, however, is too large to be explained. Perhaps the NEI effects mentioned above have influenced the abundance determination.

4.4 Wolf-Rayet binaries

Massive stars have strong stellar winds. When two such stars are in an eccentric binary, the winds of both components will show strong interaction near periastron. This is exactly what was observed in the binary HD 193793 (a WC and O star pair in an eccentric 8 year orbit). The X-ray spectrum showed a hot plasma (3 keV) that can be attributed to the shock front of the colliding winds (Koyama et al. 1994a). The spectrum is heavily absorbed through the wind of the WC star ($3\,10^{26}\,\mathrm{m}^{-2}$). The spectrum shows a strong Fe-K line with near-solar abundances. Furthermore, there is some evidence for a feature near 1.2 keV. This was interpreted as Lyman β emission from Ne X by Koyama et al., implying a Ne abundance between 50–200. Evolutionary models for WC stars predict an enhancement of Ne with respect to He corresponding to only 10–40 times solar. A part of the discrepancy may be attributed to the strong absorption cut-off in this region of the spectrum, combined with uncertainties in the Fe-L complex etc. The best-fit model also requires an overabundance of C in the absorbing wind of at most a factor of 30. This appears to be below the predicted abundances for WC stars of at least 40, but may be consistent with infrared observations.

Another WC/O binary, γ^2 Vel (Stevens et al. 1996) could also be interpreted using a colliding-wind model. The variable absorption of (3–8) $10^{26}\,\mathrm{m}^{-2}$ clearly hints to an emission region located between both stars.

5 Protostars and T Tauri Stars

5.1 Introduction

In the evolution from gas clouds to main-sequence stars several stages can be defined. In general, the young pre-main-sequence objects are divided into four infrared classes, labeled commonly as class 0 (young protostars), I (evolved protostars), II (classical T Tauri stars) and III (weak-line T Tauri stars); the typical durations of these phases are 10^4, 10^5, 10^6 and 10^7 years, respectively. The young protostars are characterized by a cold blackbody spectrum; in the evolved phase clear signatures of a disk are visible; in the classical T Tauri stars an optically thick disk develops, and in the final stages the stellar blackbody becomes dominant.

5.2 X-ray emission from protostars

Surprisingly, *ASCA* detected hard X rays from protostar candidates. Koyama et al. (1996a) studied the R CrA cloud. This nearby (130 pc) molecular cloud shows star formation. Near the center is a dense group of protostars, called the Coronet cluster. *ASCA* detected 8 X-ray sources in this group, among which 5 evolved protostars. These protostars show spectra with a temperature of 7 keV, and a strong Fe-K line feature, which are heavily absorbed by material with a column density of $4\,10^{26}\,\mathrm{m}^{-2}$. In contrast, the brightest T Tauri star in the same region shows no absorption, no clear spectral lines and a temperature of only 2.6 keV. The iron line in the brightest X-ray emitting protostar (number 8) has a centroid energy of 6.45 keV and an equivalent width of 800 eV. Interestingly, a strong flare occurred in this source, during which the iron line showed a double structure, with peaks around 6.17 and 6.81 keV. The highest-energy peak can of course be explained by a mixture of H-like and He-like emission of a hot plasma, but then the 6.17 keV feature remains unexplained. Koyama et al. suggest that both components might correspond to a bipolar jet, which then should have an average outflow velocity of $0.05c$.

The hard X-ray emission can be understood if the magnetic field of the protostar threads the protostellar disk (Hayashi et al. 1996b). The differential rotation of the disk then causes magnetic shear and eventually magnetic reconnection gives rise to heating and particle acceleration, leading finally to hard X-ray emission.

5.3 X-ray emission from T Tauri stars

Due to their young age, T Tauri stars are often found in groups near their place of birth. In the nearby (140 pc) dark cloud ρ Oph, *ASCA* detected at least 11 sources, some of which have no optical counterpart (Koyama et al. 1994b). In general, these objects show hard spectra. The brightest object has a temperature of 2 keV and is heavily absorbed (column density 1–2 $10^{26}\,m^{-2}$). The spectrum shows evidence for emission from the S-K and Fe-K complex. However the spectrum is too weak for us to distinguish between single or multiple temperature structure. Several of these sources are variable, some show flare-like events.

ASCA detected an intense X-ray flare on the weak-lined T Tauri star V 773 Tau (Skinner et al. 1997). This star is a spectroscopic binary (K2 V + K5 V) and shows strong magnetic surface activity including a spot-modulated optical light curve. The flare is one of the strongest ever recorded from a T Tauri star with a peak luminosity of 2.5 10^{25} W, 25 times brighter than the quiescent state. The total energy release is at least 10^{30} J. The temperature distribution in this object is bimodal but almost totally dominated by the hottest 3 keV component.

Another flare has been detected in V 826 Tau, a T Tauri star in the Lynds 1551 star forming cloud (Carkner et al. 1996). The quiescent emission from the T Tauri stars in this cloud shows evidence for either multi-temperature structure (0.2–1 keV) with solar abundances, or single-temperature emission with metallicity less than 0.1 times solar. *ASCA* is not sensitive enough to distinguish between these alternatives.

It is well known that cool stars with a convection zone have relatively strong magnetic fields and thereby are able to maintain a hot corona. In general, hot stars do neither have an outer convection zone nor a hot corona. The bimodal temperature structure (0.3 and 1.6 keV) and variable X-ray emission from HD 104237, a Herbig Ae object, therefore comes as a surprise (Skinner & Yamauchi 1996). Herbig Ae/Be stars are 2–10 $M_\odot$ pre-main-sequence stars. It is as yet unclear whether the X-ray emission is due to a non-solar dynamo mechanism or to an unseen normal late-type companion star.

6 Cataclysmic Variables

6.1 Introduction

Cataclysmic variables (CVs) are binary stars with a white-dwarf primary that is powered by the accretion flow from a Roche-lobe filling late-type companion (the secondary). The gravitational potential of the white dwarf is so large that infalling material can reach temperatures of 10 keV if it remains optically thin. Typical X-ray luminosities are 10^{22}–10^{27} W. Their orbital periods are, in general, 1–10 hour. CVs can be divided into two subclasses: non-magnetic CVs and magnetic CVs.

The non-magnetic CVs are subdivided into classical novae, dwarf novae and nova-like variables, according to their eruptive behaviour. In these systems, the Roche lobe overflowing cool star generates an accretion disk around the white dwarf; the X-ray emission is then produced in the boundary layer between the disk and the white-dwarf surface. The majority of the CVs belong to this class.

In magnetic CVs the white-dwarf magnetic field is so strong that the accretion disk cannot extend down to the white-dwarf surface. Instead, the material flows supersonically from the inner edge of the disk along the magnetic field towards the magnetic poles of the white dwarf. The deceleration of this polar accretion flow at the poles causes a standing shock, and heating, and thereby produces hard X-ray emission (10–30 keV). These X rays heat the white-dwarf surface and produce a blackbody-like soft X-ray component. Magnetic CVs have two subclasses. In polars (AM Hers) the magnetic field is so strong that no accretion disk forms at all; the white dwarf rotates synchroneously with the orbital period and the material flows directly from the inner Lagrangean point towards the magnetic pole. In intermediate polars (DQ Hers) the magnetic field is somewhat weaker and a partial accretion disk is present.

6.2 Non-magnetic cataclysmic variables

One of the best studied CVs, the dwarf nova SS Cyg, was observed during outburst by Nousek et al. (1994). The *ASCA* spectrum indicates the presence of at least three temperature components: 0.8, 3.5 and 18 keV. These authors needed the hottest component in order to model the relatively low line-to-continuum ratio. The spectrum shows a strong iron line, but it is not clear from the paper that the line has been modeled properly: Nousek et al. use two plasma components for the lowest temperatures plus the unphysical thermal bremsstrahlung model for the 18 keV component, but in this last case the corresponding iron emission is neglected.

Done & Osborne (1997) have reanalyzed archival *Ginga* and *ASCA* data for SS Cyg in both quiescence and outburst using multi-temperature plasma models for the continuum and line emission, together with their reflection from the X-ray-illuminated white dwarf and accretion disc. By comparing the quiescent and flare spectra, they argue that during quiescence the inner disk is optically thin or even absent. In that case the reflection should come from the white-dwarf surface. The amount of reflection during outburst is more consistent with a complete covering of the surface than with just an equatorial belt. Therefore the hard X rays originate possibly from a corona covering the entire star. The spectra are absorbed by a partially ionized medium, for which Done & Osborne propose the outflowing wind to be responsible.

The spectrum indicates abundances of 0.4 times solar for most visible elements. In quiescence, the spectrum is consistent with an isothermal plasma. Therefore the cooling material must either be reheated, or be partially hidden behind the accretion disk. In contrast, the outburst spectrum shows clearly the cooling components.

Other cases of dwarf novae where hot plasmas (8–10 keV) have been found include U Gem (Szkody et al. 1996) and HT Cas (Mukai et al. 1997). In this last case, the eclipses could be used to set an upper limit to the total size of the X-ray emitting region of 1.15 times the white-dwarf radius.

Supersoft sources Supersoft sources constitute a class of white dwarfs in binaries that are powered by steady nuclear burning of hydrogen, accreted onto the surface of the white dwarf. In general, the X-ray spectra produced by these sources are soft (hence their name), and produce no significant emission above 1 keV. Their spectra are heavily absorbed and have relatively low count rates. Supersoft sources are considered to be progenitors of type Ia supernovae.

BeppoSAX has observed CAL 87, one of these sources in the LMC (Parmar et al. 1997b). The spectrum of this source, thought to originate from the heated atmosphere of the white dwarf, can be characterized by blackbody emission with additional evidence for an O VIII absorption edge at 0.871 keV. Even better spectral fits are obtained with models that determine the atmosphere structure using LTE or NLTE models. Typical white-dwarf radii of 1000 km and luminosities of $3\,10^{29}$ W are obtained. The atmosphere has a temperature of 60–70 eV. In contrast with this, simple blackbody models yield radii and luminosities that are orders of magnitude larger.

6.3 Intermediate polars

ASCA observations of one of the brightest intermediate polars, EX Hya (Ishida et al. 1994) showed the presence of He-like and H-like emission lines from Ne to Fe (Fig. 7). The spectrum cannot be modeled by a single isothermal plasma, but instead a multi-temperature plasma is needed, with temperatures in the range of 1–8 keV. This range probably reflects the cooling of the accreting material along its path towards the white-dwarf surface. There remains a low-energy wing to the Fe-K complex that cannot be produced by the hot plasma components. Ishida et al. found a centroid energy of 6.48±0.04 keV, and explained the line as due to fluorescence at the white-dwarf surface. It cannot be due to fluorescence in the pre-shock accretion column, for the equivalent width of 80 eV would imply a column density of $10^{27}\,m^{-2}$, 100 times larger than the column density observed in absorption in the X-ray spectrum.

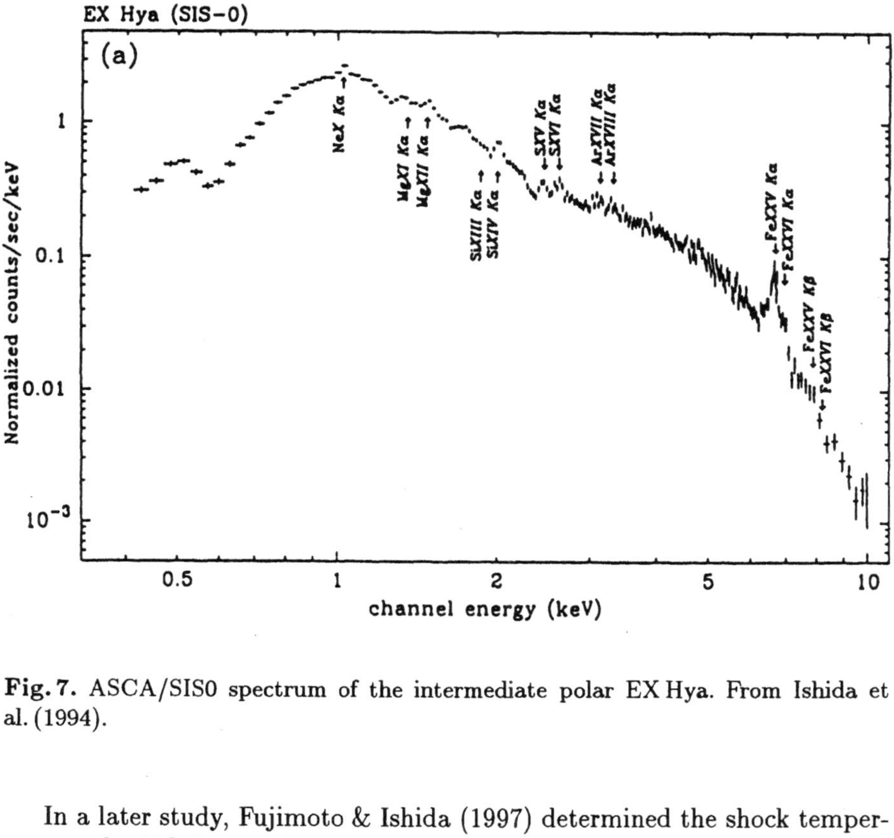

Fig. 7. ASCA/SIS0 spectrum of the intermediate polar EX Hya. From Ishida et al. (1994).

In a later study, Fujimoto & Ishida (1997) determined the shock temperature to be 15 keV, and from this they could determine the white-dwarf mass as $0.48^{+0.10}_{-0.06}\,M_\odot$, consistent with the optical mass determination. The mass can, in general, be determined from the temperature as follows. The shock temperature is directly proportional to the gravitational potential and hence to M/R. Thus, the mass-to-radius ratio can be determined directly from the shock temperature. Then by using the known M-R relation for white dwarfs, both the mass and radius can be determined. Fujimoto & Ishida have used a detailed model for the post-shock structure. In such models, the plasma temperature decreases from the shock towards the surface, while the density rapidly increases. The resulting multi-temperature plasma describes the observed spectrum well, and can account for the wide range of ionization potentials visible in the spectrum. The abundances of Si, S and Ar are consistent with solar values, but iron has a sub-solar abundance (0.6 times solar).

In another intermediate polar, FO Aqr (Mukai et al. 1994), the plasma is significantly hotter. The spectrum consists of two components. Below 1.5 keV an unabsorbed component dominates. This component is not well constrained and can be anything like 3 keV plasma emission or 0.2 keV blackbody radiation. The other component is very hot (30 keV) thermal emission absorbed by a (continuous?) distribution of cool gas with $N_{\rm H}$ ranging from 0.3–3 $10^{27}\,{\rm m}^{-2}$. The only evidence for line emission in this spectrum is Fe-K at 6.4 and 6.8 keV, probably due to fluorescence (as in EX Hya) and H-like emission from the hot component. The lack of evidence for other emission lines indicates the absence of cooler (1 keV) components, according to these authors. Note however that their spectrum shows excess emission near the O VIII Lyman α line at 0.65 keV, and that above 1.5 keV the very hot component dominates, so that their last conclusion may be doubted.

AO Psc (Hellier et al. 1996) also shows a high temperature (12 keV, implying a white-dwarf mass of 0.4 $M_{\odot}$) and is absorbed by a distribution of material with column densities in the range of 10^{26}–$10^{27}\,{\rm m}^{-2}$, typical for the accretion curtain corresponding to the flow of the material from the disk towards the poles. The H-like and He-like iron lines appear to be significantly broadened (0.15 keV), for which the authors do not give an explanation, however.

6.4 Polars

The prototype polar-type CV is AM Her. *ASCA* observations showed the tail of a soft blackbody component (T = 0.03 keV), and simultaneous *EUVE* observations confirmed this (Ishida et al. 1997). The blackbody component occupies only a fraction of 10^{-5} of the white dwarf surface area, and varies in phase with a hard X-ray component. It may be associated with the region near the poles that is heated by an accretion flow.

The spectrum during the rotational minimum phase shows a strong fluorescent Fe-K line with an equivalent width of 1000 eV. This indicates that the hard X-ray emission during the rotational minimum phase consists entirely of scattered emission, possibly in the pre-shock accretion column, which must have a column density not larger than $10^{27}\,{\rm m}^{-2}$.

Also H-like and He-like Fe-K emission lines are visible. However, the emission lines from the other light elements are generally weaker than those seen in EX Hya, in which a significant post-shock cooling flow was found from the ASCA observation. The spectrum of AM Her in the bright phase, on the other hand, is consistent with a single-temperature optically thin thermal plasma with abundances of 0.4 times solar. These facts indicate that most of the cooling part of the post-shock plasma is hidden from the observer in AM Her. A possible interpretation is that the post-shock plasma is highly inhomogeneous, and penetrates the photosphere before it has cooled significantly.

A detailed study of another polar, BY Cam, by Kallman et al. (1996) showed also three distinct components in the Fe-K complex. The component due to neutral iron was interpreted as reprocessed radiation in the white-dwarf photosphere. A medium-ionized component due to He-like to Be-like iron was attributed to emission from the accretion column, photo-ionized by the shock emission. The H-like line finally was attributed to the line created by the hot (30 keV) shock plasma.

ASCA also discovered serendipitously a polar, AX J2315–592 (Misaki et al. 1996). One of the poles in this source is probably always visible.

7 High-Mass X-ray binaries

7.1 Introduction

High-mass X-ray binaries (HMXB) are systems in which a neutron star or black hole accretes material from a massive companion star. These sources display a large variety in pulse periods (below 0.1 s – 1000 s), orbital periods (several hours to a year) and X-ray luminosities (10^{27}–10^{32} W).

Massive X-ray binaries are the product of close-binary evolution. The most massive star of the pair evolves off the main sequence and transfers material to the companion through Roche lobe overflow; it becomes a He star, which on a relatively short time scale explodes as a supernova. During the mass transfer phase, the companion star was spun up, and the result is a HMXB consisting of a rapidly rotating massive star in orbit around a neutron star. During its subsequent evolution the massive companion loses mass in a stellar wind, part of which may accrete onto the neutron star, thereby producing X-ray emission. The spectra are in general rather hard, and in several cases cyclotron lines have been seen. In the most narrow binaries, accretion may occur through Roche lobe overflow; in the wider binaries accretion occurs less efficiently through a wind; therefore the latter systems have in general lower luminosities.

7.2 Vela X-1

Vela X-1 is an eclipsing X-ray binary pulsar with an orbital period of 9 days and a pulse period of 283 s. The pulsar accretes matter from a B0.5 Ib star via capture of a stellar wind. Therefore it has a relatively low luminosity of 10^{29} W. The *ASCA* spectrum is shown in Fig. 8.

Outside of the eclipse, the pulsar spectrum is visible. It shows a heavily absorbed pulsed component and a flat spectral component which is ascribed to scattering in the stellar wind (Nagase et al. 1994). The pulsed component has a power law like spectrum with a photon index of 1.28, and shows strong

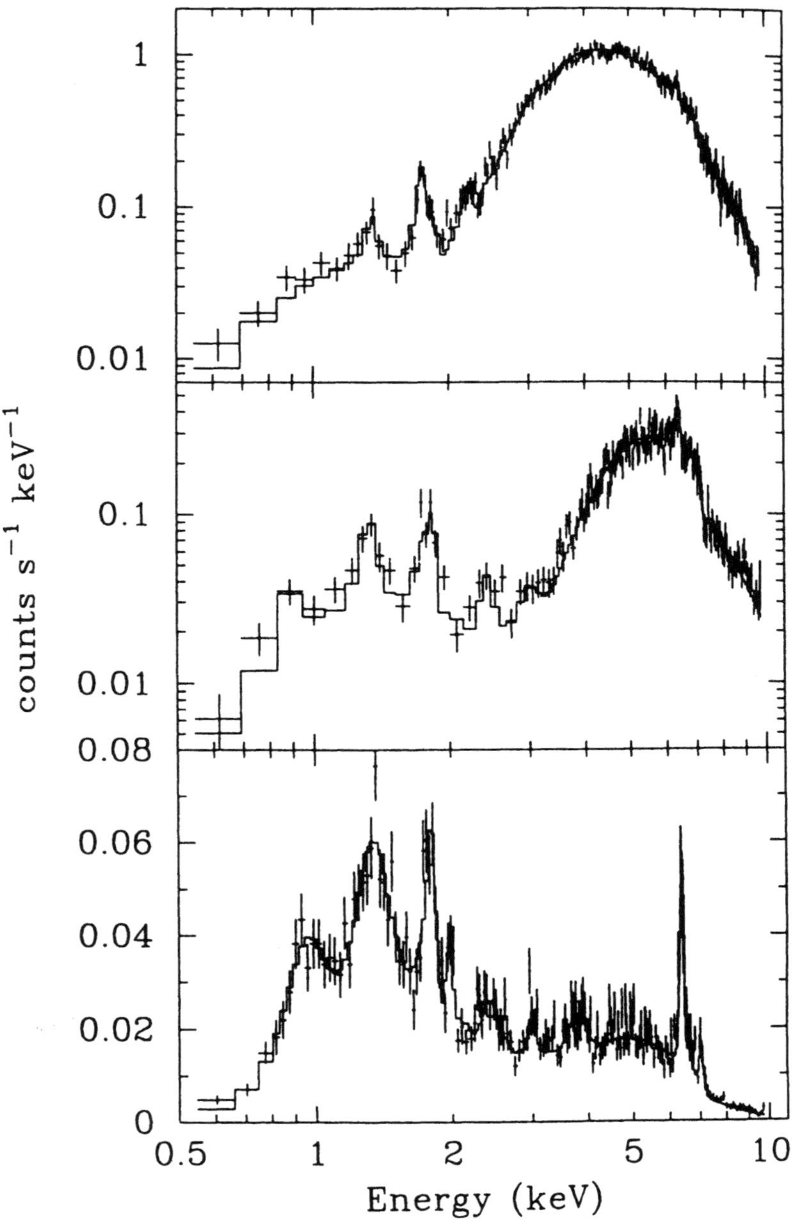

Fig. 8. *ASCA*/SIS0 spectra of the HMXB Vela X-1 obtained at post-eclipse (top), pre-eclipse (middle) and eclipse phase (bottom). From Nagase et al. (1994).

absorption ($1.3\,10^{27}\,\mathrm{m}^{-2}$). The absorption in the pre-eclipse spectrum is much stronger ($7\,10^{27}\,\mathrm{m}^{-2}$); therefore a strong Fe-K absorption edge at 7.1 keV is clearly visible in the raw data. This increased absorption has sometimes been explained by a collision between the stellar wind and the stagnant photo-ionized gas surrounding the neutron star. This causes a strong shock and creates dense sheets of gas trailing the pulsed X-ray source.

During the eclipse, however, the dominant pulsed component disappears, and the spectrum shows a flat component with strong line emission. In particular, the He-like lines of Ne, Mg, Si, S, Ar, Ca and Fe are visible. Although some lines are also visible outside the eclipse, they can be distinguished much better during the eclipse. The lines may be attributed to radiative recombination in the X-ray irradiated stellar wind; this wind is in photo-ionization equilibrium. Applying photo-ionization models, Nagase et al. estimate that the temperature in this medium should be around 0.1 keV. Remarkably, the lines of He-like Ne, Mg, S and Ar show a significant broadening of about 0.1 keV. This cannot be explained by Doppler shifts in the stellar wind, since the wind velocity is expected to be $\lesssim$1000 km/s. Therefore Nagase et al. suggest that radiative recombination continua may be responsible for the apparent line broadening.

In the Fe-K complex, not only the previously discovered component of neutral iron at 6.4 keV can be distinguished, but also the He-like component at 6.68 keV, and the Kβ component of neutral iron at 7.06 keV. However, the observed $K\beta/\mathrm{K}\alpha$ ratio of 0.30 is larger then the theoretically expected ratio of about 0.13, so that there may be a little contamination from the Lyman α line of hydrogen-like iron at 6.95 keV.

7.3 Cyg X-3

At a distance of 10 kpc and an X-ray luminosity of $2\,10^{31}$ W, Cyg X-3 is one of the intrinsically brightest X-ray sources of our galaxy. It has an orbital period of only 4.8 hour. The massive companion is a Wolf-Rayet star, and the neutron stars moves through the dense wind of this star.

The first *ASCA* observations of this source (Kitamoto et al. 1994) showed a spectrum with strong line emission. The iron line consists of components from neutral, He-like and H-like iron, at energies of 6.37, 6.67 and 6.96 keV, respectively. These last two lines show orbital modulation, in the sense that they have the strongest line flux when the continuum has a minimum. Furthermore, there is an iron absorption edge at 7.19 keV. This edge also has a maximum optical depth (0.5) at the time of minimum continuum level. Finally the spectrum shows strong emission lines from Si, S, Ar and Ca.

Liedahl & Paerels (1996) have done a proper analysis of this first ASCA spectrum of Cyg X-3 using a photo-ionization model. These authors show that the X-ray spectrum contains important contributions from radiative-recombination continua of highly ionized species at temperatures around 0.02 keV. Since for these continua the temperature is much lower than the ionization energy, the continua look like narrow line-like features. Liedahl & Paerels were able to identify the strong recombination continua from hydrogen-like Mg, Si and S. The origin of this emission is the photo-ionized wind of the Wolf-Rayet star. The ionizing photons are provided by the accreting neutron star.

7.4 Cen X-3

The X-ray binary pulsar Cen X-3 was observed with *ASCA* during eclipse (Ebisawa et al. 1996). As in Cyg X-3, the three iron emission lines at 6.4 keV, 6.7 keV, and 6.97 keV were clearly resolved. At lower energies, lines from hydrogenic Ne, Mg, Si, and S were also observed. Although the intensity of the 6.4 keV line decreased about one order of magnitude during the eclipse, the intensity decrease of the H-like and He-like lines was at most a factor of three. This indicates that the fluorescent line at 6.4 keV is emitted from the region close to the neutron star while the highly photoionized plasma is more extended than the size of the companion star ($16\,R_{\odot}$). The dense Alfvén shell or an optically thick accretion disk are likely candidates for producing the fluorescent iron line. From the intensity ratio of the 6.7 and 6.97 keV lines during the eclipse, the ionization parameter of the photoionized plasma is estimated as $\xi = 2.5\,10^{-6}\,\mathrm{kg\,m^3\,s^{-3}}$. The size and the density of the plasma could be estimated as $16\,R_{\odot}$ and $10^{17}\,\mathrm{m^{-3}}$, respectively.

7.5 SS 433

SS 433 is a spectacular X-ray binary ejecting bipolar jets at a velocity of $0.25c$ and precessing with a 163 day period. The *ASCA* spectra of this source (Kotani et al. 1994, Fig. 9) showed evidence for both red- and blueshifted lines. This was the first detection of the two Doppler-shifted beams in the X-ray spectrum, and allowed the determination of the radial jet velocity with an accuracy comparable to optical observations. The spectrum shows lines from H-like Si, S, Ar, Ca Fe and Ni. This broad range of ions is compatible with a multi-temperature structure.

In a later analysis of data obtained near precession phase 0.48, when the jet components had their maximum separation the emission could be interpreted in terms of radiatively cooling jets (Kotani et al. 1996). The jets start at a temperature of 20 keV (derived from the H/He-like iron line ratio and consistent with earlier *Ginga* measurements of the continuum temperature). The observed red/blue line ratio for He-like iron is 0.24, and this is significantly smaller than the value of 0.66 that is expected from Doppler boosting. The implication is that the red (receding) jet shows extra absorption. For other lines like Si, S and Ar, the red component of the line is even absent at this precession phase. Kotani et al. conclude that the receding component of the jet is absorbed by local gas with a column density of (0.2–3) $10^{27}\,\mathrm{m^{-2}}$, and that only the inner part of the receding jet with temperatures of at

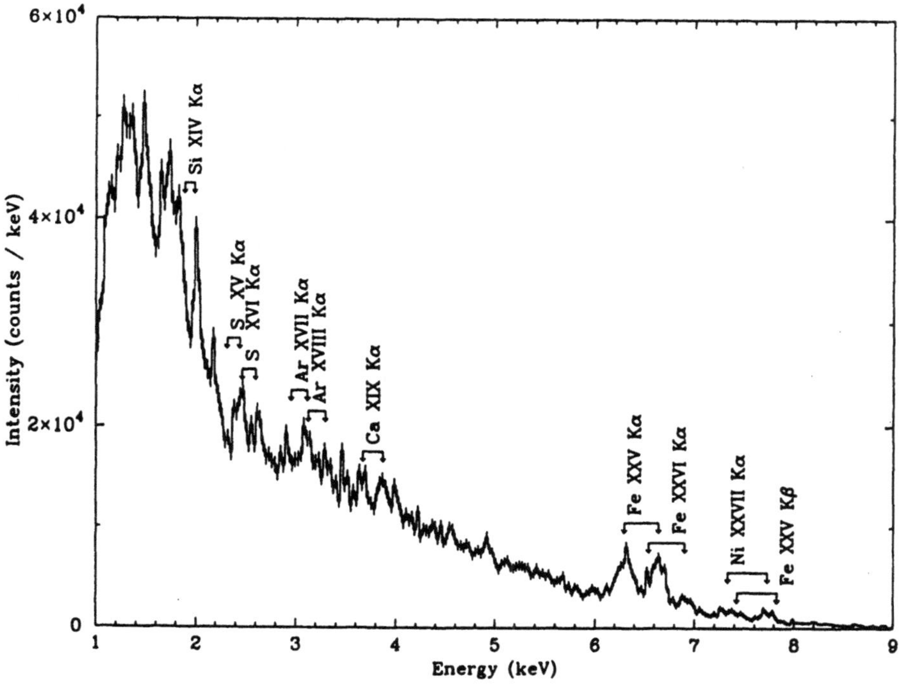

Fig. 9. X-ray spectrum of SS 433 obtained with the SIS detector of *ASCA*. Note the splitting of the lines in blue- and redshifted energies. From Kotani et al. (1994).

least 6 keV is occulted by the accretion disk. The origin of the local gas that absorbs the receding jet could be an extended thin rim to the accretion disk.

There is also a stationary component of fluorescent neutral iron present in the spectrum. Modelling showed that it cannot be due to the stellar wind or the surface of the companion star, but most likely it is due to fluorescence on the accretion disk.

7.6 Other cases

Fluorescent emission by lowly ionized iron and the corresponding absorption edge have also been found in other HMXBs, like the X-ray binary pulsar GX 301–2 (Pravdo et al. 1995; Saraswat et al. 1996). These last authors also detect fluorescent emission from (near)-neutral Ne, Si, S, Ar and Ca, originating from a plasma with a temperature of at most 10^5 K, and a thermal component (0.8 keV), which they identify as emission from shocks in the gas trailing the neutron star.

8 Low-Mass X-ray Binaries

8.1 Introduction

Low-mass X-ray binaries (LMXBs) are systems with a late-type star in a close orbit around a neutron star. Contrary to the HMXBs, LMXBs rarely show pulsations. In general, they also have weaker magnetic fields than HMXBs and softer X-ray spectra. Sometimes they show thermonuclear flashes on the neutron star that are observed as X-ray bursts. Low-mass X-ray binaries are the progenitors of radio pulsars with low-mass white-dwarf companions.

Several LMXBs have been observed with *ASCA*, among these X 1608–52 and Cen X-4 (Asai et al. 1996), and Her X-1 (Vrtilek et al. 1994; Choi 1997). The spectra of these sources can usually be described well by the sum of a few absorbed blackbody and power law components, although Mihara et al. (1994b) showed that the spectrum of Her X-1 contains a strong iron line near 6.4 keV and two unidentified features near 0.91 and 1.06 keV. These last features have now been confirmed with *BeppoSAX* (Oosterbroek et al. 1997, see Fig. 10).

A few LMXBs show more interesting X-ray spectra, and these sources are discussed in more detail in the next subsections.

8.2 4U 1626–67

4U 1626–67 is a somewhat unusual LMXB in the sense that it contains an X-ray pulsar. The pulsar has a period of 7.7 s. Surprisingly, the *ASCA* data (Angelini et al. 1995, Fig. 11) showed excess line emission above the usual blackbody plus power law spectrum that is common in these sources. A strong line at 1 keV could be identified as the Lyman α line of H-like neon. Apart from that, also other emission lines from H-like and He-like Ne and O are present. These lines do not show much variability as a function of the pulsar phase. At phase 0.2–0.4 a weak Fe-K line is present, but otherwise no significant emission from other elements seems to be present, apart from maybe some Mg near 1.41 keV, but the feature seen at that energy could also be the recombination continuum of H-like Ne. If this feature is interpreted as a recombination edge, then its observed width indicates a temperature of 0.06 keV, and hence the plasma should be photo-ionized. Both fits to models in collisional ionization equilibrium, as well as photo-ionization models, predict an overabundance for Ne of about 5 times solar and 2 times solar for O. However, both models cannot explain all details in the spectrum. The overabundance of neon might imply that the companion has gone through the helium-burning phase, since Ne is a by-product of that process. Before these *ASCA* observations, the usual assumption was that the companion is still in its hydrogen-burning phase. The presence of the 1 keV feature has been confirmed by *BeppoSAX* measurements (Owens et al. 1997), and a similar feature in Cyg X-2 was also confirmed with *BeppoSAX* (Kuulkers et al. 1997).

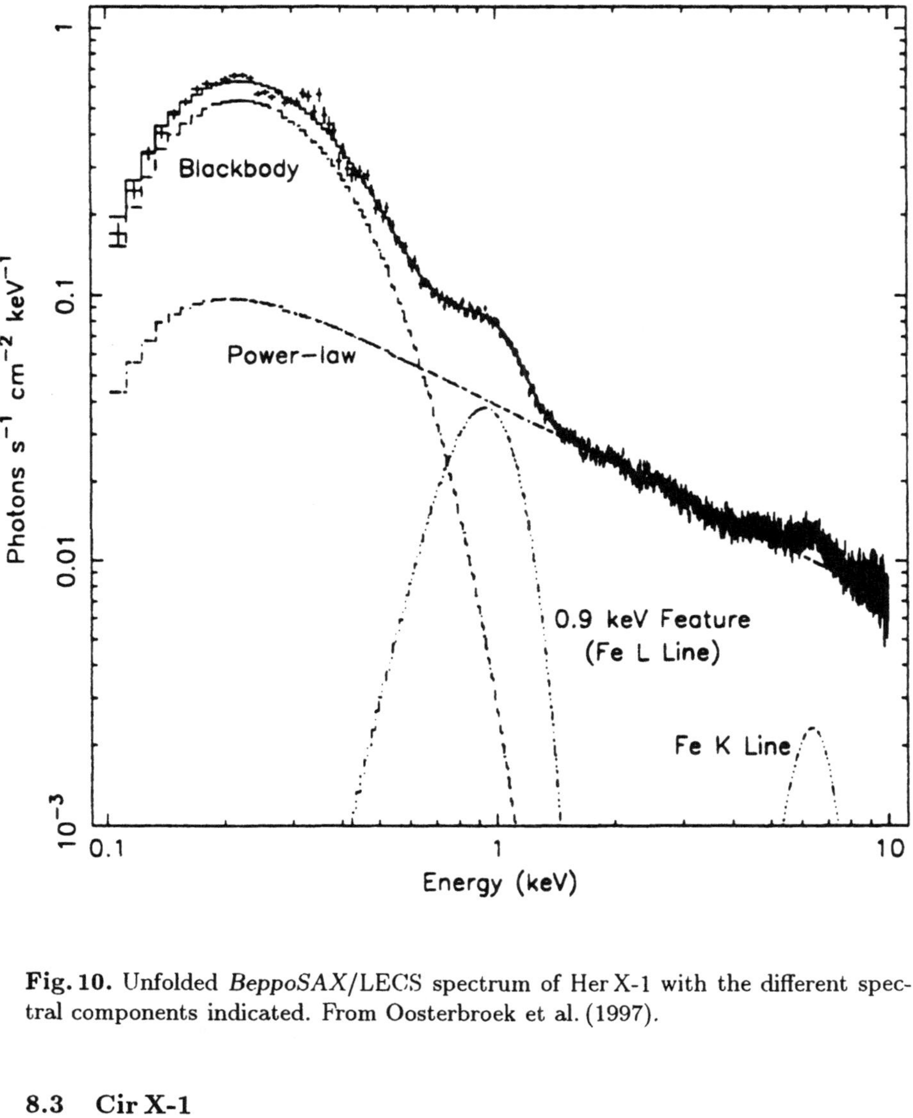

Fig. 10. Unfolded *BeppoSAX*/LECS spectrum of Her X-1 with the different spectral components indicated. From Oosterbroek et al. (1997).

8.3 Cir X-1

Cir X-1 is another LMXB with a rather unusual X-ray spectrum (Brandt et al. 1996). It shows a strong iron absorption edge near 7 keV with an equivalent column density of $1.5\,10^{28}\,\mathrm{m}^{-2}$. Despite the large column density, at low energies there is still flux visible without significant absorption above the normal galactic value. The spectrum at different time intervals could be modeled well using a constant X-ray source with a partially covering and variable absorber. The column density varies between $(0.4\text{–}1.6)\,10^{28}\,\mathrm{m}^{-2}$ and the covering factor between 40–95 %. The intrinsic spectrum is a rather fea-

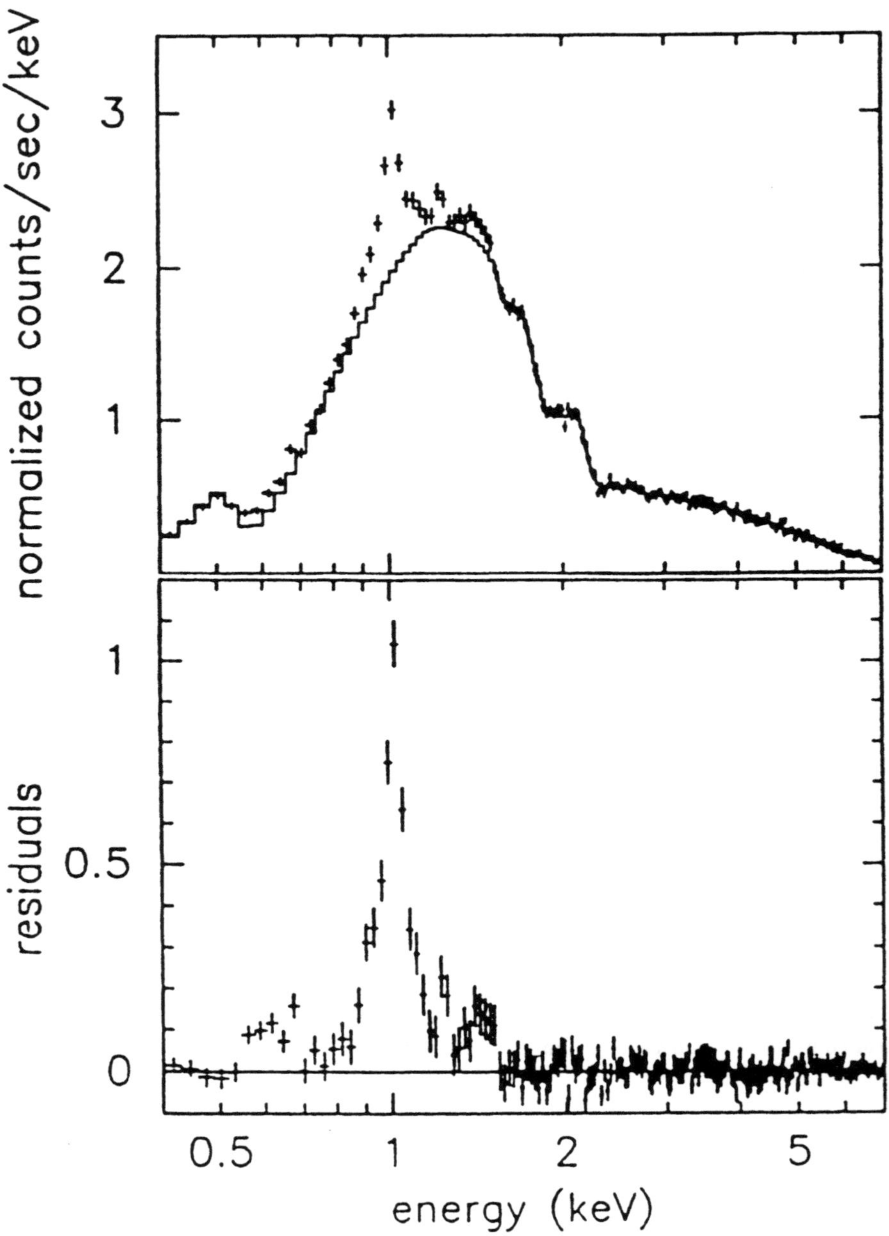

Fig. 11. *ASCA*/SIS0 spectrum of the LMXB 4U 1626–67. The best-fit model consists of a blackbody and 7 lines. The upper panel shows the data and blackbody component only, the lower panel shows the relative residuals with respect to the blackbody model. The features at 1.56, 1.7 and 2.2 keV are instrumental. Note the strong line near 1 keV. From Angelini et al. (1994).

tureless sum of two blackbodies with temperatures of 0.5 and 1.5 keV and equivalent radii of about 50 km. Brandt et al. suggest that the system might be somewhat similar to Seyfert 2 galaxies. They propose the following model. The direct emission from the hard X-ray source towards our line of sight is blocked completely by the accretion disk, which is viewed nearly edge-on. Only above 3 keV the X-rays are able to penetrate through the disk, and the Fe absorption edge is a signature of this. Apart from this, a fraction of the direct emission is scattered in the accretion disk corona, and this is the low-energy emission that is visible below 3 keV. The weak Fe-K line that is visible in the spectrum is consistent with emission from a region sustaining a solid angle of 5 % of the sky, and hence could also be produced in the accretion disk.

9 Supernova Remnants

9.1 Introduction

Supernova remnants (SNRs) can be divided, in general, in two classes: thermal and non-thermal SNRs. A prototype of non-thermal SNRs is the Crab Nebula, which has a featureless power law spectrum with photon index 2.1. Images of these SNRs often show at least two components: the pulsar and a synchrotron nebula.

These non-thermal SNRs are interesting X-ray emitters. However with instruments like *ASCA* and *BeppoSAX* there is little spectral structure to investigate, given the featureless power law spectrum. We therefore focus upon the thermal remnants.

SNRs are by definition the remnants of exploded stars. The supernova explosions leading to the SNRs can be divided in a few classes.

Type Ia supernovae are due to low-mass, old-population stars. The initial mass on the main sequence is $\sim$4–6 $M_\odot$, and the precursor is thought to be a white dwarf in a binary system that explodes due to a thermonuclear instability caused by the accumulated material accreted onto it. No stellar remnant is left, and the 1–2 $M_\odot$ that is ejected during the explosion consists for a large part of iron. Tycho (A.D. 1572) and SN 1006 may be examples of this class.

Type II supernovae are due to young and massive stars (initial main-sequence mass at least 6 $M_\odot$). The core of a red or blue supergiant collapses after it has exhausted all of its fuel, producing a neutron star and ejecting 1–10 $M_\odot$ of enriched material into the interstellar medium. The enriched material contains, in particular, carbon and oxygen burning products, like the elements from O to Ca. The circumstellar medium into which the shock develops has the signatures of the pre-explosion stellar wind. SNRs of this type may be of the non-thermal type. Examples are the Crab Nebula (A.D. 1054) and 3C 58 (A.D. 1181).

Type Ib supernovae are also due to pre-main-sequence stars with initial masses of at least $20\,M_{\odot}$. Due to the strong stellar winds during their life time the mass just before the supernova explosion is significantly smaller. Probably Wolf-Rayet stars and the helium cores of massive stars are the likely progenitors, although in this case the situation is less clear than for type Ia and type II supernovae. The stellar remnant remaining after the explosion might be a neutron star or black hole. The ejecta contain more iron than type II SNRs, but are dominated by oxygen. It is well possible that Cas A (A.D. 1680?) is an example of a type Ib remnant.

While most of the Crab-like remnants show a center-filled structure (often called a plerion), many of the thermal X-ray remnants show a shell-like structure. This shell is located behind the shockfront produced by the supernova explosion. The evolution of such shell-like remnants proceeds in general in several stages.

In the first stage (free expansion), the stellar ejecta expand freely; this phase is characterized by the fact that the mass of the swept-up interstellar medium is smaller than the ejected mass. The shock velocity is nearly constant during this stage. The duration of this phase depends strongly upon the interstellar density. It ends roughly when the shock radius is $3/n_6$ pc, where n_6 is the interstellar density in units of $10^6\,\mathrm{m}^{-3}$ ($1\,\mathrm{cm}^{-3}$).

In the next phase (adiabatic expansion) the swept-up mass is larger than the ejected mass, but the integrated energy loss due to radiation is small compared to the initial explosion energy. Due to the increasing amount of shocked material, the expansion velocity decreases during this phase. Approximately, $r \propto t^{0.4}$, $v \propto t^{-0.6}$, $T \propto t^{-1.2}$, and $L \propto t^{0.6}$ during the early phase when $T \gtrsim 2\,\mathrm{keV}$ and $L \propto t^{1.8}$ for $T \lesssim 2\,\mathrm{keV}$). This shows the rapid cooling of the shock while its luminosity increases in time. Thus, older remnants are brighter and cooler. The temperature and luminosity also depend upon the initial explosion energy E and the interstellar density n. Scaling laws show that for a fixed age $r \propto (E/n)^{0.2}$, $T \propto (E/n)^{0.4}$ and $L \propto n^{1.8}E^{0.2}$ for high temperatures, and $L \propto n^{2.2}E^{-0.2}$ for low temperatures. Thus, temperature and luminosity are not a strong function of the initial explosion energy, but the X-ray luminosity in the adiabatic phase is strongly related to the interstellar density. Moreover, when the shock front moves into denser regions, the X-ray luminosity increases considerably. Early in this phase also a so-called reverse shock may develop in the shocked ejecta (although this shock moves outwards it is called a reverse shock because it moves inwards with respect to the main shock front).

In the third phase radiative cooling contributes significantly to the reduction of the total shock energy. In this phase a dense, relatively cool (10^4–10^5 K) gas shell is formed. However, in this phase the temperature is too low to produce X rays. The same holds for the final phase, when the shock velocity decreases so strongly that it becomes comparable to the sound speed of the ISM.

In the X-ray spectra of SNRs non-equilibrium ionization effects should be taken into account as long as the product of electron density and plasma age $n_e t$ is smaller than about $10^{18}\,\mathrm{m^{-3}s}$. Moreover, non-solar abundances are almost by definition present. Finally, as shown before, there can be at least two different components: shocked ejecta and shocked interstellar material.

ASCA and *BeppoSAX* have the advantage over previous X-ray missions in that they are able to provide spatially resolved spectroscopy. Since the spectra of SNRs are in general very complex, this helps enormously in unraveling the underlying physics of the supernova remnant.

9.2 Oxygen-rich remnants: Cas A

The first *ASCA* observations of this brightest shell-like remnant (Holt et al. 1994) showed marked spatial differences over the remnant. The X-ray spectrum of its Western (W) part is significantly harder than that of the Northwest (NW) and Southeast (SE) shell. Since Cas A is such a strong X-ray source and contains a very strong He-like Si line, it was possible to measure spatial variations of the Si line centroid with a statistical accuracy of only 0.3 eV (note the unit!). Of course, the absolute energy calibration is not so good, but at least the observations showed a 2000 km/s velocity difference between the NE and SW limb of Cas A.

The analysis of Holt et al. was based upon a non-physical absorbed power law plus Gaussian lines model. A proper analysis, using two components with non-equilibrium ionization and the latest spectral code, was done by Vink et al. (1996). The cool component, associated with reverse-shock emission, was assumed to have a high oxygen abundance (10,000 times solar), in accordance with the abundances found in the optical spectra of the so-called fast-moving knots. This high oxygen abundance makes, in particular, a significant difference in the continuum spectrum: it makes the free-bound continuum of oxygen the dominant continuum component above 0.8 keV (instead of thermal bremsstrahlung). The fit also required thermal broadening of the lines of about 4000 km/s, as expected from an expanding shell. The ejected mass is $4\,M_\odot$, a factor of three smaller than earlier estimates. About 60 % of the ejected mass consists of oxygen. This mass is consistent with a late-type WN Wolf-Rayet star as progenitor, considering that these stars lose much of their initial mass due to stellar winds. Comparing the derived abundances to supernova explosion models, Vink et al. find an underabundance of a factor of 10 for Ne and Mg. The reason for this is not clear. Also iron shows an underabundance of a factor of 5; this can be explained if most of the iron layer has collapsed into a black hole.

BeppoSAX observations of Cas A (Favata et al. 1997d, Fig. 12) showed that it contains, in addition to the thermal component, a hard power law tail extending up to 80 keV. This power law has a photon index of 3, and dominates the spectrum above 20 keV. But also at lower energies it has a flux of the same order of magnitude as the thermal component. The inclusion of a power law decreases the ejected mass estimate to 2 $M_\odot$ and increases the abundances of most elements relative to oxygen by a factor of three, but the conclusions of Vink et al. concerning the progenitor and the lack of Ne and Mg are not essentially changed. The power law component can be caused by different mechanisms; it could be a synchrotron component from shock-accelerated electrons with energies of several tens of TeV as proposed for SN 1006 (see next section), or non-thermal bremsstrahlung from electrons with energies of several tens of keV, which have just been accelerated from the tail of the thermal distribution, the so-called injection spectrum, or an additional very hot component.

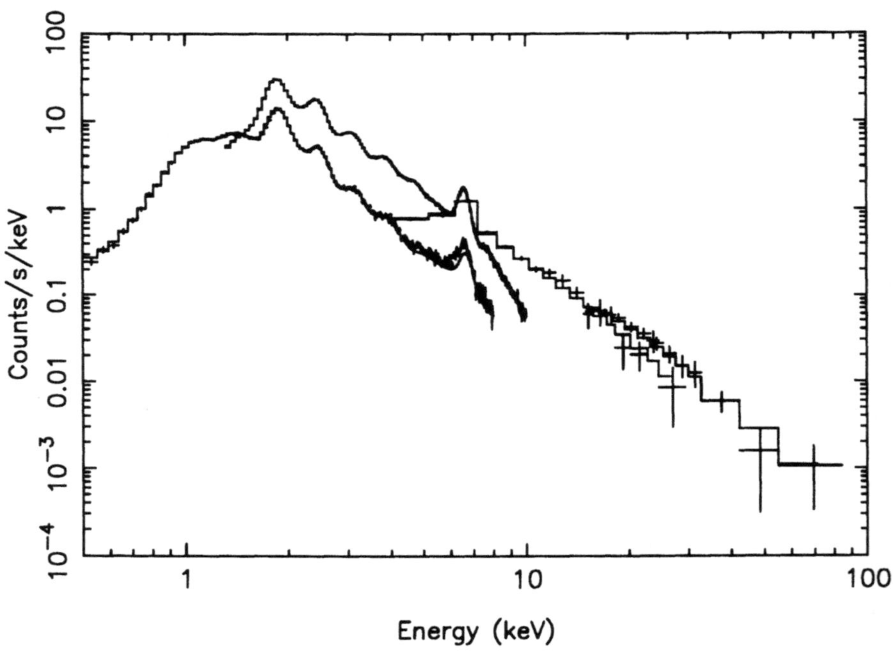

Fig. 12. *BeppoSAX* spectrum of Cas A with best fitting model. From Favata et al. (1997d).

9.3 Young type Ia remnants

Hwang & Gotthelf (1997) have studied in detail Tycho. In particular, they have attempted to create images in spectral lines that are deconvolved for the point spread function of *ASCA*. They found no systematic line centroid shifts over the remnant. Using the Si and S lines, they derived temperatures around 1 keV and $n_e t$ values around $10^{17}\,m^{-3}s$. However, the Fe-K line peaks at a smaller radius than the Si and S lines, and it has a much higher temperature (5 keV) and smaller ionization parameter (10^{16} m^{-3}s) than Si and S. Unfortunately, these authors have not tried a two-component decomposition. The azimuthal variations in line ratios seem to indicate temperature variations by a factor of 1.3–1.8.

9.4 Old shell-like remnants

Well-known examples of this class are the Cygnus loop and the Vela and Puppis A remnants. In general, these old remnants are relatively cool and therefore more difficult to study with *ASCA* due to the limited energy band.

The Cygnus loop was studied by Miyata et al. (1994). Since this is a rather large remnant (about $4° \times 3°$), the spatial distribution of the shocked material can be studied in detail. Miyata et al. fitted the spectrum as a function of the distance to the shock front with a single component NEI-model (Fig. 13). An evident rise in temperature from the shock front inwards could be observed, consistent with the predictions of the adiabatic model. Also the density decreases rapidly inwards. The $n_e t$ parameter first increases from the shock front inwards, corresponding to the aging of the plasma behind the shock, and remains constant or decreases further inwards, consistent with the decreasing density. The average $n_e t$ value combined with the estimated ISM density implies an age of about 20,000 years. The elemental abundances, however, were extremely low. For example, the O and Fe abundances are 3 % of the solar values, and Ne, Mg and Si about 10 %. This is unexpected for an old remnant like the Cygnus loop, which should show the ISM abundances. Miyata et al. suggest that either the local ISM near the Cygnus loop has sub-solar abundances, or that most metals are still contained in the dust and are slowly evaporating into the ISM. This scenario works for most elements, but not for the noble gases like neon. Similarly low abundances were found in the younger SNR RCW 86 (Vink et al. 1997).

In the shell-like remnant RCW 103 a radio-quiet X-ray point source was found. *ASCA* spectra show that the point source has a blackbody spectrum with a temperature of 0.6 keV, thereby confirming its nature as a cooling neutron star (Gotthelf et al. 1997). A similar situation holds for Kes 73. The thermal shell spectrum of this 2000 year old remnant has the characteristics of a massive progenitor (Gotthelf & Vasisht 1997). The radio-quiet pulsar rotates extremely slowly (11.8 s), perhaps due to a very strong magnetic field (10^{11} T, Vasisht & Gotthelf 1997). A third case of a neutron star within a supernova remnant is G 296.5+10.0 with a blackbody temperature of 0.3 keV (Vasisht et al. 1997). Unfortunately, these authors did not analyze the SNR spectrum.

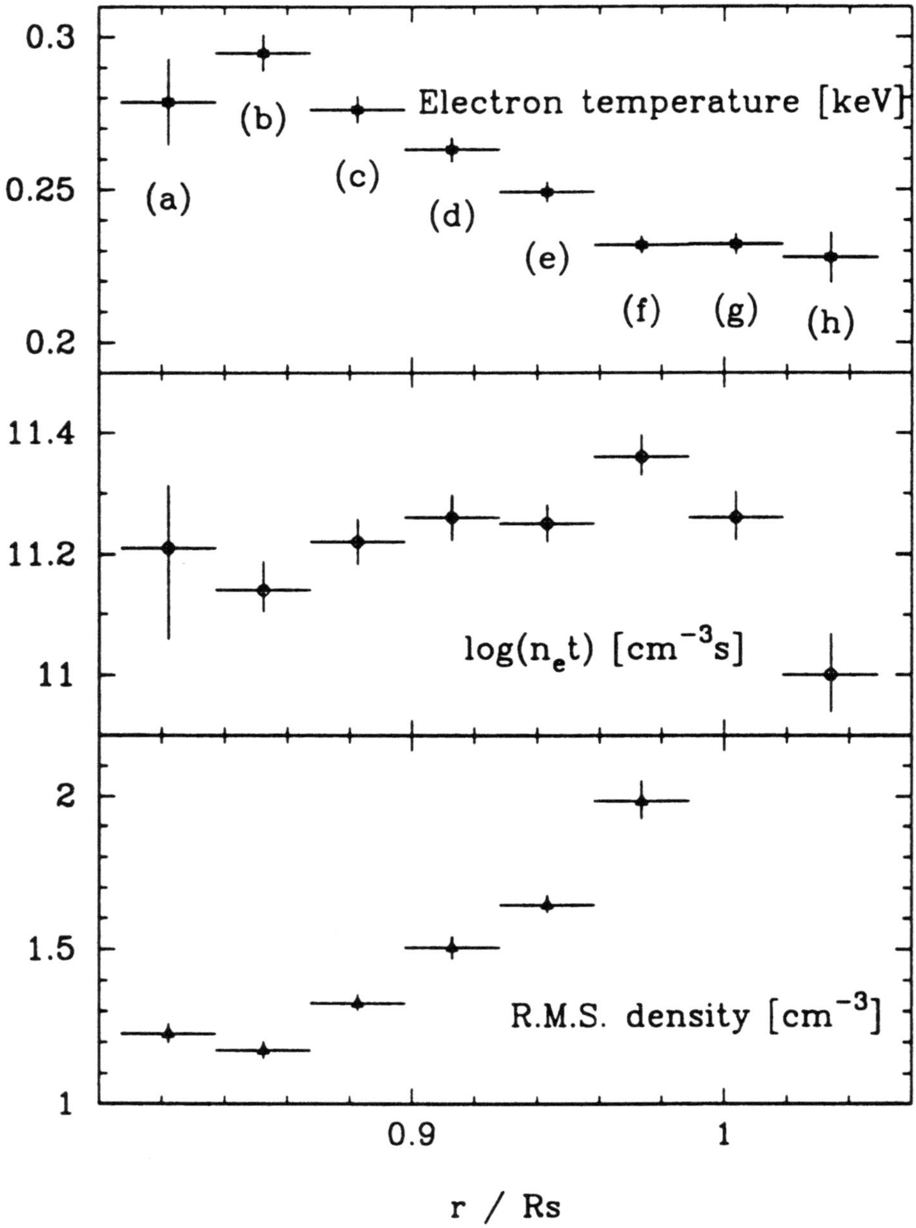

Fig. 13. Radial profile of T_e, $n_e t$ and n_e in the Cygnus loop. R_s is the radius of the shock front. From Miyata et al. (1994).

9.5 Synchrotron X-ray emission from SNRs

SN 1006 is a historical remnant with a clear shell structure. Surprisingly, the *ASCA* observations of this source (Koyama et al. 1995) showed strong and dominant non-thermal emission at the location of the bright NE and SW rim. The spectrum is characterized by a power law with a photon index of 3. In addition, a weak thermal component is visible at the rims. The interior of the remnant shows a "normal" NEI spectrum, with near-solar abundances of O, Ne and Fe, but an order of magnitude more Mg, Si and S, indicating that it is most likely the remnant of a type Ia explosion. This is also confirmed by its large height above the galactic plane and the historical light curve. The power law component in SN 1006 is significantly steeper than in the Crab nebula, which has a photon index of 2.1. Moreover, in the Crab nebula the synchrotron nebula is powered by the pulsar, while SN 1006 is shell-like and contains no neutron star. Koyama et al. explain the emission as due to X-ray synchrotron emission caused by energetic electrons accelerated at the shock front. These electrons must have energies up to several hundred TeV. This is very close to the knee in the electron spectrum of cosmic rays (1000 TeV). Therefore these observations suggest that the cosmic-ray electrons are accelerated in supernova remnant shocks. The presence of high-energy electrons is confirmed recently by TeV observations (Tanimori et al. 1998).

RX J1713.7–3946 near the galactic center also shows a shell-like structure with a featureless power law spectrum (Koyama et al. 1997). In this case the photon index is 2.5. It is perhaps a second example of synchrotron X-ray radiation from a SNR shell.

A third case is possibly IC 443 (associated with the guest star of A.D. 837?, size 45′). *ASCA* hardness maps (Keohane et al. 1997) showed a bright unresolved spot and a rim of 5′ length at the edge of the SNR, both containing hard power law like spectra, with a photon index of 2.3. The thermal component of the remnant has a much softer spectrum with a temperature of ~0.8 keV. The mechanisms proposed by the authors is slightly different from that proposed for SN 1006: for IC 443 it is thought that electrons are accelerated to energies of 20 TeV when the supernova shock impacts on dense clouds.

9.6 Crab-like remnants

ASCA has found several cases of Crab-like remnants where a central plerionic synchrotron nebula is accompanied by a thermal X-ray shell. Examples of this are G 11.2–0.3 (possibly identical to the SN of A.D. 386), showing strong Mg and Si emission lines in the shell surrounding the featureless plerion (Vasisht et al. 1996). Another example is Kes 75 (Blanton & Helfand 1996), a 1000 year old SNR at a distance of 20 kpc, the second brightest non-thermal SNR in our galaxy. This remnant is probably still in its free expansion phase. CTA 1

(Slane et al. 1997) shows a pulsar, synchrotron plerion (photon index 2.1 similar to the Crab nebula) and in addition a thermal component (0.2 keV temperature). It is an old remnant (20,000 year) in a low-energy explosion (10^{43} J) high above the galactic plane (250 pc).

The appearance of thermal emission shells in Crab-like remnants is always expected whenever the remnant gets older or the interstellar density is high. In this respect there is a problem that thermal emission from the Crab nebula has not yet been detected.

9.7 Center-filled thermal SNRs

Center-filled thermal SNRs might originate, for example, in situations where dense molecular material surrounding the progenitor evaporates after the supernova shock has passed.

W 49B is a relatively young SNR with an age of about 1000 years at a distance of 10 kpc. *ASCA* data showed that the spectrum is dominated by very strong emission lines (Fig. 14, Fujimoto et al. 1995). The spectrum cannot be described by a single NEI component. Instead, it looks like the remnant is stratified. The inner region of this center-filled remnant has a temperature of 2 keV and has the highest concentration of iron. The outer region is more patchy and shows predominantly Si, S, Ar and Ca. Moreover, these elements show an increasing degree of ionization, consistent with an increasing $n_e t$ when going inwards towards the iron-rich region.

G 166.0+4.3 is another center-filled remnant with thermal (0.7 keV) emission (Guo & Burrows 1997). It is an old remnant (24,000 years) and therefore its abundances are expected to be close to the ISM values. The observed abundances of Mg, Si and Fe are about 0.3 times solar, and this can be explained by its large distance from the galactic center: it is located at a distance of 5 kpc from the Sun towards the galactic anticenter. Assuming a galactic metallicity gradient of −0.05 dex/kpc, abundances of ∼0.5 times solar are expected.

In W 44 (age 20,000 years, size 30′) the spectrum is also thermal (0.9 keV) and out of ionization equilibrium (Harrus et al. 1996). The radio pulsar 9′ south of the dynamical center is associated with a hard X-ray source (synchrotron nebula, size smaller than 5′), but no X-ray pulses have been found.

9.8 Jets interacting with SNRs

W 50 is an interesting SNR. This remnant is related to the HMXB SS 433 discussed before. There are two spots of X-ray emission at the intersection of the relativistic jets from SS 433 with the radio shell. *ASCA* spectra (Yamauchi et al. 1994) showed that these spots have power law spectra that can be consistently modeled by synchrotron radiation from the relativistic jet electrons interacting with the magnetic field of the shocked SNR shell. Apart from these hot spots there is no detectable X-ray emission from the SNR.

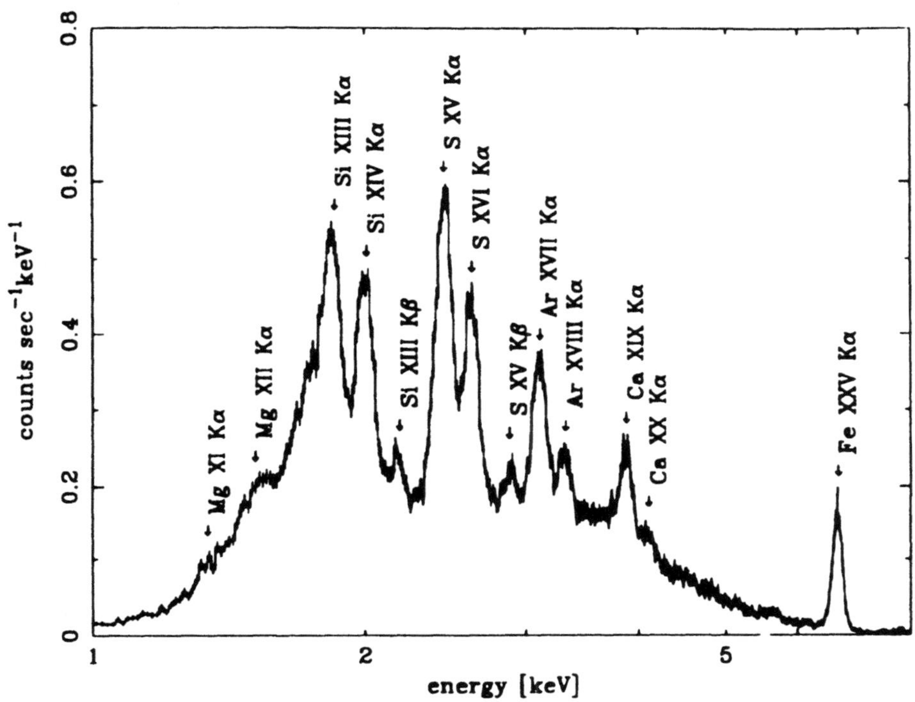

Fig. 14. *ASCA*/SIS spectrum of W 49B. From Fujimoto et al. (1995).

RCW 89 (Tamura et al. 1996) is characterized by a synchrotron nebula powered by a young radio pulsar. The nebula displays a jet-like structure, and at the end of the jet there is a thermal component with prominent lines of Ne, Mg, Si and S. The energy deposited by the jet is sufficient to supply the energy for the thermal component. Tamura et al. propose that the thermal emission is caused by the interaction of the jet with the interstellar medium.

Another case where a jet interacts with a remnant is the jet of the Vela pulsar. Surprisingly, the jet spectrum is rather similar to the surrounding SNR spectrum (Markwardt & Ögelman 1997). The spectrum shows a power law in addition to a cool (0.3 keV) thermal component.

9.9 Isolated pulsars

Since pulsars are rotating neutron stars and thereby the remains of past supernova explosions, we briefly mention them here. The X-ray spectra of isolated pulsars or pulsars in wide binaries are in general rather featureless. In spectral fits usually a single blackbody spectrum originating from the pulsar and a power law component caused by non-thermal processes in the surrounding nebula is sufficient to model the spectrum (cf. Becker & Trümper 1998). Some examples of pulsars with extended nebulae have already been treated elsewhere in this chapter. In pulsar research, most attention is focussed upon the study of the pulse profile combined with broad-band spectral information. However given the rather featureless X-ray spectra I do not discuss pulsar spectra here. As an example of what has been done with *ASCA* in this field of research I refer to papers on 1E 2259+586 (Corbet et al. 1995), PSR B 1259–63 (Kaspi et al. 1995; Hirayama et al. 1996), 4U 0142+61 (White et al. 1996), PSR 0656+14, PSR 1055–52 (Greiveldinger et al. 1996), and 1E 1048.1–5937 (Corbet & Mihara 1997).

9.10 The Magellanic Cloud SNRs

Both the LMC and the SMC contain several SNRs with detected X-ray emission. Studying those remnants has the advantage that they constitute a nearly complete sample of SNRs in various stages of evolution, all with accurately known distances. For the galactic SNRs the distances are often not well known.

Hughes et al. (1995) studied three young remnants in the LMC (younger than 1500 year). They made spectral fits with a thermal bremsstrahlung plus Gaussian lines model, and deduced the $n_e t$ values from the ratio of the H-like to He-like Si lines. The $n_e t$ values derived from this ratio are between (2.5–25) $10^{16}\,m^{-3}s$; moreover, the centroid of the Fe-L complex shifts consistently towards higher energies with increasing $n_e t$. Although Hughes et al. do not present a complete spectral fit, they do compare their spectra qualitatively to NEI model spectra with the same temperature and ionization state. A model spectrum with LMC abundances (which would be appropriate for older remnants that are dominated by swept-up ISM) predicts too much O, Ne and Mg and too little Si, S, Ar and Ca. A type II composition is even worse. However, a type Ia composition yields spectra that look similar to the observed ones. From this, Hughes et al. conclude that these three remnants (SNR 0509–67.5, SNR 0519–69.0 and N 103B) all were caused by type Ia supernovae. However, the frequency of three Ia remnants in 1500 years is hard to reconcile with the expected rate of once every 2000 years.

The first SNR in the LMC studied with *BeppoSAX* is N 132D (Favata et al. 1997c). Earlier intermediate-resolution measurements of this source were obtained with the *Einstein* SSS detector which, however, had a limited energy range (0.5–4.5 keV). The *ASCA* spectrum of this source was underexposed, so that no emission line features of the iron K complex could be seen. The better exposed *BeppoSAX* observations showed a strong Fe-K emission complex. The spectrum could be fitted well using two NEI components.

By far the brightest SNR in the SMC, and the only one studied with *ASCA*, is SN 0102–72 (Hayashi et al. 1994). The spectrum shows strong H-like and He-like lines from O, Ne and Mg. The remnant is oxygen-rich in the optical band, and is about 1000 years old. The X-ray spectrum could not be fitted with two components in non-equilibrium ionization with variable abundances (but with the abundances in both components the same). Instead, Hayashi et al. used a model with different temperature and $n_e t$ for each species O, Ne, Mg and Fe. In this fit, Ne and Mg had lower temperatures and higher $n_e t$, consistent with a reverse shock origin. On the other hand, iron should be due to the shocked ISM. The authors have difficulties in explaining the origin of the oxygen. I suggest that it would be worthwhile to try a similar model as applied to Cas A by Vink et al. (1996) to these data.

9.11 Supernova explosions in distant galaxies

Also in more distant galaxies supernova remnants emit detectable X-ray emission. I do not discuss these data in detail here. Some observations of X-ray spectra from recent supernovae are found in papers by Petre et al. (1994) for SN 1978K in NGC 1313, and by Kohmura et al. (1994) for SN 1993J in M 81.

10 Extended X-ray Emission from Normal Galaxies

10.1 The galactic ridge

The central few kpc of the Milky Way contain a hot and diffuse gas, located in a thin disk with a scale height of 0.1 kpc and cooling through thermal X-ray emission. *Ginga* spectra already showed the presence of a strong iron line in this hot plasma (4–13 keV). *ASCA* observations of a part of the ridge (the so-called Scutum arm region) showed the presence of emission lines from He-like Si, S and Fe. The temperature derived from the continuum emission is high: 7 keV, and therefore the simultaneous presence of both He-like Si and Fe means that the plasma cannot be explained by a single equilibrium ionization model. Also the centroid of the Fe-K complex is at a lower energy than expected from an equilibrium model. In fact, the plasma can be well fitted using a NEI model with an ionization parameter of $5\,10^{16}\,\mathrm{m^{-3}s}$. Yamauchi et al. (1996) argue that the emission cannot be due to discrete sources. The nature of the gas is not yet clear. For example, the pressure of the gas is 100 times higher than the pressure otherwise found in the ISM, and the temperature is so high that the sound speed exceeds the escape velocity. However, the escape time, derived from the scale height and sound velocity is shorter than the ionization time scale derived from the NEI modeling. This may either indicate that the galactic ridge plasma is non-uniform, thereby increasing the effective density, or that it is confined by, e.g., magnetic fields of the order of 10^{-9} T.

10.2 The galactic center

A black hole with a mass of $2.7\,10^6\,\mathrm{M_\odot}$ resides in the center of our Milky Way. The accretion rate onto this black hole is orders of magnitude smaller than in active galactic nuclei. Around this black hole a range of complex structures exist, visible at various wavelengths. In X rays the central 10′ contain several sources, one of which is associated with Sgr A*, the center of our galaxy.

ASCA has made several observations of the galactic-center region (Koyama et al. 1996b). The *ASCA* observations confirm the presence of diffuse hot gas out to at least 80 pc from the center. The spectrum shows distinct emission lines from H-like and He-like Si, S, Ar, Ca and Fe (Fig. 15). Moreover, there is a pronounced 6.4 keV component of neutral iron in the spectrum. The continuum temperature is 10 keV with a luminosity of 10^{30} W. For such a hot plasma, all Si–Ar ions are expected to be fully ionized, and the presence of He-like emissions shows that there must be multiple temperature components or non-equilibrium ionization effects.

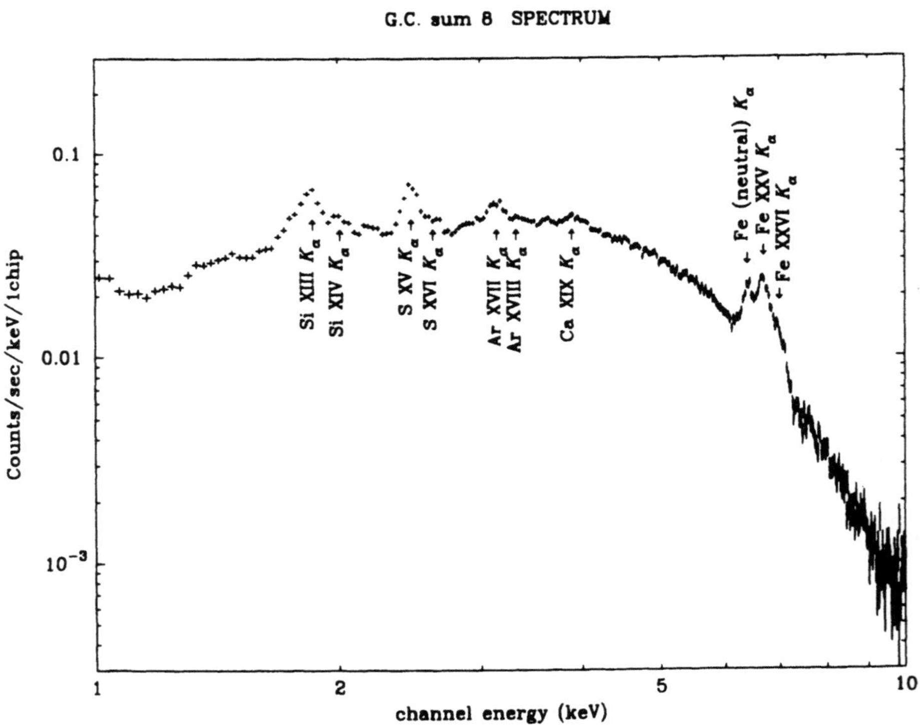

Fig. 15. Spectrum of the diffuse emission from the galactic-center region. The contributions of bright X-ray binaries have been removed. From Koyama et al. (1996b).

The electron density varies from $6\,10^6\,m^{-3}$ near Sgr A* to $0.3\,10^6\,m^{-3}$ in the outer parts. The total thermal energy content of the plasma is around 10^{46} J and corresponds to a mass of $4000\,M_\odot$. The spectrum shows no strong spatial variations, except in total intensity and the amount of absorption. The high temperature that is found implies that the hot material cannot be gravitationally bound, but using estimates of the relatively large magnetic field (10^{-7} T) the plasma might expand predominantly along the magnetic-field lines. For a sound speed of 1600 km/s (corresponding to the temperature of 10 keV), it takes 50,000 years to reach the outer parts at 80 pc from the galactic center. This is an order of magnitude smaller than the time needed to reach collisional ionization equilibrium. Hence it might be possible that the ions carry a much larger energy than the electrons, and the total energy could then easily be of the order of 10^{47} J. In fact, the measured line width of the He-like and H-like iron line of 75 eV corresponds to an ion velocity of 3300 km/s, which is of the same order of magnitude as the sound speed derived before, and most likely represents the bulk motion of the ions.

The cold 6.4 keV line, on the other hand, shows no significant broadening. A map made in this line shows two strong emission concentrations near molecular clouds (Fig. 16). Near one of these clouds, the equivalent line width is as high as 1000 eV. The X-ray emission centroid near this cloud is also shifted 2′ into the direction of the galactic center. All this suggests that the molecular cloud acts as a reprocessor of the X rays: through fluorescence in the 6.4 keV line and through Compton scattering of the continuum, thereby producing only a line with a large equivalent width at the side of the cloud facing the galactic center. From the known mass ($2\,10^6\,M_\odot$) and size (7 pc) of the cloud, the column density can be estimated ($4\,10^{27}\,m^{-2}$), and using the observed intensity of the iron line, the incident flux on the cloud can be calculated. Neither the hot gas nor other discrete sources in the neighbourhood of the clouds can provide this flux.

However, if the X rays originate from the galactic center, a luminosity of $2\,10^{32}$ W is required, much larger than what X-ray binaries can provide. This luminosity is 4 orders of magnitude larger than the current luminosity of the X-ray source at the galactic center. The presence of a hidden AGN in the galactic center with such a high luminosity can be excluded on several grounds. Therefore the luminosity of the illuminating central source must have been much larger 300 years ago. Koyama et al. suggest that this might have occurred due to a tidal-capture event of a star, although such events occur perhaps only every 10^4–10^5 years.

Finally it should be noted that not all the molecular clouds near the galactic center show such a strong iron line. This indicates that the ionizing radiation should be either beamed or that the beam is blocked in some directions.

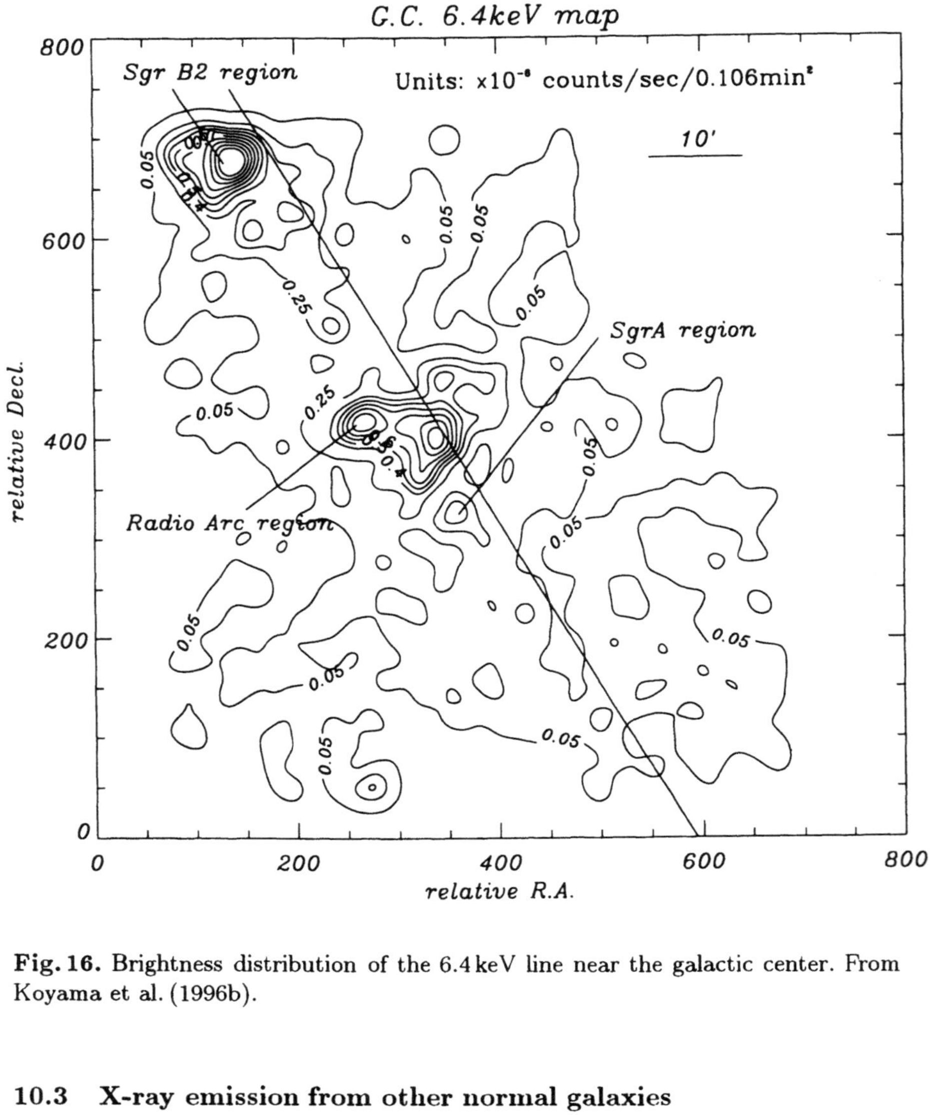

Fig. 16. Brightness distribution of the 6.4 keV line near the galactic center. From Koyama et al. (1996b).

10.3 X-ray emission from other normal galaxies

Normal galaxies show a wide variety of X-ray emission mechanisms. In the first place, emission from discrete sources such as those that are known in our galaxy play a role (X-ray binaries, SNRs). Furthermore, in some cases a weak active nucleus or starburst activity may be present. In other cases there is evidence for a diffuse hot interstellar medium.

I will limit myself here to mentioning a few useful references for further study. A major survey of M 31 was recently published by Supper et al. (1997). For early-type galaxies I mention here the work of Awaki et al. (1994), Matsushita et al. (1994), Loewenstein et al. (1994), Mushotzky et al. (1994), Kim et al. (1996), Fujita et al. (1996b) and Buote & Canizares (1996).

11 Seyfert 1 Galaxies

The first active galactic nuclei (AGN) were discovered by Carl Seyfert in 1943 as bright point-like nuclei in spiral galaxies. In their optical spectra these Seyfert galaxies show strong emission lines of ionized gas plus a non-stellar continuum. All Seyfert galaxies have a narrow emission line component, but a distinction can be made between Seyfert 1 galaxies (also broad emission lines, strong and variable X-ray emission) and Seyfert 2 galaxies (no broad emission lines, heavily absorbed X-ray emission). There is a smooth transition between Seyfert galaxies and the more luminous quasars. In quasars, the central nucleus is often so bright and the distance is so large that it is difficult or impossible to detect the emission from the underlying galaxy. The boundary between both classes cannot be specified precisely but occurs approximately at a redshift of 0.1, or at a 2–10 keV X-ray luminosity of $5\,10^{37}$ W.

The power in these active galactic nuclei is provided by accretion onto a massive (10^6–$10^{10}\,M_\odot$) black hole. Signs of the accretion disk can be seen in the presence of a strong blue/UV component and a soft X-ray excess above the dominant power law component of the X-ray spectrum. The soft excess is thought to be produced either as direct emission from the warm accretion disk or as hard X-ray emission that is reprocessed in the disk. Perhaps both mechanisms play a role, but to what extent is yet unknown. The hard X rays are thought to be produced by inverse Compton emission originating in a hot corona surrounding the inner parts of the accretion disk-black hole system, but in some cases a large contribution might also arise from other mechanisms like synchrotron-self-Compton emission in a jet or outstream along the rotation axis of the system. Around the nucleus also cooler material is present, showing up in the form of fluorescent line emission from iron ions near an energy of 6.4 keV. Furthermore, in many cases the X-ray spectrum is absorbed partially by warm (10^5–10^6 K) photo-ionized material surrounding the nucleus. This leads to clearly detectable absorption edges, the most prominent being the K-edges of O VII at 0.739 keV and O VIII at 0.871 keV. At higher column densities also the absorption edges of iron in its various ionization stages may be seen in the 7.11–9.28 keV energy range. The edge energy increases monotonically with the ionization stage, thereby providing a useful diagnostic tool. Finally, at energies above 10 keV in several cases the reprocessed and reflected X-ray spectrum from the accretion disk is visible in the form of an apparent spectral hardening.

11.1 The iron line

If the iron line in Seyfert 1 galaxies were produced by, e.g., fluorescence in the broad- or narrow-line region, we would expect to see a narrow line at

6.4 keV. Surprisingly, the iron line in MCG −6-30-15[2] is broad with a FWHM of 400 eV or 20,000 km/s (Fabian et al. 1994a).

A year later, Mushotzky et al. (1995) showed that also in two other cases the iron line is broadened. In IC 4329A and NGC 5548, the lines have a FWHM of at least 20,000 and 35,000 km/s, respectively. This is much larger than the width of the optical and UV lines in these sources. Moreover, the line centroids are not located at 6.4 keV, but at 6.19 and 6.15 keV respectively. It can be argued that models with Compton scattering cannot reproduce the line properties. Instead, the line profiles can be explained using the models for relativistic rotating accretion disks as presented, e.g., by Laor (1991). In such models, the line profile for a narrow emission line in an accretion disk rotating around a spinning black hole are calculated, including the effects of gravitational redshift, Doppler shifts and boosting, etc. The models predict asymmetric lines with a steep blue wing and an extended red wing. The line peak is blueshifted for most cases but for nearly face-on disks it can be redshifted by no more than 5 %. Mushotzky et al. concluded that the disks in IC 4329A and NGC 5548 have inclinations $\lesssim 24°$ and 15°–38°, respectively, and that the emission should originate very close to the black hole, within 100 Schwarzschild radii ($R_{\rm S} = 2GM/c^2$).

A breakthrough was achieved with the 4-day observation of MCG −6-30-15 (Fig. 17; Tanaka et al. 1995). The signal-to-noise ratio near the iron complex was now much better. The line appeared to be broad (between 5–7 keV, full width of $0.3c$) and asymmetric. The profile shows a narrow peak at 6.4 keV with an equivalent width of 120 eV and a skewed reddened wing with an equivalent width of 200 eV. The effective line centroid is near 5.9 keV.

Again Compton scattering can be excluded as a possible broadening mechanism. The argument goes as follows. Compton down-scattering of Fe-K photons requires low temperatures (less than 1.6 keV). Each scattering yields on average a 80 eV redshift. For the centroid to shift from 6.4 to 5.9 keV, about $(500/80)^2$ scatterings are needed, hence the optical depth should be about 6. If the scattering material would be in our line of sight, this would produce strong absorption that is not observed; if it is not located in the line of sight (i.e. if all observed emission is scattered) the continuum should show a break near 20 keV that is not observed.

Also in the case of MCG −6-30-15 the relativistic disk model is well able to reproduce the observed line profile. For a Schwarzschild black hole, the best fit indicates an inclination angle of 30°. The emission is produced between the inner edge of the disk ($3R_{\rm S}$) and a radius of about $10R_{\rm S}$. For a Kerr geometry (rapidly rotating black hole) similar conclusions can be drawn.

The rather high equivalent line width of about 400 eV can be explained if either iron is 1.5–3 times overabundant, or if the material is highly ionized.

[2] The minus sign before the 6 in the name *is* significant; MCG +6-30-15 is a Seyfert 2 galaxy.

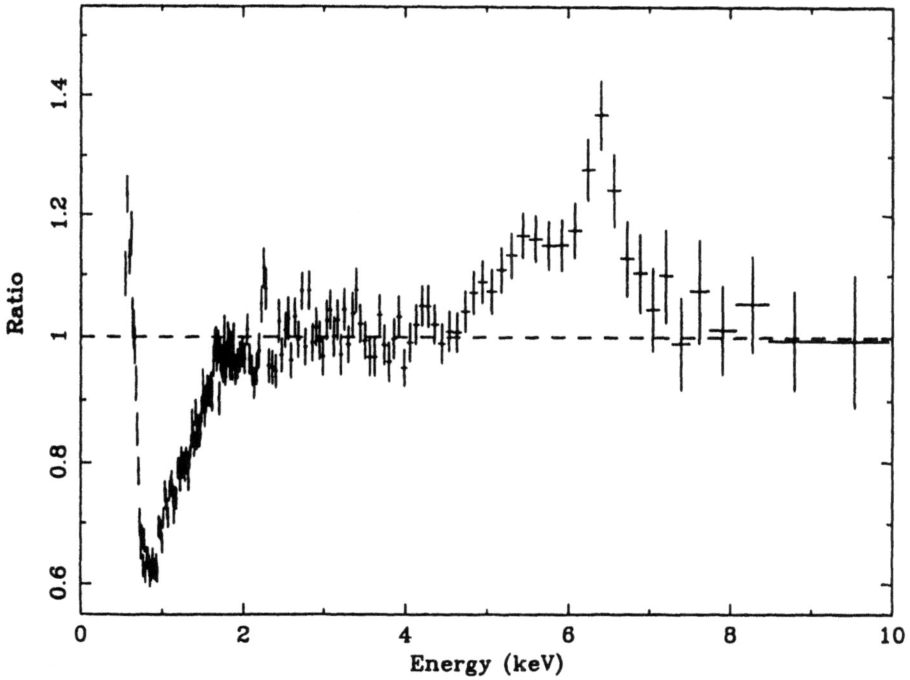

Fig. 17. The iron line in MCG −6-30-15. Plotted is the ratio between the *ASCA*/SIS0 data and a simple power law spectrum. Note the strong oxygen absorption complex below 1 keV and the broad, asymmetric iron line between 5–7 keV. From Iwasawa et al. (1996b).

A significant narrow contribution from a molecular torus can be excluded (it has an equivalent width less than 45 eV).

In a later study of MCG −6-30-15, Iwasawa et al. (1996b) analyzed the variability of the iron line using the data from Tanaka et al. They found a complex behaviour, where the narrow and broad component of the line respond differently to continuum variations on long and short time scales. We do not repeat their arguments here, but summarize how the data can be explained.

A Kerr black hole is needed. The broad red tail of the line profile is produced mainly in the innermost regions of the accretion disk, between 0.62–$5R_\mathrm{S}$ (Contrary to a non-rotating hole, the last stable circular orbit around a rotating black hole is smaller than $3R_\mathrm{S}$ and can be nearly equal to $0.6R_\mathrm{S}$). The red tail is red and broadened, because very close to the black hole even the blue wing of the emitted line profile is redshifted due to the enormous gravitational potential. The large equivalent width (1 keV) of this red component during the minimum continuum flux phase could be due to light bending around the spinning black hole or due to an ionized disk: a 6.7 keV He-like line should have an equivalent width twice as large as a neutral 6.4 keV line.

The narrow-line component (the blue horn) is mostly produced in regions somewhat farther outward than the broad red wing, typically at 3–5 R_S. Flares with a short duration (less than a few hours) tend to occur in the innermost regions where the rotation periods, and related time scales are shortest. Consequently, only the broad red wing of the line can respond fast enough to these short-duration flares. The larger size of the blue-horn region prevents it from reacting fast enough. On the other hand, slower events arising at larger distances from the black hole, e.g., near 3.5 R_S, mainly cause a response from the narrow blue component, because this region is nearest and has the largest Doppler boosting factor.

In many other cases evidence for broadened lines have been reported, like MCG−2-58-22 (Weaver et al. 1995), Mrk 1040 (Reynolds et al. 1995a), NGC 4051 (Mihara et al. 1994a; Guainazzi et al. 1996), Mrk 290 (Turner et al. 1996b), NGC 3516 (Nandra et al. 1997a), and EXO 055620–3820.2 (Turner et al. 1996a). In several of these sources disk inclination angles around 30° have been found. In some cases (like NGC 4051) the peak of the narrow-line component is significantly redshifted to 6.1 keV. In NGC 3516 two observations one year apart showed that the line flux had varied in accordance with the continuum, showing that the line must originate from a region within 1 light year distance from the nucleus.

Nandra et al. (1997c) have studied systematically the iron line for a sample of 18 Seyfert 1 galaxies. In 14 of these, the line is resolved. The average line energy is at 6.34±0.04 keV, corresponding to a weak average redshift of 1 %, although a few individual sources have a larger redshift of 5 %. The equivalent line width ranges from 50–600 eV with a mean value of 160 eV. If the lines are modeled using the relativistic disk model, the average inclination appears to be 29°±3°, consistent with orientation-dependent unification schemes. Differences in the line profiles from source to source imply slight variations in the geometry, which cannot be accounted for solely by inclination effects. In most cases, it is required that the line emission arises from a range of radii. Although a small contribution to the emission from a region other than the accretion disk is not ruled out, in general it is not required and it has little effect on the conclusions regarding the disk line. The data are fit equally well using rotating (Kerr) and non-rotating (Schwarzschild) black-hole models.

Thus it appears that in most Seyfert 1s the inner regions of the accretion disk produce the skewed iron line, and the study of its shape and variability offer a unique opportunity to investigate the behaviour of cold material very close to a black hole.

In NGC 4151, however, the disk model is problematic (Yaqoob et al. 1995), although it provides a good fit to the observed line profile. The reason is that the upper limit to the disk inclination of 20° is hard to reconcile with the inclination of at least 40° derived from the extended soft X-ray emission image, the ionization cones and the radio jet.

Not all Seyfert 1 spectra contain broad X-ray emission lines. For example, in NGC 7469 Guainazzi et al. (1994) only find a narrow-line component with an equivalent width of 120 eV. Perhaps this line is produced in the outer regions of the nucleus and not in the inner disk.

11.2 Warm absorbers

The presence of warm absorbing material in the line of sight towards the nucleus of Seyfert 1 galaxies was already deduced from *Ginga* observations of an ionized iron K absorption edge, and from spectral fits to *ROSAT* PSPC data, although in the latter case the resolution is insufficient to directly make a distinction between the O VII edge at 0.739 and the O VIII edge at 0.871 keV. With *ASCA* both edges can be resolved, and the broad energy band allows a good determination of the underlying continuum. Moreover, also the Fe-K edge is observable with the same instrument.

In several AGN both oxygen edges have indeed been observed. The first example reported is again MCG –6-30-15 (Fabian et al. 1994a). Two spectra were obtained at a time difference of three weeks . A strong O VII edge was seen. Its optical depth increased from 0.57 to 1.17 between the observations, while the shape of the continuum did not change. This implies a doubling of the column density from 0.6 to 1.3 $10^{26}\,m^{-3}$. Further analysis by Reynolds (1995) indicated variability of the absorber on time scales of 10^4 s.

The four-days observation of MCG –6-30-15 that was used for determining the iron line profile, showed rapid variability of the O VIII edge (Otani et al. 1996). The O VIII column density increased inversely proportional to the luminosity on time scales of 10^4 s. This provides evidence for photoionization of the warm absorber by continuum photons: increasing X-ray flux produces more ionizations, thereby decreasing the O VIII population relative to the O IX population. Otani et al. argue that the O VIII edge should be produced at distances smaller than 10^{15} m, near or within the broad-line region. This estimate is based on the assumption that the response time of the edge is the recombination time scale for O VIII. Interestingly, the O VII edge does not respond at all to the continuum variations (slower than 10^6 s). This indicates that this less ionized component originates in the more outward regions of the nucleus (at least 1 pc), perhaps near the molecular torus or in the narrow-line region.

The first *BeppoSAX* observations of MCG –6-30-15 showed clear evidence for an unresolved line near 0.59 keV with an equivalent width of 45 eV, attributed to a blend of O VII and O VIII emission lines (Orr et al. 1997). *BeppoSAX* confirmed the more rapid response of the O VIII edge to continuum variations, but also found more complex variability behaviour than the inverse proportionality to the luminosity.

In NGC 4051 (Mihara et al. 1994a) there is a hint for the presence of the Ne X edge at 1.36 keV. A more detailed study of this source was done by Guainazzi et al. (1996). The continuum in this source varied rapidly on time scales down to 500 s. However, the O edges responded to continuum variations opposite to what was found in MCG -6-30-15. In NGC 4051, the O VII edge responded rapidly within 10^4 s to the continuum variations, while the O VIII edge did not vary.

From the above observations it is evident that the warm absorber has at least two components originating from different regions. This is confirmed by simultaneous far-UV/*ASCA* observations of NGC 3516 as reported by Kriss et al. (1996). They need two absorption components with a factor of 8 difference in ionization parameter; the more highly ionized component has a column density twice as large as the less ionized component.

A statistical study of 24 type 1 sources (among which 18 Seyfert 1 galaxies) showed that in at least half of them O VII or O VIII edges are present (Reynolds 1997). The higher-luminosity objects tend to be less ionized. The O VII optical depths are typically in the range of 0–1, the O VIII optical depths are in general somewhat smaller, 0–0.4. Why no objects with much larger optical depth are seen is yet unclear.

Ionized warm absorbers not only absorb hard X rays, but they also produce line emission, like any other warm medium. Consequently, when an oxygen edge is observed, there should also be oxygen line emission. Theoretical models (e.g., Netzer 1993) predict that the strongest line emitted by the warm absorber is the O VIII Lyman α line at 0.653 eV. While, in principle, the line could have an equivalent width up to 530 eV, several effects (e.g., photoabsorption by competing ions, and absorption of the line in the source itself) limit its strength, so that the observed equivalent line width is expected to be in the range of 5–50 eV. Other emission lines (O VII, Ne) are also predicted to have comparable or slightly smaller strengths. Unfortunately, the resolution (50 eV) of *ASCA* near the oxygen features makes it rather difficult to demonstrate the presence of these lines, also since the strong absorption edges are nearby. Nevertheless, George et al. (1995) deduced from their spectral fits some evidence for the presence of an O VII line at 0.57 keV and the O VIII line at 0.65 keV. The equivalent line widths of 45 and 19 eV, respectively, are in the range predicted by the photo-ionization models.

11.3 The power law component

The power law component in Seyfert galaxies dominates the total X-ray luminosity. However, since a power law is by definition featureless, it produces less detailed information than other spectral components. Most progress should come from variability studies or statistical studies comparing the photon index or luminosity with other properties of the nucleus.

Nandra et al. (1997b) studied the variability of 18 Seyfert 1s observed with *ASCA*. As expected from the results of earlier X-ray missions, all but one source showed variability on time scales ranging from minutes to hours. At least 5 sources showed variability, on time scales shorter than 1000 s. The general tendency of less variability for the more luminous sources could be confirmed. Typically, the r.m.s. variations behave like $(\Delta L/L)^2 L_{34} \simeq 1$, where L_{34} is the 2–10 keV luminosity in units of 10^{34} W. This relation can be understood in terms of models where the variability time scale is proportional to the size of the source, the central black-hole mass, the accretion rate, and the bolometric luminosity.

In MCG –6-30-15 some very rapid X-ray variations have been found: in one case 50 % intensity variations in only 100 s hve been observed (Reynolds et al. 1995b). The corresponding luminosity change of $10^{34}\,\mathrm{W s^{-1}}$ implies an accretion efficiency of at least 5 %, approaching the maximum for the Schwarzschild geometry (6 %). Since the efficiency for accretion onto a rotating black hole can be much larger (up to 42 %) this is additional evidence (apart from the iron line diagnostics) that the black hole in MCG –6-30-15 is rapidly rotating.

Most Seyfert 1s observed with *ASCA* have photon indices around 1.9 (Nandra et al. 1997c). Some sources observed with *ASCA*, however, have very steep photon indices. An example is RE 1034+39 (Pounds et al. 1995) with a photon index of 2.6 in the 2–10 keV range. This source has also a very strong soft component, as observed with *ROSAT*.

Brandt et al. (1997) studied the correlation between the photon index and the width of the optical Hβ line. They found that the photon index is enhanced up to values of 2.2–2.6 for a FWHM of the Hβ line less than 2000 km/s. Above a FWHM of 2000 km/s, no enhancement or decrement is seen. These systems with steep photon indices correspond to the so-called ultra-soft narrow-line Seyfert 1s, which in the *ROSAT* band have steep photon indices of about 3. RE 1034+39 mentioned above is an example of this class. These sources often show extreme variability (over a factor of 30). The strong soft component may cause enhanced inverse Compton cooling of the accretion disk corona, thereby lowering its temperature and increasing its photon index.

11.4 Soft components

The loss of sensitivity and resolution, and the calibration uncertainties of *ASCA* below 0.5 keV, make an analysis of the softest spectral components difficult. Moreover, at the lowest-energy bands visible with *ASCA*, the warm absorber plays an important role, and the spectral properties of any intrinsic soft X-ray emission component are easily hidden in the details of the warm absorber. Several cases where no soft excess is needed to model the *ASCA* data have been reported, like NGC 3227 (Ptak et al. 1994) and IC 4329A (Cappi et al. 1996b); however, in other cases a soft component is definitely needed (NGC 4051, Mihara et al. 1994a; NGC 7469, Guainazzi et al. 1994). The statistical study of Reynolds (1997) cited earlier shows that at least 5 out of 24 sources have a significant soft excess.

It is evident that in this field of research the broader energy band of *BeppoSAX*, extending down to 0.1 keV, will help enormously. Up to now, however, no results have become available.

11.5 Low-luminosity AGN

At present, no lower limit to the activity level of AGN is known. In fact, deep observations often reveal weak active nuclei in a large fraction of all galaxies. Even our own galaxy has a very-low-luminosity nucleus around its $2.6\,10^6 M_\odot$ black hole. It is often difficult to disentangle the active nucleus from a starburst component. I do not treat these cases here in detail, but mention only a few references.

One of the nearest AGN is Cen A at 5 Mpc distance. This radio galaxy has a complicated structure with an extended jet and diffuse emission, plus a strongly absorbed nucleus. *ASCA* observations of this source have been presented by Sugizaki et al. (1997) and Turner et al. (1997). Possible synchrotron X-ray emission from For A has been described by Kaneda et al. (1995).

Other low-luminosity AGN are found in M 106 (Makishima et al. 1994), M 81 (Ishisaki et al. 1996), and NGC 3147 (Ptak et al. 1996). An example of *ASCA* data of starburst galaxies is found in Della Ceca et al. (1996) for NGC 1569.

11.6 Broad-line radio galaxies

Broad-line radio galaxies (BLRGs) are in many respects similar to Seyfert 1 galaxies. The main difference is that BLRGs are located within giant elliptical galaxies instead of spiral galaxies, and contain broad extended radio lobes. In addition, the optical lines often have very large widths, up to several 10^4 km/s. Sometimes it is argued that this class of objects is similar to radio-loud quasars, but then seen at another inclination.

The number of BLRGs studied with *ASCA* is limited. Eracleous et al. (1996) analyzed 3C 390.3. Like Seyfert 1s, this galaxy shows a broad iron line with an equivalent width of 300 eV. In 3C 109 Allen et al. (1997) found an extremely broad iron line, with a FWHM of 120,000 km/s. Applying the relativistic disk model, they derive a lower limit of 35° for the inclination of the disk. In adition, VLBI observations have shown the inclination to be smaller than 56°. This inclination range is compatible with the unification schemes, where the BLRGs are viewed at intermediate viewing angles near the edge of the torus.

In the sample of Reynolds (1997) two more BLRGs are found, 3C 120 and 3C 382. Both of them have iron lines with a large equivalent width (more than 900 eV). They show significant line broadening of 1.5–1.8 keV (this is the Gaussian σ, corresponding to 0.25–0.30c).

None of the four BLRGs discussed above have strong absorption features from a warm absorber. In 3C 382, one of the most evident cases with absorption features, the O VII edge has an optical depth of only 0.16 (Reynolds 1997).

12 Seyfert 2 Galaxies

12.1 Introduction

Seyfert 2 galaxies differ from the classical Seyfert 1 galaxies in the sense that they only have narrow emission lines in the optical band, contrary to Seyfert 1 galaxies which both have narrow and broad lines. In both cases an active nucleus is present; however, in Seyfert 2s the broad-line region and most of the other nuclear emission is hidden behind a dust torus or other absorbing structure; the narrow-line region is located outside of this torus, and hence is always visible.

Currently a popular theory is the so-called unification scheme for active galactic nuclei. In this scheme, Seyfert 1s are those objects where it is possible to look more or less along the axis of the obscuring torus towards the inner nucleus. Therefore the broad-line region, the X-ray emission and all other emission from the accretion disk are visible without much attenuation. Seyfert 2 galaxies are the same objects, but since they are viewed more along the plane of the torus, they are heavily absorbed.

However, at present several facts still remain that are difficult to reconcile with the unification scheme, and it is not yet clear whether (a part of) the Seyfert 2 population has intrinsically different properties from the Seyfert 1 population.

The earlier generation of X-ray satellites detected many Seyfert 1 galaxies and fewer Seyfert 2 galaxies. This is due to the strong absorption that essentially deletes most of the X-ray flux in the classical 2–10 keV band.

In several Seyfert 2 galaxies a strong iron line was already discovered with *Ginga* (Awaki 1992). The paper of Awaki gives a nice overview of the *Ginga* results for a large set (21 objects) of Seyfert 2 galaxies.

12.2 NGC 1068

The prototype Seyfert 2 galaxy is NGC 1068 (=M 77, strangely enough this older name is not often used). At a redshift of 0.0034, its distance of 20 Mpc is similar to that of the Virgo cluster. *ROSAT* observations (Wilson et al. 1992) showed that half of the soft X-ray emission is extended. A core with a radius of 1 kpc contains hot gas that emits a quarter of the total soft X-ray flux. It is possibly associated with the hot medium that confines the narrow-line region. A more extended component, most likely associated with the starburst activity, emits another quarter of the soft X rays, within a radius of 10 kpc. The iron line as detected with *Ginga* (Awaki 1992) is much stronger than for all the other Seyfert 2 galaxies of his sample. The equivalent line width is 1300 eV, compared to typical values of a few hundred eV for most other Seyfert galaxies. This has been explained by the blocking of the direct X-ray emission, so that only scattered hard X rays can be observed; this reduces the continuum emission and hence increases the equivalent line width.

ASCA confirmed the existence of the iron line (Ueno et al. 1994), and resolved it into three components at energies of 6.40, 6.62 and 6.87 keV, corresponding approximately to the fluorescent line of lowly charged iron, He-like and H-like iron, respectively, with equivalent widths of 1600, 1000 and 600 eV, respectively. This shows that there are contributions from fluorescence in cool material, as well as emission by very hot gas.

More importantly, the spectrum below 3 keV can be decomposed into the extrapolated power law underlying the Fe-K complex, and an additional thermal emission, as is evident from the strong lines of H-like and He-like O, Ne, Mg, Si and S, as well as Fe-L emission that is present in the spectrum. Ueno et al. fitted the thermal component using two CIE-plasmas with temperatures of 0.6 and 0.15 keV, producing 40 and 50 % of the soft X-ray flux below 3.5 keV, respectively. It is not yet clear how these components are related to the three spatial components as discovered with *ROSAT*. Moreover, the abundances of 0.3 and 0.03 times solar found by Ueno for the 0.6 and 0.15 keV components, respectively, are rather low. Complicating factors in the analysis are the difficulty in subtracting the power law component, the old (*meka*) plasma code used and the resolution and calibration of *ASCA* below 0.5 keV, affecting the coolest component. More importantly, the presence of H-like and He-like iron ions hints to the presence of multi-temperature structure, or more likely to the presence of photo-ionized plasmas.

BeppoSAX detected for the first time hard X rays in the 20–100 keV band in this source (Guainazzi & Piro 1997). This confirms the predominance of the cold reflector that is needed to explain the low-energy scattering.

12.3 NGC 6552

This Seyfert 2 galaxy was discovered serendipitously near the north ecliptic pole (Fukazawa et al. 1994). Fitted to a partial absorber model, its spectrum shows a covering factor of 98 % with a column density of $6\,10^{27}\,\mathrm{m}^{-2}$, as is typical for Seyfert 2s. More importantly, it shows fluorescent lines of neutral Ne, Mg, Si, S, Ar, Ca and Fe, although the significance of each individual line except iron is only at the $1{-}2\,\sigma$ level. He-like or H-like emission could be excluded, contrary to the case of NGC 1068. The lines thus originate in a medium that is thick and dense enough not to suffer from significant photoionization. This medium is also expected to produce some Compton scattering of hard X-ray continuum.

12.4 NGC 4945

This Seyfert 2 is one of the brightest at an energy of 100 keV. Nevertheless it is relatively weak in the classical X-ray band (Done et al. 1996). It shows a heavily absorbed power law, with only 3 % of the continuum leaking through or being scattered at low energies. The absorbing column density of $5\,10^{28}\,\mathrm{m}^{-2}$ is so large, that the calculated optical depth for Thomson scattering is about 7.5. This, however, implies that each photon should scatter several times before being absorbed or escaping, thereby effectively increasing the path length of the photon in the absorbing medium. Approximating the number of scatterings by τ^2, the true optical depth is about 3. Since this is still larger than 1, a proper analysis should include the (energy dependent) Compton scattering of photons through the absorber. In Fig. 18 I show the model spectrum for this source.

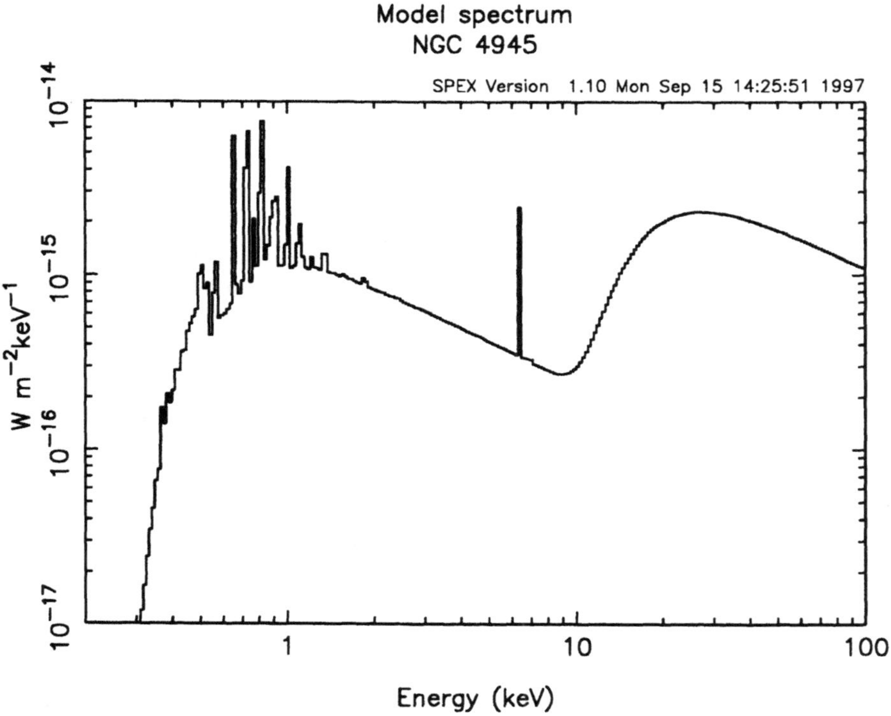

Fig. 18. Model spectrum of NGC 4945. Note the 30-fold flux decrease between 10–20 keV, the strong iron line and the thermal emission near 1 keV.

12.5 NGC 1808

In NGC 1808 (Awaki et al. 1996) there is both a starburst and an active-nucleus component. The presence of the active nucleus follows from the time variability of the hard X-ray flux. Apart from the absorbed power law component, there are at least two thermal components visible (0.35 and 0.8 keV temperature), with a spatial extent of at least 0.6 kpc (from *ROSAT* imaging). The cooling time of the hotter component is only a few million years and hence much shorter than the duration of the starburst; it requires continuous energy supply. Its flux is consistent with interstellar matter heated by supernova shocks, provided that most of the supernova remnants are in their radiative phase, therefore producing a relative large luminosity per remnant. The near-solar abundances are higher than in other starburst galaxies (e.g., M 82 has 0.1 times solar), and this might indicate that the galaxy is approaching the end of its starburst phase, with a heavily enriched ISM. The cooler component could be consistent with the hot gas confining the narrow-line region.

12.6 Other cases

Other well-studied Seyfert 2s include Mrk 3 (Iwasawa et al. 1994), NGC 4388 (Iwasawa et al. 1997), and NGC 5252 (Cappi et al. 1996a).

Mrk 3 has both a strong cool (6.4 keV) and a weaker hot (7 keV) iron emission line, and the continuum is variable on time scales of years, showing that the hard X rays are due to the active nucleus and not to a starburst component. NGC 4388 near the core of the Virgo cluster has a soft thermal, spatially extended component with low abundances (iron less than 0.4 times solar). It has no strong starburst, and most likely its thermal emission is due to ram pressure stripping of the ISM in the cluster: the host galaxy moves with 1300 km/s with respect to the Virgo cluster. NGC 5252 shows a flat spectrum without emission lines, hence any thermal component (if present) should have less than 0.01 solar abundances.

Mrk 463 is an interesting case: this is a pair of interacting galaxies with two nuclei separated by only 4″ (6 kpc). Both nuclei appear to be Seyfert 2 nuclei. *ASCA* has observed this system (Ueno et al. 1996) and confirmed the Seyfert 2 nature of the X-ray spectrum. Of course, it was not possible to separate both components.

BeppoSAX observations of two Seyert 2s, selected solely on the basis of their optical properties, showed that these sources have strong iron lines, with equivalent widths up to 1000–2000 eV (Salvati et al. 1997).

12.7 Intermediate cases: narrow emission line galaxies and others

Narrow emission line galaxies (NELG) are most likely Seyfert 1 nuclei where the broad-line region is shielded by the optically thick molecular torus. In their optical spectra they resemble Seyfert 2 galaxies, while their X-ray properties are similar to Seyfert 1s. In general, the absorbing columns found in their hard X-ray spectra are of the order of 10^{26}–10^{27} m^{-2}, i.e. 100 times less than the column density of the molecular torus. In the context of unification schemes a likely explanation is that these sources are viewed through the rim of the torus, at a viewing angle between those of Seyfert 1s and Seyfert 2s.

Perhaps the best-studied case with *ASCA* is NGC 2992 (Weaver et al. 1996). This source shows a narrow emission line of neutral iron. There are no indications for the presence of highly ionized iron. The line has a small width, with a FWHM of less than 6600 km/s. This implies that, unlike for Seyfert 1s, the line cannot be produced in an accretion disk, for in those cases the lines are much broader. However, the line can be produced in the scattering region around the torus, the same region where the warm absorbers in Seyfert 1s may be present. The only problem with the *ASCA* spectrum of NGC 2992 is the large equivalent width of the iron line of 500 eV. This is rather large for an otherwise Seyfert 1-like X-ray spectrum. However, by comparing to previous X-ray missions Weaver et al. showed that the continuum decreased by a factor of 20 in 16 years while the line flux decreased only by a factor of 3. At a distance of 3 pc (10 light years), the expected lag of the line with respect to the continuum may artificially produce this large equivalent width. Such a location would be compatible with the unification scheme; a location in the broad-line region would yield a much smaller equivalent width.

Also in NGC 2110 (Hayashi et al. 1996a) a narrow line from cold iron was found. This NELG shows evidence for an Fe-K absorption edge from cold absorbing material with a column density of 10^{27} m^{-2}, with about 5 % scattered radiation leaking through.

The Circinus galaxy (Matt et al. 1996) shows a very flat continuum (photon index less than 1). When fitted to a power law plus reflection model, the reflected component is more then hundred times brighter than the direct component. This picture is consistent with reflection off the torus, with the direct continuum mostly suppressed. Besides this, the spectrum shows strong emission lines from Ne, Mg, Si, S (at mildly ionized stages) and Fe, both in its neutral phase and H-like, although the line blend at 7 keV might also contain a significant contribution from the Kβ line of neutral iron, because the Kα line at 6.4 keV has an extremely large equivalent width of 2000 eV.

Another intermediate case is IRAS 18325–5926 (alternative name H 1829–591) (Iwasawa et al. 1996a). It is optically classified as a Seyfert 2. However it has distinctive spectral differences with respect to NELGs, although the optical lines do show weak broad wings. The hard X-ray spectrum is highly variable on a time scale of 10^4 s, showing that it is possible to see the active nucleus directly (i.e., in X rays it is like a Seyfert 1). There is a soft excess, but *ASCA* is not able to distinguish here between thermal or non-thermal emission. More importantly, there is a complex iron line with a sharp peak at 6.9 keV, a large equivalent line width of 500–800 eV, and skewed wings down to 4 keV. Iwasawa et al. interpret the line profile with similar models as used for Seyfert 1s, namely the relativistic inclined-disk model. They need a cold disk with emission up to 7 Schwarzschild radii (perhaps a Kerr black hole), at an inclination of about 42°, which is slightly larger than the typical inclination of 30° as found in some Seyfert 1s (e.g., Mushotzky et al. 1995; Tanaka et al. 1995). The inclination angle can be explained in terms of the unification scheme if IRAS 18325–5926 is a system just at the limit of the Seyfert 1 viewing domain. In this model, the sharp peak near 6.9 keV is due to a blue-shifted wing of the line profile. The large equivalent line width can only be explained if the iron abundance is at least three times solar.

In another NELG, NGC 7314 (Yaqoob et al. 1996), the continuum emission varies on time scales of a few minutes; the broadened red wing of the iron line responds very rapidly, within 30,000 s, to these continuum variations. However, closer to the peak of the emission lines the response time is larger. If the line is due to a disk, the disk should have an inclination of at least 33°. However, the line core could also originate in, e.g., the torus, given the slow response of the line core. Assuming a combined disk/torus origin, the disk inclination should be larger than 38°, a typical value for NELGs.

The brightest Seyfert galaxy in the sky is NGC 4151. It is often thought to be an intermediate case between Seyfert 1s and Seyfert 2s, and one of the best studied Seyfert galaxies, but it has some peculiarities. Here I only mention that *ASCA* observations of this source (Weaver et al. 1994; Warwick et al. 1996) can be well understood by dual cold-absorber models with covering factors varying between 30–75 % and variable column densities of $0.3–3\,10^{27}\,\mathrm{m}^{-2}$, plus a scattered component of 2–4 % of the original continuum, and an additional soft excess corresponding probably to the extended X-ray component seen in high-spatial-resolution observations.

13 Quasars

Quasars constitute the most powerful class of active galactic nuclei. At cosmological distances, they contain important information about the early history of the universe. It is often assumed that quasars are the more distant and luminous versions of Seyfert galaxies. In their X-ray properties they show also many similarities to Seyferts.

Quasars are commonly subdivided into two classes: radio-loud and radio-quiet, according to the presence or absence of a strong central radio source. Contrary to the radio-quiet objects, the radio-loud objects are thought to be beamed into the line of sight.

13.1 Radio-quiet quasars

The X-ray spectra of radio-quiet quasars look most similar to those of Seyfert 1 galaxies. In radio-loud objects, a relativistic beamed X-ray component dominates the hard X-ray spectrum. Due to the different viewing angle or intrinsically weaker beams, such a contamination is less important in radio-quiet quasars.

The first radio-quiet quasar studied with *ASCA*, PG 1211+143, showed a soft excess that varied together with the hard X rays on a typical time scale of 5 hours, indicating that both spectral components originate from nearby locations (Yaqoob et al. 1994b). There is only marginal evidence for an Fe-K line of about 200 eV equivalent width.

In PG 1116+215 (Nandra et al. 1996) there is clearer evidence for an iron line, with an equivalent width of 260 eV. However, contrary to many Seyfert 1 galaxies, the line centroid is not consistent with (near)-neutral iron. Instead, the line centroid is indicative for H-like iron (6.9 keV), although He-like emission cannot be ruled out. The line can be attributed to reprocessing of radiation in the innermost strongly ionized parts of the accretion disk. This has further implications. For example, highly ionized gas is less able to reprocess X rays than neutral gas. Therefore, it seems unlikely that the large blue bump that is observed in this source is caused by reprocessing.

Lines with intermediate ionization (6.54–6.58 keV in the rest frame) were detected in E 1821+643 (redshift 0.297, Yamashita et al. 1997) and in IRAS 13349+2438 (redshift 0.107, Brinkmann et al. 1996). In both cases the lines have an equivalent width of about 200 eV. The line centroids are clearly not consistent with neutral iron, but rather with Be–C-like iron. Alternatively, Yamashita et al. propose that the observed line centroid at 6.58 keV may also be due to emission from He-like iron, if the models for rapidly rotating accretion disks are used. In that case, the disk should have an inclination of about $10°$.

There are also examples of radio-quiet quasars that show no significant line emission. In PG 1634+706 and PG 1718+481, both at a redshift of about 1, neither a line nor a reflection component is visible (Nandra et al. 1995). Since in these objects no substantial beaming is expected, other explanations need to be sought. Nandra et al. suggest that either the disk is highly ionized or that it is optically thin. In both cases no line or only a weak iron line is expected. Both scenarios could be consistent with a quasar luminosity close to the Eddington limit; in that case the accretion disk is expected to become geometrically thick.

A special class of radio-quiet quasars are the so called broad-absorption line quasars (BAL-quasars). These radio-quiet objects show in their optical spectra broad absorption lines redward of the system velocity with a width up to 0.1–0.2 c, caused by ionized absorbing material with column densities of 10^{25}–$10^{26}\,m^{-2}$. Some of these objects show in their optical emission line spectra over-abundances of the metals by a factor of 10–100. PHL 5200 is the prototype source. It has a redshift of 1.98. Mathur et al. (1995) measured the extremely weak spectrum of this source, and found that it should be absorbed by material in the source with a column density of $10^{27}\,m^{-2}$ if it has a solar composition. For higher metallicities, the required column density scales inversely proportional to the metallicity. Thus, both in the optical and in X rays this source shows strong absorption.

13.2 Radio-loud quasars

Radio-loud quasars are the best studied population of quasars with *ASCA*. These objects are dominated by a power law component that often has a rather hard spectrum, and in general they show no significant iron line emission.

A spectral analysis of the first two radio-loud quasars observed with *ASCA* showed a spectral steepening above 7 keV (Serlemitsos et al. 1994). However, this was not confirmed in a later re-analysis of these data (Cappi et al. 1997), and it is probably due to the preliminary calibration data used. The prototype quasar, 3C 273, showed no iron line (Yaqoob et al. 1994a), and in several other cases also neither significant line emission nor the presence of a reflection component could be detected (Elvis et al. 1994; George et al. 1994; Turner et al. 1995; Siebert et al. 1996; Kubo et al. 1997). In some cases very strict upper limits could be set to the equivalent width of the iron line, such as 38 eV (NRAO 140, Turner et al. 1995) or 19 eV (PKS 2149–306, Siebert et al. 1996). It is possible that in some of these sources the lack of an iron line is due to the absence of a reflecting medium, but more likely it is due to the dominance of the beamed X-ray component, increasing the continuum and thereby hiding any iron line of normal strength that would be otherwise visible. From the upper limits to the equivalent line width, it can be deduced that this beamed component may be an order of magnitude stronger than the normal unbeamed component that is visible in radio-quiet quasars.

Cappi et al. (1997) have analyzed a set of 9 radio-loud quasars, among which several of the sources analyzed before by others. All these sources have power law spectra, and the six brightest show evidence for excess absorption above the galactic foreground value. The average 2–10 keV photon index of 1.53 with a spread of 0.12 is much lower than that for radio-quiet objects and most Seyfert 1 galaxies. This, and the absence of iron line emission, points again to the important role of beaming in these sources. In S5 0836+71 the absorbing column density is variable. In S5 0014+81 there is marginal evidence for an Fe-K absorption edge at the redshifted energy. Both facts strongly support the idea that the excess absorption in these systems is intrinsic to the quasars, rather than caused somewhere along the line of sight. The derived excess column densities are in the range of (0.2–50) $10^{26}\,\mathrm{m}^{-2}$.

BeppoSAX observations of the prototype quasar 3C 273 showed the presence of a narrow absorption feature near 0.5 keV, which can be interpreted as due to O VIII Lyman α absorption at 0.65 keV, blueshifted by 60,000 km/s (Guainazzi & Piro 1997). Therefore, it might be associated with the outflowing jet, similar to what has been reported in the past for the BL Lac object PKS 2155–304.

13.3 Type 2 quasars

For low-luminosity AGN like Seyfert nuclei, a clear distinction between type 1 and type 2 nuclei can be made, as was shown in the previous chapters. Before the launch of *ASCA*, it was not known whether type 2 nuclei (i.e., objects with only narrow optical emission lines) of high luminosity also had detectable X-ray emission, and what spectral characteristics these objects have.

Fabian et al. (1994b) detected X rays from the galaxy IRAS P09104+4109. This galaxy at a redshift of 0.442 has narrow optical emission lines similar to Seyfert 2 galaxies. Interestingly, the X-ray spectrum showed strong Fe-K emission with an equivalent width of 450 eV. Such large equivalent widths are usually found only in Seyfert 2 galaxies. It indicates the scattering of light into our line of sight from a hidden nucleus. The intrinsic luminosity of this galaxy is very high: $5\,10^{40}$ W. Thus, this galaxy represents a new class of Seyfert 2-like nuclei, emitting at quasar luminosities.

A similar case is IRAS 20460+1925 at a redshift of 0.181, with a 260 eV equivalent width iron line in a moderately absorbed ($3\,10^{22}\,\mathrm{m}^{-3}$) intermediate luminosity (10^{37} W) nucleus (Ogasaka et al. 1997). Such systems are also detected serendipitously with *ASCA*, like the redshift 0.9 object detected in the Lynx field by Ohta et al. (1996).

13.4 BL Lac objects

BL Lac objects are active nuclei that show a strong beaming into our line of sight. The X-ray spectra of these interesting objects are rather dull, however, showing mostly power laws with sometimes a break. The main contribution of *ASCA* to this field consists of participation in multi-wavelength campaigns, with the aim of deriving the source structure and spectrum using correlated timing analysis of the data taken at different energies. I mention here only Mrk 421 (Macomb et al. 1995; Takahashi et al. 1996), AO 0235+164 (Madejski et al. 1996), EXO 055625–3838.6 (George & Turner 1996), H 1426+428 (Sambruna et al. 1997) and last, but not least, PKS 2155–304 (Urry et al. 1997).

14 Clusters of Galaxies

Clusters of galaxies constitute the most massive objects in the universe. They are known to emit X rays from hot thermal plasmas. The X-ray emitting mass is generally larger than the mass of the visible galaxies, although the total mass of the cluster is dominated by invisible dark matter. From their X-ray morphology, clusters can be divided into two classes. The first class consists of clusters with a dominant central cD galaxy and a relatively small X-ray core, which often show a cooling flow. The prototype is the Virgo cluster, with the giant elliptical M 87 in its core. The other class consists of clusters that do not have a single dominant galaxy. Among these are clusters with two bright ellipticals in the center (like the Coma cluster), and more irregular clusters. These less well concentrated clusters often show no cooling flows.

The X-ray brightness of non-cooling-flow clusters or the brightness outside the cooling-flow region in clusters, is often well described by the so-called isothermal β models, in which the X-ray surface brightness S is given by

$$S = S_0[1 + (r/a)^2]^{-3\beta+0.5} , \tag{5}$$

with a the core radius and β a dimensionless variable, related to the ratio of the galaxy velocity dispersion in the cluster to the X-ray temperature. A typical value is $\beta = 2/3$. The gas density profile corresponding to Eq. (5) is given by

$$n = n_0[1 + (r/a)^2]^{-3\beta/2} , \tag{6}$$

showing that for $\beta = 2/3$ the density at large distances is proportional to r^{-2} and hence that the integrated gas mass $M(r)$ is proportional to r.

The core radius a of clusters with a dominant central galaxy is 200 kpc, with a typical spread of a factor of 2. For the other clusters, the core radius is generally larger, about 600 kpc. In the cases where a cooling flow is present it is usually found within one core radius distance from the center.

14.1 Temperature distribution of the hot medium

In most cases, the *ASCA* data of clusters are consistent with an isothermal plasma at all radii, except for a possible cooling-flow component. However, a few exceptions to this general rule have been published. Arnaud et al. (1994) report a slight temperature increase outside the cooling radius for the Perseus cluster, but it is not clear how much this temperature rise is influenced by the substructures with different temperatures that are also present in this cluster.

Markevitch et al. (1994) studied A 2163, one of the hottest known clusters (temperature 16 keV) and concluded that within the central 3–4 core radii the temperature varies, indicating that the gas is not in equilibrium. In this cluster there is also evidence that the galaxies are not in equilibrium. Markevitch (1996) later reported a temperature decrease beyond 2 Mpc in four non-cooling-flow clusters. In two of these clusters (A 2163 and A 665) the temperature gradient is so large that the cluster atmosphere should be convectively unstable and transient. Also, such a steep temperature gradient cannot be due to heating by the release of gravitational energy during infall. Therefore, Markevitch proposed that there should be an additional heat source in the inner parts of the cluster, such as merger shocks. The other two cases (A 2256 and A 2319) were considered to be pre-mergers by Markevitch. In a later study, Markevitch & Vikhlinin (1997) reported also a case with a marginal temperature decrease from 6 to 5 keV beyond 1 Mpc (A 3558), but in AWM 7 no temperature variations outside the cooling-flow radius were seen.

ASCA's sensitivity allows also the measurement of temperatures of more distant clusters. In a statistical analysis of 38 clusters at redshifts larger than 0.14, Mushotzky & Scharf (1997) found the same luminosity-temperature relation as the one that is known for nearby clusters. This strongly limits those cosmological models that predict substantial changes in this relation due to the recent rapid growth of clusters. Either the cluster core is already pre-heated when it starts to grow, or the cluster structure does not evolve strongly with time.

14.2 The cooling flow and central temperature distribution

The presence of cooling flows in clusters of galaxies has been deduced from X-ray spectra as well as X-ray imaging. In the latter case, the excess emission above the isothermal β models is considered to be evidence for a low-temperature component. The first cooling-flow spectra gave rise to difficulties in the interpretation (Fabian et al. 1994c, Fig. 19). In the Centaurus cluster, the strength of the Fe-L 4–2 transitions could not be explained by the plasma codes that were available at that time. Stimulated by these observations, the plasma codes have since been improved substantially with calculations by Liedahl, and resulted in, e.g., the *mekal* model.

There are several clusters that indeed show a classical cooling flow. For example, the nearby ($z = 0.01$) Centaurus cluster (Fukazawa et al. 1994) is isothermal outside a radius of 10′ (200 kpc) with a temperature of 4 keV. Within that radius there is also a 1 keV component, clearly visible through its strong Fe-L complex, that is absorbed by cool gas in the emitting region with a column density of $10^{25}\,m^{-2}$. However, Fukazawa et al. argue that the brightness profile indicates that 70 % of the high-temperature flux within a projected radius of 5′ is actually emitted within the corresponding radius of 100 kpc. This implies that the gas does not cool uniformly, and that the core contains a mixture of hot and cooling gas. Assuming pressure equilibrium between both components, one finds that in the central zone the cool component occupies about 5 % of the total volume, hence it is probably in the form of dense filaments.

In Virgo (Matsumoto et al. 1996) there are also two temperature components (1.3 and 3 keV). Within the central 10′, both temperatures remain constant, although the cooler component is more centrally concentrated than the hot component. The low-luminosity AGN of the central M 87 galaxy only contributes 3–5 % to the total X-ray flux.

Ikebe et al. (1996) claimed evidence that the hot component in the Fornax A cluster also extends down to the core, despite the fact that the central brightness enhancement was previously attributed to a cooling flow. They decomposed the X-ray image into two components, one corresponding to the central cD galaxy (core radius 5 kpc, central gas density $2\,10^4\,m^{-3}$) and the

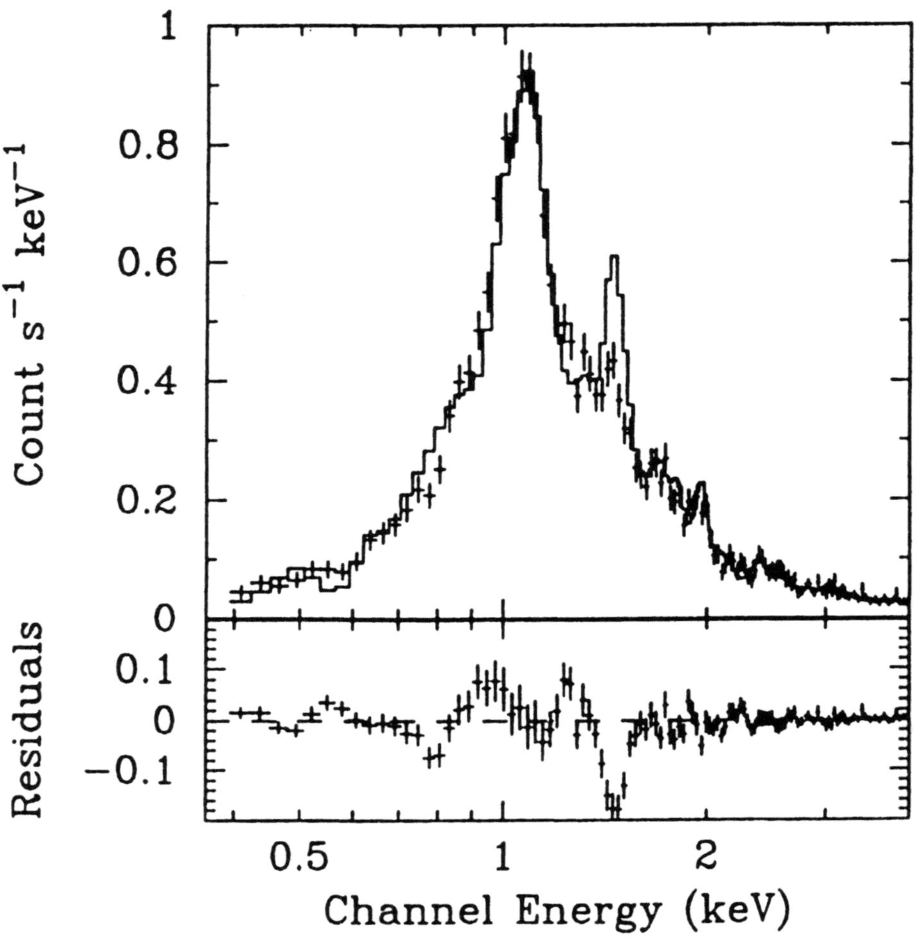

Fig. 19. The best-fit *meka* cooling-flow model for the *ASCA*/SIS0 spectrum of the Centaurus cluster. The Fe 4–2 transition blend at 1.4 keV was not included in the fit. From Fabian et al. (1994c).

other to the cluster (core radius 130 kpc, central gas density $8\,10^2\,\mathrm{m}^{-3}$). Interestingly, the optical velocity dispersion in the cD galaxy is similar to the cluster velocity dispersion: both are 300 km/s. Ikebe et al. conclude that this provides evidence that the dark matter in the cluster (which essentially determines the gravitational potential and thereby the gas distribution) must have a hierarchical distribution.

In yet another case, the nearby Hydra A cluster, Ikebe et al. (1997) found that the strong excess above the single β-model, that was interpreted previously as a cooling-flow component, is also present at energies above 4 keV. A joint *ROSAT/ASCA* fit to the data is consistent with two interpretations. First, a two-temperature fit to the spectrum of the central regions yields components of 3.5 keV (equal to the temperature of the outer regions) and 0.6 keV. The hot component occupies 90 % of the volume within 1.′5; the total mass of the cool component is $4\,10^9 M_\odot$, comparable to the mass of the hot halos seen in non-cD elliptical galaxies. In this interpretation, the cooler component is therefore due to the normal ISM of the central cD galaxy. Alternatively, the spectrum can be fitted by a cooling-flow spectrum with a mass flow rate of $60\,M_\odot\,yr^{-1}$, an order of magnitude smaller than previously found.

In the hot (10 keV) and nearby cD cluster Ophiuchus, Matsuzawa et al. (1996) claim to see no spectral indications for a cooling flow, contrary to imaging observations. It should be noted, however, that especially in hot clusters like this one, there is a very strong dominant bremsstrahlung continuum, which makes detection of soft X-ray features more difficult than in cooler clusters.

Some have claimed to have observed the strongest cooling flow. Allen et al. (1996) report a cooling rate of $1000\,M_\odot\,yr^{-1}$ within 200 kpc for PKS 0745–191. The champion of all clusters up to now is RX J1747.5–1145, the most luminous X-ray cluster known (Schindler et al. 1997). At a redshift of 0.451 it has a luminosity of $2\,10^{39}$ W, a temperature of 9 keV and a total mass of $10^{15}\,M_\odot$. The cooling flow of $3000\,M_\odot\,yr^{-1}$ within 30″ (200 kpc) emits half of the cluster luminosity.

An interesting result is the claim by Lieu et al. (1996) that in several clusters there are additional cool plasma components with an extended distribution. Evidence for this comes from observations with the *EUVE*/DS photometer, sensitive around 0.1 keV, and additional support from *ROSAT* PSPC data. In the Coma cluster, the DS detector shows a soft-excess flux of 50 % above the 8.7 keV component as determined from *ASCA* observations. The Coma cluster has no known cooling flow. The soft halo extends out to 15′ (600 kpc), and can be modeled by the sum of 0.07 and 0.2 keV components. The origin of these components is as yet unclear. However, these observations can be easily tested with *BeppoSAX* observations, the results of which are not yet available, however.

14.3 Mass distribution

Using the observed density and temperature distribution derived from X-ray observations, the mass distribution in the cluster can be mapped using the hydrostatic-equilibrium assumption. This has now been done in a standard way for several clusters. In those cases where the mass can also be deduced from the gravitational-lensing properties of the cluster, in general an agreement within a factor of two is found. Hydrostatic equilibrium gives the *total* mass distribution; the gas density profile itself gives the *hot gas* mass distribution; and from galaxy counts and assumed mass-to-light ratios of the individual galaxies, the *galaxy* mass distribution is derived. In general, the total galaxy mass is smaller than the gas mass, and this is often an order of magnitude smaller than the total mass. Subtracting the hot-gas mass and the galaxy mass from the total mass yields the mass of the dark matter.

The baryonic fraction is not always the same for all clusters. For example, Loewenstein & Mushotzky (1996) determined the baryonic fraction in two poor clusters. AWM 7 has twice as large a baryonic fraction (25 %) as A 1060 (15 %). There are also more extreme cases, like MS 0451.6–0305 (Donahue 1996), where the hot gas mass is only 4 % of the total mass. Loewenstein & Mushotzky propose that during the proto-cluster phase baryons may be depleted due to strong supernova-driven winds; the degree to which this happens determines the final baryon fraction.

14.4 Groups of galaxies

Groups of galaxies with membership ranging from a few to a few hundred can be regarded in some aspects as the low-mass end of the cluster distribution. Also in such loose groups hot plasma is found. Fukazawa et al. (1996) report a temperature of 1 keV and 0.3–0.5 times solar abundances for two groups studied by them, showing thereby similar abundances as in rich clusters of galaxies.

14.5 Cluster mergers and dynamical evolution

In several clusters evidence is found for merging of sub-clusters. This evidence is not only derived from optical velocity dispersion measurements, but also from the X-ray spectra and morphology. Cluster merging is an important process in the formation of the large-scale structure of the universe. Also *ASCA* observations have contributed significantly to our insight into the merger physics.

Fujita et al. (1996a) have studied the neighbouring pair of cD clusters A 399 and A 401. These clusters are likely to be gravitationally bound, but *ASCA* spectra do not show much evidence for interaction: the region in between the clusters has definitely excess X-ray emission, but it shows no temperature rise; instead, the temperature varies smoothly between the clusters. This is not expected from numerical models, which predict that there should be a shock front with enhanced temperatures between the clusters. The absence of a temperature rise either means that the pair is still a pre-merger pair, or that both clusters have interacted already once, and that the shock has cooled down already. Recent support for the latter hypothesis comes from *ROSAT* HRI observations, showing evidence for a tidal tail between the clusters (Fabian et al. 1997).

In several other clusters there is evidence for non-azimuthal temperature differences, attributable to merging processes. Examples are the Perseus cluster (Arnaud et al. 1994), showing temperature variations from 5–9 keV, and also some iron abundance variations associated with it. Another case is Coma, where temperature differences of 4.5–10 keV over 40′ (1.6 Mpc) have been found. A 754 (Henriksen & Markevitch 1996) shows temperature differences from 5–12 keV with no significant abundance variations.

14.6 Optical-depth effects

In general, it is assumed that the X-ray emission from clusters of galaxies is due to thermal emission from hot optically thin plasmas. This is certainly true for the continuum radiation, but not always for the line emission, as is shown below. Although the densities in clusters are extremely small (10^2–$10^4\,\mathrm{m}^{-3}$ or even lower), their size is so large that the total column density of hot gas can be as large as $10^{26}\,\mathrm{m}^{-2}$ in some cases. The strongest resonance lines of the plasma then have an optical depth of order unity. Resonance scattering of line emission then has the effect that photons from the central regions of the cluster moving towards us are scattered out of the line of sight. Since clusters are more or less spherically symmetric, the net effect on the projected line profile is that near the center of the cluster the line intensity is suppressed, while in the outer regions it is enhanced. Of course, resonance scattering does not destroy photons. Thus the total line intensity does not change.

In the cases with the largest column density, the H-like or He-like iron line transitions 2–1 (erroneously often called Kα) may have optical depths larger than unity, while the 3–1 transitions (Kβ) have a smaller optical depth. Thus, the Kα/Kβ line ratio should increase from the center of the cluster towards the outer regions, even for isothermal clusters. This is what was found in a sample of clusters by Tawara (1996). In several cases, he obtaied Kα/Kβ line ratios that are a factor of two below the prediction for an optically thin plasma. Although also the equivalent width of the resonance line should change as a function of radius, this is not a good diagnostic tool since it is difficult to disentangle these scattering-induced variations from possible abundance variations. That is the reason why line ratios should be used. However, it should be noted that recent work by myself and R. Mewe shows that the present-day plasma codes may significantly under-estimate the strength of the 3–1 transitions for hydrogen-like iron; the newer calculations decrease the predicted Kα/Kβ line ratio.

BeppoSAX observations of the Perseus cluster (Molendi et al. 1998) show clearly the effect of resonance scattering. The Kα/Kβ line ratio at the center of the cluster is only half of the predicted value for the optically thin case; it increases smoothly up to the CIE value near 7′ away from the core; beyond that radius the statistics are not good enough to obtain useful constraints. These data imply that the iron abundance near the center of the cluster is nearly solar and twice as large as previously thought. This is due to the fact that the He-like resonance line of iron yields the dominant contribution to the iron abundance estimate.

14.7 The quest for the Hubble constant

Using X-ray data, it is possible to determine the Hubble constant. This can be done in several ways. First, by inferring optical depths from resonantly scattered lines as indicated in the previous subsection, essentially the product of density times size (nr) is determined. Together with the observed emission measure (essentially n^2r^3/d^2) and the angular size (r/d) the distance d can be calculated directly and the Hubble constant can be deduced. At present, however, there are too many uncertainties involved in this method to be reliable.

A better method is to use the Sunyaev-Zeldovich effect to measure the quantity nr. The Sunyaev-Zeldovich effect is the decrease of the apparent 3 K background radiation temperature due to inverse Compton scattering on the hot electron gas in the cluster. Note that although the 3 K photons gain energy through the scattering, the effective temperature is decreased because the photons are scattered out of the line of sight. Temperature measurements with *ASCA* combined with imaging from *ROSAT* and observations of the 3 K radiation have been used to determine H_0. At this moment, however, systematic effects make this method not more accurate than other competing methods. For example, Myers et al. (1997) report a Hubble constant of $54 \pm 14\,\mathrm{km/s\,Mpc^{-1}}$. Other more or less successful attempts and measurements can be found in Holzapfel et al. (1997) and Matsuura et al. (1996).

14.8 Abundances in nearby clusters

For the hottest clusters, bremsstrahlung dominates the spectrum and only the Fe-K complex is strong enough to be useful for abundance determinations. In the cooler clusters, also lines of Si, S and others have better statistics, as well as the Fe-L complex. Because of the difficulties associated with the atomic physics of the Fe-L complex, some people have doubted its usefulness in abundance estimates. However Hwang et al. (1997) showed that both iron complexes lead to consistent abundances.

Near the center of the Virgo cluster, the iron abundance is about 0.6 times solar, Si and S are 0.9 times solar (Matsumoto et al. 1996). Surprisingly, the oxygen abundance in Virgo is rather small: 0.4 times solar. This is difficult to understand, since both oxygen and Si are thought to originate in massive stars.

Mushotzky et al. (1996) have studied the abundances in four bright clusters. For A 496, A 1060, A 2199 and AWM 7 they find average abundances of 0.48 (O), 0.62 (Ne), 0.65 (Si), 0.25 (S) and 0.32 times solar for Fe. They argue that this composition (a high Ne/Fe and Si/Fe ratio) is typical for the enrichment by type II supernova explosions. They suggest that, therefore, almost all metals are produced in type II supernovae during the early star formation phase of the galaxies in the cluster. In order to explain the abundances quantitativly, they need a rather flat initial mass function. The mass-weighted average progenitor of type II SNe is 25 $M_\odot$. Each 25 $M_\odot$ SN II produces enough metals to enrich 225 $M_\odot$ of primordial material to solar abundances. Therefore, some 10^{12} type II supernovae are required in order to produce the total metal mass of the cluster.

The above model is subject to several uncertainties, however. First, the abundances of S, Ca and Ar are lower than expected. Furthermore, the amount of iron produced by type II supernovae is a strong function of the initial mass and is subject to uncertainties in the physics of the explosion. Also type I supernovae produce mainly iron. This is elaborated somewhat by Ishimaru & Arimoto (1997). They argue that Mushotzky et al. use solar photospheric abundances for iron ($4.68\,10^{-5}$ by number), while the meteoritic abundance of iron is $3.24\,10^{-5}$, about 30 % smaller. Ishimaru & Arimoto state that in chemical-evolution models often this last meteoritic abundance is used. Correcting Mushotzky's results, they find that a type Ia contribution of at least 50 % is not ruled out.

14.9 Abundances in distant clusters

Also in distant clusters relatively high iron abundances are found. In A 370, a Coma-like cluster with two giant ellipticals in its core, at a redshift of 0.37, Bautz et al. (1994) measured an iron abundance of 0.5. This indicates that the enrichment occurred already early, before $z = 0.37$. Later, Donahue (1996) reported an iron abundance of 0.12 in the $z = 0.54$ cluster MS 0451.6–0305.

Mushotzky & Loewenstein (1997) have systematically studied the *ASCA* spectra of 21 distant clusters (redshift larger than 0.14). The average iron abundance for this sample is 0.29, within the error bars consistent with the average value for low-redshift clusters (0.27). The conclusion is that the iron abundance shows little or no evolution out to $z = 0.3$. Mushotzky & Loewenstein suggest that the lack of evolution can be explained if most cluster enrichment is due to the precursors of elliptical galaxies at $z > 1$. From their analysis of nearby cluster abundances, they conclude that the mass function in these early galaxies is skewed towards high-mass stars, that are able to produce a massive outflow of metal-enriched gas. The global metal production is 4 times greater than is inferred from recent studies of galaxies at high redshift.

14.10 Abundance gradients

Abundance gradients have been reported in a number of clusters. In the nearby Centaurus cluster Fukazawa et al. (1994) report an increase of the iron abundance from 0.3 times solar outside 200 kpc to solar in the core.

This abundance estimate is solely based upon the He-like iron line, hence is dominated by the high-temperature component. Fukazawa et al. explain this by the relatively low gas density of the Centaurus cluster, which makes it easier to enrich the core by winds from the central cD galaxy.

In Virgo (Matsumoto et al. 1996) the abundances of Si, S and Fe all rise towards the center by a factor of 3, with about the same ratio. Surprisingly, the oxygen abundance does not seem to be that strongly concentrated.

Also in the poor cD cluster AWM 7 (Xu et al. 1997) the iron abundance increases from 0.4 times solar in the outer parts to 0.6 times solar in the center. Abundance gradients thus appear to be a common feature in several clusters.

Acknowledgement
I thank Wouter Hartmann and Jacco Vink for critically reading parts of the manuscript.

References

Allen, S.W., Fabian, A.C. & Kneib, J.P. (1996): MNRAS 279, 615
Allen, S.W., Fabian, A.C. & Idesawa, E. et al. (1997): MNRAS 286, 765
Angelini, L., White, N.E. & Nagase, F. (1995): ApJ 449, L41
Antunes, A., Nagase, F. & White, N.E. (1994): ApJ 436, L83
Arnaud, K.A., Mushotzky, R.F., Ezawa, H. et al. (1994): ApJ 436, L67
Arnaud, K.A. (1996): in *Astronomical data analysis software and systems V*, Eds G. Jacoby & J. Barnes, ASP Conf Ser. 101, p. 17
Arnaud, M. & Raymond, J. (1992): ApJ 398, 394
Arnaud, M. & Rothenflug, R. (1985): A&AS 60, 425
Asai, K. Dotani, T., Mitsuda, K. et al. (1996): PASJ 48, 257
Awaki, H. (1992): in *Frontiers of X-ray astronomy*, Eds Y. Tanaka & K. Koyama, Universal Ac. Press, Tokyo, p. 537
Awaki, H., Mushotzky, R., Tsuru, T. et al. (1994): PASJ 46, L65
Awaki, H., Ueno, S., Koyama, K., Tsuru, T. & Iwasawa, K. (1996): PASJ 48, 409
Bautz, M.W., Mushotzky, R., Fabian, A.C. et al. (1994): PASJ 46, L131
Becker, W., Trümper, J. (1998): in *The Many Faces of Neutron Stars*, R. Buccheri, J. van Paradijs, M.A. Alpar (Eds), Kluwer Acd. Publishers, p. 531
Blanton, E.L. & Helfand, D.J. (1996): ApJ 470, 961
Boella, G., Butler, R.C., Perola, G.C. et al. (1997a): A&AS 122, 299
Boella, G., Chiappetti, L., Conti, G. et al. (1997b): A&AS 122, 327
Brandt, W.N., Fabian, A.C., Dotani, T. et al. (1996): MNRAS 283, 1071
Brandt, W.N., Mathur, S. & Elvis, M. (1997): MNRAS 285, L25
Brickhouse, N., Edgar, R., Kaastra, J. et al. (1995): Legacy 6, 4
Brinkmann, W., Kawai, N., Ogasaka, Y. & Siebert, J. (1996): A&A 316, L9
Buote, D.A. & Canizares, C.R. (1996): ApJ 474, 650
Cappi, M., Mihara, T. & Matsuoka, M. (1996a): ApJ 456, 141
Cappi, M., Mihara, T., Matsuoka, M. et al. (1996b): ApJ 458, 149
Cappi, M., Matsuoka, M., Comastri, A. et al. (1997): ApJ 478, 492

Carkner, L., Feigelson, E.D., Koyama, K., Montmerle, T. & Reid, I.N. (1996): ApJ 464, 286
Choi, C.S., Seon, K.I., Dotani, T. & Nagase, F. (1997): ApJ 476, L43
Cohen, D.H. (1996): PASP 108, 1140
Corbet, R.H.D., Smale, A.P., Ozaki, M., Koyama, K. & Iwasawa, K. (1995): ApJ 443, 786
Corbet, R.H.D. & Mihara, T. (1997): ApJ 475, L127
Corcoran, M.F., Waldron, W.L., MacFarlane, J.J. et al. (1994): ApJ 436, L95
Della Ceca, R., Griffiths, R.E., Heckman, T.M. & Mackenty, J.W. (1996): ApJ 469, 662
Donahue, M. (1996): ApJ 468, 79
Done, C., Madejski, G.M. & Smith, D.A. (1996): ApJ 463, L63
Done, C. & Osborne, J.P. (1997): MNRAS 288, 649
Drake, S.A., Singh, K.P., White, N.E. & Simon, T. (1994): ApJ 436, L87
Ebisawa, K., Day, C.S.R., Kallman, T.R. et al. (1996): PASJ 48, 425
Elvis, M., Matsuoka, M., Siemiginowska, A. et al. (1994): ApJ 436, L55
Eracleous, M., Halpern, J.P. & Livio, M. (1996): ApJ 459, 89
Fabian, A.C., Kunieda, H., Inoue, S. et al. (1994a): PASJ 46, L59
Fabian, A.C., Shioya, Y., Iwasawa, K. et al. (1994b): ApJ 436, L51
Fabian, A.C., Arnaud, K.A., Bautz, M.W. & Tawara, Y. (1994c): ApJ 436, L63
Fabian, A.C., Peres, C.B. & White, D.A. (1997): MNRAS 285, L35
Favata, F., Mewe, R., Brickhouse, N.S. et al. (1997a): A&A 324, L37
Favata, F., Mineo, T., Parmar, A.N. & Cusumano, G. (1997b): A&A 324, L41
Favata, F., Vink, J., Parmar, A.N., Kaastra, J.S. & Mineo, T. (1997c): A&A 324, L45
Favata, F., Vink, J., Dal Fiume, D. et al. (1997d): A&A 324, L49
Fleming, T.A. & Tagliaferri, G. (1996): ApJ 472, L101
Fujimoto, R., Tanaka, Y., Inoue, H. et al. (1995): PASJ 47, L31
Fujita, Y., Koyama, K., Tsuru, T. & Matsumoto, H. (1996a): PASJ 48, 191
Fujita, Y., Fukumoto, J. & Okoshi, K. (1996b): ApJ 470, 762
Fukazawa, Y., Ohashi, T., Fabian, A.C. et al. (1994): PASJ 46, L55
Frontera, F., Costa, E., Dal Fiume, D. et al. (1997): A&AS 122, 357
Fujimoto, R. & Ishida, M. (1997): ApJ 474, 774
Fukazawa, Y., Makishima, K., Ebisawa, K. et al. (1994): PASJ 46, L141
Fukazawa, Y., Makishima, K., Matsushita, K. et al. (1996): PASJ 48, 395
George, I.M., Nandra, K., Turner, T.J. & Celotti, A. (1994): ApJ 436, L59
George, I.M., Turner, T.J. & Netzer, H. (1995): ApJ 438, L67
George, I.M. & Turner, T.J. (1996): ApJ 461, 198
Gotthelf, E.V., Jalota, L., Mukai, K. & White, N.E. (1994): ApJ 436, L91
Gotthelf, E.V., Petre, R. & Hwang, U. (1997): ApJ 487, L175
Gotthelf, E.V. & Vasisht, G. (1997): ApJ 486, L133
Greiveldinger, C., Camerini, U., Fry, W. et al. (1996): ApJ 465, L35
Guainazzi, M., Matsuoka, M., Piro, L., Mihara, T. & Yamauchi, M. (1994): ApJ 436, L35
Guainazzi, M., Mihara, T., Otani, C. & Matsuoka, M. (1996): PASJ 48, 781
Guainazzi, M. & Piro, L. (1997): Astron. Nachr. 318, 223
Güdel, M., Guinan, E.F., Mewe, R., Kaastra, J.S. & Skinner, S.L. (1997a): ApJ 479, 416

Güdel, M., Guinan, E.F. & Skinner, S.L. (1997b): ApJ 483, 947
Guo, Z. & Burrows, D.N. (1997): ApJ 480, L51
Harrus, I.M., Hughes, J.P. & Helfand, D.J. (1996): ApJ 464, L 161
Hayashi, I., Koyama, K. & Ozaki, M. (1994): PASJ 46, L121
Hayashi, I., Koyama, K., Awaki, H., Yamauchi, S. & Ueno, S. (1996a): PASJ 48, 219
Hayashi, M.R., Shibata, K. & Matsumoto, R. (1996b): ApJ 468, L37
Hellier, C., Mukai, K. Ishida, M. & Fugimoto, R. (1996): MNRAS 280, 877
Henriksen, M.J. & Markevitch, M.L. (1996): ApJ 466, L79
Hirayama, M., Nagase, F. & Tavani, M. (1996): PASJ 48, 833
Holt, S.S., Gotthelf, E.V., Tsunemi, H. & Negoro, H. (1994): PASJ 46, L151
Holzapfel, W.L., Arnaud, M., Ade, P.A.R. et al. (1997): ApJ 480, 449
Honda, H., Hirayama, M., Watanabe, M. et al. (1996): ApJ 473, L71
Hughes, J.P., Hayashi, I., Helfand, D. et al. (1995): ApJ 444, L81
Hwang, U. & Gotthelf, E.V. (1997): ApJ 475, 665
Hwang, U., Mushotzky, R.F., Loewenstein, M. et al. (1997): ApJ 476, 560
Ikebe, Y., Ezawa, H., Fukazawa, Y. et al. (1996): Nature 379, 427
Ikebe, Y., Makishima, K., Ezawa, H. et al. (1997): ApJ 481, 660
Ishida, M., Mukai, K. & Osborne, J.P. (1994): PASJ 46, L81
Ishida, M., Matsuzaki, K., Fujimoto, R., Mukai, K. & Osborne, J.P. (1997): MNRAS 287, 651
Ishimaru, Y. & Arimoto, N. (1997): PASJ 49, 1
Ishisaki, Y., Makishima, K., Iyomoto, N. et al. (1996): PASJ 48, 237
Iwasawa, K., Yaqoob, T., Awaki, H. & Ogasaka, Y. (1994): PASJ 46, L167
Iwasawa, K., Fabian, A.C., Mushotzky, R.F. et al. (1996a): MNRAS 279, 837
Iwasawa, K., Fabian, A.C., Reynolds, C.S. et al. (1996b): MNRAS 282, 1038
Iwasawa, K., Fabian, A.C., Ueno, S. et al. (1997): MNRAS 285, 683
Iyomoto, N., Makishima, K., Fukazawa, Y. et al. (1996): PASJ 48, 231
Kaastra, J.S. (1992): An X-ray spectral code for optically thin plasmas, Internal SRON-Leiden report, version 2.0.
Kaastra, J.S., Mewe, R., Liedahl, D.A. et al. (1996): A&A 314, 547
Kaastra, J.S., Mewe, R. & Nieuwenhuijzen, H. (1996): in *UV and X-ray Spectroscopy of Astrophysical and Laboratory Plasmas*, Eds K. Yamashita & T. Watanabe, Univ. Acad. Press, p. 411
Kallman, T.R., Mukai, K., Schlegel, E.M. & Paerels, F.B. (1996): ApJ 466, 973
Kaneda, H., Tashiro, M., Ikebe, Y. et al. (1995): ApJ 453, L13
Kaspi, V., Tavani, M., Nagase, F. et al. (1995): ApJ 453, 424
Keohane, J.W., Petre, R., Gotthelf, E.V., Ozaki, M. & Koyama, K. (1997): ApJ 484, 350
Kim, D.W., Fabbiano, G., Matsumoto, H., Koyama, K. & Trinchieri, G. (1996): ApJ 468, 175
Kitamoto, S., Kawashima, K., Negoro, H. et al. (1994): PASJ 46, L105
Kitamoto, S. & Mukai, K. (1996): PASJ 48, 813
Kohmura, Y., Inoue, H., Aoki, T. et al. (1994): PASJ 46, L157
Kotani, T., Kawai, N., Aoki, T. et al. (1994): PASJ 46, L147
Kotani, T., Kawai, N., Matsuoka, M. & Brinkmann, W. (1996): PASJ 48, 619
Koyama, K., Maeda, Y., Tsuru, T. et al. (1994a): PASJ 46, L93
Koyama, K., Maeda, Y., Ozaki, M. et al. (1994b): PASJ 46, L125

Koyama, K., Hamaguchi, K., Ueno, S., Kobayashi, N. & Feigelson, E.D. (1996a): PASJ 48, L87
Koyama, K., Maeda, Y., Sonobe, T. et al. (1996b): PASJ 48, 249
Koyama, K., Kinugasa, K., Matsuzaki, K. et al. (1997): PASJ 49, L7
Kriss, G.A., Krolik, J.H., Otani, C. et al. (1996): ApJ 467, 629
Kubo, H., Makishima, K., Takahashi, T. et al. (1997): MNRAS 287, 328
Kuulkers, E., Parmar, A.N., Owens, A., Oosterbroek, T. & Lammers, U. (1997): A&A 323, L29
Landini, M. & Monsignori Fossi, B.C. (1990): A&AS 82, 229
Laor, A. (1991): ApJ 376, 90
Liedahl, D.A. & Paerels, F. (1996): ApJ 468, L33
Lieu, R., Mittaz, J.P.D., Bowyer, S. et al. (1996): Science 274, 1335
Loewenstein, M., Mushotzky, R.F., Tamura, T. et al. (1994): ApJ 436, L75
Loewenstein, M. & Mushotzky, R.F. (1996): ApJ 471, L83
Macomb, D.J., Akerlof, C.W., Aller, H.D. et al. (1995): ApJ 449, L99
Madejski, G., Takahashi, T., Tashiro, M. et al. (1996): ApJ 459, 156
Makishima, K., Fujimoto, R., Ishisaki, Y. et al. (1994): PASJ 46, L77
Makishima, K., Tashiro, M., Ebisawa, K. et al. (1996): PASJ 48, 171
Manzo, G., Giarrussu, S., Santangelo, A. et al. (1997): A&AS 122, 341
Markevitch, M., Yamashita, K., Furuzawa, A. & Tawara, Y. (1994): ApJ 436, L71
Markevitch, M. (1996): ApJ 465, L1
Markevitch, M. & Vikhlinin, A. (1996): ApJ 474, 84
Markwardt, C.B. & Ögelman, H.B. (1997): ApJ 480, L13
Masai, K. (1984): Ap&SS 98, 367
Masai, K. (1997): A&A 324, 410
Mathur, S., Elvis, M. & Singh, K.P. (1995): ApJ 455, L9
Matsumoto, H., Koyama, K., Awaki, H. et al. (1996): PASJ 48, 201
Matsushita, K., Makishima, K., Awaki, H. et al. (1994): ApJ 436, L41
Matsushita, K. (1997): in *ASCA/ROSAT Workshop on Clusters of Galaxies*, Ed. Ohashi, T., Jap. Soc. for the Promotion of Science, p. 207
Matsuura, M., Miyoshi, S.J., Yamashita, K. et al. (1996): ApJ 466, L79
Matsuzawa, H., Matsuoka, M., Ikebe, Y., Mihara, T. & Yamashita, K. (1996): PASJ 48, 565
Matt, G., Fiore, F., Perola, G.C. et al. (1996): MNRAS 281, L69
Mewe, R., Gronenschild, E.H.B.M. & Van den Oord, G.H.J. (1985): A&AS 62, 197
Mewe, R., Heise, J., Muller, J.M. et al. (1997): IAU Circ. 6551
Mewe, R., Kaastra, J.S. & Liedahl, D.A. (1995b): Legacy 6, 16
Mewe, R., Kaastra, J.S., White, S.M. & Pallavicini, R. (1996): A&A 315, 170
Mewe, R., Kaastra, J.S., Van den Oord, G.H.J., Vink, J. & Tawara, Y. (1997): A&A 320, 147
Mewe, R., Lemen, J.R. & Van den Oord, G.H.J. (1986): A&AS 65, 511
Mewe, R., Kaastra, J.S., Schrijver, C.J., Van den Oord, G.H.J. & Alkemade, F.J.M. (1995a): A&A 296, 477
Mihara, T., Matsuoka, M., Mushotzky, R.F. et al. (1994a): PASJ 46, L137
Mihara, T., Soong, Y., *ASCA* team (1994b): in *New Horizon of X-ray Astronomy*, Eds F. Makino & T. Ohashi, Univ. Ac. Press, Tokyo, p. 419
Misaki, K., Terahima, Y., Kamata, Y. et al. (1996): ApJ 470, L53
Miyata, E., Tsunemi, H., Pisarski, R. & Kissel, S.E. (1994): PASJ 46, L101

Molendi, S., Matt, G., Antonelli, A. et al. (1998): ApJ Lett., accepted for publication
Mukai, K., Ishida, M. & Osborne, J.P. (1994): PASJ 46, L87
Mukai, K., Wood, J.H., Naylor, T., Schlegel, E.M. & Swank, J.H. (1997): ApJ 475, 812
Mushotzky, R.F., Loewenstein, M., Awaki, H. et al. (1994): ApJ 436, L79
Mushotzky, R.F., Fabian, A.C., Iwasawa, K. et al. (1995): MNRAS 272, L9
Mushotzky, R., Loewenstein, M., Arnaud, K.A. et al. (1996): ApJ 466, 686
Mushotzky, R.F. & Loewenstein, M. (1997): ApJ 481, L63
Mushotzky, R.F. & Scharf, C.A. (1997): ApJ 482, L13
Myers, S.T., Baker, J.E., Readhead, A.C.S., Leitch, E.M. & Herbig, T. (1997): ApJ 485, 1
Nagase, F., Zylstra, G., Sonobe, T. et al. (1994): ApJ 436, L1
Nandra, K., Fabian, A.C., Brandt, W.N. et al. (1995): MNRAS 276, 1
Nandra, K., George, I.M., Turner, T.J. & Fukazawa, Y. (1996): ApJ 464, 165
Nandra, K., Mushotzky, R.F., Yaqoob, T., George, I.M. & Turner, T.J. (1997a): MNRAS 284, L7
Nandra, K., George, I.M., Mushotzky, R.F., Turner, T.J. & Yaqoob, T. (1997b): ApJ 476, 70
Nandra, K., George, I.M., Mushotzky, R.F., Turner, T.J. & Yaqoob, T. (1997c): ApJ 477, 602
Netzer, H. (1993): ApJ 411, 594
Nousek, J.A., Baluta, C.J., Corbet, R.H.D. et al. (1994): ApJ 436, L19
Ogasaka, Y., Inoue, H., Brandt, W.N. et al. (1997): PASJ 49, 179
Ohashi, T., Ebisawa, K., Fukazawa, Y. et al. (1996): PASJ 48, 157
Ohta, K., Yamada, T., Nakanishi, K et al. (1996): ApJ 458, L57
Oosterbroek, T., Parmar, A.N., Martin, D.D.E. & Lammers, U. (1997): A&A 327, 215
Orr, A., Molendi, S., Fiore, F. et al. (1997): A&A 324, L77
Otani, C., Kii, T., Reynolds, C.S. et al. (1996): PASJ 48, 211
Owens, A., Oosterbroek, T. & Parmar, A.N. (1997): A&A 324, L9
Parmar, A.N., Martin, D.D.E., Bavdaz, M. et al. (1997a): A&AS 122, 309
Parmar, A.N., Kahabka, P., Hartmann, H.W. et al. (1997b): A&A 323, L33
Petre, R., Okada, K., Mihara, T., Makishima, K. & Colbert, E.J.M. (1994): PASJ 46, L115
Pounds, K.A., Done, C., Osborne, J.P. (1995): MNRAS 277, L5
Pravdo, S., Day, C., Angelini, L. et al. (1995): ApJ 454, 872
Ptak, A., Yaqoob, T., Serlemitsos, P.J., Mushotzky, R. & Otani, C. (1994): ApJ 436, L31
Ptak, A., Yaqoob, T., Serlemitsos, P.J., Kunieda, H. & Terashima, Y. (1996): ApJ 459, 542
Raymond, J.C. & Smith, B.W. (1977): ApJS 35, 419
Reynolds, C.S. Fabian, A.C. & Inoue, H. (1995): MNRAS 276, 1311
Reynolds, C.S. (1997): MNRAS 286, 513
Salvati, M., Bassani, L., Della Ceca, R. et al. (1997): A&A 323, L1
Sambruna, R.M., George, I.M., Madejski, G. et al. (1997): ApJ 483, 774
Saraswat, P., Yoshida, A., Mihara, T. et al. (1996): ApJ 463, 726
Schindler, S., Hattori, M., Neumann, D.M. & Boehringer, H. (1997): A&A 317, 646
Schmitt, J.H.M.M., Drake, J.J. & Stern, R.A. (1996): ApJ 465, L51

Schrijver, C.J., Van den Oord, G.H.J. & Mewe, R. (1994): A&A 289, L23

Serlemitsos, P., Yaqoob, T., Ricker, G. et al. (1994): PASJ 46, L43

Serlemitsos, P.J., Jalota, L., Soong, Y. et al. (1995): PASJ 47, 105

Siarkowski, M., Preś, P., Drake, S.A., White, N.E. & Singh, K.P. (1996): ApJ 473, 470

Siebert, J., Matsuoka, M., Brinkmann, W. et al. (1996): A&A 307, 8

Singh, K.P., Drake, S.A. & White, N.E. (1995): ApJ 445, 840

Singh, K.P., White, N.E. & Drake, S.A. (1996): ApJ 456, 766

Skinner, S.L., Güdel, M., Koyama, K. & Yamauchi, S. (1997): ApJ 486, 886

Skinner, S.L. & Yamauchi, S. (1996): ApJ 471, 987

Slane, P., Seward, F.D., Bandiera, R., Torii, K. & Tsunemi, H. (1997): ApJ 485, 221

Stevens, I.R., Corcoran, M.F., Willis, A.J. et al. (1996): MNRAS 283, 589

Sugizaki, M., Inoue, H., Sonobe, T., Takahashi, T. & Yamamoto, Y (1997): PASJ 49, 59

Supper, R., Hasinger, G., Pietsch, W. et al. (1997): A&A 317, 328

Szkody, P., Long, K.S., Sion, E.M. & Raymond, J.C. (1996): ApJ 469, 834

Tagliaferri, G., Covino, S., Fleming, T.A. et al. (1997): A&A 321, 850

Takahasi, T., Tashiro, M., Madejski, G. et al. (1996): ApJ 470, L89

Tamura, K., Kawai, N., Yoshida, A. & Brinkmann, W. (1996): PASJ 48, L33

Tanaka, Y., Inoue, H. & Holt, S.S. (1994): PASJ 46, L37

Tanaka, Y., Nandra, K., Fabian, A.C. et al. (1995): Nature 375, 659

Tanimori, T., Hayami, Y., Kamei, S. et al. (1998): ApJ 497, L25

Tawara, Y. (1996): in *UV and X-ray Spectroscopy of Astrophysical and Laboratory Plasmas*, Eds K. Yamashita & T. Watanabe, Univ. Acad. Press, p. 145

Tsuboi, Y., Koyama, K., Sakano, M. & Petre, R. (1997): PASJ 49, 85

Turner, T.J., George, I.M., Madejski, G.M., Kitamoto, S. & Suzuki, T. (1995): ApJ 445, 660

Turner, T.J., Netzer, H. & George, I.M. (1996a): ApJ 463, 134

Turner, T.J., George, I.M., Kallman, T., Taqoob, T. & Zycki, P.T. (1996b): ApJ 472, 571

Turner, T.J., George, I.M., Mushotzky, R.F. & Nandra, K. (1997): ApJ 475, 118

Ueno, S., Mushotzky, R.F., Koyama, K. et al. (1994): PASJ 46, L71

Ueno, S., Koyama, K., Awaki, H., Hayashi, I. & Blanco, P.R. (1996): PASJ 48, 389

Urry, C.M., Treves, A., Maraschi, L. et al. (1997): ApJ 486, 799

Vashisht, G., Aoki, T., Dotani, T., Kulkarni, S.R. & Nagase, F. (1996): ApJ 456, L59

Vashisht, G. & Gotthelf, E.V. (1997): ApJ 486, L129

Vashisht, G., Kulkarni, S.R., Anderson, S.B., Hamilton, T.T. & Kawai, N. (1997): ApJ 476, L43

Vink, J., Kaastra, J.S. & Bleeker, J.A.M. (1996): A&A 307, L41

Vink, J., Kaastra, J.S. & Bleeker, J.A.M. (1997): A&A 328, 628

Vrtilek, S.D., Mihara, T., Primini, F.A. et al. (1994): ApJ 436, L9

Warwick, R.S., Smith, D.A., Yaqoob, T. et al. (1996): ApJ 470, 349

Weaver, K.A., Yaqoob, T., Holt, S.S. et al. (1994): ApJ 436, L27

Weaver, K.A., Nousek, J., Yaqoob, T., Hayashida, K. & Murakami, S. (1995): ApJ 451, 147

Weaver, K.A., Nousek, J., Yaqoob, T. et al. (1996): ApJ 458, 160

Wheaton, W.A., Dunklee, A.L., Jacobson, A.S. et al. (1995): ApJ 438, 322
White, N.E., Arnaud, K.A., Day, S.R. et al. (1994): PASJ 46, L97
White, S.M., Pallavicini, R., Lim, J. (1995): in *Flares and Flashes*, Eds R. Gershberg, H. Duerbeck & J. Greiner, Springer, Berlin, p. 168
White, N.E., Angelini, L. & Ebisawa, K. (1996): ApJ 463, L83
Wilson, A.S., Elvis, M., Lawrence, A. & Bland-Hawthorn, J. (1992): ApJ 391, L75
Xu, H., Ezawa, Y., Fukazawa, Y. et al. (1997): PASJ 49, 9
Yamauchi, S., Kawai, N. & Aoki, T. (1994): PASJ 46, L109
Yamauchi, S., Kaneda, H., Koyama, K. et al. (1996): PASJ 48, L15
Yamashita, A., Matsumoto, C., Ishida, M. et al. (1997): ApJ 486, 763
Yaqoob, T., Serlemitsos, P., Mushotzky, R. et al. (1994a): PASJ 46, L49
Yaqoob, T., Serlemitsos, P., Mushotzky, R. et al. (1994b): PASJ 46, L173
Yaqoob, T., Edelson, R., Weaver, K.A. et al. (1995): ApJ 453, L81
Yaqoob, T., Serlemitsos, P.J., Turner, T.J., George, I.M. & Nandra, K. (1996): ApJ 470, L53

Future X-ray Spectroscopy Missions

Frits Paerels

SRON Laboratory for Space Research, Sorbonnelaan 2,
3584CA Utrecht, The Netherlands,

and

Columbia Astrophysics Laboratory, Columbia University,
538 West 120th Street, New York, NY 10027, USA

Abstract. This article provides a description of the high-resolution X-ray spectrometers for astrophysics that will become operational in the near future. The emphasis is on the physical principles of operation.

1 Introduction

The title of this Chapter refers to future missions for astrophysical X-ray spectroscopy. At the time of this writing (Spring 1998), the future is especially close, and we expect it to be bright. Within the next two years, high-resolution X-ray spectrometers will be placed in orbit on *AXAF*, *XMM*, *Spectrum X/γ*, and *Astro-E*. For the first time, we will have the sensitivity to spectroscopically detect a wide variety of diagnostic physical effects, from essentially all types of cosmic X-ray source. This article therefore concentrates on this immediate future, although at the end I will briefly discuss what is under study for the 21st century.

In selecting topics for this chapter, I have tried to emphasize the physics of the instruments, especially of the grating spectrometers on *AXAF* and *XMM*, rather than enumerating their properties and giving examples of simulated spectroscopy for various types of X-ray source (which might suggest that this field is intellectually already pretty much covered). A brief description of some X-ray spectroscopic diagnostics provides an estimate for astrophysically interesting ranges of resolving power, and I largely leave the astrophysical applications to your imagination.

Careful quantitative spectroscopy requires an understanding of the operation of the spectrometers at various levels of detail. A description of the physical principles may help to appreciate the properties of the instruments, the reasons for various design choices, and the physical and practical limits to their resolving power and efficiency (which will be pushed in future generations of instruments). As such, this chapter aims to provide an introduction to what is sometimes referred to as 'the theory of the experiment' in experimental physics.

2 Resolving Powers of Interest in Astrophysical X-ray Spectroscopy

In this section, I will collect spectroscopic diagnostics of interest in the X-ray band, and the spectral resolving power required to use the diagnostic. You have seen a lot of this material before. I have added some more, and I will put it in the definite context of the spectroscopic capabilities of the X-ray spectrometers to be put in orbit in the near future, in the next section.

2.1 Ionization stage spectroscopy

At the lowest level of spectrosopic analysis is the ability to distinguish the emission spectra of the various elements and their ionization stages. The most basic parameters to be measured with this 'diagnostic' are of course the elemental abundances themselves, but an accurate abundance measurement must rely on an accurate measurement of the ionization balance as well. The ionization balance itself is an important diagnostic for the physical conditions in the emitting gas. It contains information on the excitation mechanism (collisional or recombination), the temperature and density, and the thermal history of the gas. You can find a nice set of scaling relations in George Fraser's contribution, one of which gives the relative energy splittings between the strongest transitions in the hydrogenic and helium-like ions as a function of nuclear charge Z, but I will repeat the numbers here for a few astrophysically important elements. Table 1 lists the energies of the principal $n = 2 - 1$ transitions in the hydrogenic and helium-like ions of O, Si, and Fe, which span the $0.8 - 7$ keV band. I also list the resolving power $\mathcal{R} \equiv E/\Delta E$ required to distinguish the H- and He-like transitions (which scales like Z).

Table 1. Energies of the principal $n = 2 - 1$ lines in H- and He-like ions

element	H-like $1s - 2p$	He-like $1s^2 - 1s2p$	$\mathcal{R}$
O	654 eV	574 eV	8
Si	2005	1865	14
Fe	6960	6701	27

2.2 Excitation mechanism

Another very basic diagnostic is the dominant line excitation mechanism. As discussed extensively in Duane Liedahl's contribution, there are very clear spectroscopic differences between the emission spectra arising from collisionally or photoionized media. In the first case, line emission results primarily from radiative decays following collisional excitation, in the second case, radiative decays following recombination dominate. For a gas of cosmic composition in photoionization equilibrium, this leads to enhanced emission from the low- and mid-Z hydrogenic and He-like ions with respect to the Fe L ions, as compared to emission from a collisional plasma of comparable mean ionization.

In addition, for low enough electron temperatures the radiative recombination continua appear as narrow, line-like features at the series limits. To detect these features clearly separated from (at least) Lyβ in hydrogenic ions (to take the simplest case) requires resolving power $\mathcal{R} = 9$.

Another, subtler diagnostic is based solely on the spectroscopy of the Fe L ions, and it therefore also works in cases where the electron temperature is too high to make the narrow recombination continua detectable. As first described by Liedahl et al. (1990), the recombination spectra of the Fe L ions are markedly different from the collisionally excited spectra. This is shown in Fig.1, which shows the spectra of Fe XVI-XIX under conditions of coronal and photoionization equilibrium. The differences between these two different spectra should be evident even at resolving powers of order $\mathcal{R} \gtrsim 10 - 20$.

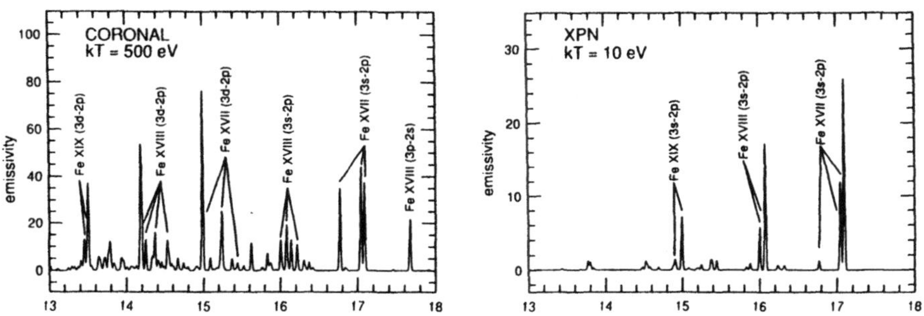

Fig. 1. Line spectra emitted by Fe XVI-XIX under conditions of coronal and photoionization equilibrium. From Liedahl et al. (1990).

2.3 Density diagnostics

He-like ions The intensities in the $n = 2 - 1$ transitions in the helium-like ions are sensitive to the density of the plasma. The physical principle underlying this diagnostic is treated in the Chapters by Rolf Mewe and Duane Liedahl, but I include a brief description for completeness. The standard reference is Gabriel & Jordan (1969). The diagnostic rests on the competition between spontaneous and collisional transition rates out of the various $n = 2$ levels. The first are determined purely by atomic structure, the second scale with the electron density, and a measurement of the excitation balance in the $n = 2$ sublevels must therefore be sensitive to density. Four transitions are used in this diagnostic: $1s^2\,^1S_0$ to $1s2p\,^1P_1, 1s2p\,^3P_{1,2}$ and $1s2s\,^3S_1$. The first is an allowed transition, the second two involve a change in the total spin-angular momentum of the two electrons ('spin-flip'), the fourth is dipole-forbidden. The transitions are therefore commonly referred to as the resonance line ('R'), the 'intercombination' (between the triplet and singlet systems) lines ('I' – the two lines are too closely spaced to be resolved in practical cases), and the forbidden line ('F').

At low densities, the intensities in the lines are purely determined by the collisional excitation rates into the various upper levels: every collisional excitation is followed by a spontaneous radiative decay (in the case of a photoionized plasma, the intensities would be determined by the rate at which the upper levels are populated by recombination and cascades, instead). But at high densities, the collisional rate from the upper level of the F line to the upper level of the I lines becomes comparable to or exceeds the slow spontaneous radiative decay rate out of the upper level of the F line, and the ratio between the I and the F lines changes. The dividing line occurs when the two rates out of $1s2s\,^3S_1$ are equal, and the density for which this happens is referred to as the critical density for the ion. Critical densities span the range 10^9 to 10^{17} cm^{-3} for the helium-like ions from O to Fe.

In order to apply this diagnostic, the I and F lines need to be resolved. Line energies and resolving powers for three important helium-like ions are listed in Table 2.

Table 2. Energies of the $n = 2 - 1$ R, I, and F lines in He-like ions

element	R	I	F	$\mathcal{R}$
O	574 eV	569 eV	561 eV	72
Si	1865	1853	1840	155
Fe	6701	6673	6634	240

Fe L Ions Again, the Fe L ions provide an alternative diagnostic in a photoionized plasma (Liedahl et al. 1992). At sufficiently high density, low-lying metastable excited states of the ions develop a significant population through collisional excitation. These excited ions have recombination spectra that are different from those in the ground state, as illustrated in Fig.2 for Fe XX. To separate the strongest of the many lines that appear with increasing density from the lines present at low density requires $\mathcal{R} \gtrsim 150 - 200$ (for the specific but representative case of Fe XX).

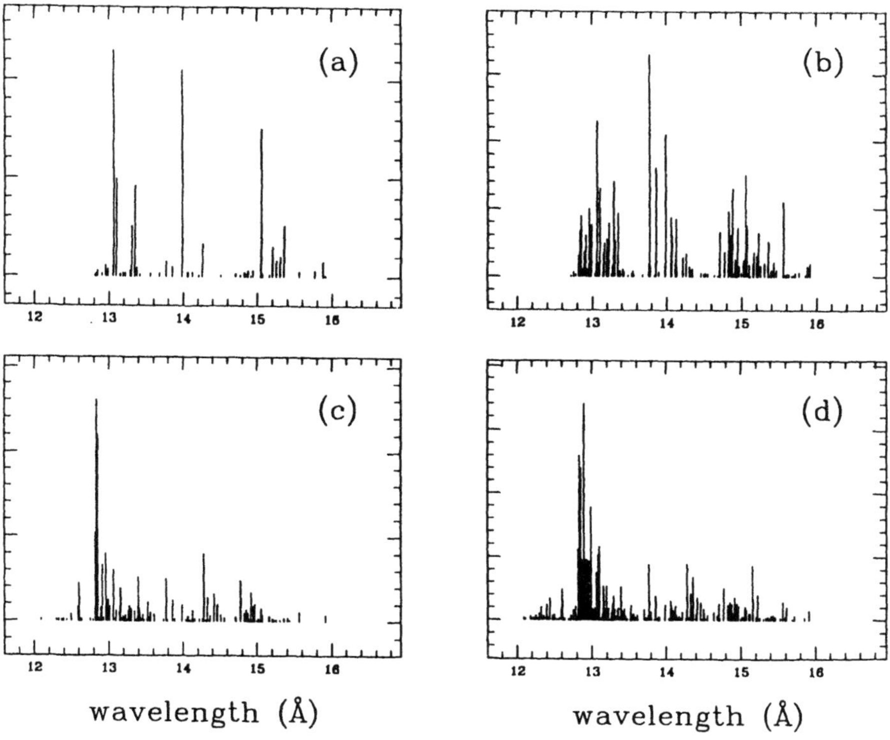

Fig. 2. Line spectra of Fe XX, under conditions of photoionization [(a),(b)] and coronal equilibrium [(c),(d)]. Panel (a) shows the recombination spectrum at low density, panel (b) the spectrum above the critical density of $n \sim 10^{13.5}$ cm^{-3}. From Liedahl et al. (1992).

2.4 Satellite line spectroscopy

A powerful electron temperature diagnostic, proven in terrestrial plasma experiments and solar spectroscopy, is spectroscopy of dielectronic satellite lines (Rolf Mewe, Duane Liedahl, this volume). Dielectronic recombination onto a target ion of (net) charge $z + 1$ leaves an ion in charge state z in a doubly excited state, with one electron, the 'spectator', in a high Rydberg level. This spectator provides some shielding of the nuclear field, and so the stabilizing downward radiative transition involving the other electron has an energy that is slightly lower than the corresponding transition in an ion of charge $z+1$ (with no spectator); there will be a satellite for each possible value of the principal quantum number of the spectator. The ratio of the intensities in the satellite line to that of the 'parent' transition (in charge state $z+1$) is sensitive to the electron temperature, because the satellite intensity is proportional to the dielectronic recombination rate onto charge state $z+1$, while the parent intensity is equal to its collisional excitation rate, and both rates have different temperature dependence. The strength of the method is that it does not depend on knowing the ionization balance: the target charge state for both the parent transition and the dielectronic satellites is charge state $z+1$.

For definiteness, we examine the difference in the energies of the dielectronic satellites in the lithium-like ions of Fe and O, and the resonance transitions in the corresponding helium-like ions. A direct measurement of the satellites to the helium-like Fe resonance line at 6701 eV (1.850 Å) by Beiersdorfer et al. (1992) finds the $n = 3$ satellite at a distance of $\Delta\lambda = 1.82 \times 10^{-3}$ Å towards longer wavelengths, so to unambiguously resolve at least this resonance requires high resolving power, $\mathcal{R} \sim 1000$. For the $n = 3$ satellite to the He-like O resonance line at 574 eV (21.60 Å), Mewe, Gronenschild & van den Oord (1985) list a wavelength difference of 0.034 Å, so a resolving power of $\mathcal{R} = 640$ is required to apply the diagnostic.

2.5 Radiative recombination continuum spectroscopy

As explained in detail in Duane Liedahl's contribution, the characteristic that most distinguishes a plasma in photoionization equilibrium from one in collisional ionization equilibrium is the comparatively low electron temperature. As a consequence, the radiative recombination continuum photons will all be 'piled up' just above the ionization limit, because the free electrons all have kinetic energies much smaller than the ionization energy. The resulting narrow, quasi-discrete recombination continuum is a strong spectroscopic signature of photoionization. It can be used to straightforwardly measure the electron temperature in the zone where the radiative recombination rate onto a given ion species peaks: the emissivity depends on photon energy E as $\exp(-\Delta E/kT_{\rm e})$, with $\Delta E = E - \chi$, χ the ionization potential, $T_{\rm e}$ the electron temperature. The recombination continuum therefore has a characteristic width of order $kT_{\rm e}$, and to resolve it requires roughly $\mathcal{R} \gtrsim \chi/kT_{\rm e}$. As a practical example, you need $\mathcal{R} \sim 140$ to resolve the continuum in hydrogenic Neon, at a (representative) electron temperature of ~ 10 eV. Higher temperatures are easier to resolve, but the contrast at the ionization edge goes down proportionally as well (which is why this diagnostic does not work for the hot plasmas in collisional equilibrium).

2.6 Thermal Doppler broadening

A straightforward ion temperature diagnostic is the thermal Doppler width of emission lines. The radial velocity distribution in a thermal plasma is Gaussian, with a variance[1] $\sigma^2 = kT_i/m_i$, with T_i and m_i the ion temperature and mass, respectively. For the resolving power corresponding to this width, we find $\mathcal{R} = \lambda/\Delta\lambda = c/\sigma = (m_i c^2/kT_i)^{1/2} = 1000 A^{1/2} T_7^{-1/2}$, with A the atomic weight of the ions, T_7 the ion temperature in units 10^7 K. This indicates that this is not a trivial diagnostic—for all ions of interest, $A > 12$, so a high spectral resolving power is required.

2.7 Compton scattering effects

Monochromatic photons of wavelength λ_0 scattering off of stationary electrons suffer an energy loss due to the Compton recoil effect. The maximum corresponding wavelength shift $\Delta\lambda_{\rm max}$ in a single scattering interaction is equal to twice the Compton wavelength of the electron, $\lambda_C = h/m_e c$, so $\Delta\lambda_{\rm max}/\lambda_0 = -2h/m_e c\lambda_0 = -2E_0/m_e c^2$ (for complete backscattering). To resolve the Compton-downscattered photons, you therefore need a resolving power of order $\mathcal{R} \sim 511/E_{\rm keV}$, with $E_{\rm keV}$ the photon energy in keV.

A measurement of the intensity of downscattered photons yields an estimate of the electron column density through which the photons passed. In order for the scattering probability to be non-negligible, you need to have the Thomson depth $\tau_T \sim 1$, or $N_H \sim \sigma_T^{-1} = 1.5 \times 10^{24}$ cm^{-2} (with N_H the total ionized Hydrogen column density, and σ_T the Thomson cross section). At such large column densities, the photoelectric opacity at lower energies is large, so that the usefulness of this spectroscopic diagnostic is probably limited to the Fe K lines.

Photons passing through a large column density τ_T will suffer on average $\overline{n} = \tau_T^2/2$ collisions, so a line at energy E_0 will appear downshifted by $\Delta E/E_0 \sim \overline{n} E_0/m_e c^2$. The line will also appear broadened, due to statistical fluctuations both on the number of scattering interactions, as well as on the energy shift per scattering. After passing through a column equivalent to an average number of $\overline{n}$ scatterings, the r.m.s. width of the line is $\Delta E/E_0 = \overline{n}^{1/2}(7/5)^{1/2} E_0/m_e c^2$ [the fundamental papers on Comptonization in cold gas are Ross et al. (1978) and Illarionov et al. (1979); one factor $\overline{n}^{1/2}$ comes from the variance in the number of scattering interactions, the distribution of energy shifts per scattering supplies the other factor $(\frac{2}{5}\overline{n})^{1/2}$].

If the scattering electrons have a finite temperature, the Doppler shifts due to the velocities of the electrons have to be taken into account. For a Maxwellian velocity distribution of temperature T_e, the characteristic width of the scattered line radiation, expressed as a variance or r.m.s. width, is

[1] not to be confused with the usual, 'official' definition of the frequency 'Doppler width', $\Delta\nu_D \equiv \sqrt{2}\nu_0\sigma/c$ (Rybicki & Lightman 1979, p.288)

$\Delta E/E_0 \approx (2kT_e/m_ec^2)^{1/2}$ (Illarionov et al. 1979; compare this with the r.m.s. width of ionic emission lines given above—the additional factor two arises from two successive Lorentz transformations, into the electron rest frame and back to the observer's frame). The net line *shift* due to scattering from a finite-temperature medium can be both positive and negative, depending on whether the average electron energy is larger or smaller than one-fourth of the line photon energy.

The characteristics of the Compton scattering effects can conveniently be summarized as follows (cf. McCray 1984, and correcting a typo): passing through a column density of free electrons of temperature T_e, equivalent to an average number of scatterings $\overline{n} = \tau_T^2/2$, line photons of energy E_0 will experience a net shift

$$\Delta E_{\text{shift}}/E_0 \approx \overline{n}(4kT_e - E_0)/m_ec^2, \tag{1}$$

and an r.m.s. broadening of the scattered radiation

$$\Delta E_{\text{broadening}}/E_0 \approx \overline{n}^{1/2}(7E_0^2/5 + 2kT_em_ec^2)^{1/2}/m_ec^2, \tag{2}$$

and so the relevant resolving power is still set by $\mathcal{R} \gtrsim 511/E_{\text{keV}}$, except for warm to hot plasmas ($kT_e \gtrsim 10$ eV).

2.8 Raman scattering

If X rays are absorbed in a neutral medium, you may observe, instead of Compton scattering, the effects of scattering by bound electrons, that is, Rayleigh scattering and Raman scattering. Since hydrogen is the most abundant element, you would in practice only expect to see Rayleigh and Raman scattering off hydrogen atoms, although very sensitive experiments might also detect the effects of scattering off helium atoms.

Rayleigh scattering involves excitation of the bound electron into an intermediate state (not a stationary state of the atom), from which it decays back to the initial state. The scattered photon therefore emerges with unchanged energy (coherent scattering), only its direction has been changed. If the photon has an energy larger than the excitation energy of a discrete transition between stationary states in the atom, however, the excited atom may deexcite to the excited level, and the outgoing photon has less energy than the incoming photon, the difference being the excitation energy of the atomic transition (Raman scattering). Since most Raman scattering events will involve a transition $n = 1-2$ in hydrogen (Lyα), the signature of Raman scattering of X-ray lines would be the presence of scattered photons at an energy 10.2 eV (the energy of the Lyα transition) below an emission line.

Again, since the scattering cross section is small (a fraction of the Thomson cross section), only absorbers with very large neutral column densities will have finite Raman scattering optical depth, and these will be photoelectrically opaque at low energies. Raman scattering will therefore probably only be detected in Fe K photons, and it requires a resolving power $\mathcal{R} \sim 6400\text{eV}/10\text{eV} = 640$ to do so.

A careful study of the Raman spectrum (including the angular dependence, and the effects of the distribution of the velocity of the bound scattering electron—on average of order $\alpha c \sim 2000$ km s^{-1} in the ground state of Hydrogen! [α is the fine structure constant]) will reveal detailed information on the geometrical distribution and amount of scattering hydrogen atoms with respect to the emission line source and the observer. Sunyaev & Churazov (1996) give a (characteristically) complete calculation of the effect, and describe an application to the study of the Galactic Center Region. For more general applications of Raman spectroscopy in astrophysics, see Nussbaumer, Schmid & Vogel (1989).

2.9 Fluorescence spectroscopy

From ions with a filled K-shell, and at least a partially filled L-shell, you may get fluorescent X rays. The radiative transitions follow an inner-shell ionization, either collisional or by photoionization, with a probability y, the fluorescence yield, which is a strong function of nuclear charge. For instance, $y_{\text{K}} \approx 0.3$ for Fe Kα, but only a few percent for O Kα. L-shell fluorescence is very inefficient. Despite the low yield, fluorescent emission from low-Z elements may still be detectable, from photoionized plasmas. The fluorescence emissivity j_{fl} is the product of the yield with the photoionization rate: $j_{\text{fl}} = y_{\text{K}} n_i \int_{\chi}^{\infty} dE\, \sigma(E) S(E)$ photons cm^{-3} s^{-1}, with n_i the density of element i, $\sigma(E)$ the photoionization cross section, $S(E)$ the ionizing radiation field (photons cm^{-2} s^{-1} keV^{-1}), and χ the ionization potential. The low yield y_{K} for the light elements is offset by higher abundances (as compared to Fe), by larger photoabsorption cross sections, and by the fact that for a typical AGN ionizing radiation field (for instance) the integral actually favors low-Z elements [$S(E)$ a steeply falling powerlaw in E, and lower χ's for low-Z elements].

Careful spectroscopy of the fluorescent spectrum may provide information on the ionization state of the fluorescing gas, its ionization history, and its physical and chemical state. To date, this has only been attempted at Fe K, although fluorescent emission from lighter elements has now also been detected (e.g., the buried active nucleus in NGC 6552, cf. Fukazawa et al. 1994).

Ionization state Precise determination of the (rest frame) energy of fluorescent emission lines or absorption edges in principle allows the determination of the ionization stage in which the transitions arise. This is no different in principle from the the Ionization Stage Spectroscopy mentioned above, except that in ions more neutral than helium-like, the transition energies vary only slowly with charge state, requiring medium to high spectral resolution, and a very reliable calibration of the spectrometer energy scale to uniquely separate them. Also note that in any astrophysical source, charge states that differ little in ionization potential (like the Fe M shell ions) are likely to coexist, so that the emission spectrum shows the superposition of the individual spectra of the separate charge states. Rather than subtle energy shifts, the major effect may be to broaden the transitions by superposition.

To get some idea of which distinctions become important or measurable at what resolving power, let us look at the Fe K spectrum in particular. The dependence of the energies of the various Kα transitions on ionization stage is fairly steep for the Fe L ions. For instance, it takes $\mathcal{R} \approx 200$ to distinguish the $1s - 2p_{3/2}$ transitions arising in the various L-shell ions (Decaux et al. 1997). Higher resolving power is required to do the same for the more neutral species. The average separation between the $1s - 2p_{3/2}$ transitions arising in the ions Fe X-XVII is approximately $\Delta\lambda = 1.5 \times 10^{-3}$ Å per ionization stage (Decaux et al. 1995), so a resolving power $\mathcal{R} \gtrsim 1300$ is required to distinguish these stages uniquely (the wavelength of all Fe Kα transitions is approximately $\lambda \approx 1.93$ Å). At this resolution, a lot of the intrinsic complexity of the Fe K spectra of individual charge states will also be resolved (e.g. the fine structure split between Kα_1 and Kα_2).

Finally, at high resolving power, other peculiar effects become important. For the ionization stages below Fe XI, accidents of atomic structure make the energies of the Kα transitions actually go slightly below those for neutral Fe, with decreasing ionization stage, before going up again (Decaux et al. 1995). This underlines the importance of having an accurate and stable wavelength scale calibration for the spectrometer, as well as accurate and reliable rest frame energies for the transitions under study, or else the small effect I just mentioned might easily masquerade as a spurious velocity field in the source with an amplitude of several hundred km s^{-1}!

Physical state of the fluorescing material This issue is actually almost completely unexplored in astrophysics: astrophysical X-ray spectrometers never had the required resolving power. The following remarks are therefore somewhat sketchy—time will tell how important these issues really are. Because this field is only now becoming of interest to general astrophysics, accurate calculations and laboratory data that would be of peculiar interest to astrophysics are not always available (this applies to the requirements of pure astrophysical spectroscopy, as well as to laboratory calibration of X-ray astrophysical spectrometers!). The situation is changing with the advent of new experiments expressly designed to meet this need, though.

Precise measurement of the wavelengths of the fluorescent emission lines provides information on the physical state of the fluorescing atoms. In principle, there is a difference in the energy of a given transition depending on whether it occurs in a free atom, or in an atom bound to other atoms, as in a molecule or a solid. There are small shifts and broadenings caused by the interaction of atomic electrons with the other charges in the molecule or solid. Moreover, the shifts and broadenings depend on the chemical constitution as well. These issues are already important for the ground calibration of the grating spectrometers on *AXAF* and *XMM*. These have resolving powers ranging from several hundred to several thousand, at which point the finite (but poorly characterized) *width* of the popular (for calibration purposes) Al and Mg Kα characteristic X-ray lines excited in a solid (approximately 1 eV, at 1.49 and 1.25 keV, respectively) is already resolved. Information on transition *shifts* has been tabulated by Sevier (1979), from which I take the data listed in Table 3.

Table 3. Energy Levels and Line Energies in Oxygen and Iron: Physical and Chemical Shifts

		Oxygen			Iron	
energy	gas	oxide	$\mathcal{R}$	gas	oxide	$\mathcal{R}$
$1s$	545.4 eV	532.0 eV	40	7124 eV	7113 eV	650
$2p_{1/2}$	16.40	7.1		733	721.2	
$2p_{3/2}$	—	—	—	720	708.1	
Kα_1	529.0	524.9	130	6391	6392	6400
Kα_2	—	—	—	6404	6405	6400

First note that in situations where near-neutral gas is viewed against a sufficiently strong background X-ray continuum source, a precise measurement of the energy of absorption edges already provides information on the physical state of the absorbing atoms. At least for the low-Z elements, only modest resolving power is required (e.g. $\mathcal{R} = 40$ to distinguish absorption by monatomic gaseous oxygen, from oxygen bound in oxide). For this technique to work, however, you require a bright 'backlighter', and the column density of the absorbing element should be such that the optical depth at the absorption edge is of order unity, to have good contrast. By contrast, a measurement of the energy of fluorescent line emission may be more widely applicable, but requires higher resolving power, and will only be practicable for low Z elements.

Fluorescence spectroscopy is relevant for all elements for which there is evidence that they are easily bound in dust grains in free space [see for instance Snow & Witt (1996), and references therein]. Just to be able to show that material is indeed locked in dust, and how much of it is locked in dust as opposed to being in the gas phase, is already important information when studying certain environments (e.g. the cold gas near the nuclei of AGN).

2.10 EXAFS spectroscopy

Continuum photoelectric absorption by atoms bound in a solid or in molecules may show the curious EXAFS effect ('Extended X-ray Absorption Fine Structure'), which appears as a slow, wavelike modulation of the atomic photoelectric absorption cross section just above an ionization threshold. This modulation is the result of quantum interference, and the physical mechanism works as follows. A photon of energy E ionizes an atom, and produces a photoelectron of momentum $p = [2m_e(E - \chi)]^{1/2}$, with χ the ionization potential. If E is only slightly larger than χ, the de Broglie wavelength of the photoelectron is large, and may be as large as the lattice period of the absorbing solid (or the interatomic distance, in molecules). The outgoing photoelectron probability wave is scattered by neighboring atoms, and if the de Broglie wavelength matches the lattice period, strong interference may result between the outgoing wave and the scattered waves. If the resulting interference is such that the photoelectron wavefunction amplitude is suppressed near the origin, the photoabsorption transition probability is reduced with respect to that for a free atom, and the opposite for constructive interference.

To estimate the characteristic energy scale involved in the effect, consider the following simplified argument [for a complete quantitative description, see Lee et al. (1981); astrophysical implications are discussed by Martin (1970), Evans (1986), and Woo (1995)]. The photon energies corresponding to successive interference resonances are given roughly by the condition $2ka = 2n\pi$, with $k = p/\hbar$ the photoelectron wavevector, a the lattice period or interatomic distance of the absorbing material, and n a positive integer. This neglects any phase shifts in the photoelectron wave function associated with the Coulomb interaction of the photoelectron with the ion, as well as with the scattering atoms. Using $\hbar^2 k^2/2m_e = E - \chi$, we have for the resonance energies

$$E_n \sim \chi + \frac{h^2}{8m_e a^2} n^2 \qquad (3)$$

so the spacing between the first two peaks in the ionization cross section is, very roughly,

$$\Delta E \sim \frac{h^2}{8m_e a^2} = 38(a/1\ \text{Å})^{-2}\ \text{eV}. \qquad (4)$$

To see the effect at the oxygen K edge, assuming a characteristic lattice spacing $a = 1$ Å, therefore requires resolving power $\mathcal{R} = \chi_{\mathrm{O\,K}}/\Delta E = 14$. Similarly, at the silicon edge, you need $\mathcal{R} = 50$.

The precise shape and amplitude of the modulation is sensitive to the exact structure and composition of the absorbing crystals or molecules, through the phase shifts incurred by Coulomb scattering of the photoelectron waves by the atoms neighboring the absorbing atom. That is, the effect can tell, for instance, whether photoelectrically absorbing Si atoms are all surrounded by other Si atoms, or by O atoms. The exact crystalline properties of interstellar dust is an unsolved problem, and EXAFS spectroscopy is a possible technique to address the issue. Again, just to be able to detect the presence of dust or molecular material, and its abundance relative to material in the gas phase, may in itself provide important information.

Given the fact that photoelectric absorption is the physical effect employed in many types of X-ray detector, it is not surprising that EXAFS is a major concern for instrument calibration. In fact, many detectors contain two elements likely to show the astrophysical effect: silicon and oxygen—and the instrumental effect may mask the astrophysical effect (for examples of the instrumental effects, see Owens et al. 1997).

2.11 Radial-velocity spectroscopy

Radial-velocity spectroscopy is one of the oldest applications of spectroscopy in astronomy, and is obviously also of great interest to X-ray astronomy. With sufficiently high sensitivity, one could for instance detect binary motion (both in binaries containing coronal X-ray sources, as well as in classical X-ray binaries), motion in accretion flows, shock velocities in supernova remnants, bulk relative motion in merging clusters, etc.

Despite the conceptual simplicity of the measurement, accurate radial-velocity measurements are actually not trivial in practice. As a rough criterion, you need resolving power $\mathcal{R} \sim c/v$ to detect a velocity field of amplitude v, and this criterion says that velocity spectroscopy is hard: to detect $v \sim$ few hundred km s^{-1}, characteristic of many astrophysical dynamical situations, would require $\mathcal{R} \gtrsim 1000$.

To beat this limit, the following argument is often invoked: if an emission line has been detected with N photons, one can determine its centroid wavelength with an accuracy of order $\Delta\lambda = \Delta\lambda_{\rm sp}/\sqrt{N}$, with $\Delta\lambda_{\rm sp}$ the wavelength resolution of the spectrometer. Translated into velocity, this gives $v = c\Delta\lambda/\lambda = c/(\mathcal{R}_{\rm sp}\sqrt{N})$, so in principle you can do better in v/c than the spectrometer resolution by a factor $\sqrt{N}$. From the previous discussions, it will be clear that you have to be careful with this argument. For it to work, you need to know the laboratory wavelengths to high precision, and you need an accurate and stable wavelength calibration of the spectrometer.

In Figure 3, I have plotted representative values of the resolving powers we just discussed, as a function of X-ray photon energy. As you can see, the diagnostics are literally all over the diagram. Unfortunately, as we will see in the next section, there is no physical effect that will allow you to construct a spectrometer with a nearly energy-independent (and high) resolving power across the entire wavelength band. No single spectrometer will probably ever cover all of these diagnostics simultaneously, and so either compromises are made, or observatories are designed to carry multiple, complementary instruments.

This was only a very rough overview, which will mainly serve as an aid to appreciate the capabilities of the spectrometers to be put into orbit over the next few years. Several important classes of diagnostic have not been discussed, mainly in order for the scope of the article not to blow up out of all proportion. For instance, nothing has been said about spectroscopy of hot, high-density plasmas (atmospheres of neutron stars and white dwarfs). It is also reasonable to expect that there will be surprises: unsuspected physical effects with clear spectroscopic signatures, waiting to be discovered.

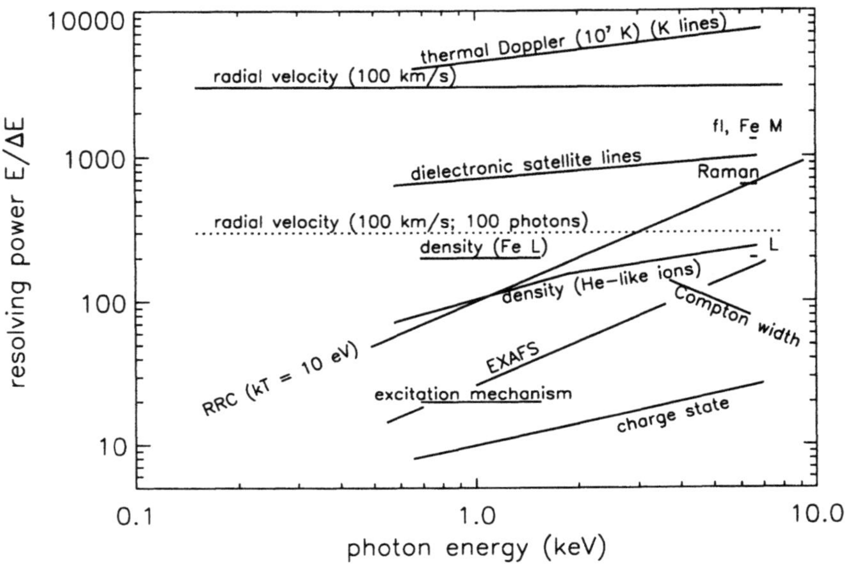

Fig. 3. Resolving power required for some spectroscopic diagnostics, as a function of the photon energy of the diagnostic feature. 'fl, Fe M' indicates the location of the resolving power required to distinguish between the Kα lines in the different Fe M-shell ions, and 'L' indicates the same for the Fe L-shell ions. 'RRC' stands for 'radiative recombination continuum'.

3 X-ray Astrophysical Spectrometers

To set the stage for the discussion of the real instruments on future observatories, we very briefly review the principles of the main types of spectrometer in use in X-ray astronomy.

X-ray spectrometers are usually divided into diffractive and non-diffractive instruments; the terms 'dispersive' and 'non-dispersive' are also used, as well as 'wavelength-dispersive' and 'energy-dispersive', or, less appropriately, 'constant-$\Delta\lambda$' and 'constant-ΔE' spectrometers. The first rely on diffraction of X rays, and comprise grating and crystal spectrometers. The second comprise ionization detectors, calorimeters, and superconducting tunneling junctions, and as yet to be invented other devices that rely on conversion of photon energy into some other measurable quantity.

3.1 Diffractive spectrometers

Grating spectrometers Grating spectrometers are in many ways the simplest of all possible X-ray spectrometers. In a standard application, a grating or set of gratings, either transmission or reflection, is placed behind a focusing telescope. This is usually referred to as an 'objective grating spectrometer', although that term is actually, strictly speaking, reserved for configurations with the dispersing element located *in front* of the telescope. Figure 4 shows a schematic of the arrangement. First, let us look at transmission gratings. From the condition for constructive interference for a grating of period d, illuminated by light of wavelength λ incident at an angle χ, you derive the dispersion equation

$$m\lambda = d(\sin\theta - \sin\chi) \tag{5}$$

where θ is the dispersion angle, and m the spectral order (Fig. 4). If the incident beam has an intrinsic angular spread $\Delta\chi$, the diffracted beam will have a corresponding angular spread, $\Delta\theta = \Delta\chi$, at fixed wavelength and order. That $\Delta\theta$ corresponds to a wavelength width, according to the dispersion equation:

$$\Delta\lambda = \frac{d}{m}\cos\theta\Delta\theta, \tag{6}$$

and, for small θ, this is independent of λ for a fixed telescope angular resolution $\Delta\theta$—hence the name 'constant-$\Delta\lambda$ spectrometer'. The resolving power is $\mathcal{R} = \lambda/\Delta\lambda = \tan\theta/\Delta\theta \approx \theta/\Delta\theta$. You can increase the resolving power by either increasing θ (increasing the line density of the grating, or going to higher order m), or by decreasing $\Delta\theta$ (better telescope). We will return to this issue when we discuss the relative merits of transmission and reflection gratings.

Transmission gratings were flown on the *Einstein* and *EXOSAT* observatories (Brinkman et al. 1980; Seward et al. 1982). Spectra of a variety of cosmic X-ray sources were obtained, of which the results on stellar coronal emission deserve special mention (see Mewe 1991, for a review). Transmission gratings are also at the heart of the high-resolution spectrometers on *AXAF*, to be discussed later.

The corresponding geometry for a reflection grating is shown in Figure 5. Defining α as the angle of incidence with respect to the grating plane, and β as the dispersion angle, you derive the dispersion equation

$$m\lambda = d(\cos\beta - \cos\alpha). \tag{7}$$

For $m = 0$ (zero order) you obviously have $\alpha = \beta$, i.e. the grating acts as a mirror. For $m < 0$, the dispersed ray is inside the triangle defined by

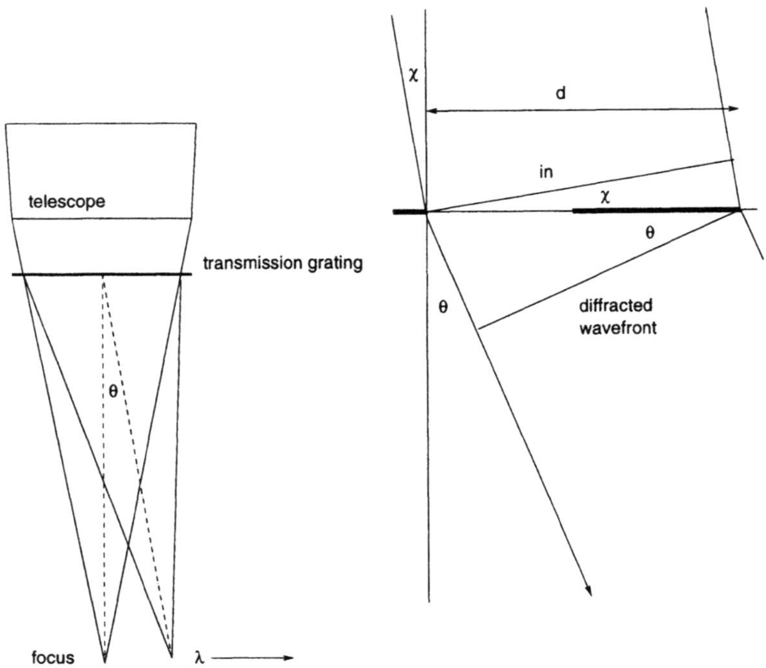

Fig. 4. Schematic arrangement for an X-ray transmission grating spectrometer. A grazing-incidence telescope focuses radiation, which is diffracted by a transmission grating placed in the focused beam (the grating bars are oriented perpendicular to the plane of the page). Focused X-rays are dispersed in the plane of the page, and detected by a position-sensitive detector in the focal plane.

the incoming ray and the zero order ray; these orders are referred to as the 'inside orders'. Outside orders ($m > 0$) don't always exist; for certain ranges of λ and given d, m and α, the dispersion equation may not have a solution ($\cos\beta > 1$).

When illuminated by a beam of finite angular resolution $\Delta\alpha$, the diffracted beams have an angular spread as well, which according to the dispersion relation corresponds to a wavelength resolution

$$\Delta\lambda = \frac{d}{m}\sin\alpha\Delta\alpha. \tag{8}$$

This shows that a reflection grating instrument is also a 'constant-$\Delta\lambda$' spectrometer. The resolving power is

$$\mathcal{R} = \lambda/\Delta\lambda = \frac{\cos\alpha - \cos\beta}{\sin\alpha\Delta\alpha} \tag{9}$$

and this shows that you can increase the resolving power by using a good telescope (small $\Delta\alpha$), or by lowering the angle of incidence. What counts is

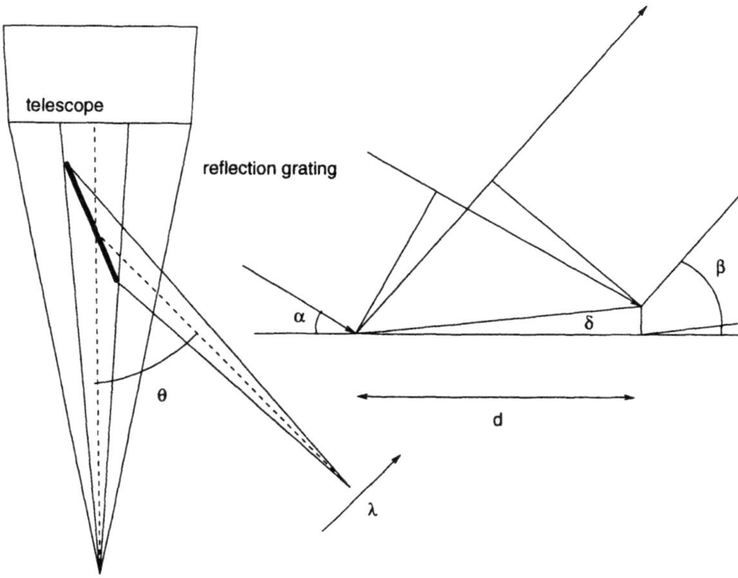

Fig. 5. Diffraction geometry for a reflection grating spectrometer.

the 'projected grating period', $d/\sin\alpha$. At grazing incidence, you can reach fantastically high effective ruling densities this way; for example, at $\alpha =$ 1 degree, and a moderate line density of 500 lines/mm, the effective line density is $\sim 29,000$ lines/mm! Reflection gratings therefore have the potential advantage over transmission gratings of delivering high resolving power at moderate, easy to fabricate ruling density.

But note that there is a price to pay: the grating must be extremely flat. A piece of the grating with the wrong slope, off by $\Delta\alpha$ from the nominal incidence angle α, will produce diffracted light in the wrong direction for a fixed λ, thereby broadening the outgoing diffracted beam, which implies a decrease in resolving power. A change in α, to $\alpha + \Delta\alpha$, produces a different dispersion angle, β', with $\cos\beta' = \cos(\alpha + \Delta\alpha) + m\lambda/d$, but β' refers to the *rotated* plane, not the nominal plane of the grating (Figure 6). To get the change in β with respect to the nominal dispersion direction, you have to add $\Delta\alpha$: $\Delta\beta = \beta' + \Delta\alpha - \beta$. Using the dispersion relation, you find $\Delta\beta = (1 + \sin\alpha/\sin\beta)\Delta\alpha$, and the apparent wavelength shift is $\Delta\lambda = d(\sin\alpha + \sin\beta)\Delta\alpha/m$. To get an idea, use $\lambda = 15$Å, $d = 15,000$ Å (666 lines/mm), $\alpha = 1$ degree. To keep $\Delta\lambda < 0.03$ Å(or resolving power 500), you need $\Delta\alpha \lesssim 6$ arcsec. Similarly, when using an array of gratings to cover the telescope beam, you need to align the gratings with respect to each other to similar precision.

The other obvious disadvantage of having to use very small angles of incidence on the grating is that its area projected to the incoming beam is

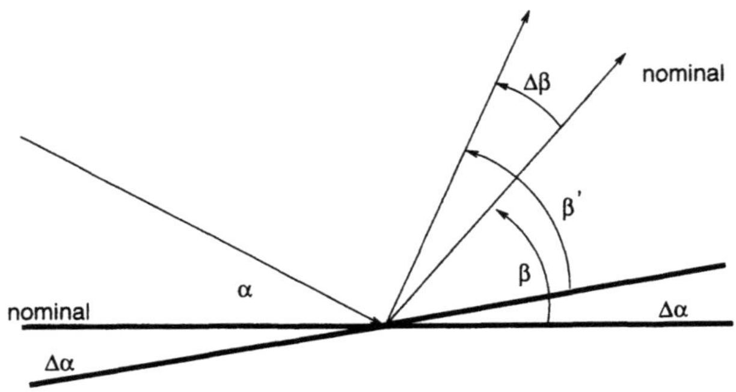

Fig. 6. A small piece of a reflection grating has the wrong orientation, and the incoming beam makes an angle $\alpha + \Delta\alpha$ with the local grating plane. The ray is dispersed to β' according to the dispersion equation, but with respect to the tilted plane. The change in angle with respect to the nominal dispersion angle β is $\Delta\beta = \beta' + \Delta\alpha - \beta$.

very small: you either have low throughput, or else you have to make many gratings to cover the telescope beam.

X-ray reflection gratings will be flown for the first time on *XMM*, and we will describe the Reflection Grating Spectrometer in more detail later. The Extreme Ultraviolet Explorer (*EUVE*) carried three reflection gratings, which provided spectra longward of $\lambda 70$Å (Bowyer & Malina 1996). A reflection grating spectrometer is also currently being considered for NASA's fleet of spectroscopic X-ray observatories *Constellation-X* (most recent information: `http://constellation.gsfc.nasa.gov`), to provide high resolving power in the soft X-ray band.

Crystal spectrometers The most venerable of all X-ray spectrometers is the crystal spectrometer, which was used by Friedrich, Knipping, and von Laue in 1912 to demonstrate the diffraction of X rays by periodic structures, proving conclusively that X rays are electromagnetic waves [for a historical account, read Compton & Allison (1935), pp. 20-38]. The reason this works is the coincidence that typical crystal lattice spacings are of order 1 Å, of the same order of magnitude as X-ray wavelengths.

Referring to Figure 7, it is straightforward to show that constructive interference between rays reflected off two planes spaced by the crystal lattice spacing d will occur if

$$2d \sin\theta = m\lambda, \tag{10}$$

the *Bragg condition*. Note that this is not true dispersion: for a chosen Bragg angle θ, only certain wavelengths are reflected: λ, $\lambda/2$ (in second order), etc. The crystal acts more like a narrow-bandpass interference filter, whereas a grating will simultaneously diffract all wavelengths, so you can record a complete spectrum all at once. To obtain a spectrum in a finite-width band, a crystal spectrometer has to scan through a range of Bragg angles.

The resolution of a crystal spectrometer is

$$m\Delta\lambda = 2d\cos\theta\Delta\theta, \tag{11}$$

with $\Delta\theta$ the combined angular spread in the Bragg angle due to the finite angular resolution of the incoming beam, convolved with misorientations of different pieces of the crystal, and the effects on the diffraction pattern of absorption and dispersion within the crystal. The combined contributions from the crystal are referred to as the 'rocking curve'. If you illuminate the crystal with monochromatic light, and slowly rotate ('rock') the crystal through the Bragg condition, you will see the intensity sharply peak around the Bragg angle. The finite width of the intensity vs. angle curve is partly due to the fact that the crystal is not perfect, and the narrower this 'rocking curve', the higher the resolution of the crystal. With good crystals, very high resolving powers can be achieved ($\mathcal{R} \gtrsim$ several thousand). Crystal spectrometers are also 'constant $\Delta\lambda$' spectrometers, for a small range in θ.

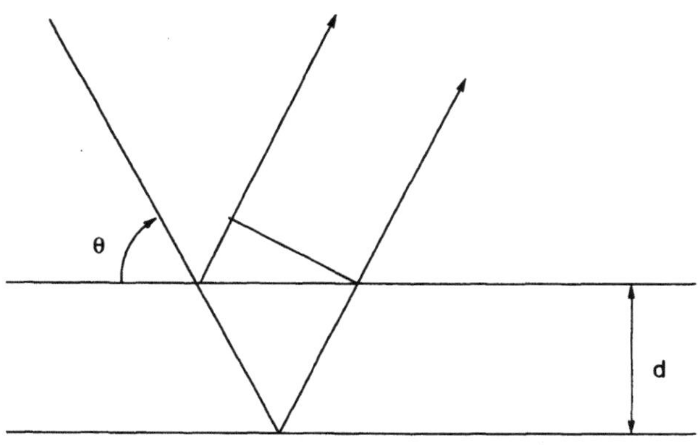

Fig. 7. Geometry for constructive interference between waves diffracted by successive crystal planes, with distance d.

There have been a small number of crystal spectrometers on rockets, and the Focal Plane Crystal Spectrometer (FPCS) on *Einstein* provided the highest resolution X-ray spectra of cosmic sources to date. The first astrophysical detection of an emission line (O VIII Lyα from hot gas in the supernova remnant Puppis A) was in a crystal experiment performed by Zarnecki & Culhane (1978). For a review of the beautiful FPCS results, see Canizares (1990). A crystal spectrometer has been flown on the Shuttle to investigate the emission from the hot ISM in the 44-85 Å range (Diffuse X-ray Spectrometer; Sanders, Edgar & Liedahl 1996). A crystal specrometer of novel design (OXS) will be flown on the *Spectrum X/γ* observatory (see below).

3.2 Non-diffractive spectrometers

Ionization detectors Ionization detectors rely on conversion of the energy of a photon into free electrons, through photoelectric absorption. They include various kinds of proportional counters and solid-state devices. We will briefly recall the properties of CCD detectors only, because the other detectors have too low resolving power to be of interest in the present context. A review of semiconductor detectors can be found in George Fraser's Chapter, and in Fraser (1989).

A photon produces a primary photoelectron, which produces more electrons by collisional processes. There is no amplification in a CCD, so that, very roughly, the number of electrons is

$$N = E/w \tag{12}$$

with E the photon energy, and w the average energy needed to produce one secondary electron (more appropriately, N is the number of electron-hole pairs in the semiconductor). For Silicon, $w = 3.62$ eV. The charge due to a single absorbed photon is collected and measured, and the (apparent) photon energy can be calculated. The energy resolution is set by the statistical fluctuation on N, for a given photon energy. The ionization event also produces other excitations, with their own expectation values and fluctuations. The fluctuation on the number of electron-hole pairs and on the other excitations are correlated, because energy must be conserved in the conversion process, and this implies that the fluctuation on N is smaller than that given by Poisson statistics, $\sigma = N^{1/2}$. This reduction can be described by a 'Fano factor' F, such that the variance in the number of electrons is

$$\sigma^2 = FN \quad (F < 1). \tag{13}$$

For Si, $F \approx 0.1$, and the maximum ('Fano-limited') energy resolution is $\Delta E(\mathrm{FWHM}) = 2.35w(FE/w)^{1/2} \approx 45(E/1\ \mathrm{keV})^{1/2}$ eV (for a Gaussian distribution, FWHM $= 2.35\sigma$). This goes like $E^{1/2}$, so a CCD is an 'almost-constant-ΔE' detector—$E^{1/2}$ varies by a factor ~ 3 between 1 and 10 keV, and other, small sources of noise tend to weaken the dependence of resolution on energy even further.

The first CCD's to be used extensively for X-ray astrophysics are the detectors in the Solid-state Imaging Spectrometers on *ASCA* (Tanaka, Inoue & Holt 1994). Jelle Kaastra's chapter provides a comprehensive overview of results obtained with these instruments.

Superconducting tunneling junctions The fundamental idea behind STJ's is to use a physical process with a very small w, in order to create a large number N of 'countable objects', with correspondingly small statistical fluctuations. Instead of photoionization, the STJ relies on the breaking of Cooper pairs in a superconductor, which have a binding energy of order 10^{-3}

eV, instead of the ~ 1 eV atomic binding energies that are characteristic of photoionization.

The general mood among experimenters, however, seems to be one of pessimism that the early promises of these detectors will be realized in practical high-resolution devices, at least in the near future, for a variety of technical reasons. STJ's still have great potential as superb photon-counting detectors for IR/optical/UV applications (Rando et al. 1996).

Microcalorimeters *Very* loosely speaking, a microcalorimeter counts not photoelectrons or broken Cooper pairs, but phonons—heat. X rays are absorbed by a tiny cooled sensor with very low heat capacity, and the resulting rise in temperature is sensed with a thermometer. The limiting energy resolution corresponds to the fundamental statistical fluctuations in the total energy of the sensor volume, and can be made almost arbitrarily small by reducing the heat capacity. The expectation is that with high-sensitivity thermometers and clever electronic readout schemes, microcalorimeters can be made to deliver $\Delta E \sim 2$ eV (approximately constant), within the next few years.

The microcalorimeter to be flown on *Astro-E* will have $\Delta E \approx 12$ eV. We will come back to this instrument in a later section.

One astrophysical spectrum has already been obtained with a microcalorimeter. A detector flown on a rocket by the Wisconsin/NASA Goddard collaboration has recently detected (expected) emission lines in the spectrum of the hot ($\sim 10^6$ K) phase of the interstellar medium (Deiker et al. 1997). Microcalorimeters are also planned for NASA's future *Constellation-X* observatories, and for the European observatory *XEUS* (Turner et al. 1996).

3.3 Comparison with astrophysically significant resolving powers

Let us choose some representative parameters for each type of spectrometer, and overlay the resulting resolving power on the resolving powers required for the diagnostics discussed in the previous section. This is a crude procedure, in the sense that sensitivity to a particular effect is not just a function of resolving power, but also of signal-to-noise, crowdedness of the spectral region of interest (confusion), etc., but it will at least allow us to establish a broad overview.

For a Fano-limited CCD we had $\Delta E \approx 45(E/1\ \mathrm{keV})^{1/2}$ eV, so $\mathcal{R} \approx 22(E/1\ \mathrm{keV})^{1/2}$ (see Figure 8). The microcalorimeter on *Astro-E* has $\Delta E \approx 12$ eV, so $\mathcal{R} \approx 83(E/1\ \mathrm{keV})$. The 'microcalorimeter of the future' (*Constellation X* and *XEUS*) may have $\Delta E \approx 2$ eV, so $\mathcal{R} \approx 500(E/1\ \mathrm{keV})$. With a telescope response of $\Delta\theta \approx 1$ arcsec, and a line density of 5000 lines/mm, a transmission grating spectrometer has $\Delta\lambda = d\Delta\theta \approx 0.01$ Å, so $\mathcal{R} = 1240(E/1\ \mathrm{keV})^{-1}$. With a line density of 1000 lines/mm, the resolving power is $\mathcal{R} = 250(E/1\ \mathrm{keV})^{-1}$. These numbers are roughly representative of the two grating spectrometers on *AXAF*. We will discuss the performance of these instruments, and of the Reflection Grating Spectrometer on *XMM* in more detail later. Finally, a crystal spectrometer could reach $\mathcal{R} \sim$ several thousand, but given the fact that each crystal can only scan a fairly narrow band (the arrangements on a satellite don't allow large, variable ranges of Bragg angles), an extensive array of crystals with different *d*-spacings would be needed to cover a wide energy band. In practice, crystals are therefore targeted for a certain narrow band of interest.

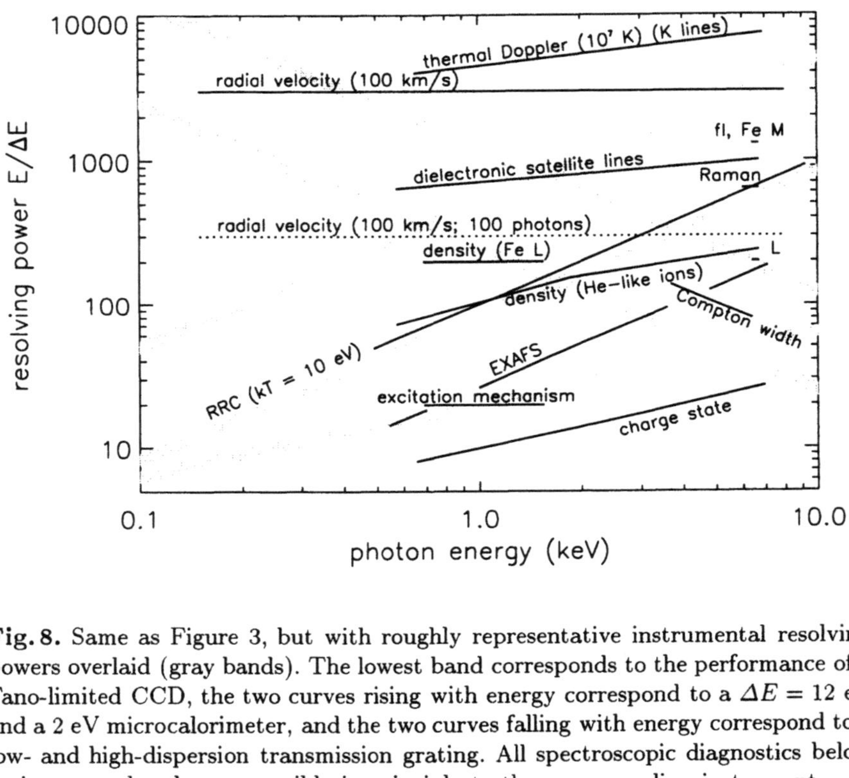

Fig. 8. Same as Figure 3, but with roughly representative instrumental resolving powers overlaid (gray bands). The lowest band corresponds to the performance of a Fano-limited CCD, the two curves rising with energy correspond to a $\Delta E = 12$ eV and a 2 eV microcalorimeter, and the two curves falling with energy correspond to a low- and high-dispersion transmission grating. All spectroscopic diagnostics below a given gray band are accessible in principle to the corresponding instrument.

As you can see, there is not a single type of instrument that covers the entire 0.1 – 10 keV band with uniformly high resolving power. Diffractive and non-diffractive spectrometers 'cross over' in $\mathcal{R}$ in the range around 1 keV, and the crossover is likely to remain around those energies to within a factor $\sim$ 2, for a variety of practical reasons. On a fundamental level, the competition is between microcalorimeters and diffractive spectrometers, because the concept underlying either type of instrument does not contain a fundamental energy scale (like the ionization energy of the absorbing material). Decisions on what type of instrument to use will therefore be based on practical considerations (need for large numbers of diffracting elements vs. need for cryogenic equipment, etc.). But note that microcalorimeters, by their nature, in principle also allow for imaging spectrometers, a concept that does not come natural to diffractive spectrometers.

As a final remark, note that for any finite redshift z of the sources of interest, the loci of the spectroscopic diagnostics in Fig. 8 shift to the left by a factor $1 + z$, and at significant redshifts $z \gtrsim 1$, this would emphasize the low-energy performance of any given spectrometer as its crucial characteristic. The result would be a complicated set of tradeoffs. At lower energies, it is easier to manufacture high resolving power diffractive spectrometers. On the other hand, some of the most interesting objects to study at such redshifts are clusters of galaxies, whose finite angular extent would complicate the problem (requires either large dispersion angles with diffractive spectrometers to preserve spectral resolution, or imaging *arrays* of non-diffractive spectrometers, coupled with good low-energy response).

3.4 The Rowland circle

In this section, I want to briefly discuss the concept of the Rowland circle. It is integral to the design of modern diffractive spectrometers (though not fundamental to the concept underlying their operation), so we will save time when we discuss the grating spectrometers on *AXAF* and *XMM*. The reason I am putting it in here at all is that you will not usually find the idea explained in modern texts [but see Michette (1986) for an algebraic derivation], and it is tricky enough to cause confusion.

Nineteenth-century spectroscopists were faced with the problem that you can not get the full resolving power from a flat grating (usually a reflection grating in this case) if it is not illuminated with collimated light (plane parallel waves). In order to collimate the light from a source at a finite distance away, you either need lenses or mirrors, or a system of narrow slits. The first cause large losses of light, especially at short wavelengths, while the latter are also an obvious waste of precious light in the instrument. After being diffracted by the grating, the beam has to be refocused onto a detector, again at the expense of light. The problem is how to design a spectrometer that will work with a diverging beam of light. The idea is to use a curved grating that automatically corrects for the aberrations associated with the diverging beam, and also 'refocuses' the dispersed light onto the detector. Of the various possible solutions proposed, the one devised by Rowland is the best known.

The idea is the following (see Figure 9). A source at S illuminates a grating, placed at a distance L. A ray of wavelength λ strikes the grating at A and is dispersed into an angle θ. Place a detector at F, on a circle of radius $R = L/2$ (the Rowland circle). You can show that a different ray of the same wavelength and spectral order, which strikes the grating at a different place (A') will intersect with the first ray *almost* at F, provided (1) the grating has a radius of curvature $L = 2R$, (2) the grating period is constant along a plane tangent to the grating at the apex A. The residual aberrations are of order l^2/R^2 (l the length of the grating), *and are to first order independent of the wavelength*. The proof of these statements is messy; you can find it, for instance, in the massive classical review of grating spectrometers by Stroke (1967). You can appreciate that the aberrations are small as long as l/R is small, if you mentally tilt triangle SAF around S. Point A describes a circle of radius $2R$ centered on S, which should be the grating surface, and triangle $SA'F$ is almost congruent with triangle SAF as long as the tilt angle remains small.

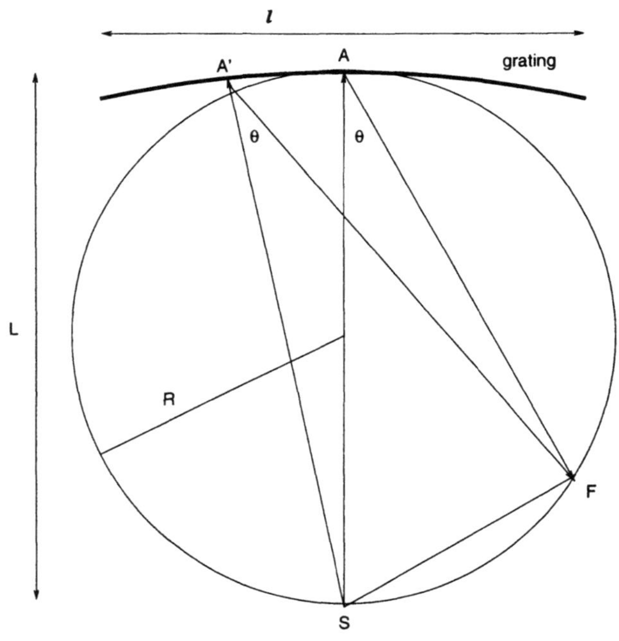

Fig. 9. Schematic of the Rowland circle geometry. A source is placed at S, rays of arbitrary wavelength strike the grating at A and A', and come to a common focus at F, if the grating has radius of curvature $L = 2R$, and the source, grating, and focus are placed on a circle of radius R.

You could reduce or eliminate the residual aberrations further, by choosing a clever line space variation along the grating face, or a more complicated geometry than the circles in Figure 9. But such improvements will turn out to only work for one particular wavelength and spectral order. The importance of the compromise Rowland circle geometry is that it is independent of wavelength, so that a focused spectrum at *all* wavelengths is obtained simultanously. The focus is not perfect, but since the aberrations scale like l^2/R^2, you can limit the resulting blurring by scaling up R for given l (at the obvious expense of loss of light—the larger R, the smaller the fraction of light from S that will strike the grating).

It is now clear how to incorporate these ideas into the design of X-ray spectrometers. Figure 10 gives the equivalent Rowland geometries for both transmission and reflection grating spectrometers, with the gratings placed behind a grazing incidence focusing X-ray telescope —by reversing the direction of the rays SA and SA' in Fig. 9.

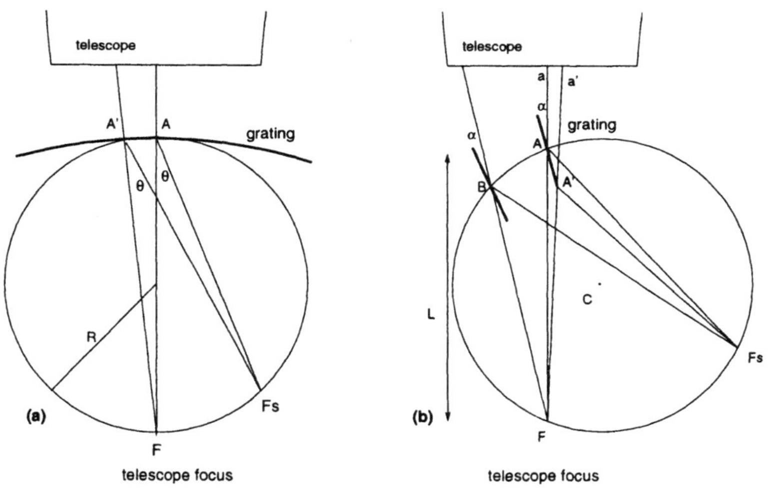

Fig. 10. Relation of the Rowland circle geometries for X-ray transmission grating spectrometers *(a)* and reflection grating spectrometers *(b)*, to the classical Rowland geometry of Figure 9.

Figure 10(a) displays the transmission grating case. Rays from the telescope converge on the telescope focus F (the equivalent of the source in the classical Rowland geometry), the grating bars are perpendicular to the plane of the page, and dispersion is in the plane of the page. Again, the grating and the focus are placed on a circle of radius R, and the grating has radius of curvature $2R$; the optimum spectroscopic focus F_s appears on the circle, for arbitrary wavelength and spectral order.

Figure 10(b) shows the equivalent geometry for the reflection grating case. The grating grooves are perpendicular to the plane of the page. Ray a strikes the grating a A, and is dispersed towards the spectroscopic focus F_s. Ray a' strikes at A', and, just like in the classical geometry, the grating has to be curved in order to make both rays converge on a common spectroscopic focus. Instead of using a physically curved grating, it is easier to introduce a variation in the grating period along its face, which has the same effect (Hettrick & Bowyer 1983). The telescope and spectroscopic foci, F and F_s, and the center of the grating, A, are again placed on a Rowland circle, but the radius of the circle now depends on the chosen line density gradient—the smaller the line density gradient, the further away F_s will be from A, and the larger the radius of the Rowland circle. The optimum-focus spectrum again appears along the circle. In order to cover the focused telescope beam, other gratings have to be placed with their centers on the same circle, like the grating at B, at the same angle of incidence α for the focused ray passing through their centers. In principle, each grating should have its own unique line density gradient (because each grating has its own unique distance to F_s), but in practice the aberrations resulting from having identical gratings are small (as long as typical distances AB are small compared to AF_s), and are typically less important than the effect of the finite angular resolution of the telescope, and the finite accuracy of alignment of the gratings with respect to each other.

Finally, note that all Rowland circle grating spectrometers are intrinsically astigmatic (i.e. rays from a monochromatic point source do not pass through a single point in the focal plane, but instead have different foci in the dispersion, and in the cross-dispersion directions). The telescope has a finite extent out of the plane of the paper, and the telescope beam is 'filled' with grating elements by 'rocking' the Rowland circle back and forth around an axis passing through the telescope focus, lying in the plane of the paper, perpendicular to the optical axis. According to Fermat's principle, the telescope focuses on a cirle centered at A, of radius equal to the length of AF. Focused rays traveling in different planes rotated around F out of the plane of the paper will therefore converge in the *cross-dispersion* direction on that circle. As you can see from Fig. 10(a), that circle is always outside the Rowland circle for a transmission grating spectrometer (the two circles intersect at the telescope focus), which implies that light will focus in the dispersion direction on the Rowland circle, but is defocused in the cross-dispersion direction. The monochromatic image of a point source appears at F_s as a 'stripe' perpendicular to the dispersion direction. Similarly, a circle of radius $|AF|$ centered on A will not coincide with the Rowland circle for the reflection grating case either [Fig 10(b)], so the reflection grating spectrometer is also astigmatic.

This astigmatism does not affect the resolving power of the spectrometer, but the actual two-dimensional size of the image does determine the detector and diffuse sky background level in a resolution element, so you want to keep the astigmatism to a minimum, if possible.

4 The High Resolution X-ray Spectrometers on *AXAF*

4.1 Introduction

NASA's *Advanced X-ray Astrophysics Facility* (*AXAF*) is currently scheduled for launch on December 3, 1998. It is built around a high-resolution grazing-incidence telescope. The overal angular resolution of the telescope, across the X-ray band up to 10 keV ($\lesssim$ 1 arcsec FWHM, driven by the desire to explicitly resolve the point source contribution to the 2-10 keV diffuse X-ray background), implies superb high-resolution imaging and, with transmission gratings placed in the focused X-ray beam, high-resolution spectroscopy. In the following, we will briefly look at the various instruments on *AXAF*, and then discuss the grating spectrometers in detail.

Scientific operation of the observatory is the responsibility of the *AXAF* Science Center in Cambridge, MA (**http://asc.harvard.edu**). A good source of information on the observatory and instruments is the *AXAF Observatory Guide* (**http://asc.harvard.edu/USG/docs/docs.html**). An older standard reference is Weisskopf et al. (1987).

The heart of the observatory is the High Resolution Mirror Assembly (HRMA; Telescope Scientist: Leon van Speybroeck, Smithsonian Astrophysical Observatory). It consists of 4 pairs of iridium-coated paraboloid- hyperboloid shells in a Wolter I configuration. The focal length of the telescope is 10.066 m. The figure and the surface smoothness of the mirror shells (glass) were very accurately controlled to ensure the very high angular resolution and high-quality focus.

There are four separate detectors in the focal plane, two designed for imaging observations, and two designed for reading out the spectra obtained with the gratings. One imaging and one spectroscopic detector consist of CCD's [the *AXAF* CCD Imaging Spectrometer detectors, ACIS-I (imaging), and ACIS-S (Spectroscopy), Instrument PI Gordon Garmire, Pennsylvania State University], and the other imaging and spectroscopic detector are microchannel plate detectors [the High Resolution Camera detectors HRC-I (imaging) and HRC-S (spectroscopy), Instrument PI Stephen Murray (SAO)]. All four instruments can be moved into the telescope focus by means of a mechanism for lateral motion. In addition, the detectors can be moved along the optical axis of the telescope to optimize the focusing (the focal depth of the telescope is only 200 micron! and a small change in the length of the 10 m telescope tube, if uncorrected, would easily defocus the system). A schematic of the focal plane is shown in Figure 11. The full field of view of ACIS-I is 16.9×16.9 arcmin, of HRC-I is 31×31 arcmin.

The ACIS-I and HRC-I cameras were designed with different goals in mind. ACIS obviously offers spatially resolved CCD spectroscopy, whereas the HRC-I offers higher spatial resolution (ACIS slightly undersamples the telescope response), large field of view, high time resolution, and an extended

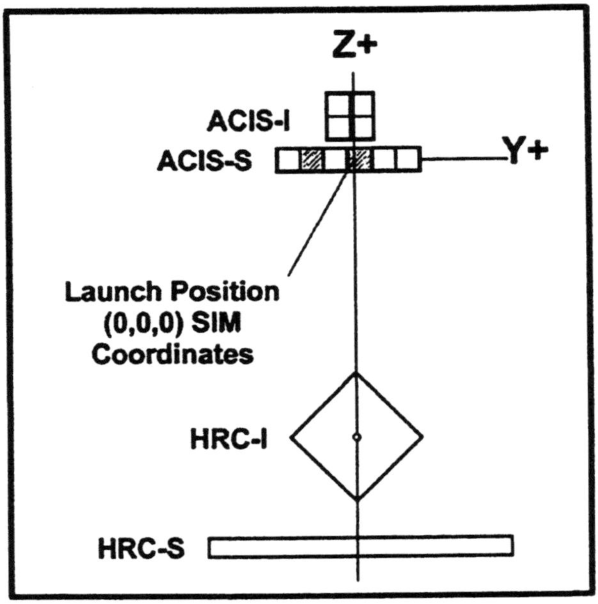

Fig. 11. Schematic of the *AXAF* focal plane, looking down on the focal plane from the position of the mirrors. The SIM (Scientific Instrument Module) can be moved along the vertical line; the focus can be moved in the perpendicular direction by offset-pointing the entire telescope. The dispersion direction for the two gratings is the $\pm Y$ direction (source: *AXAF Observatory Guide*, Ch. 1).

sensitivity to soft photons (down to $\sim$ 100 eV), but no intrinsic energy resolution.

The ACIS-S camera was specifically designed to read out the spectra obtained with the High Energy Transmission Grating (HETG) in the beam, while the HRC-S was designed specifically for use with the Low Energy Transmission Grating (LETG), although either detector can be used with either grating for specific non-standard applications. For extensive information on the detectors and their predicted performance in combination with the HRMA, I refer you to the *AXAF Observatory Guide*.

Either one of two sets of transmission gratings can be rotated into the focused X-ray beam, behind the HRMA, and together with the HRMA and the focal-plane detectors, these make up the High Energy Transmission Grating Spectrometer [HETGS, Instrument PI Claude Canizares (MIT)], and the Low Energy Transmission Grating Spectrometer [LETGS, Instrument PI Albert Brinkman (SRON/Utrecht)]. These will be the first astrophysical *truly* high-resolution X-ray spectrometers, and from even just a cursory glance at Figure 8, you can see what a tremendous increase in sensitivity to diagnostic physical effects they represent. Both grating spectrometers take advantage of the very high angular resolution of the mirrors to attain high spectral resolution, even at moderate dispersion angles.

In the following we will discuss both spectrometers, work out the efficiency of a transmission grating and how it can be optimized for a chosen range of photon energies, and calculate the effect of random fluctuations in the grating properties on the performance.

4.2 The High Energy Transmission Grating Spectrometer

The HETGS (Canizares et al. 1987; Markert et al. 1994; and **http://space.mit.edu/HETG**) consists of two different sets of gratings, with different periods. These form the High and the Medium Energy Transmission Gratings (HETG and METG). Due to the finite size of the ACIS-S detector, the HETG's first-order bandpass is limited to photon energies $E > 800$ eV. The METG has lower dispersion and complements the HETG down to 400 eV. The gratings actually consist of small rectangular flat elements arranged on four separate annuli, covering the four hollow-cone shaped focused beams emerging from the HRMA's four mirror shells. The HETG gratings are arranged behind the inner two shells, the METG gratings behind the outer two shells, because the inner shells have higher throughput at the highest photon energies (smaller diameter, hence smaller graze angle [for fixed focal length], which implies high reflectivity up to higher photon energies). This arrangement is shown in Figure 12.

The gratings are mounted in a Rowland configuration slightly different from the one shown in Figure 10(a): all elements are placed on the Rowland circle itself (diameter 8633.69 mm), perpendicular to the focused rays, instead of on a surface of radius equal to twice the Rowland circle radius. As long as the diameter of the Rowland circle is large compared to the diameter of the grating (approximately 1000 mm), and the grating elements are kept small, the aberrations arising from the approximation of the toroidal surface by finite, flat elements of constant period are small (Beuermann, Bräuniger & Trümper 1978).

The gratings themselves are made of gold, have rectangular bars, and are supported on thin polyimide films. The period is 2000.81 Å (HETG) and 4001.41 Å (METG). The bar widths are 1200 Å (HETG) and 2080 Å (METG).

At high photon energies, the photoelectric absorption coefficient of gold drops approximately as $E^{-3.5}$, and eventually any grating becomes transparent to the radiation, and doesn't diffract at high energies. The optical depth at wavelength $\lambda \equiv 2\pi/k$ through a gold bar of thickness z_0 is $\tau = k\gamma z_0 \approx (z_0/5000\ \text{Å})(E/3\ \text{keV})^{-2.5}$ for $E \gtrsim 2$ keV. Here, γ is the imaginary part of the complex index of refraction, $n \equiv 1 - \delta + i\gamma$ [tables of optical constants as a function of photon energy for all elements can be found at **http://xray.uu.se/hypertext/henke.html**, which contains the tables

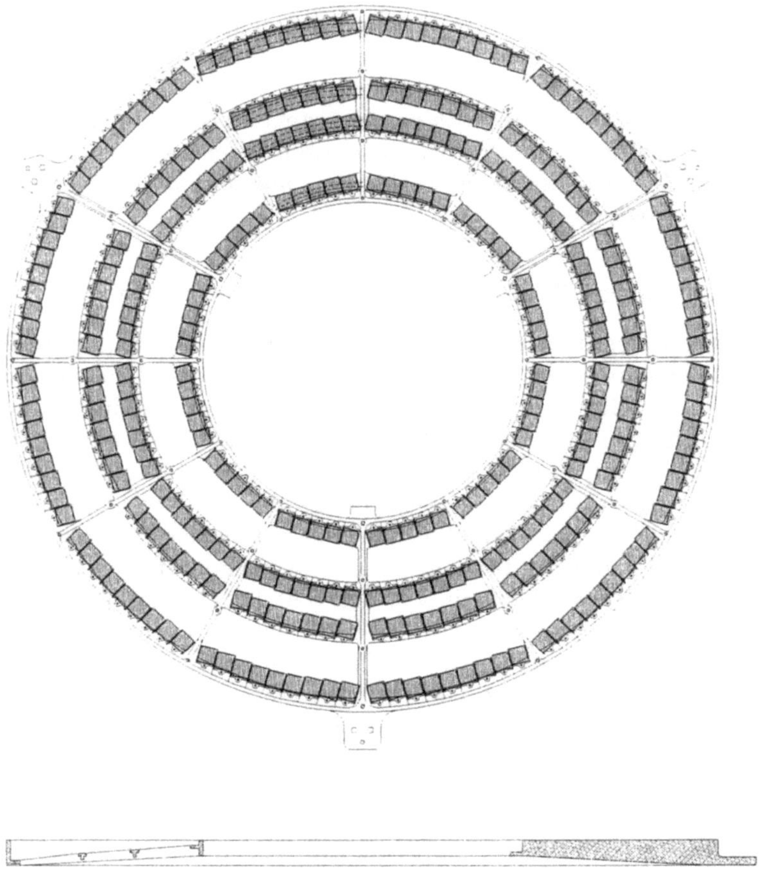

Fig. 12. The High Energy Transmission Grating assembly. The four annuli cover the beams from the four mirror shells; the high dispersion gratings are on the inner two annuli, the low dispersion gratings on the outer two shells (source: *AXAF Observatory Guide*, Ch. 7).

published by Henke, Gullikson & Davis (1993) in digital form]. The HETG has a bar thickness $z_0 = 5100$ Å, so it starts to become optically thin above 3 keV. Fortunately, the real part of the index is still non-zero in this regime, so in addition to small attenuation, the radiation suffers a significant phase shift on passing through the bars. Interference between this phase-shifted part of the wavefronts with those parts that passed through the slits is considerable, leading to significant diffraction efficiency. In this regime, the grating operates effectively as a phased array. The phase shift through the bars depends on wavelength as $\Delta\phi = kz_0 \cdot Re(n-1) = kz_0\delta$, so by optimizing the bar thickness you can optimize the diffraction efficiency for a chosen energy band, as we will explicitly calculate in the next section.

The requirements of high dispersion (small period), and significant phase shift at energies up to 10 keV, leads to the extreme aspect ratio of $\sim 4/1$ of the HETG grating bars. The manufacture of these novel gratings by a photolithographic process is described by Schattenburg et al. (1994). Laboratory efficiency measurements (which cleverly use the interference properties of the large aspect ratio of the grating bars, as a function of angle of incidence) are described by Nelson et al. (1994). Preliminary results on the ground calibration of the HETGS are given by Dewey et al. (1997) and Marshall et al. (1997).

As discussed above, the HETGS was designed to take advantage of the very high angular resolution of the HRMA, so its resolving power in first order is very nearly given by the expression we used to produce Figure 8. The actual resolving power, as predicted from the ground calibrations, is shown in Figure 13. At large dispersion angles, the effect of small grating-to-grating variations Δd in grating period slightly widen the resolution. From the dispersion relation, we have $\Delta\lambda/\lambda = \Delta d/d$, so this effect grows linearly with wavelength. At short wavelengths (small dispersion angles), on the other hand, slight errors in the telescope attitude reconstruction and focusing contribute to the width of the spectral image.

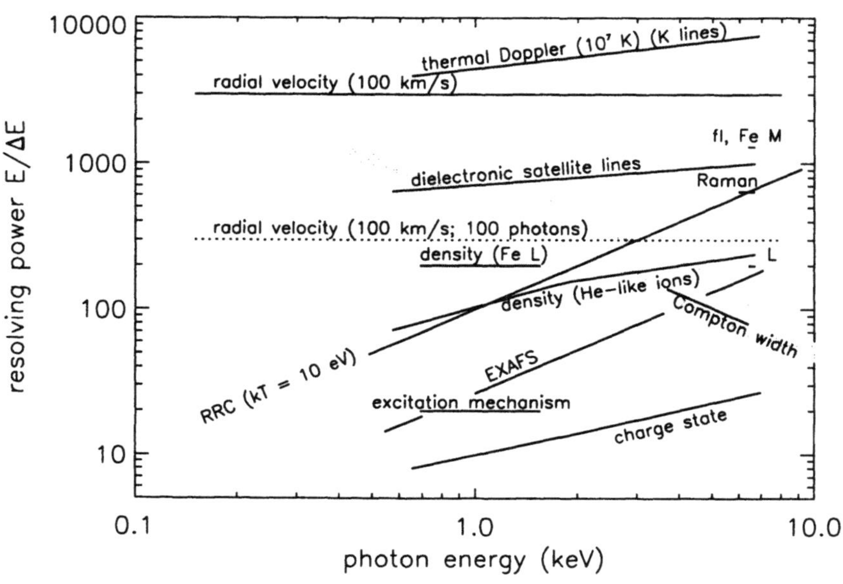

Fig. 13. Same as Figure 8, but with the predicted resolving power for the *AXAF* HETGS overlaid (gray bands). The upper gray band corresponds to the HETG, the lower band to the METG (source: *AXAF Observatory Guide*, Ch. 7).

The spectrum can be read out by either the ACIS-S or HRC-S detector, but the HETGS was designed for use with ACIS-S. The six individual chips that make up ACIS-S are 'folded around' the Rowland circle. Four chips are conventional 'front-illuminated' (FI) devices, two are back-illuminated (BI), which gives them higher quantum efficiency at low energies (but poorer energy resolution). One BI chip was placed at the on-axis position of the zero order, one in the middle of the spectral range on one side. The dispersion directions of the METG and HETG are offset from each other by a small angle, so that the two spectra (high dispersion and low dispersion) form a shallow 'X' on the detector.

From the dispersion equation you conclude that different spectral orders will overlap in the spectral image: the first-order position for wavelength λ coincides with the mth order for wavelengths λ/m. These higher orders 'contaminate' the first order, but they also contain interesting information, because they have resolving powers m times higher than first order (the higher orders are faint though). Separation, or at least unique identification, of the higher orders is therefore desirable, and for this, the intrinsic energy resolution of ACIS-S can be used. The CCD resolution is good enough to uniquely assign events in the focal plane to the correct spectral order to which they belong. This is illustrated in Figure 14, which shows a ground calibration spectrum obtained at the *AXAF* X-ray Calibration Facility (XRCF) at Marshall Space Flight Center. The calibration spectrum has been summed over the cross-dispersion direction, and each event has been plotted in a diagram of CCD event pulse height vs. position along the dispersion direction. For any given position along the dispersion direction, photons belonging to mth order dispersed light of wavelength λ/m are offset vertically with respect to each other. Conversely, the different spectral orders for a given wavelength are offset horizontally with respect to each other. The resulting curved spectra are the different spectral orders of the same incident spectrum. The CCD therefore acts in the same way as the 'cross-disperser' of an echelle grating.

In addition, the pulse height-dispersion plane nicely visualizes the effects of 'pileup' of events in the CCD. If two or more photons arrive in neighboring pixels during a single CCD frame exposure time, the so-called event reconstruction algorithm combines them into a single event in the output, with the sum of the individual charges assigned to this single, 'reconstructed' event. The reason the event reconstruction has to be applied to the CCD frames is that charge from individual photons may physically leak to neighboring pixels, and if this is not corrected for (each pixel treated as a separate event), the energy resolution of the CCD is degraded, and a fraction of fake events with low apparent charges is generated. The unavoidable side effect, though, is the unwanted 'pileup' of true single pixel photons referred to above. But since pileup is likely to occur only in intense parts of the spectrum (emission lines), the piled-up events will appear distinctively patterned and easily recognizeable in the pulse height-dispersion plane, as a series of dots offset

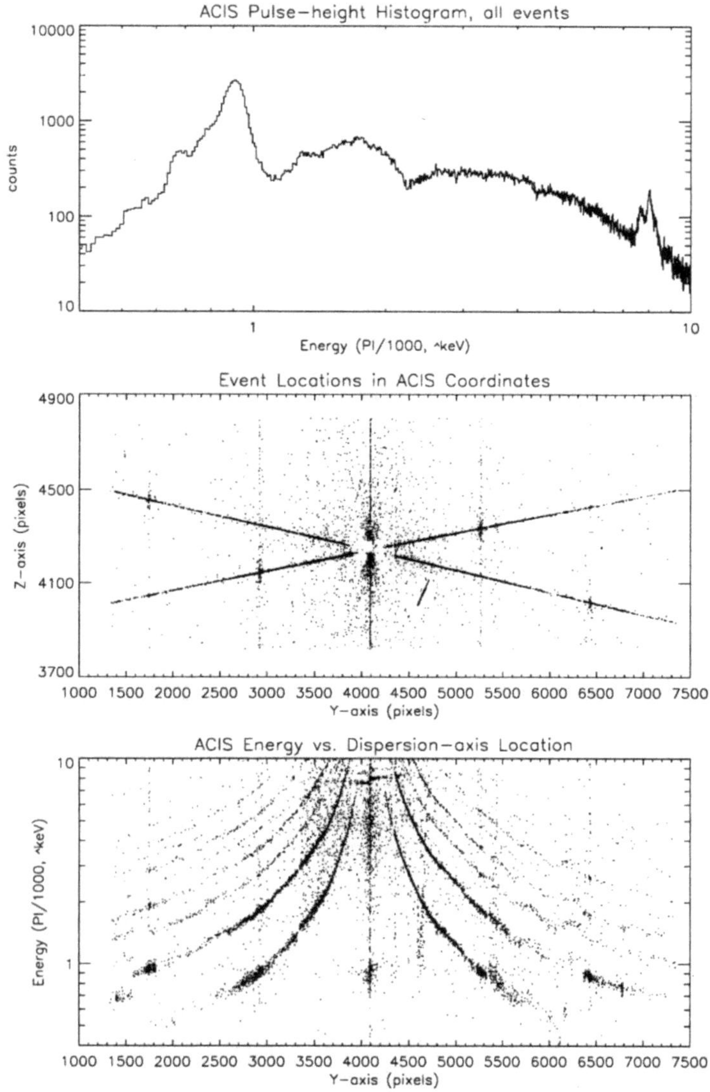

Fig. 14. HETGS spectrum of a Cu target bombarded by electrons, consisting of a bremsstrahlung continuum and fluorescent Cu L and K lines (0.93 and 8.0 keV, respectively). The top panel shows the CCD spectrum for all events combined, the middle panel the HETGS spectral image (you can see the 'X' shape formed by the HETG and METG spectra). The dark spots are the emission line photons (the vertical lines are artefacts from photons arriving at the detector during a 'frame transfer' of the CCD). The bottom panel shows the location of all events on the CCD pulse height vs. dispersion angle plane. The different spectral orders are clearly separated in the pulse height direction. The bright zero-order image has been suppressed (source: *AXAF Observatory Guide*).

in the vertical direction, at one, two, three, etc. times the true single photon charge. In actual flight operation of the HETGS, this effect is expected to be small, because of the low expected photon rates.

The pulse height-dispersion plot also clearly separates dispersed and non-dispersed (background) photons. Non-dispersed photons are sky and intrinsic detector background, and light scattered by the telescope (and possibly by the gratings). In other words, by using the CCD pulse height information, you can reduce the background in the spectrum by requiring that the pulse height and the dispersion angle match, for a given wavelength. Any events that don't match are background, and are smoothly distributed in between the orders in Figure 14.

For accurate quantitative analysis of actual spectra, it will probably be a good idea to take advantage of these powerful features of the HETGS, and perform both data analysis and spectral simulation in this 2D pulse height-dispersion space.

All statements so far regarding the resolving power assume that the source is pointlike. If the source has a finite extent, like a cluster or supernova remnant, the angular width $\Delta\theta$ in the expression for the resolving power becomes the angular size of the source, and the resolving power in principle degrades more or less linearly by a factor (source extent/telescope response). This is a severe effect with *AXAF*, because the telescope resolution is so high. For moderately extended sources, however, that radiate primarily in a small number of distinct, well-separated strong emission lines, the spectral image will look like a discreet set of partially overlapping monochromatic images of the source. Under these circumstances, it may be possible to perform spectrally resolved imaging, i.e. imaging in isolated emission lines. Figure 15 shows an example of this idea, the HETGS spectral image of the supernova remnant N132D in the Large Magellanic Cloud.

Let us discuss one more specific example that nicely illustrates the novel spectroscopy that will become available with the HETGS, the predicted 1-10 keV spectrum of the galactic X-ray binary Cygnus X-3. Cyg X-3 is a highly unusual X-ray binary, with properties unique among the galactic population. It is currently suspected to be the only binary we happen to catch in a rapid, short-lived evolutionary phase of massive X-ray binaries, where the compact object rapidly 'spirals into' the extended atmosphere of its massive companion (van den Heuvel 1994). The ultimate outcome of the binary evolution may be the formation of a white dwarf/neutron star binary, or a neutron star/neutron star binary.

Near-infrared spectroscopy of the IR counterpart shows distinctive emission lines from He I and II, characteristic of Wolf-Rayet stars (van Kerkwijk et al. 1992). The compact object is therefore likely accreting from a massive stellar wind, if the primary really is a WR star. Direct evidence for this conjecture was provided by the 1-10 keV spectrum of Cyg X-3 obtained with *ASCA* (Kitamoto et al. 1994). The spectrum shows a rich discrete emission

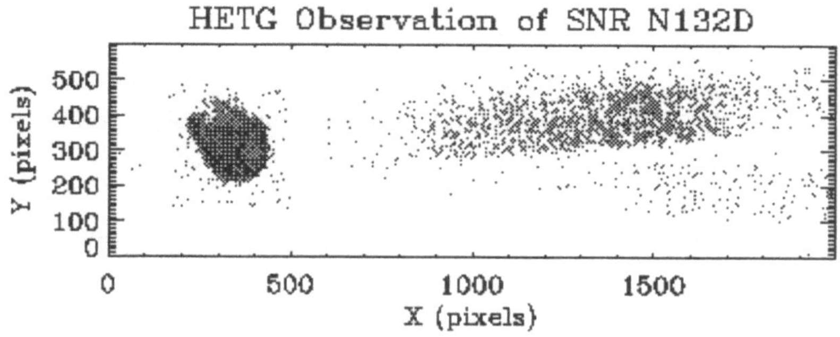

Fig. 15. Partial HETGS spectral image of the SNR 132D in the LMC (about 2 arcmin diameter). The dark image at left is the zero-order image of the remnant (a straight image). The dark band is the METG spectrum (the HETG spectrum below it is very faint because the remnant spectrum is very soft). You can see monochromatic emission line images of the SNR superimposed on the continuum in the MEG spectrum (courtesy of John Houck and the HETGS group, MIT).

spectrum superimposed on a strong, hard continuum which originates near the compact object. The discrete spectrum shows strong line emission from H- and He-like Mg, Si, S, Ar, Ca, and Fe. The spectrum also clearly shows narrow radiative recombination continua (RRC), of which the H-like Si, and especially the H-like S RRC can be seen directly in the data (Liedahl & Paerels 1996). The presence of these narrow features in the spectrum provides unambiguous evidence for the presence of X-ray photoionized gas in the system, probably relatively tenuous material in the WR wind, ionized by the strong central continuum.

With *ASCA*, the RRC's were distinguishable, and possibly just barely resolved, which indicates electron temperatures in the photoionized wind of order $kT_e \lesssim 80$ eV (Liedahl & Paerels 1996; Kawashima & Kitamoto 1996). With the HETGS, these features will be clearly resolved, and a detailed analysis of the spectrum will provide constraints on the conditions in various parts of the ionized stellar wind (density, temperature, abundances, possibly flow velocity). To illustrate this, I show a simulation in Figure 16, obtained by folding the recombination spectra for the He- and H-like ions of Ne, Mg, Si, S, Ar, Ca, and Fe, with values for the free parameters that fit the *ASCA* spectrum, through the HETGS response. The RRC's are clearly resolved, which will give us accurate measurements of the electron temperature in the various ionization zones. The 'triplets' (resonance, intercombination, and forbidden lines) of the He-like ions should also be resolved in principle, up to He-like Ca, which will provide limits on the electron density in the various ionization regions in the wind.

Unfortunately, heavy interstellar and circumsource absorption preclude the detection of low-energy photons from Cyg X-3. A measurement of the abundances of the low-Z elements in the wind would have given us an independent constraint on the low-Z photospheric abundances, and a direct window on the nucleosynthesis and mass loss history of at least this WR binary.

4.3 The diffraction efficiency of an X-ray transmission grating

We will calculate the diffraction efficiency of a transmission grating, show how it is optimized to yield maximum efficiency in a chosen wavelength band, and indicate how the efficiency calibration of the spectrometer is established in terms of a physical model for the instrument.

As long as the wavelength of the radiation is much smaller than the physical size of the diffracting elements (the grating period, for a grating), and at large distance away from the grating, Fraunhofer diffraction applies (Born & Wolf 1959), and the diffraction pattern is simply calculated by Huygens' principle: you just add up the complex phases of waves originating from the various parts of the grating, for a given dispersion angle. It is straightforward to show that the intensity of the angular pattern is

$$I(p,q) = \frac{1}{D^2}\frac{\sin^2((p-q)D/2)}{\sin^2((p-q)d/2)}|f(p,q)|^2 \tag{14}$$

with

$$f(p,q) = a\frac{\sin((p-q)a/2)}{(p-q)a/2}\exp(-i(p-q)a/2) + \int_0^b ds \exp i\Delta(s). \tag{15}$$

Here, d is the grating period, D is the total width of the grating, b is the width of the grating bars, and a is the width of the slits between the bars. The variables p and q are defined as $p = k\sin\chi$, $q = k\sin\theta$, where χ and θ are the angle of incidence and the dispersion angle, respectively, and k the radiation wave number. The first factor in the expression for $I(p,q)$ represents the rapid modulation due to the large number of periods in the grating. The two terms in the expression for $f(p,q)$ are the contributions of the slits and the bars to the amplitude, respectively. The phase shift $\Delta(s)$ in the contribution of the bars is

$$\Delta(s) = (p-q)s + k(n-1)z(s) \tag{16}$$

with s a coordinate in the plane of the grating, perpendicular to the bars, and $n \equiv 1 - \delta + i\gamma$ is the complex index again. The function $z(s)$ describes the cross-sectional shape of the bars: it is the bar thickness, as a function of s, measured perpendicular to the grating plane.

The efficiency of the grating is defined as the intensity for a given wavelength and spectral order, integrated over an angular region containing the sharp interference peak plus the weak, rapidly modulated 'wings' due to the

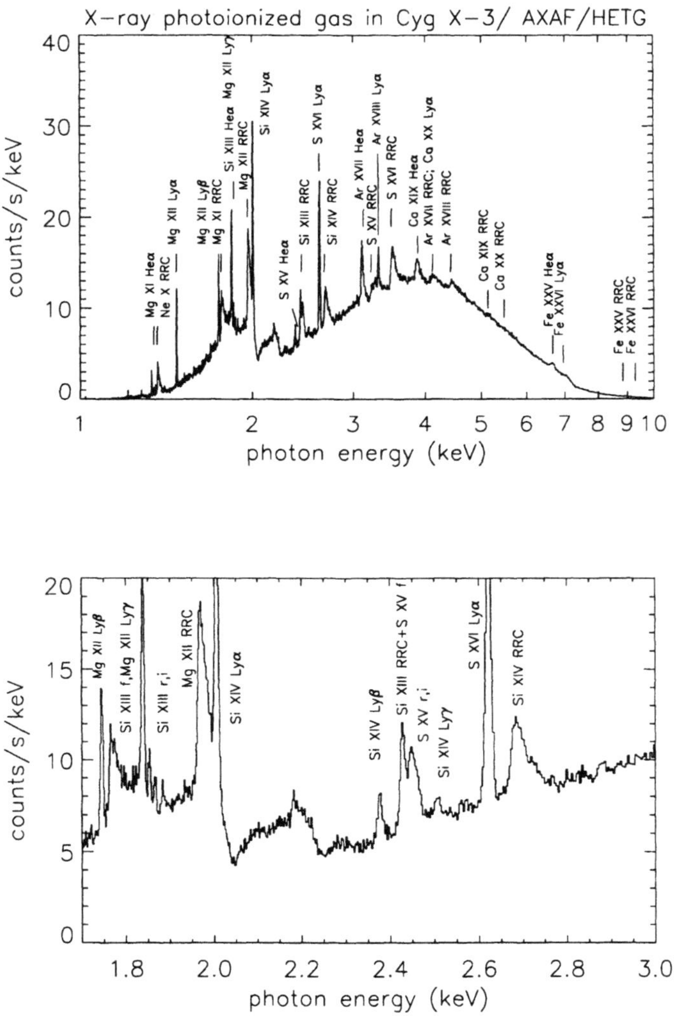

Fig. 16. Simulated spectrum of Cyg X-3, in a 40,000 sec exposure with the HETGS; the HETG spectrum is shown (positive and negative spectral orders summed). Data have been binned in approximately 0.005 Å bins (corresponding to one ACIS pixel). The lower panel shows a blow-up of the 1.8-3.0 keV region.

finite size of the grating, normalized to the intensity of the incoming radiation.

Let us examine the efficiency of the simplest possible grating: square bar cross sectional shape, bar thickness z_0. Inserting the condition for constructive interference, $p - q = -2\pi m/d$, it is easy to show that with the proper normalization, Eq.(14) reduces to

$$\text{for } m = 0: \; \eta = \frac{a^2}{d^2}\left\{1 + \frac{b^2}{a^2}e^{-2\gamma k z_0} + 2\frac{b}{a}e^{-\gamma k z_0}\cos\delta k z_0\right\}, \tag{17}$$

$$\text{for } m \neq 0: \; \eta = \frac{a^2}{d^2}\frac{\sin^2(m\pi a/d)}{(m\pi a/d)^2}\left\{1 + e^{-2\gamma k z_0} - 2e^{-\gamma k z_0}\cos\delta k z_0\right\} \tag{18}$$

These expressions were first given by Schnopper et al. (1977); note that for $m \neq 0$ they refer to the diffraction efficiency into *one* (i.e., positive or negative) spectral order only.

The efficiency of a grating with rectangular bars has several interesting properties. First, note that for a particular choice of bar width ($a/d = b/d = 1/2$, or bar width equal to slit width), diffraction into the even orders disappears. This is usually a desirable feature—more light is diffracted into the odd (first !) orders, and there is less overlap of orders. Second, the efficiency is a sensitive function of wavelength (through the wavelength dependence of the complex index n) at wavelengths short enough that the bars are partially transparent. This behavior is contained in the factor $\cos\delta k z_0$. By tuning the thickness z_0 of the bars, the efficiency can be enhanced over its value for an opaque grating, for a chosen range of wavelengths. The grating is said to be 'blazed' for that particular wavelength range (the origin of the term 'blazing' will become clearer when we discuss reflection gratings, below). Moreover, the dependence of the efficiency on wavelength, apart from an overall order-dependent constant, is identical for all orders. At long wavelengths, the bars are opaque, and the diffraction efficiency depends only on the bar-to-slit ratio. All these properties are illustrated in Fig.17, where I show the diffraction efficiency for a grating with gold bars, of rectangular cross-sectional shape, slit-to-period ratio 1/2, and two different bar thicknesses (0.35 and 0.60 μm). Notice how the thinner grating has much higher efficiency in the 1-2 keV band, while the thicker grating is (of course) superior above 2 keV. The resonance feature at 100 Å is due to a local minimum in the absorption coefficient of gold at those wavelengths, causing interference between the bars and the slits.

The *AXAF* LETG was optimized to provide enhanced efficiency in the 1-2 keV band, while the thicknesses of the bars of the HETG and METG were chosen to provide optimum efficiency at energies $E \gtrsim 2$ keV.

Things change when the bar cross sectional shape is no longer strictly rectangular (either by design or because of small imperfections in the manufacturing process). Once you realize that the edges of such bars don't have the same optical depth as the centers of the bars (usually smaller—the bar cross

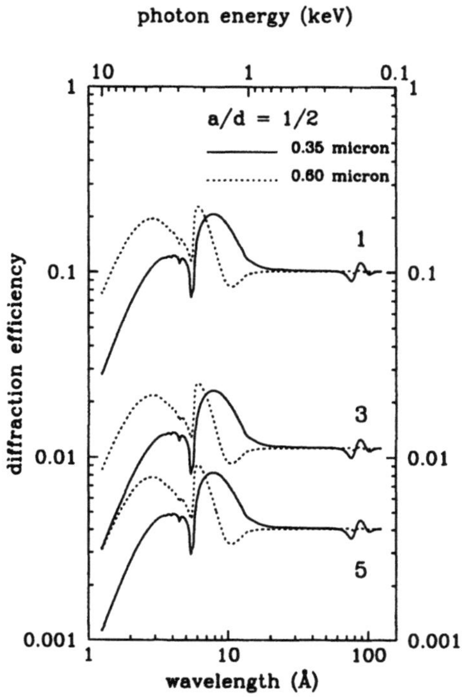

Fig. 17. Diffraction efficiency of a gold transmission grating, of slit-to-period ratio 1/2, rectangular bar cross sectional shape, for two different bar thicknesses.

section may look like a trapezoid, or even an ellipsoid), it is obvious that, for instance, the exact cancellation of the even-order diffraction no longer works for wavelengths at which the bars are partially transparent, even for bar-to-slit ratio equal to unity. In fact, the even order diffraction efficiency becomes extremely sensitive to bar shape and thickness at short wavelengths. Also, the ratio of the efficiencies in the various orders becomes a function of wavelength, even in the odd orders. These properties are illustrated in Fig.18, where I show efficiencies for two gratings with identical parameters (slit-to-period ratio 1/2, bar thickness 0.35 μm), but different bar cross-sectional shape (either rectangular or ellipsoidal). The 'ellipsoidal' grating shows the expected diffraction into the even orders (in fact, the even orders become more efficient than the odd orders at high energies). The higher the order is, the more sensitive is the efficiency to the details of the bar properties.

Finally, if there is a distribution in the properties of the bars across the grating (as will always be the case to some extent), the simple properties of a strictly uniform grating obviously no longer apply. The simplest example is a

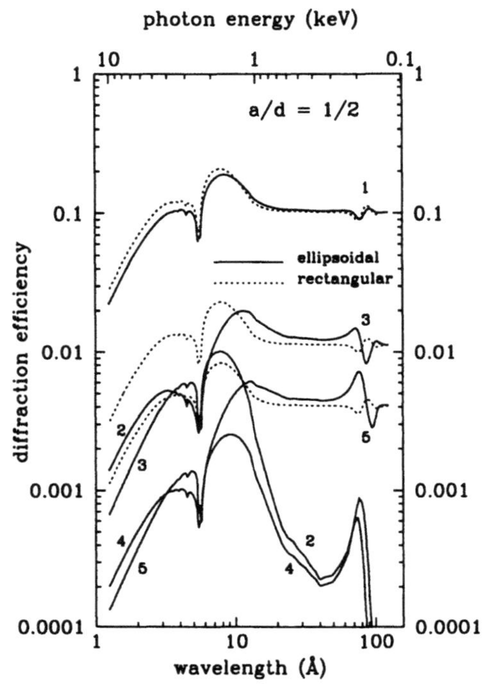

Fig. 18. Diffraction efficiency of a gold transmission grating, of slit-to-period ratio 1/2, bar thickness 0.35 μm, and two different bar cross sectional shapes (rectangular or ellipsoidal). The grating period is 1.0 μm.

grating designed with $a/d = 1/2$, which exhibits finite even-order diffraction due to the fact that the bar-to-slit ratio varies slightly around the mean over the grating.

With these features in mind, you will appreciate the complications associated with trying to uniquely establish the efficiency calibration of an X-ray grating. Calibrating all different parts of the grating, at all wavelengths (and possibly different angles of incidence) is obviously out of the question for practical reasons, so one typically relies on a physical model for the grating efficiency, with a (small) number of free parameters, whose values are constrained by comparing calibration data with model calculations. This is a difficult and not necessarily unique procedure. Past experiments (the gratings on *Einstein* and *EXOSAT*) for practical reasons had to rely almost exclusively on a finite set of efficiency measurements, averaged over the entire grating, at a discrete set of characteristic X-ray wavelengths (Seward et al. 1982; Paerels, Kahn & Wolkovitch 1998), which did not always admit of a fully consistent, simple and unique model for the grating and it efficiency. The *AXAF* grating spectrometer calibration relies on a much wider range of data, including optical and IR period measurements and synchrotron X-ray efficiency measurements for many or all individual grating elements, and is ultimately expected to produce agreement between data and the best physical model for the gratings on the order of or better than a few percent.

4.4 The Low Energy Transmission Grating Spectrometer

The Low Energy Transmission Grating Spectrometer (LETGS; Brinkman et al. 1987) consists of four annular sets of free-standing gold transmission grating elements, mounted behind the HRMA mirror shells in a Rowland configuration identical to that of the HETGS (cf. Section 4.2). It is designed for use with the HRC-S imaging camera, which consists of three separate microchannel plate detectors, placed tangent to the Rowland circle. The grating period is 9912.5 Å, the bar thickness 0.5μ. The low dispersion, as compared to the HETGS, implies that, given the size of the HRC-S, the LETGS provides spectroscopy out into the EUV band, out to 170 Å. At the lowest photon energies, the LETGS reaches resolving powers in excess of $\mathcal{R} \sim 2000$, the highest resolving power on *AXAF*. The resolving power is shown overlaid on the spectroscopic diagnostics we discussed at the beginning, in Fig.19. The somewhat wiggly appearance of the resolving power curve at the lowest energies is due to the fact that the HRC-S camera elements are flat, and do not precisely follow the Rowland circle, which leads to a small amount of defocusing at some wavelengths.

The LETGS has unique capabilities for performing high-resolution spectroscopy of (very) soft sources (either intrinsically soft or sources with low interstellar absorption). Soft coronal emitters can be studied at high spectral resolution, which may actually allow the detection of bulk velocity fields, or binary motion in short-period active binaries (RS CVn stars). Other prime targets are cataclysmic binaries, white-dwarf stars, and thermal emission from hot neutron stars. The LETGS may also provide unique access to the spectra of the low-Z elements in sources with a significant redshift.

With these goals in mind, the gratings were again (like their predecessors on *Einstein* and *EXOSAT*) designed to be free-standing (no support film), and the design of the HRC-S was optimized with the lowest photon energies in mind. Instead, the grating is supported by two gold support grids, one (the fine support grid) at right angles to the X-ray grating bars, the other (the coarse support) actually consists of three grids at 120 degree angles. These support grids have very large periods, and very large slit-to-period ratios a/d, so while they do disperse light at an angle to the X-ray dispersion direction out of the X-ray spectral image, they do so at very low efficiency.

The actual gratings were produced at the Max Planck Institut für Extraterrestrische Physik in Garching, Germany, under the direction of Peter Predehl. Details on the design, manufacturing, and laboratory calibration,

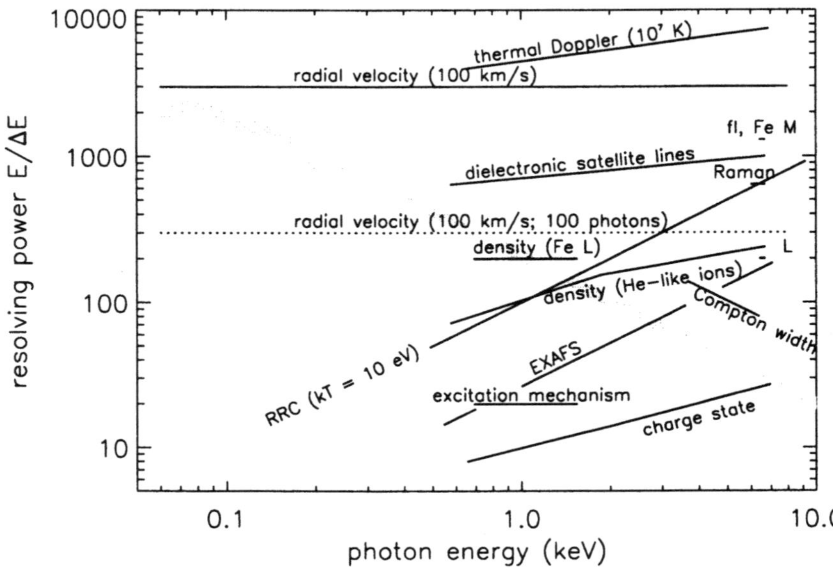

Fig. 19. Same as Figures 8 and 13, but with the predicted resolving power of the LETGS overlaid (gray band) (source: *AXAF Observatory Guide*, Ch. 8)

as well as a stunning color image of the complete, assembled LETG with all the facets shining like the rainbow in diffracted optical light, can be found at `http://www.rosat.mpe-garching.mpg.de/axaf/LETG_description.html`. A preliminary analysis of the ground calibration can be found in Brinkman et al. (1997) and Predehl et al. (1997).

Many of the considerations mentioned earlier that apply to the HETGS, apply equally to the LETGS, with one important exception. Since the HRC-S has no intrinsic energy resolution, the overlapping spectral orders cannot be separated by applying the detector energy resolution to the individual photons, as is the case with the HETGS/ACIS-S combination [imaging detectors with sufficient energy resolution and high quantum efficiency at low energies ($E \lesssim 500$ eV) did not yet exist at the time the instrument was designed]. This situation has received considerable attention (somewhat surprising, since the *Einstein* and *EXOSAT* gratings exhibited the same feature, so experience with this type of data in astrophysics exists), and a separate optical element was designed, the High Energy Suppression Filter (HESF), that can be inserted into the optical path to provide additional data in case spectral-order ambiguity precludes quantitative interpretation of a given dataset. It must be emphasized that the relative efficiencies of the various orders of the LETGS have accurately and extensively been characterized, so that the relative intensities in the various orders, for an assumed incident spectrum, can always be predicted with precision, even if the individual photons cannot be sorted by order using detector energy information. No quantitative information is ever lost by the overlapping of orders.

It is generally agreed that order confusion may arise in sources with bright lines at relatively high energies ($E \gtrsim 1$ keV). The third orders of these lines may contribute significantly to the spectrum at energies below 1 keV, especially if the first-order low-energy spectrum is faint (for instance, due to heavy interstellar absorption). The HESF was designed with this situation in mind. It cleverly uses the notion of the critical angle for total external reflection (cf. Dick Willingale's chapter), by having the focused and dispersed light bounce off a special mirror made of low-Z material. This mirror is oriented along the dispersion direction, and is slightly offset from the HRC-S in the cross dispersion direction. By tilting the entire telescope slightly in the cross dispersion direction, the dispersed light can be made to bounce off the HESF before reaching the detector in the focal plane. This additional reflection suppresses photons with energies higher than the critical energy that corresponds to the angle of incidence on the HESF, and their contaminating effect in higher order is reduced. The HESF was incorporated in the *AXAF* instrument package at a late date, and there has not been much time to carry out measurements to characterize its performance, so experience with astrophysical data will probably have to tell how useful it will turn out to be in practice.

Ironically, the cross dispersion properties of the fine support grid could have supplied the additional information required to assign spectral orders to bright emission line images—the extent of the faint cross dispersion image scales linearly with wavelength, and in principle this allows another measurement of the wavelength of the photons in the image (a true cross disperser). However, this cross dispersion image will usually be very faint in astrophysical spectra.

As an example of the power of spectroscopy with the LETGS, Figure 20 displays a simulated LETGS spectrum of the famous Seyfert 1 galaxy NGC 5548, obtained in 40,000 sec exposure. This object has a complex soft X-ray/EUV spectrum. Continuum emission originating in the innermost regions of the AGN is seen through a partially ionized absorber (Fabian et al. 1994; Mathur, Elvis & Wilkes 1995), which itself also emits detectable amounts of X-ray line radiation (Kaastra, Roos & Mewe 1995). These spectral components are present in the simulation as well, at parameter values consistent with recent observational constraints. The line emission, specifically, has been represented by a source in collisional equilibrium at a temperature of $\sim 6 \times 10^5$ K, consistent with the discrete emission seen in the EUV spectrum.

All features in the $40 - 100$ Å band are due to the L-shell ions of the low- and mid-Z elements, which are formed in a range of ionization conditions that has not yet extensively been probed spectroscopically. Not much is known about the properties of the medium that gives rise to the emission, and it has been speculated that this is the same medium that gives rise to the highly-ionized Ne emission lines seen in the UV spectra of high-redshift quasars (Hamann et al. 1998). Spectroscopy with the LETGS will also allow further quantitative investigation of the $\sim$ 3800 km s^{-1} line broadening seen in the *EUVE* spectrum, as well as the $\sim$ 1200 km s^{-1} bulk velocity field seen in the UV absorption spectrum of the absorbing medium.

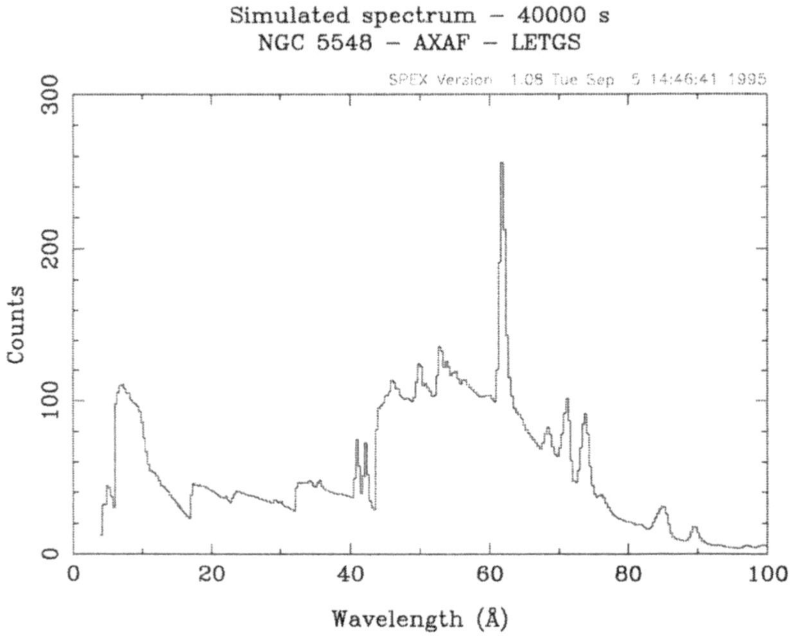

Fig. 20. Simulated 40,000 sec exposure *AXAF* LETGS spectrum of NGC 5548, illustrating the novel spectroscopy that can be expected longward of the interstellar Carbon edge (44 Å) in very soft-spectrum objects, such as AGN (courtesy Jelle Kaastra, SRON).

4.5 In Von Laue and Debye's footsteps: scattering by random fluctuations in the properties of a transmission grating

As another example of how basic physical principles can be applied directly to a calculation of the properties of a diffraction grating, I discuss the presence of small amounts of scattered light in the diffraction patterns of the HETG and LETG. The calculation also provides yet another illuminating example of the physics of scattering mechanisms, which gives rise to very similar concepts in widely differing physical systems.

During calibration measurements of the *AXAF* grating spectrometers, it was found that the HETG and LETG exhibit small amounts of scattering, visible as a very faint, nearly continuous distribution of light in between the various diffracted orders of a given emission line. You can see this light in Figure 21, which shows a LETG calibration spectrum of the Al Kα line, as recorded with ACIS. From top to bottom, I plot the two-dimensional spectral image (each photon is represented by a single dot; the zero-order position is at the origin), followed by the ACIS pulse height spectrum of all events. The third panel shows the position of all photons in the dispersion-pulse height plane, and here you can clearly see the separation between true continuum radiation (the curved shapes running away from the lowest spectral orders) and monochromatic, Al Kα radiation spread out horizontally between the orders (all photons the same pulse height). Finally, the bottom panel shows a monochromatic grating spectrum, obtained by integrating the spectral image over the cross dispersion coordinate, and including only events in the $1.4-1.8$ keV CCD pulse height range.

The Al Kα light detected between the orders is the puzzle: light scattered by the telescope should be distributed circularly symmetrically around each spectral order in the two-dimensional spectral image, which is not what is seen. The dispersion-pulse height diagram shows that the inter-order light is not continuum radiation, so scattering by the grating is logically the only possible explanation. What causes the scattering ?

Small random fluctuations in the grating properties (the period, bar thickness, bar cross sectional shape) cause scattering. The simplest way to understand this is the following; we will specialize to scattering by period fluctuations, because it is the easiest case.

Arguing from Huygens' principle, you find that the diffraction pattern of a grating is formally equal to the squared modulus of the Fourier transform of the grating pattern (the pattern of slits and bars), as you can see as follows. Adding the complex amplitudes from all parts of the grating (assuming an opaque grating for simplicity) gives

$$I = \left| \int_{\text{grating slits}} ds \exp i(p-q)s \right|^2 = $$
$$= \left| \int_{-\infty}^{\infty} ds\; G(s)\; \exp i(p-q)s \right|^2 \qquad (19)$$

where $G(s)$ is the grating pattern [$G(s) \equiv 1$ for the slits, $G(s) = 0$ otherwise], which shows that I is the square of the transform of G.

Now suppose that the grating period were itself slowly modulated, by a slow sinusoidal deviation in the period. That means that $G(s)$ contains another period, other than the average grating period, which causes additional interference in the diffraction pattern. Right next to each bright diffraction

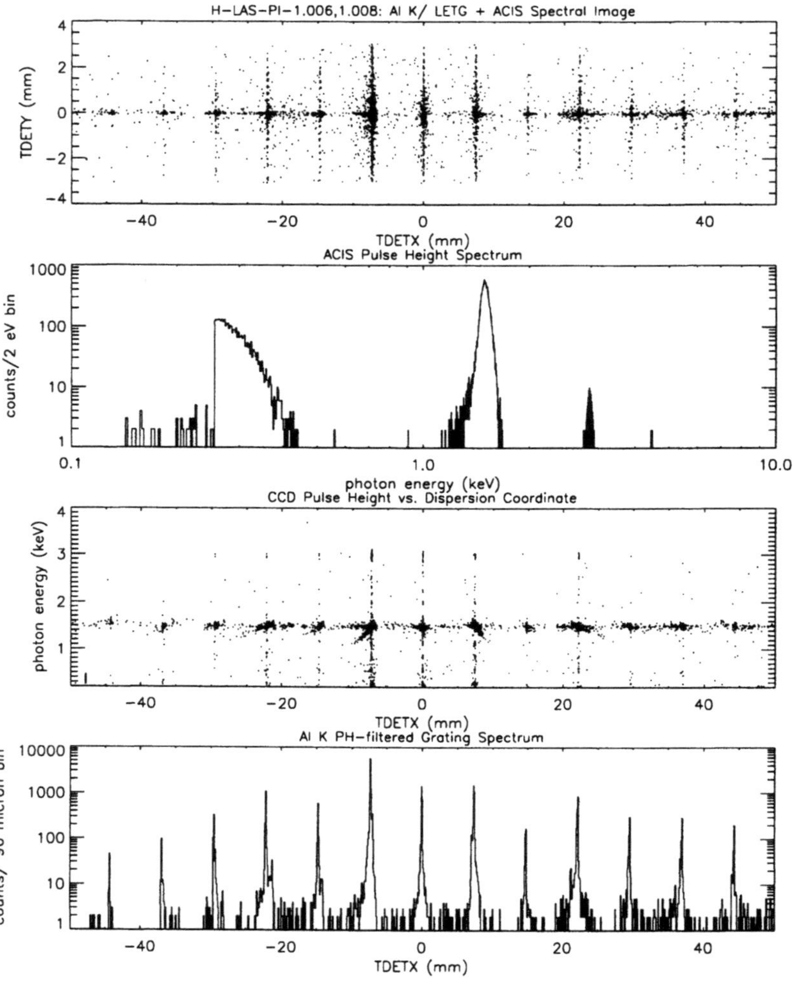

Fig. 21. LETG Al Kα calibration spectrum, recorded with ACIS-S. Top to bottom: two-dimensional spectral image, CCD pulse height spectrum, position of all photons in the dispersion-pulse height plane, and grating spectrum with the continuum suppressed by PHA filtering. The dispersion-pulse height plane shows faint scattered light in between the spectral orders as a horizontal band.

peak you will see smaller peaks, corresponding to 'beats' with the main grating period. Such subsidiary images are called 'ghost images', and the ones due to slow sinusoidal modulation of the grating period are called 'Rowland ghosts', after Henry Rowland, who first gave the correct explanation for their existence, and managed to produce ghost-free gratings.

The general case of small random fluctuations in the grating period can be represented by a Fourier decomposition, with random phases, for the fluctuations. Each of the Fourier components produces a set of ghosts outside the main diffraction peak, at an angular separation determined by the spatial wavelength of the Fourier component. The entire superposition of Fourier components therefore produces a continuous distribution of ghosts, or a scattering 'wing' on the main diffraction peaks. And as long as the amplitude of the period fluctuations is small, you intuitively guess that the shape of the scattering wing will reflect the distribution of power over the various Fourier components in the grating pattern. This is the basic mechanism behind scattering by diffraction gratings.

It is easy to estimate the total amount of light thus scattered. For simplicity, assume an opaque grating with bar/slit ratio equal to unity (it is easy to relax these conditions). Slit number n contributes a complex amplitude

$$\begin{aligned} A_n &= \int_{\text{slit } n} \mathrm{d}s \exp(-iks\sin\theta) \\ &= \qquad\qquad (q \equiv k \sin\theta) \\ &= \int_{(2n+1)d/2}^{(2n+2)d/2} \mathrm{d}s \exp(-iqs) \\ &= \frac{1}{-iq}\Big(\exp(-iq(2n+2)d/2) - \exp(-iq(2n+1)d/2)\Big), \end{aligned} \tag{20}$$

using the same notation as earlier (d is the grating period). The intensity is (we are not worrying about the correct normalization because we will be interested only in the fraction of light scattered)

$$\begin{aligned} I = AA^* &= \sum_{n=-N}^{N} \sum_{m=-N}^{N} A_n A_m^* \\ &= \frac{4\sin^2(qd/4)}{q^2} \sum_{n=-N}^{N} \sum_{m=-N}^{N} \exp(-iq(n-m)d) \end{aligned} \tag{21}$$

The intensity has its main peaks at the zeros of $1 - e^{iqd}$ (or $d \sin\theta = m\lambda$, the dispersion relation). This factor appears in the denominator in the final expression for I, and arises from the sums over $\exp(inqd)$.

Now apply a small random perturbation Δd_n to the location of slit n; this amounts to applying a small phase shift to A_n:

$$A_n = \frac{1}{-iq}\Big(\exp(-iq[(2n+2)d/2 + \Delta d_n]) - \exp(-iq[(2n+1)d/2 + \Delta d_n])\Big) \tag{22}$$

(Here you see that the result will not depend on period-to-slit ratio; the phase shift is the same regardless of the slit-to-period ratio). The diffraction pattern is

$$I = AA^* = \frac{4\sin^2(qd/4)}{q^2} \sum_{n=-N}^{N} \sum_{m=-N}^{N} \exp(-iq(\Delta d_n - \Delta d_m)) \exp(-iq(n-m)d) \tag{23}$$

The observable quantity is the statistical average of I, averaged over all possible realizations of the Δd_n:

$$\langle I \rangle = \frac{4\sin^2(qd/4)}{q^2} \sum_{n=-N}^{N} \sum_{m=-N}^{N} \Big\langle \exp(-iq(\Delta d_n - \Delta d_m)) \Big\rangle \exp(-iq(n-m)d) \tag{24}$$

Now assume that the perturbations are small: $q\Delta d_n \ll 1$, and expand the exponential in $\langle\ \rangle$:

$$\begin{aligned} &\Big\langle \exp(-iq(\Delta d_n - \Delta d_m)) \Big\rangle = \\ &= \langle 1 - iq(\Delta d_n - \Delta d_m) - \frac{1}{2} q^2 (\Delta d_n - \Delta d_m)^2 + \ldots \rangle \\ &\cong 1 - q^2\sigma^2 \end{aligned} \tag{25}$$

since $\langle \Delta d_n \rangle = \langle \Delta d_m \rangle = 0$, $\langle \Delta d_n \Delta d_m \rangle = 0$, and we define the variance of the perturbations $\langle \Delta d_n^2 \rangle \equiv \sigma^2$. Therefore, to first order, the diffraction pattern is reduced in intensity by a factor $1 - q^2\sigma^2$, which implies that a fraction $q^2\sigma^2 = k^2\sigma^2 \sin^2\theta$ of the light has been scattered out of the main diffraction peak. Recalling that $\sin\theta = m\lambda/d$ (m the order number), you find that the total amount of scattered light, as a fraction of the light in a given diffraction peak, scales like $f_m = 4\pi^2\sigma^2 m^2/d^2$; there is no scattering in zero order.

This calculation closely parallels Debye's famous calculation for scattering of X-rays by thermal vibrations of the ions in a crystal lattice in crystal diffraction experiments (Debye 1914). You can show that if the Δd_n are normally distributed, then the statistical average of $\exp(-iq(\Delta d_n - \Delta d_m))$ (the reduction in the intensity of the diffraction pattern, or the fraction of unscattered light) is exactly equal to $\exp(-q^2\sigma^2)$, and this is the so-called Debye-Waller factor, by which the intensity of the sharp diffraction peaks decreases, and which occurs in all kinds of scattering calculations. Our factor $1 - q^2\sigma^2$ is of course the first-order approximation to this expression. Anecdote has it that when von Laue first discussed with Debye his idea of looking for diffraction of X-rays by crystals, the latter raised the objection that thermal vibrations of the ions around the lattice sites would destroy the interference pattern, and that von Laue wouldn't see anything. Von Laue proposed he would do the experiment anyway, if Debye would calculate the effect of the vibrations. That led to the Debye-Waller factor and the conclusion that scattering would not significantly affect the sharp diffraction peaks, while von Laue got the Nobel prize for the experiment ...

John Davis (MIT) has given an elegant and mathematically powerful alternative derivation of the scattering effect, where he also calculates the angular distribution of the scattered radiation explicitly, to arbitrary order of approximation, by assuming a normal distribution of completely uncorrelated (bar-to-bar) period fluctuations (Davis 1997).

How does this compare to the observed scattering properties of the LETG? In order to measure the intensity of the scattered light in the lowest few orders, I took the Al K spectrum shown in Fig. 21(d), and divided the spectral range into equally sized subranges centered on the spectral orders. I approximately fitted a copy of the telescope point response function, integrated over the cross-dispersion direction, to the diffraction peaks, plus a Gaussian to represent the scattered light. The Gaussian was integrated over the relevant detector subrange, to obtain the scattered light associated with each order. These estimated scattered intensities, normalized to the intensities in the sharp diffraction peaks, are shown in Fig. 22, as a function of spectral order. The dotted line in this figure shows the prediction based on the expression derived above, $f_m = 4\pi^2(\sigma/d)^2 m^2$, with the r.m.s. fluctuation in the period σ normalized at $m = 4$.

As you can see, the qualitative behavior of the scattered light in low order is roughy consistent with our prediction, but two features stand out. First, there is some scattering in $m = 0$, which should not be there, and second, there seems to be an even/odd asymmetry in m of the measured f_m, with respect to the predicted f_m. Both these features have a natural explanation. Scattering near $m = 0$ is partially an artefact of my analysis, beacuse some light scattered through large angles from $m = \pm 1$ has leaked into the analysis region centered on $m = 0$. I can correct for this effect if I knew the theoretical angular distribution of the scattered light. To obtain some idea of how large this correction might work out, I used the explicit shape of the full diffraction pattern (including scattering) given by Davis (1997), which was derived under the assumption that the period fluctuations can be described by completely uncorrelated, Gaussian-distributed shifts of the bar locations. I integrated the intensity of this pattern over the spatial regions centered on the spectral orders, and normalized to the theoretical efficiencies of the orders. The shape of the scattering distribution turns out to depend on a/d, just like the diffraction efficiencies themselves, and so I adjusted a/d and σ/d to approximately fit the measured scattering fractions. The result is displayed in Fig. 22 as the solid histogram. The agreement is improved, and the physical interpretation of the scattering as being due to random fluctuations in the period appears to be validated. The parameters a/d and σ/d are not very well constrained, though, and the fact that the best-fitting a/d does not entirely agree with the value determined from IR

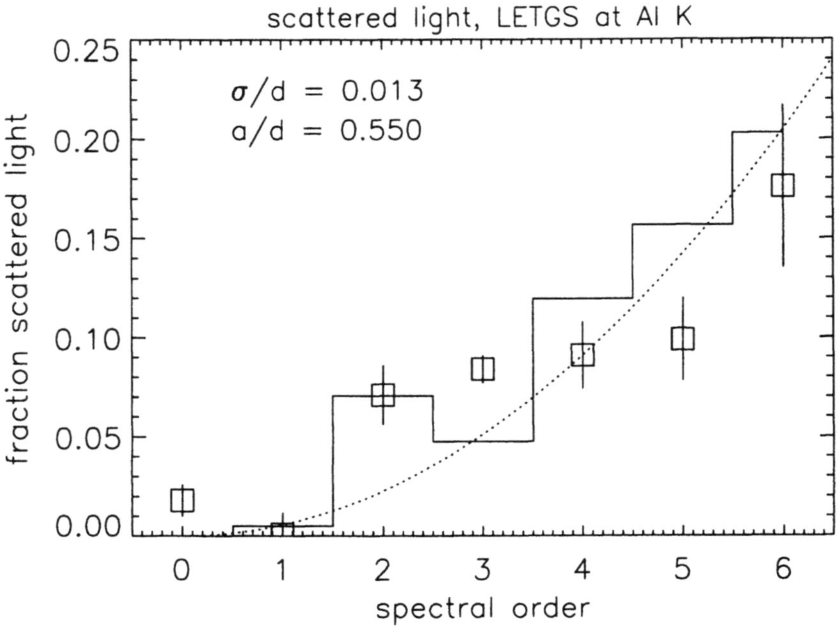

Fig. 22. Measured fraction scattered light, as a function of spectral order, for low-order Al K radiation. The dotted line is the simple first-order estimate of the scattering fraction ($f_m = 4\pi^2(\sigma/d)^2m^2$), normalized at $m = 4$, while the solid histogram is based on an explicit shape for the distribution of scattered light.

reflectivity measurements of the individual grating elements is therefore not a serious concern. Also, the exact shape of the scattering distribution I have used is reasonable, but has not yet been independently verified.

The 'leakage' correction at $m = 0$ turns out to be small, and we have to conclude that there appears to be true scattered light associated with $m = 0$. This could be due to fluctuations in the bar thickness, which also cause scattering, and, in contrast to the period fluctuations, also cause scattering in zero order. The calculation of the bar thickness fluctuation effect is much nastier than the period fluctuation effect, because a perturbation to the bar thickness is not equivalent to a simple phase shift to the complex amplitude contributed by a given bar+slit. In addition, there is true absorption in the bars, so you cannot repeat the simple calculation leading up to Eq. (25): the apparent decrease in the diffraction efficiency can be due both to scattering as well as true absorption. In order to derive the fraction scattered light, you therefore have to calculate the diffraction pattern explicitly, to extract the scattered light. If you are curious, the details of the lowest-order approximate calculation are given in Paerels (1997).

It turns out that the bar-thickness-fluctuation scattered light depends strongly on wavelength (as it should—the effect should disappear at long wavelengths, where the bars are opaque), occurs in zero order as well as in orders $m \neq 0$, and does not depend on order for the orders $m \neq 0$. Taken at face value, the measured fraction scattered light in $m = 0$ at Al K indicates a relative r.m.s. fluctuation in bar thickness of ~ 0.08, and the theory predicts that the scattering by this effect should be very small ($f_m \lesssim 0.6\%$) in the orders $m \neq 0$ at Al K.

You may wonder why this scattering is a concern at all, given that it is measured to be a sub-1% effect in $m = 1$. The reason is, of course, that if the scattering would turn out to be strongly wavelength- or order-dependent, then the scattered light could significantly contaminate the true continuum light (and these components cannot be separated with the HRC-S detector, which has no intrinsic energy resolution), especially in line-dominated spectra, such as the cooler coronal plasmas ($T \lesssim 10^7$ K). Since absolute abundance measurements depend on line intensities relative to the continuum, the scattering could systematically bias abundance measurements.

And this turns out to be the case. From our calculation we find that $f_m \propto m^2$, whereas Eq.(18) shows that the diffraction efficiencies are approximately proportional to m^{-2}. So in the lowest orders, each order has the *same* absolute amount of scattered light! The higher orders of intense short-wavelength lines can therefore contaminate the continuum at longer wavelengths. At some high m, f_m must of course saturate to $f_m = 1$ through the Debye-Waller factor, and all the diffracted light is scattered, at which point the scattered light must again decrease like η_m itself. The actual degree of contamination and its effect on abundance measurements is still under investigation.

5 The Reflection Grating Spectrometers on *XMM*

5.1 Introduction

ESA's X-ray Multimirror Mission (*XMM*; `http://astro.estec.esa.nl/XMM/xmm_top.html`) was defined to be the high-throughput X-ray spectroscopy cornerstone of the Horizon-2000 strategic scientific program. Launch is currently scheduled for August 1999. The core of the observatory are three identical high-throughput, medium angular resolution ($\lesssim 15$ arcsec half-power spot diameter [HPD, or half energy width HEW], and approximately 6 arcsec FWHM) X-ray telescopes, of peak effective area ~ 1500 cm^2 at 1 keV, ~ 500 cm^2 at 5 keV (per telescope), and sensitivity out to 10 keV (Telescope Scientist: Bernd Aschenbach, Max Planck Institut für Extraterrestrische Physik). The focal length is 7500 mm. All three telescopes have CCD focal-plane cameras, collectively referred to as the European Photon Imaging Camera (EPIC; PI: M. Turner, Leicester University). Two CCD cameras are MOS-type devices, the third is a pn-junction device. The mirror assemblies represent a breakthrough in the fabrication of lightweight X-ray optics: each assembly consists of 58 separate, densely nested gold-coated thin Ni shells of Wolter-I geometry. Details and pictures can be found at the site mentioned above; a set of technical references is given on `http://astro.estec.esa.nl/XMM/user/xmmpub_top1.html`.

Two of the telescopes will be equiped with an array of reflection gratings each, which, together with two dedicated CCD cameras at the spectroscopic focus, constitute the Reflection Grating Spectrometer (RGS) experiment (PI: Albert Brinkman, SRON). The Reflection Grating Arrays (RGA) will be mounted permanently behind their mirror assemblies, where they intercept approximately half the light for medium- to high-resolution spectroscopy in the 0.3-2.5 keV band; the other half of the light goes to the EPIC detectors at the prime focus. Therefore, an RGS spectrum will be obtained for each target observed with *XMM*.

Finally, the Optical Monitor (OM) experiment (PI: Keith Mason, Mullard Space Science Laboratory) will simultaneously cover the X-ray field of view in the UV/optical band (1700 − 6000 Å) with a 30 cm aperture Ritchey-Chrétien telescope.

In order to ensure a uniform, optimum, and timely analysis of the 'Serendipitous Survey' formed by the data content of all of the imaged fields, an *XMM* Survey Science Center has been formed (PI: Mike Watson, Leicester University), a collaboration of institutes in the UK, France, and Germany. The SSC will analyze all fields, and survey the *XMM* X-ray sky.

5.2 Properties of reflection gratings, and design of a grazing-incidence reflection grating spectrometer

As stated above, *XMM* was designed around high-throughput, medium angular resolution optics. The desired instrument package was to include a high-resolution spectrometer for the soft X-ray band, with high throughput and a resolving power of at least $\mathcal{R} = 250$ (also in the HEW sense, just like the telescope angular resolution specification) at 1 keV. This resolving power was chosen with the density diagnostics in the He-like ions in mind, among other things. At the time these recommendations were formulated (mid-eighties), the only way to achieve these goals was with diffractive spectrometers.

But with a moderate telescope angular resolution, the dispersion angles need to be large in order to achieve $\mathcal{R} \sim 250$; for instance, with a $20''$ telescope blur, a transmission grating needs to disperse to $250 \times 20'' = 1.4°$, which, for 1 keV photons, implies a line density of $\sim 20,000$ lines mm^{-1} ! Instead, as we have seen before, it is much easier to achieve these high dispersions with grazing-incidence reflection gratings, a solution first advocated for *XMM* by Kahn & Hettrick (1985).

In addition, the reflectivity properties of reflection gratings have certain attractive features. First, the efficiency in the soft X-ray band can be high (> 30% in first order), and second, the wavelength at which this maximum occurs can again be tuned by optimizing the grating groove shape. To understand these properties, let us look at how to calculate the reflectivity of a reflection grating, an exercise in the application of electrodynamics to the solution of a real problem (as opposed to a textbook EM problem).

There is a fundamental distinction between the operation of a transmission grating and a reflection grating, in the regime $\lambda/d \ll 1$ that applies to the X-ray band for moderate line density structures. All a transmission grating does in that case is block and possibly phase shift parts of the incoming wave fronts; Huygens' principle is sufficient, and you don't need Maxwell's equations to work out the diffraction efficiency. But with a reflection grating, you have to apply the correct boundary conditions to the fields at the reflective surface in order to obtain the intensity of the reflected waves in terms of the intensity of the incident waves. For a flat mirror surface of given material properties, this is trivial, and the calculation is in every textbook. But there is no mathematically simple way to impose general boundary conditions at a periodically modulated (as opposed to a flat) surface. Hence, a general solution in closed analytical form to the problem of determining the reflectivity of a reflection grating does not exist.

Only approximate solutions to the problem exist. Let us look at a well-known, but as it turns out for our application physically unrealistic choice of boundary condition. If you ad-hoc impose the condition of perfect reflectivity at the grating surface, that is, the amplitudes of the fields inside the grating and at the surface are zero everywhere, each part of the grating surface radiates in phase with the incident wavefronts, and the efficiency calculation reduces again to applying Huygens' principle to the periodic surface. This is also referred to as 'scalar diffraction theory', because the actual dynamics of the electromagnetic field (its tensor character and the interaction with the atoms of the reflecting material) is entirely ignored.

As a benchmark example, a straightforward application of Huygens' principle to the case of a grating of period d, illuminated at incidence angle α, with triangular-shaped grooves of tilt angle δ (see Figure 5 for the geometry) yields the following expression for the scalar diffraction efficiency as a function of wavelength and order m:

$$\epsilon_m = (4\sin^2\theta\sin\alpha\sin\beta_m)^{-1}g^2P_m^2(\sin Q_m/Q_m)^2 \tag{26}$$

with

$$g = \sin\alpha/\sin(\alpha+\delta) \tag{27}$$

$$P_m = \sin\theta\{\sin(\alpha+\delta)+\sin(\beta_m-\delta)\} \tag{28}$$

$$Q_m = (\pi g d/\lambda)\sin\theta\{\cos(\alpha+\delta)-\cos(\beta_m-\delta)\} \tag{29}$$

with β_m the dispersion angle in order m, and θ the angle of the incident ray with the groove direction.

By inspecting this expression, you see that it peaks at wavelengths satisfying

$$\alpha + \delta = \beta - \delta \tag{30}$$

because $Q_m \to 0$. When this condition is fulfilled, the incident and dispersed ray make the same angle with the groove surface, and the grating facets act kind of like tiny mirrors—the grating literally 'blazes' in reflected light, and is said to 'be blazed' at this wavelength and spectral order. The blaze can be tuned by varying the facet angle δ.

But we know that mirror reflectivities in the soft X-ray band are definitely finite, and less than unity. As a stopgap measure, one therefore multiplies ϵ_m by the appropriate mirror reflectivity R_λ to 'correct' for this effect. This is physically not consistent, because it violates the choice of boundary conditions underlying Eq.(26), and the conceptual crisis is evident once again when it comes to choosing the incidence angle at which to evaluate the reflectivity correction factor: except at the blaze, the angle of incidence on the facets $\alpha + \delta$ and the 'reflection' angle $\beta - \delta$ are not equal. Often, the mean of R_λ evaluated at both angles is used: $R_\lambda \equiv [R_\lambda(\alpha + \delta) R_\lambda(\beta_m - \delta)]^{1/2}$, and the full grating efficiency is defined as $\eta_m = R_\lambda \cdot \epsilon_m$.

Theories that incorporate solutions to the full Maxwell equations are designated 'vector theories', to distinguish them from the scalar theory outlined above. All vector theories of grating reflectivity (Stroke 1967, Sect. V, and the references therein) employ the periodicity in the direction along the plane of the grating, and start with a Fourier expansion, with constant coefficients for the far fields; the components of the reflected field are the various diffraction orders. Following Rayleigh, one makes the assumption that this expansion is valid everywhere (that is, that the coefficients do not depend on the coordinate direction perpendicular to the grating plane), including near and at the grating surface. Imposing the boundary conditions on the grating surface leads to an infinite set of coupled linear equations, which, when suitably cut off at some maximum harmonic, can in principle be solved numerically. A further traditional simplification arises when one assumes that the grating material has infinite conductivity (i.e., imaginary part of the complex index of refraction equal to zero).

But if you have to resort to using a computer to evaluate the reflectivities in these approximate vector theories, you might as well solve the Maxwell equations directly and exactly, and apply the correct boundary conditions with no restrictions on the properties of the grating material. Various schemes have been worked out to accomplish this (Petit 1980). In the conceptually simplest of these, a Fourier expansion is again substituted in Maxwell's equations, but this time one allows for the general dependence of the expansion coefficients on the vertical distance to the grating. The resulting set of coupled ordinary differential equations for the Fourier coefficients can again be solved numerically with a suitable cutoff (see for instance Jark & Nevière 1987). These numerical procedures also naturally allow for arbitrary grating groove shapes.

Figure 23 displays an example comparison between the scalar diffraction and the EM theory efficiencies for a Gold grating of parameters very similar to those for the gratings in the RGS. The general features of both calculations are roughly the same (such as the blazing behavior - the blaze was chosen to be at 15 Å in $m = -1$), but for accurate predictions the scalar theory is clearly inadequate.

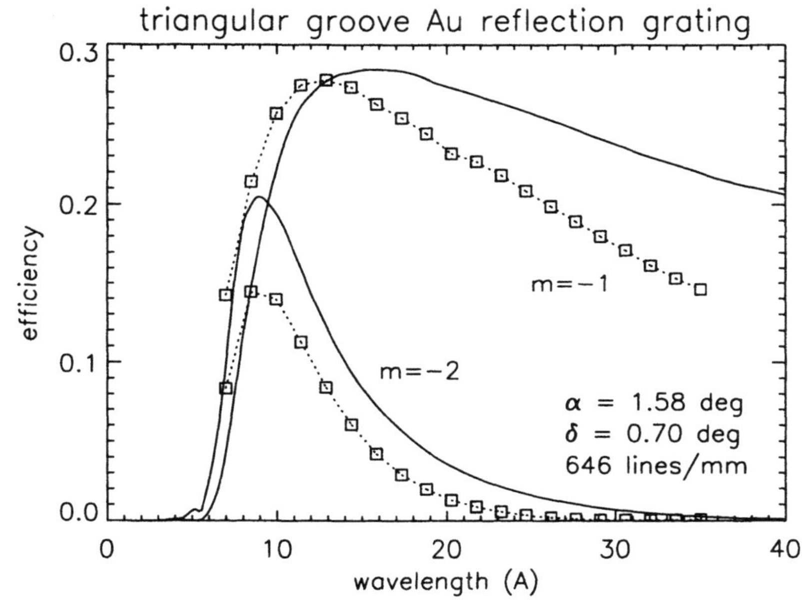

Fig. 23. Absolute efficiency of a gold reflection grating of triangular profile, illuminated at an angle $\alpha = 1.58$ deg; the line density is 646 lines/mm, the groove facet tilt angle $\delta = 0.70$ deg. The solid line gives the scalar diffraction result, the datapoints were calculated from a numerical solution to Maxwell's equations developed for the RGS project.

Reflection grating spectrometers have a few more free parameters than transmission grating spectrometers (angle of incidence α, groove shape, spacing between adjacent gratings in the array), so optimizing the design is a little more complicated. We will discuss the optimization of the design for the RGS as an example.

The primary design drivers are of course the chosen wavelength band of operation, and the resolving power at some chosen fiducial wavelength. You require that the design deliver the optimum throughput. For the RGS, the (minus) first-order wavelength band was chosen to be $5 - 35$ Å. The short wavelength cutoff is dictated by the properties of suitable grating materials.

All elements near $Z = 80$ (Au, Pt, Ir) have deep M-shell absorption edges near 2.5 keV, which provides a natural boundary. The upper limit was loosely chosen with atomic physics, and the absorbing properties of the interstellar medium in mind; many sources are very faint well beyond the O K edge at 23 Å. Also, a much wider band sampled at resolving powers of at least $\mathcal{R} = 250$ would have required a physically much larger focal-plane detector. The $5 - 35$ Å band contains the K-shell spectra of the abundant elements C through Si (for C, only the Hydrogenic species is in the band), as well as the diagnostic-rich Fe L spectra. The blaze wavelength was chosen in the middle of the band, $\lambda_B = 15$ Å, roughly where the Fe L $n = 3 - 2$ emission lines occur.

The resolving power at blaze, for given telescope angular resolution $\Delta\alpha$, depends of course on the angle of incidence α and the grating period d. It will be convenient to rewrite it as (Eq. 9):

$$\begin{aligned}\mathcal{R}_B = \lambda_B/\Delta\lambda &= \frac{\cos\alpha - \cos\beta_B}{\sin\alpha\Delta\alpha} = \\ &= \frac{(1-\sin^2\alpha)^{1/2} - (1-\sin^2\beta_B)^{1/2}}{\sin\alpha\Delta\alpha} \approx \\ &\approx \frac{(\sin\alpha + \sin\beta_B)(\sin\beta_B - \sin\alpha)}{2\sin\alpha\Delta\alpha} \approx \\ &\approx \frac{\gamma}{\Delta\alpha}\left(\frac{1}{\eta} - 1\right)\end{aligned} \tag{31}$$

where we have defined the graze angle γ on the groove facets, $\gamma \equiv \alpha + \delta = \beta_B - \delta$, used small angle approximations, and defined the ratio

$$\eta \equiv \sin\alpha/\sin\beta_B \tag{32}$$

(don't confuse this η with the efficiency of a grating; the coincidence in nomenclature is unfortunate, but the use of η for both is common in papers on the RGS). For a given telescope angular resolution $\Delta\alpha$, the resolving power at blaze therefore depends on η and γ.

Next, we calculate the throughput at blaze. From Eq.(26), you see that the diffraction efficiency ϵ_B at blaze is equal to $\epsilon_B = \sin\alpha/\sin\beta_B \equiv \eta$, and to obtain the grating reflectivity we multiply by the mirror reflectivity at the graze angle γ, at the chosen blaze wavelength, $R_{\lambda_B}(\gamma)$ (we use scalar theory for analytical convenience). The gratings cannot be placed infinitely close together—rays dispersed at large enough angles will be intercepted at the back of the next grating in the array, so there is a limit to how densely you should cover the focused beam in order to intercept the optimum amount of light. Packing the gratings more densely intercepts a larger fraction of the beam, but also increases the vignetting. For the RGS, the gratings were spaced such that all rays up to the first-order blaze wavelength escape shadowing (ray dispersed from the top of one grating barely grazes the bottom of the back side of the next grating in the array). That implies that the fraction of the beam that is intercepted is equal to η, and the balance $1 - \eta$ passes unhindered through the grating array and can be used by the detectors at the prime focus. So the total throughput at blaze, t_B, is the product of the fraction light intercepted times the reflectivity of the gratings:

$$t_B = \eta^2 R_{\lambda_B}(\gamma). \tag{33}$$

You can see that there is a tradeoff between resolving power and throughput: lowering η increases the resolving power, but decreases the throughput, and *vice versa*.

The optimization might proceed as follows. Choose the resolving power at blaze; the telescope resolution $\Delta\alpha$ is given. Now examine the throughput at blaze as a function of η: for every value of η, Eq.(31) gives the corresponding value of γ, and η and γ together determine the throughput at blaze. For $\Delta\alpha = 30''$ (the original optimization for the RGS was based on a figure of $\Delta\alpha = 30''$), a resolving power at blaze of 300, and a 15 Å blaze, I plot this throughput as a function of η in Figure 24; the grating material is gold. With increasing η, the throughput first rises roughly as η^2. As η approaches unity, γ rapidly increases, and therefore the reflectivity $R_{\lambda_B}(\gamma)$ sharply decreases. Therefore, there is a maximum in the curve, in this case at $\eta = 0.50$. Given this optimum η, γ follows from Eq.(31); for $\eta = 0.50$, we find $\gamma = 2.50°$. From $\alpha + \delta = \beta_B - \delta = \gamma$, we find

$$\delta = \gamma \frac{1 - \eta}{1 + \eta} \tag{34}$$

so in our case, the facet tilt angle $\delta = 0.833$ deg. The grating period follows from the dispersion equation at blaze, $\lambda_B = 2d \sin\gamma \sin\delta$: $d = 11823$ Å, or a density of 846 lines mm^{-1}. Finally, the angle of incidence is $\alpha = \gamma - \delta = 1.67°$.

The actual optimization for the RGS was done slightly differently; there, the physical length of the detector (nine CCD chips long) was held constant instead of the resolving power at blaze. This yields an alternative relation between η and γ. The optimum value for η was chosen as a compromise between throughput in $m = -1$ and $m = -2$, and lies at $\eta = 0.53$, consistent with the overall requirement that approximately half the light should go to the EPIC detectors. The resulting actual design parameters for the RGS are listed in Table 4. Note that the zero order does not fall on the detector, so that the exact wavelength scale will have to be derived from accurate knowledge of the geometry, and the pointing direction of the satellite.

The design is completed by choosing the arrangement of the gratings in the focused beam. In the RGS, the gratings are placed on a Rowland circle as shown in Figure 10(b). As discussed earlier, the actual radius of the Rowland circle is arbitrary for the grazing-incidence reflection grating instrument, as long as the line density gradient on the gratings matches the Rowland circle radius. For the RGS, the center of the grating array was placed at 6700 mm from the telescope focus. The distance to the spectroscopic focus at the

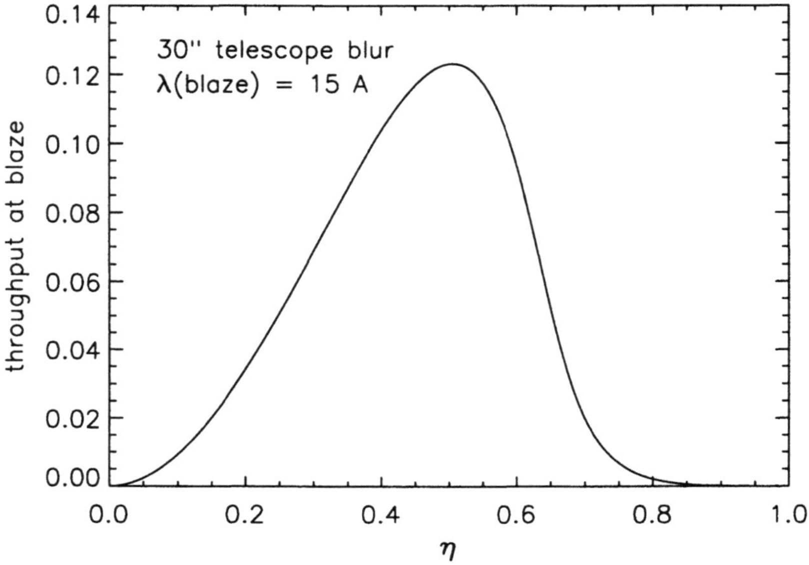

Fig. 24. Throughput at blaze, for an array of gold reflection gratings, with blaze wavelength $\lambda_B = 15$ Å, a resolving power at blaze of 300 with a 30″ telescope blur, as a function of the parameter $\eta = \sin\alpha / \sin\beta_B$.

blaze wavelength was chosen to be equal to this same distance, which implies that the Rowland circle geometry is nicely symmetric between the focused telescope and first-order blaze beams. The entire set of gratings lies on the toroidal surface generated by rotating the geometry around an axis passing through the telescope focus and first-order blaze focus; this arrangement minimizes the astigmatism of the spectrometer. In all, with a grating size of 100×200 mm, of order 200 gratings per array are needed to cover the beam.

5.3 Implementation of the design, and actual performance of the RGS

Implementation The gratings were produced by replication in gold on an epoxy layer from a mechanically ruled master grating. The master grating was ruled directly into gold with an interferometrically controlled ruling machine, which also allowed direct implementation of the required line density variation (using the same density gradient on all gratings introduces only a negligible aberration, as it turns out). The groove shape is approximately triangular, with a tilt angle close to the design value of $\delta = 0.70°$. X-ray performance tests of master gratings are described in Bixler et al. (1991). Performance of replicated gratings, and a preliminary physical model for the

Table 4. Design Parameters of the Reflection Grating Spectrometer on *XMM*

Mean Line Density	646 lines mm^{-1}
Blaze Angle δ	0.70 deg
Mean Graze Angle γ	2.28 deg
Angle of Incidence α	1.58 deg
Dispersion Angle at Blaze β_B	2.97 deg
Blaze Wavelength λ_B	15 Å
Fraction of Beam Intercepted η	0.53
Resolving Power at Blaze	290

grating efficiency, is decribed in Kahn et al. (1996). The primary variable in this model is the actual average groove shape, loosely characterized by the mean slope δ of the grating facets. From the measured angular, wavelength, and order dependence of the efficiency, a more detailed shape has been derived, with turns out to have small but noticeable departures from the triangular shape.

As discussed in Sect. 3.1, the gratings need to be very flat, and very accurately aligned in order to preserve the full resolution implied by the angular width of the telescope beam. In practice, this amounted to a combined flatness and alignment requirement of $\Delta\phi \lesssim 5$ arcsec. Flatness was achieved by replicating gratings onto lightweight, very flat SiC substrates, which were stiffened in the dispersion direction with thin support 'ribs' on the back of the substrates. In practice, the flatness of the substrates is such that this contribution to the spectrometer resolution is almost negligible. The grating-to-grating alignment requirement (both with respect to position and orientation) was one of the drivers on the design of the overall 'integrating structure' in which the gratings are held, directly behind the mirror module in the focused X-ray beam. The required angular precision is achieved by a 'multi-scale' design, where the actual 'grating box' provides low-precision mounts for high-precision elements (which in turn hold the gratings), which are interferometrically aligned to micron accuracy (Kahn et al. 1996). Figure 25 shows a picture of a complete RGA. The RGA's were designed and built by Columbia University (PI: Steven M. Kahn).

Resolving power and shape of the response In practice, the telescope blur and finite accuracy of grating-to-grating angular aligment contribute approximately equally to the resolution of the RGS. The actual performance of the telescopes has steadily and significantly improved over its original specification, whereas the grating alignment is fundamentally limited by the accuracy of the interferometric aligment procedure. The contribution of the

Fig. 25. Schematic of a single Reflection Grating Array for the RGS. The focused X-ray beam is incident from the top. The 182 gratings are placed on a Rowland toroidal surface. The lightweight structure holding the gratings is manufactured from beryllium, and is mounted with the triangular mounts directly behind the mirror module. The reflective faces of the gratings face forward; the five stiffening ribs on the back of each grating are also visible.

telescope blur according to Eq.(8) scales like

$$\Delta\lambda = (d/m)\sin\alpha\Delta\alpha, \tag{35}$$

which is strictly independent of wavelength, for given α and telescope blur $\Delta\alpha$. Relative misalignments, characterized by a angular spread $\Delta\phi$ produce a weak dependence on wavelength (cf. Section 3.1):

$$\Delta\lambda = (d/m)\sin\alpha(1+\sin\beta/\sin\alpha)\Delta\phi. \tag{36}$$

and the telescope contribution becomes progressively less important with increasing dispersion angle. As it is, the telescope blur and misalignment contributions are about equal at the short wavelength end of the band. Figure 26 shows the predicted resolving power of the RGS (based on the FWHM of the profile, not the HEW, to facilitate comparison with the other instruments), calculated with a dedicated raytrace code. Both first- and second-order resolving power are plotted this time, because the second-order efficiency of the instrument is actually appreciable.

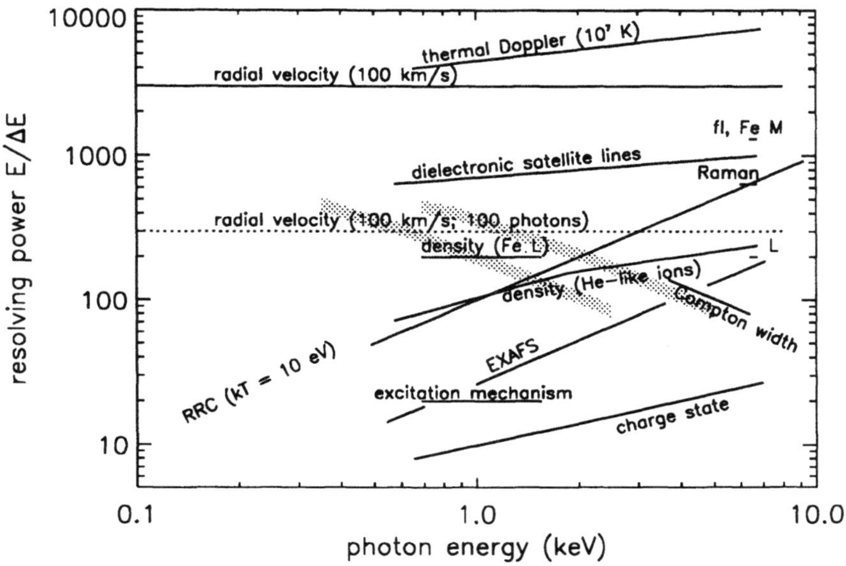

Fig. 26. Predicted resolving power of the RGS, in orders $m = -1$ (*lower gray band*) and $m = -2$ (*upper gray band*), overlaid on the set of spectroscopic diagnostics. The second-order bandpass formally extends up to 5 keV, but the effective area is small above 2.5 keV. (courtesy Jean Cottam, Columbia University).

The gratings exhibit scattering, in this case (just like for a mirror surface) by low-amplitude microroughness on the surface, which gives rise to a separate, wider component to the spectrometer profile (period fluctuation scattering is negligible, because the spacing of the grooves was actively controlled during the ruling of the master grating). The calculation of the effect of the roughness on a grating surface is entirely analogous to the calculation for grazing-incidence X-ray scattering by a mirror (e.g., Church, Jenkinson & Zavada 1977, and the references given by Klos 1985). In that case, one imagines the perturbations to the flat surface to be decomposed into their equivalent Fourier series. Each sinusoidal component in this expansion basically represents a sinusoidal grating on the mirror surface, which diffracts light out of the specularly reflected beam, into a series of ghost images on either side of the specular beam. Again, a distribution of sine waves with different spatial frequencies will produce a distribution of ghosts, i.e., a continuous scattering distribution, with the higher spatial frequency sine waves scattering light further from the core (through the dispersion relation for the light diffracted by the sine waves). The shape of the scattering distribution will again reflect the power spectral density of the surface perturbations, an intrinsic property of the mirror surface. One often characterizes this power spectral density by a spatial-frequency bandwidth, which is then inversely proportional to the corresponding correlation length of perturbations on the mirror surface. The total amount of scattered light is again completely determined by the r.m.s. roughness of the surface (r.m.s. amplitude of the perturbations), just as with the transmission grating scattering.

Just as with calculating the reflectivity of a reflection grating, one has to face the fundamental question of how to impose the boundary conditions to the electromagnetic field on the scattering surface, and there are again 'scalar' and 'vector' scattering theories. The simplest calculation is of course the scalar calculation, assuming very-low-amplitude long spatial-wavelength perturbations, such that only the first-order diffraction by the Fourier components of the roughness has to be taken into account. This calculation is straightforward (although the standard papers on the subject are somewhat confusing). The extension to a periodically modulated surface (a grating) is also straightforward, and is decribed in Spodek et al. (1998), Kahn et al. (1996), and Paerels et al. (1994).

The expression for the fraction of light scattered in the lowest-order scalar diffraction scattering theory is closely analogous to the expression we found for the fraction of light scattered by period fluctuations in a transmission grating (σ is the r.m.s. amplitude of surface perturbations):

$$f_m = k^2\sigma^2(\sin\alpha + \sin\beta_m)^2 \tag{37}$$

which exhibits a rather strong wavelength- and spectral-order dependence. Just as with the transmission grating scattering, we expect that you can prove (under the usual assumptions) that this factor is just the lowest-order approximation to a Debye-Waller factor of the form

$$f_m = 1 - \exp\left(-k^2\sigma^2(\sin\alpha + \sin\beta_m)^2\right) \tag{38}$$

The characteristic angular width Γ of the scattering distribution scales inversely proportional to the correlation length l of the perturbations:

$$\Gamma \equiv (kl\sin\beta_m)^{-1} \tag{39}$$

As a general rule of thumb, with decreasing wavelength or increasing spectral order, the fraction scattered light increases, while the angular distribution narrows.

The RGS gratings, we found, exhibit scattering on two distinct angular scales, corresponding to correlation lengths of order $l \sim 10 - 12$ μm and $l \lesssim$ few grooves, respectively. The first gives rise to a component of scattered light close to the core of the spectrometer profile (the angular scale corresponds to $\sim$ 8 resolution elements at blaze), the second basically produces a faint, almost constant background of inter-order light. The long-correlation length roughness has an amplitude of $\sim 11 - 13$ Å, which implies a fraction scattered light of $\sim$ 13% at blaze. The effect of this scattered light has not been included in the resolving power plot (Fig. 26). Because the profile consists of a sharp core, and fainter, more extended scattering wings, the width of this core is the determining factor when deciding whether a particular set of spectral features can be resolved.

The small-correlation length scattering still poses some conceptual problems. The small-angle approximation is not valid when you try to calculate its exact angular distribution—in a sense, the scattered photons do not 'belong' to a particular order anymore, which means that you really have to calculate the statistical average of the entire diffraction pattern, assuming a statistical distribution for the high-spatial-frequency surface perturbations. However, the fraction of light removed from the sharp diffraction peaks is well calibrated, and corresponds to a surface roughness of $\lesssim 11$ Å. Since the RGS is read out with CCD's, this scattered light is neatly separated from true continuum radiation in a pulse height vs. dispersion coordinate plot, and so the lack of a precise calculation for its angular distribution may not matter very much in practice. The large-angle scattering wil show up mostly as a 'lost light' correction factor to the grating efficiencies (and that is how it was found in the first place).

Effective area Fig. 27 shows the effective area of the RGS (mirror + RGA + detector), for the two modules combined. The Figure also shows the effective areas for three other high-resolution spectrometers, the *AXAF* HETGS and LETGS and the microcalorimeter on *Astro-E* ('X-ray Spectrometer', or XRS, to be discussed below). For the RGS we show both first and second order, because the second order actually has appreciable area. All curves are based on data and models which were available before the final instrument calibrations were performed, so they may (have) become slightly inaccurate.

All of the above instrument properties have been verified and characterized at the long beam X-ray testing facility 'Panter' of the Max Planck Institut für Extraterrestrische Physik near Munich (Rasmussen et al. 1998). The same physical models we have described here were also used to predict the performance of the instruments in this facility; the free parameters in the models were set to the values derived from sub-assembly calibrations, and fine tuned until agreement of the entire instrument performance with the data was reached. This roundabout 'calibration' is necessary, because the laboratory configuration is not fully representative of the flight configuration; with a 7.5m telescope focal length, you can still tell the difference between a source at a finite distance in the laboratory (125 m), and a source at infinity (the same was true for the *AXAF* calibration, with the additional complication of the effect of terrestrial gravity on the shape of the mirrors).

5.4 Examples

As a first example of the RGS performance, we show a laboratory calibration spectrum, obtained with a gold-anode electron impact X-ray source. Figure

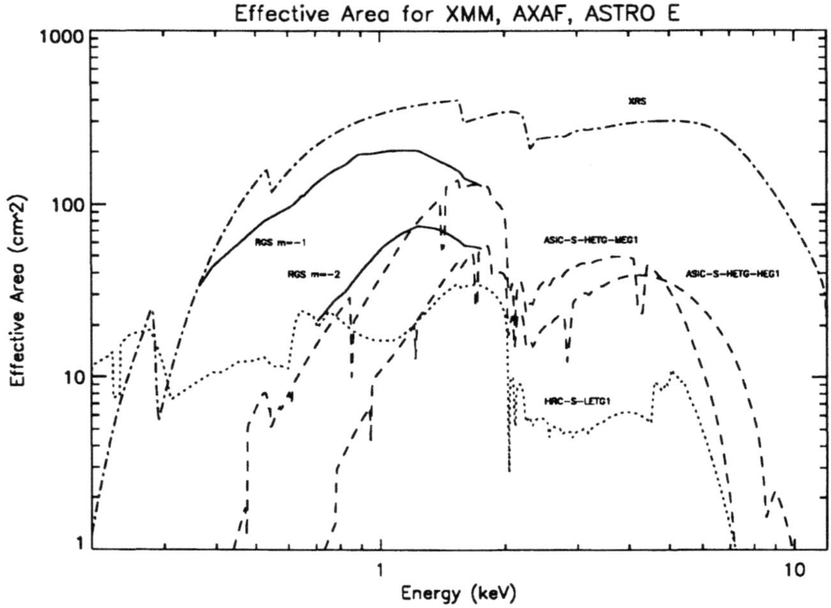

Fig. 27. Effective area of the RGS, compared with three other high-resolution spectrometers (courtesy Jean Cottam, Columbia University).

28 shows the CCD pulse height vs. dispersion coordinate diagram. This spectrum is interesting, because it actually looks a little bit like an astrophysical spectrum, and because the pulse height/dispersion plot once again demonstrates the power of performing analysis in that plane. CCD pulse height increases vertically, dispersion runs horizontally, with wavelength increasing to the left. The left edge of the image corresponds approximately to 35 Å in first order, the right edge to approximately 5 Å and the horizontal scale is approximately linear in dispersion angle β. You can see the small gaps between the nine individual CCD chips if you look closely. The curved dark bands are the continuum again (compare Fig. 14 for the *AXAF* HETGS); spectral orders $m = -1$ through -4 are visible in the continuum. Superimposed are Au fluorescence lines, the strongest being an $n = 4-3$ transition at approximately 6 Å (at the right edge of the plot). In the horizontal direction you can distinguish five spectral orders of this line, as well as the large-angle scattering referred to above. Notice how the fourth order nearly coincides in position along the dispersion direction with the first order O K emission line. In the vertical direction, centered on each emission line, you see the CCD response to monochromatic radiation, in addition to a small number of piled-up events.

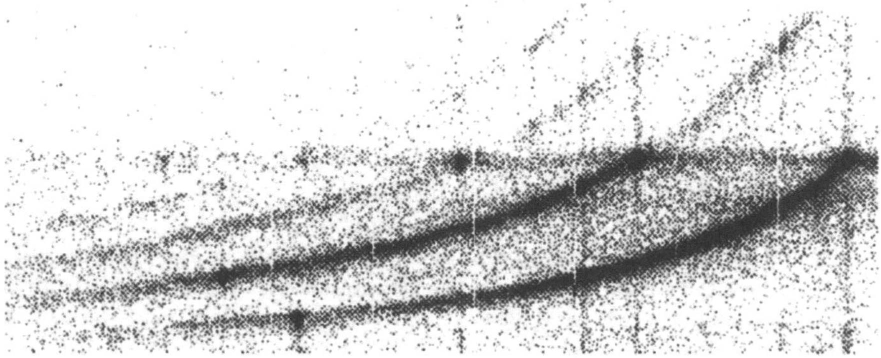

Fig. 28. Pulse height/dispersion coordinate diagram, for a laboratory calibration Au continuum spectrum. CCD pulse height increases vertically, dispersion runs horizontally, with wavelength increasing to the left. The left edge of the image corresponds approximately to 35 Å in first order, the right edge to approximately 5 Å (courtesy Masao Sako, Columbia University).

Finally, an astrophysical example. With the advent of CCD spectroscopy on *ASCA*, it was found, as had been suspected, that the soft X-ray spectra of Seyfert 2 galaxies are dominated by emission lines (e.g., Kunieda et al. 1994; Matt 1996; Ueno et al. 1994; Iwasawa et al. 1994; Iwasawa et al. 1997). The current view on the different properties of Seyfert 1 and 2 objects is that these are merely due to different viewing angles into a highly intrinsically anisotropic environment. The nuclear region may be surrounded by an absorbing torus of cool material, and in Seyfert 2's, the strong X-ray continuum arising in the region near the central compact object is blocked from direct view, leaving a large equivalent width discrete emission spectrum evidently associated with a region still in view. Such regions should also exist in Seyfert 1's, but their line emission is harder to detect against the bright nuclear continuum.

There is great interest in determining the location and origin of the line emission: it could be associated with hot, shocked gas in a strong circumnuclear starburst (detected at IR wavelengths), or in a cooler X-ray photoionized region associated with the very nucleus of the object, or a combination of both. A complete understanding of the relative contributions of both emission mechanisms would shed light on the nature of the currently suspected connection between Seyfert-type nuclear activity in galaxies, and starbursts. As we have seen in Section 2, the soft X-ray line spectrum is highly sensitive to the ionization and excitation mechanism that drives the emission.

Figure 30 shows a simulated EPIC spectrum of the famous Seyfert 2 NGC 4945 (Done, Madejski & Smith 1996), which basically looks like a much brighter version of the *ASCA* spectrum. You can see the continuum, scattered into our line of sight (the intrinsic nuclear continuum does not start to become detectable through the obscuring material until you reach photon energies $E \gtrsim 10$ keV), a strong fluorescent Fe K complex, and unresolved soft X-ray line emission in the 500-1500 eV band. At this resolution, it is impossible to uniquely distinguish between the various excitation mechanisms for these lines, let alone measure average abundances, velocity fields, or diagnose radiative-transfer effects expected to be associated with an origin in X-ray photoionized gas.

Figure 29 shows the corresponding RGS spectrum, where the discrete emission detected with *ASCA* has been represented with emission from a collisional plasma at $kT_e = 0.35$ keV and solar abundances, normalized to produce the correct amount of line power. With spectra of this quality, we can finally address the questions mentioned above directly, and in detail. For faint objects, like the Seyfert 2's, the large throughput of the RGS will be required to obtain well-exposed spectra.

Some of the most interesting objects for soft X-ray spectroscopy are, or may be, extended: supernova remnants, clusters, the circumnuclear regions in nearby AGN, ordinary galaxies. As mentioned in Section 4.2, the resolution of a diffraction grating spectrometer degrades more or less linearly with source extent, if the source is resolved by the telescope. But since the RGS relies on a medium resolution telescope (15″ HEW), it is relatively insensitive to this effect. Sources up to a sizeable fraction of an arcminute still yield an RGS spectrum with much higher resolving power than a CCD; examples are the compact supernova remnants in the LMC, and cooling flows in clusters.

6 The Objective Crystal Spectrometer on *Spectrum X/γ*

The Objective Crystal Spectrometer (OXS) will be flown on the *Spectrum X/γ* observatory, which is a collaboration between Russia and a large number of institutions in Europe and the US. The experiments on it cover the entire UV through hard X-ray bands. Of these, the OXS is interesting in the present context, and in fact will reach the highest resolving power of all experiments discussed. The launch date for *Spectrum X/γ*, unfortunately, is highly uncertain. You can find information on the project at `http://hea.iki.rssi.ru/SXG/ SXG-home.html` (Institute for Space Research (IKI), Moscow, Russia).

The OXS (PI: H. Schnopper, Danish Space Research Institute) consists of a 4000 cm^2 panel of flat crystals, which can be placed in front of the aperture of the SODART (Soviet-Danish X-ray Telescope), such that the diffracted radiation is sent into the focusing telescope for imaging at the focal plane. A number of different energy bands are covered with different crystals, centered on different important emission line complexes. LiF crystals cover the Fe K band ($5.0 - 7.4$ keV at resolving power $\mathcal{R} = 1250$), Si crystals cover the S and Ar K complexes ($2.3 - 4.6$ keV at resolving power $\mathcal{R} = 3200$), and RAP crystals cover the O K band ($0.55 - 0.81$ keV at resolving power $\mathcal{R} = 770$).

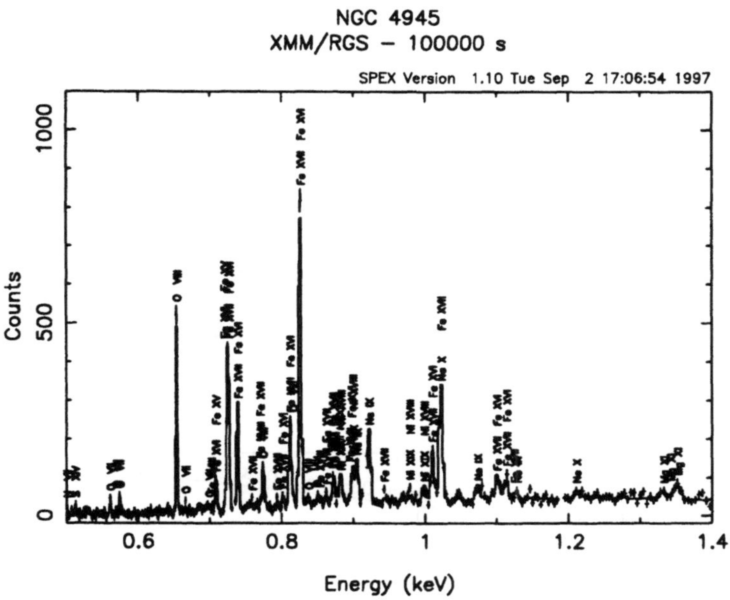

Fig. 29. Simulated RGS spectrum for the Seyfert 2 galaxy NGC 4945 (courtesy Jelle Kaastra, SRON).

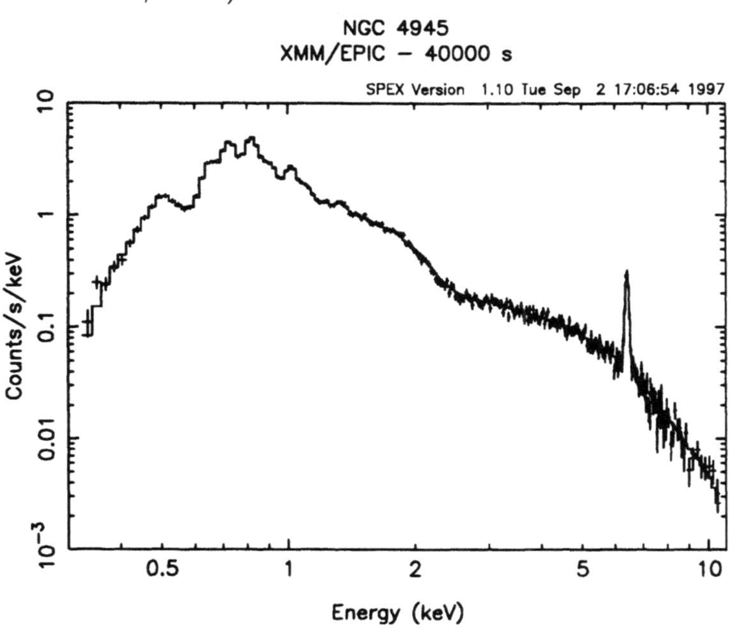

Fig. 30. Simulated EPIC spectrum for the Seyfert 2 galaxy NGC 4945 (courtesy Jelle Kaastra, SRON).

Finally, there is a Co/C multilayer which covers the band below the C K edge (0.175 – 0.280 keV, at resolving power $\mathcal{R} = 50$).

These resolving powers apply to observations of point sources, where one has to scan through a range of Bragg angles by moving the crystal panel with respect to the source direction; the telescope plus focal plane detector merely act as a very sensitive photometer. The resolving powers for the three highest-energy bands are plotted in Figure 31. As you can see, they are sufficient to detect almost all spectroscopic diagnostics we have discussed. The observations will be very slow, though: to scan the He-like S 'triplet' requires scanning a $\gtrsim$ 30 eV band at 2500 eV. At a resolving power of 3200, that requires $\gtrsim 30 \times (3200/2500) = 38$ separate exposures (Bragg angle steps)!

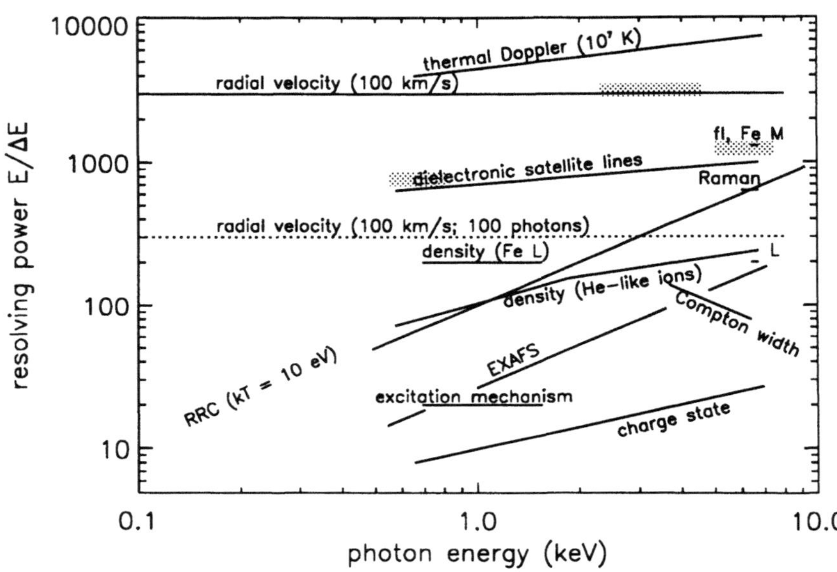

Fig. 31. Resolving powers for the three highest-energy bands of the OXS.

For an extended source, the OXS has an interesting multiplexing property. Light from different parts of the source strikes the crystals at different angles, which means that different photon energies are reflected and imaged from different parts of the source simultaneously (Figure 32). For a given orientation of the crystal/telescope, a weird image appears in the focal plane: it is monochromatic, but in different wavelengths in different spots! By changing the orientation, you can scan through a range of Bragg angles, and obtain a spectrum for every spot in the source. You cannot do this with a grating—since a grating diffracts all wavelengths at the same time, the image of an extended source in a slitless spectrometer like the grating spectrometers on *AXAF* and *XMM* has the wavelength and spatial information completely scrambled. But note that now that the source is no longer a point source, your ability to separate two different wavelengths is effectively limited by the angular resolution of the system (crystal rocking curve plus telescope resolution), whereas the spectral resolving power for a point source is limited by the rocking curve only.

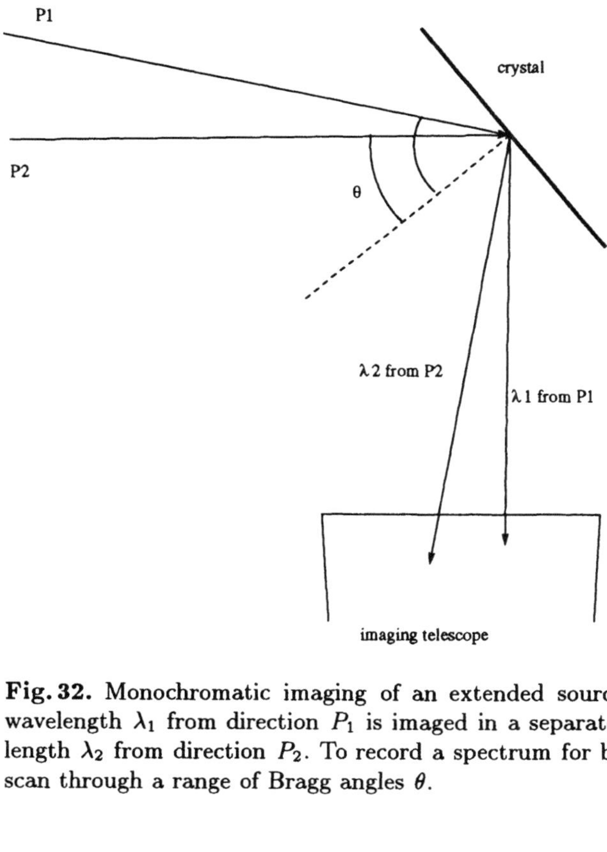

Fig. 32. Monochromatic imaging of an extended source with the OXS: light of wavelength λ_1 from direction P_1 is imaged in a separate spot from light of wavelength λ_2 from direction P_2. To record a spectrum for both P_1 and P_2, you again scan through a range of Bragg angles θ.

7 The Microcalorimeter Experiment on *Astro-E*

7.1 Introduction

The principle behind the operation of a microcalorimeter is simple: a photon is absorbed in a piece of material. Its energy is converted into heat, and the temperature rise is a measure of the photon energy (Moseley, Mather & McCammon 1984). An idealized detector consists of an absorber with a thermometer, connected to large heat bath at constant temperature.

A simplified theory of the device proceeds as follows. Assume the photon energy is instantly and completely converted into heat, and thermalized. The thermometer reacts instantly to this change in temperature. In the absence of incident X-ray photons the total energy of the system formed by the heat reservoir and absorber plus thermometer is constant, and the statistical properties of the entire system can be described with the microcanonical ensemble (Kittel 1958, Ch. 4). From this, the energy of any small subsystem, such as the X-ray absorber, is found to fluctuate spontaneously, and evidently there is a continuous exchange of energy between the subsystem and the heat reservoir. The subsystem is described by the canonical ensemble, and the calculation of the amplitude of the energy fluctuations is straightforward. For completeness, we will do the calculations in the next subsection. In this idealized instrument, the energy resolution is determined by the spontaneous thermodynamic fluctuations in the energy of the absorber.

But first, a subtle point. The photon energy is determined from the temperature rise in the absorber, so, strictly speaking, it is spontaneous fluctuations in the *temperature* of the absorber we want to know about. In a real microcalorimeter, energy flows back and forth across the link to the reservoir, and the total thermal energy in the absorber fluctuates. If the time constant associated with the heat conduction is long compared to the time scale for thermalization of the energy in the absorber, the temperature in the absorber will fluctuate. This situation is not described by the canonical ensemble, which assumes complete thermodynamic equilibrium. However, Mather (1982) has shown that as long as the temperature gradients are small, the simple thermodynamic result still applies, multiplied by a factor of order unity, which depends on the temperature dependence of the thermal conductivity of the link. For simplicity, we will ignore this factor here.

It will turn out that at very low temperatures the fluctuations in the energy of the absorber become very small (in fact, they vanish at zero temperature) due to the quantized nature of the vibrational motions (heat) in the absorber crystal.

7.2 Thermodynamic fluctuations

A system in thermodynamic equilibrium in contact with a large heat reservoir at constant temperature can be described by the canonical ensemble. The probability of finding a member of the ensemble, or a single system at a given instant in time, in a state n of total energy E_n is given by

$$p_n = \frac{e^{-\beta E_n}}{\sum_n e^{-\beta E_n}} \tag{40}$$

where $\beta \equiv 1/kT$. The quantity $\mathcal{Z}$,

$$\mathcal{Z} \equiv \sum_n e^{-\beta E_n} \tag{41}$$

is the partition function, which carries all the information about the thermodynamic properties of the system through the enumeration of the E_n.

We can derive an expression for the variance of the energy fluctuations as follows. Taking the derivative with respect to β, we find

$$\frac{\partial \mathcal{Z}}{\partial \beta} = -\sum_n E_n e^{-\beta E_n} \tag{42}$$

so that the average energy $\langle E \rangle$ of the system is

$$\begin{aligned} \langle E \rangle &= \frac{\sum_n E_n e^{-\beta E_n}}{\sum_n e^{-\beta E_n}} = \\ &= -\frac{1}{\mathcal{Z}} \frac{\partial \mathcal{Z}}{\partial \beta} \end{aligned} \tag{43}$$

The variance of the fluctuations in the energy is defined as

$$\begin{aligned} \langle \delta E^2 \rangle &\equiv \langle (E - \langle E \rangle)^2 \rangle = \\ &= \langle E^2 \rangle - \langle E \rangle^2 \end{aligned} \tag{44}$$

Taking another derivative, you find

$$\langle E^2 \rangle = \frac{1}{\mathcal{Z}} \frac{\partial^2 \mathcal{Z}}{\partial \beta^2} \tag{45}$$

so that

$$\langle \delta E^2 \rangle = \frac{1}{\mathcal{Z}} \frac{\partial^2 \mathcal{Z}}{\partial \beta^2} - \frac{1}{\mathcal{Z}^2} \left(\frac{\partial \mathcal{Z}}{\partial \beta} \right)^2 \tag{46}$$

which can be rewritten

$$\langle \delta E^2 \rangle = -\frac{\partial \langle E \rangle}{\partial \beta}, \tag{47}$$

or in terms of temperature,

$$\langle \delta E^2 \rangle = kT^2 \frac{\partial \langle E \rangle}{\partial T}. \tag{48}$$

Since the specific heat at constant volume, c_V, is defined as $c_V \equiv \partial \langle E \rangle / \partial T$, this can also be written as

$$\langle \delta E^2 \rangle = kT^2 c_V. \tag{49}$$

This is a general thermodynamic relation, valid regardless of the precise nature of the system, provided it is in equilibrium.

Now we have to calculate c_V for our system. As a model, we will assume a pure insulating crystal. In that case, the only contribution to the specific heat is due to elastic vibrations of the ions in the crystal lattice. As you remember, at high temperatures such that classical physics applies, c_V is a constant because the average energy per degree of freedom for each particle is equal to $\frac{1}{2}kT$. Each vibrating particle has six degrees of freedom, so for N particles

$$c_V = \frac{\partial \langle E \rangle}{\partial T} = \frac{\partial (N \cdot 3kT)}{\partial T} = 3Nk. \tag{50}$$

But at low temperatures, the quantized nature of the vibrations becomes evident and with decreasing temperature, eventually you start noticing that more and more oscillators are in the ground state, and that the amount of heat absorbed or given up per degree of temperature rapidly declines with decreasing temperature. This happens when kT becomes comparable to or goes below the energies $\hbar\omega_0$ associated with the highest eigenfrequencies ω_0 of the vibrations of the solid. The associated temperature is referred to as the Debye temperature, Θ:

$$\Theta \equiv \hbar\omega_0/k. \tag{51}$$

Well below the Debye temperature, the heat capacity must rapidly decrease with decreasing temperature. It is this fact that makes the calorimeter an attractive spectrometer at low temperatures: the specific heat can be made very small at low temperatures, and hence the thermodynamic fluctuations in the energy can be made very small.

The standard calculation for the specific heat associated with the lattice vibrations of a solid at low temperatures goes as follows (e.g., Peierls 1955). First, you write down the coupled equations of motion for the particles in the crystal, and you apply boundary conditions to find the normal modes. One usually applies periodic boundary conditions, in which case the allowed values of the wavevector $\mathbf{f}$ form a lattice, with a density of $V/(2\pi)^3$, where V is the volume of the crystal. The eigenfrequencies are $\omega(\mathbf{f}, s)$. For a lattice containing r particles in the unit cell, there are in principle $3r$ solutions with the same $\mathbf{f}$ to the boundary value problem, each with a different frequency $\omega(\mathbf{f}, s)$. In the case of a simple lattice with only one particle in the unit cell, the three solutions belonging to the same $\mathbf{f}$ are two transverse and one longitudinal wave, in which case s simply labels the three possible polarizations. This remains true for more complicated unit cells, for very-long-wavelength (low-frequency) vibrations, which correspond to relative motions of entire, undistorted unit cells with respect to each other.

You can write the total energy (kinetic plus potential) of the crystal in terms of the canonical coordinates of the normal modes. You then interpret these coordinates as operators and apply the commutation rules to them. The resulting expression for the total energy now looks like a sum over discrete excitations of a set of independent harmonic oscillators:

$$E = \sum_{\mathbf{f}} \sum_{s} \left(n(\mathbf{f}, s) + \frac{1}{2} \right) \hbar\omega(\mathbf{f}, s), \tag{52}$$

with the oscillator of eigenfrequency $\omega(\mathbf{f}, s)$ excited into its n'th excited state. You can also interpret this expression as saying that there are $n(\mathbf{f}, s)$ discrete quanta of vibration, phonons, with energy $\hbar\omega(\mathbf{f}, s)$.

By calculating the average energy $\langle E\rangle$ for this set of phonons, we get the expression for c_V. The partition function, $\mathcal{Z}_1$, for a single harmonic oscillator is

$$\mathcal{Z}_1 = \sum_{n=0}^{\infty} e^{-\beta(n+\frac{1}{2})\hbar\omega} = \\ = \frac{e^{\frac{1}{2}\beta\hbar\omega}}{e^{\beta\hbar\omega}-1} \tag{53}$$

so that the average energy, $\langle E_1\rangle$, for one oscillator is

$$\langle E_1\rangle = -\frac{1}{\mathcal{Z}_1}\frac{\partial \mathcal{Z}_1}{\partial\beta} = \frac{1}{2}\hbar\omega + \frac{\hbar\omega}{e^{\beta\hbar\omega}-1} \tag{54}$$

and for the entire crystal

$$\langle E\rangle = \sum_{\mathbf{f},s}\left\{\frac{1}{2}\hbar\omega(\mathbf{f},s) + \frac{\hbar\omega(\mathbf{f},s)}{e^{\beta\hbar\omega(\mathbf{f},s)}-1}\right\}. \tag{55}$$

The first term is the familiar zero point energy of a harmonic oscillator summed over all modes, the second term is the thermal energy E_T of the crystal (kinetic plus potential energy of vibration):

$$E_\mathrm{T} = \sum_{\mathbf{f},s}\frac{\hbar\omega(\mathbf{f},s)}{e^{\beta\hbar\omega(\mathbf{f},s)}-1}. \tag{56}$$

The evaluation of this sum is difficult, because it requires knowing $\omega(\mathbf{f},s)$ for every value of $\mathbf{f},s$. You can get an approximate solution, valid at very low temperatures, as follows (Peierls 1955). First, you argue that at very low T only very low frequency modes are excited, and those are the long-wavelength sound wave like excitations we mentioned above. Assume that the sound velocity c_s is constant at very-low-frequency, $\omega(\mathbf{f},s) = c_\mathrm{s}(\theta,\phi)f$, where θ,ϕ indicate the direction of $\mathbf{f}$ (for each of the three polarizations $s = 1,2,3$). Replace the summation in the expression for E_T by an integral over $\mathbf{f}$-space:

$$E_\mathrm{T} = \frac{V}{(2\pi)^3}\sum_{s=1}^{3}\int_0^\infty f^3 df \int\int d\Omega \frac{\hbar c_\mathrm{s}(\theta,\phi)}{e^{\beta\hbar c_\mathrm{s} f}-1} \tag{57}$$

where $d\Omega$ is an element of solid angle, and the integral has been extended up to infinity, because at low temperatures only vibrations with low frequencies are excited, and the high-frequency part of the normal mode spectrum does not contribute. Change variables to $x = \beta\hbar c_\mathrm{s}(\theta,\phi)f$:

$$E_\mathrm{T} = \frac{V(kT)^4}{(2\pi\hbar)^3}\sum_{s=1}^{3}\int\int\frac{d\Omega}{c_\mathrm{s}^3}\int_0^\infty dx\frac{x^3}{e^x-1}. \tag{58}$$

The integral over x is equal to $\pi^4/15$. The integral and sum of c_s^{-3} still require knowing the elastic constants of the crystal. Peierls introduces a mean effective sound velocity by

$$\sum_{s=1}^{3} \int\int \frac{d\Omega}{c_s^3} \equiv \frac{12\pi}{c_{\text{eff}}^3} \tag{59}$$

and with this

$$E_{\text{T}} = \frac{\pi^2 (kT)^4 V}{10\hbar^3 c_{\text{eff}}^3}. \tag{60}$$

Since E_{T} depends on temperature as T^4, the specific heat depends on T as T^3, indeed a steep function of T, and c_V goes to zero as T goes to zero.

To do a practical calculation, we need an estimate for c_{eff}. Debye has suggested a way of expressing this parameter in terms of the empirically easier to characterize parameter Θ. In transforming Eq.(56) to an integral over $\mathbf{f}$-space, he also took c_s to be constant, even for high frequencies, and independent of polarization. The number of independent normal modes in a crystal can never be larger than three times the number of particles in the system. So, given that three polarizations are associated with each allowed $\mathbf{f}$, the integral over $\mathbf{f}$-space that replaces the sum over normal modes should only be extended up to a value of f such that the volume in $\mathbf{f}$-space contains just Nr $\mathbf{f}$-space lattice points, with N the number of unit cells (and r the number of ions in the unit cell). The corresponding radius f_0 of this sphere in $\mathbf{f}$-space is thus found from the condition

$$\frac{4\pi}{3} f_0^3 \frac{V}{(2\pi)^3} = Nr, \tag{61}$$

and f_0 is then the highest frequency in the problem. With the definition of the Debye temperature, Θ,

$$\begin{aligned} \Theta = \frac{\hbar\omega_0}{k} = \frac{\hbar c_s f_0}{k} &= \\ = \frac{\hbar c_s}{k} \left(\frac{6\pi^2 Nr}{V} \right)^{\frac{1}{3}}. \end{aligned} \tag{62}$$

Expressing c_s in terms of Θ, and substituting for c_{eff} in Eq.(60), we get for the thermal energy at low temperature

$$E_{\text{T}} = \frac{3\pi^4}{5} NrkT \left(\frac{T}{\Theta} \right)^3, \tag{63}$$

and for the specific heat

$$c_V = \frac{12\pi^4}{5} kNr \left(\frac{T}{\Theta} \right)^3. \tag{64}$$

The Debye temperature is an empirical parameter, characteristic of the crystal material, to be determined experimentally from a fit to measured values of c_V as a function of T.

For the r.m.s. fluctuation in the energy of the crystal we finally obtain

$$\langle \delta E^2 \rangle^{\frac{1}{2}} = \left(\frac{12\pi^4}{5} \frac{k^2 N r}{\Theta^3} \right)^{\frac{1}{2}} T^{\frac{5}{2}}. \tag{65}$$

To get a very rough idea what this could work out to in practice, consider a piece of Si, of volume 1 mm^3, at a temperature of 0.1 Kelvin. The density of Si is 2.33 g cm^{-3}, its Debye temperature is $\Theta = 640$ K. It is customary to quote the FWHM energy resolution, which is equal to 2.35 times the r.m.s. resolution. In this sense, the energy resolution is 4.2 eV! And notice there is no fundamental reason it couldn't be smaller—just lower the number of particles and/or the temperature.

Let us take a short break, and examine the expression for the r.m.s. fluctuation in the energy of the quantized harmonic oscillator. With very little work, we can demonstrate a result of profound significance for bosonic quantum systems, too beautiful not to highlight here. From Eq.(47) we had $\langle \delta E^2 \rangle = -\partial \langle E \rangle / \partial \beta$, which you can write as an equation for the mean square fluctuation $\langle \delta n^2 \rangle$ in the average occupation number $\langle n \rangle$ of a harmonic oscillator:

$$\langle \delta n^2 \rangle = -\frac{1}{\hbar\omega} \frac{\partial \langle n \rangle}{\partial \beta}. \tag{66}$$

Substitute

$$\langle n \rangle = \frac{1}{e^{\beta\hbar\omega} - 1} \tag{67}$$

and you get

$$\begin{aligned} \langle \delta n^2 \rangle &= -\frac{1}{\hbar\omega} \frac{\partial}{\partial \beta} \left(\frac{1}{e^{\beta\hbar\omega} - 1} \right) = \\ &= \frac{e^{\beta\hbar\omega}}{(e^{\beta\hbar\omega} - 1)^2} = \\ &= \frac{e^{\beta\hbar\omega} - 1}{(e^{\beta\hbar\omega} - 1)^2} + \frac{1}{(e^{\beta\hbar\omega} - 1)^2} = \\ &= \langle n \rangle + \langle n \rangle^2. \end{aligned} \tag{68}$$

This seemingly innocent result has a profound implication. In the classical regime, we have $\hbar\omega \ll kT$, so $\langle n \rangle \approx kT/\hbar\omega \gg 1$, and in that case evidently $\langle \delta n^2 \rangle \approx \langle n \rangle^2$. The fractional fluctuation in the number of excitations is then $\langle \delta n^2 \rangle / \langle n \rangle^2 \approx 1$, which is the result for the fractional mean square fluctuation in the energy density of a random superposition of plane waves (e.g., Longair 1984, p. 245). In the quantum regime, however, the other term dominates, and the expression for the mean square fluctuations reduces to the usual Poisson counting statistics expression for discrete particles. In general, therefore, the thermodynamic properties of vibrations in a crystal indicate that the vibrations exhibit both a wave- and a particle-like character!

Since this is just a property of the quantized harmonic oscillator, the same is true for a box filled with pure radiation in thermodynamic equilibrium. What produces the result is that the harmonic oscillator in thermodynamic equilibrium has a Planck spectrum (Eq.(67)). Einstein used Eq.(48) in 1909 to point out that the empirical fact that radiation in thermodynamic equilibrium has a Planck spectrum implies that, thermodynamically, radiation behaves as if it has both a wave- and a particle-like character (Pais 1982; Longair 1984). In fact, in the 1905 paper on the photoelectic effect (as it is usually referred to), he had already used a different thermodynamic relation to show that the Wien spectrum led to the inference that 'monochromatic radiation of low density behaves thermodynamically as though it consisted of a number of independent energy quanta of magnitude [$h\nu$]'. Equation (68) holds for the Bose-Einstein distribution in general, and, even more miraculously, in 1925, before the advent of wave mechanics, Einstein showed that the thermodynamic fluctuation properties of a Bose gas of material particles also implied that the 'particles' exhibit both particle- and wave-like characteristics!

The above digression is actually relevant to the theory of the microcalorimeter. The qualitative argument is sometimes made that you can estimate the amplitude of the thermodynamic energy fluctuations in the absorber as follows. There are of order $N \sim c_V/k$ modes in the crystal, the typical mode has occupation number $\langle n\rangle \sim 1$ and variance $\langle \delta n^2\rangle \sim 1$, and each mode carries of order kT energy. Hence,

$$\begin{aligned}\langle \delta E^2\rangle &\sim N \cdot \langle \delta n^2\rangle \cdot (kT)^2 \sim \frac{c_V}{k} \cdot 1 \cdot k^2T^2 = \\ &= kT^2 c_V\,. \end{aligned} \tag{69}$$

This is somewhat misleading. At high T, the average occupation number of each mode is large, but the variance in the number is also large, and equal to $\langle n\rangle^2$, so the squared fractional fluctuation is equal to unity, and the correct estimate is

$$\begin{aligned}\langle \delta E^2\rangle &\sim N \cdot \frac{\langle \delta n^2\rangle}{\langle n\rangle^2} \cdot (kT)^2 \sim \frac{c_V}{k} \cdot 1 \cdot k^2T^2 = \\ &= kT^2 c_V\,. \end{aligned} \tag{70}$$

But at low T, $\langle \delta n^2\rangle \approx \langle n\rangle$, and the average energy per mode is *not* kT, but rather $\sim \hbar\omega_i \exp(-\beta\hbar\omega_i)$ (with ω_i the eigenfrequency of the mode), and you don't get the right answer this way because the oscillations are quantized with most oscillators in the ground state, and because the particle oscillations are not statistically independent.

This also brings us to the question of the ultimate energy resolution of an ideal microcalorimeter. At very low T, or with a very small device (small number of modes), there are only very few phonons present in equilibrium, and the fractional (quantum) fluctuations are large. But Eq.(49) still holds, that is just thermodynamics (provided the time scale for thermalization of the phonon population is short compared to other timescales in the problem). Since c_V is now very small, the fluctuations only carry tiny amounts of energy, and the resolving power is still high (and given by Eq.(49)). Presumably, however, at some point you will start to notice the effect of the statistical spread in the number of phonons excited by the absorption of a single photon, and how fast a given phonon population evolves (e.g., relaxes to a Planck distribution). The resolving power averaged over a large number of identical photons will still be given by thermodynamics (if the thermalization is sufficiently rapid), but for a single photon of unknown energy (and that is what you build the spectrometer for, to determine uniquely the energy of each photon), the resolution will degrade.

7.3 An alternative derivation

The thermodynamic fluctuations we worked out in the previous section provide a useful measure of the energy resolution of the microcalorimeter. In this section we will work out an alternative derivation of the energy resolution that is actually physically more correct, in that it explicitly models how the actual signal of the temperature pulse induced by the absorption of a photon is detected against the noise in the absorber. This treatment is also the starting point for the calculation of the properties of actual devices (including amplifier noise, etc.). As it turns out, the derivation yields a surprise: the 'thermodynamic limit' we derived above is not a real limit on the energy resolution!

We start again from the fluctuation formula Eq.(49). We argue that the energy fluctuations $\langle \delta E^2 \rangle$ correspond to temperature fluctuations:

$$\langle \delta T_{\mathrm{n}}^2 \rangle = \frac{\langle \delta E^2 \rangle}{c_V^2} = \frac{kT^2}{c_V} \tag{71}$$

where the subscript 'n' indicates that δT_{n} is the noise fluctuation in the temperature due to the spontaneous energy fluctuations. This is a somewhat questionable step, as mentioned earlier—the fluctuation formula for $\langle \delta E^2 \rangle$ only holds in thermodynamic equilibrium, but we will assume again that the equality Eq.(71) is good to order unity. For a complete treatment, thermodynamics doesn't work, and you would have to resort to solving the Boltzmann equation to study the energy transport and thermalization processes!

When analyzing the signal pulse shape and the noise properties, it will be convenient to work in frequency space, rather than in the time domain. The principles of this type of analysis, common in detector physics, can be found in various texts; I have used Houghton & Smith (1966, Ch. 5). The calculation goes as follows: we will derive the power spectral density of the temperature noise due to the thermodynamic fluctuations, and the power spectral density of the temperature signal induced by the absorption of an X-ray photon. The ratio, at each frequency, of these two spectra, is a 'monochromatic' signal-to-noise ratio. Integrating this signal-to-noise ratio over all frequencies, we obtain the total signal to noise ratio, and the ultimate energy resolution obtainable in the presence of noise.

We will analyze the noise first. We start by writing down the energy conservation law for the absorber:

$$\frac{d(\delta E)}{dt} = c_V \frac{d(\delta T)}{dt} = W - G\delta T \tag{72}$$

where δT is the difference in temperature of the absorber and the heat reservoir, and W is any external source of power. The term $G\delta T$ describes the conduction of heat from the absorber to the heat reservoir; G is the thermal conductance of the link (in units Watt Kelvin^{-1}).

Eq.(72) is most conveniently solved in frequency space. But if we apply the equation to the fluctuations $\delta T_{\rm n}$ caused by a source of noise power $W_{\rm n}$, we have to be careful with the Fourier transforms; the usual definition of the transform, $\widehat{\delta T_{\rm n}}(\omega)$, of $\delta T_{\rm n}(t)$,

$$\widehat{\delta T_{\rm n}}(\omega) \equiv \frac{1}{\sqrt{2\pi}} \int_{-\infty}^{\infty} {\rm d}t\ e^{-i\omega t} \delta T_{\rm n}(t), \tag{73}$$

does not converge. Instead, assume that $\delta T_{\rm n}(t)$ is non-zero only for a finite time interval, $t = -a \to +a$, where a is chosen large, but finite. Then *define* the function (see for example Bennett 1956)

$$\widehat{\delta T_{\rm n}}(\omega, a) \equiv \frac{1}{\sqrt{2\pi}} \int_{-a}^{a} {\rm d}t\ e^{-i\omega t} \delta T_{\rm n}(t). \tag{74}$$

Then, for $-a < t < a$,

$$\delta T_{\rm n}(t) = \frac{1}{\sqrt{2\pi}} \int_{-\infty}^{\infty} {\rm d}\omega\ \widehat{\delta T_{\rm n}}(\omega, a) e^{i\omega t}. \tag{75}$$

The definite integral of $\delta T_{\rm n}^2(t)$ is

$$\begin{aligned}
\int_{-a}^{a} {\rm d}t\ \delta T_{\rm n}^2(t) &= \int_{-a}^{a} {\rm d}t\ \delta T_{\rm n}(t) \cdot \frac{1}{\sqrt{2\pi}} \int_{-\infty}^{\infty} {\rm d}\omega\ \widehat{\delta T_{\rm n}}(\omega, a) e^{i\omega t} = \\
&= \int_{-\infty}^{\infty} {\rm d}\omega\ \widehat{\delta T_{\rm n}}(\omega, a) \cdot \frac{1}{\sqrt{2\pi}} \int_{-a}^{a} {\rm d}t\ \delta T_{\rm n}(t) e^{i\omega t} = \\
&= \int_{-\infty}^{\infty} {\rm d}\omega\ \widehat{\delta T_{\rm n}}(\omega, a) \cdot \widehat{\delta T_{\rm n}}^*(\omega, a) = \\
&= \int_{-\infty}^{\infty} {\rm d}\omega\ \left|\widehat{\delta T_{\rm n}}(\omega, a)\right|^2,
\end{aligned} \tag{76}$$

so the variance of the temperature noise can be written as the integral over a spectral density,

$$\langle \delta T_{\mathrm{n}}^2 \rangle = \frac{1}{2a} \int_{-a}^{a} \mathrm{d}t \; \delta T_{\mathrm{n}}^2(t) = \int_{-\infty}^{\infty} \mathrm{d}\omega \; \frac{\left|\widehat{\delta T_{\mathrm{n}}}(\omega, a)\right|^2}{2a}. \tag{77}$$

You can find a logically rigorous definition of the spectral density in Kittel (1958, Ch. 28).

Now we are ready to solve Eq. (72) for the noise; in terms of the transforms $\widehat{\delta T_{\mathrm{n}}}(\omega, a)$ and $\widehat{W_{\mathrm{n}}}(\omega, a)$, we get

$$\widehat{\delta T_{\mathrm{n}}}(\omega, a) = \frac{\widehat{W_{\mathrm{n}}}(\omega, a)}{G + i\omega c_V}, \tag{78}$$

from which we find

$$\begin{aligned} \left|\widehat{\delta T_{\mathrm{n}}}(\omega, a)\right|^2 &= \frac{\left|\widehat{W_{\mathrm{n}}}(\omega, a)\right|^2}{G^2 + \omega^2 c_V^2} = \\ &= \frac{\left|\widehat{W_{\mathrm{n}}}(\omega, a)\right|^2}{G^2(1 + \omega^2 \tau^2)} \end{aligned} \tag{79}$$

with $\tau \equiv c_V/G$. The interpretation of τ is obvious from solving Eq.(72) in the absence of an external power source: the equation

$$c_V \frac{\mathrm{d}(\delta T)}{\mathrm{d}t} = -G\delta T \tag{80}$$

has the solution

$$\delta T(t) = \delta T_0 \; e^{-t/\tau} \tag{81}$$

with $\tau = c_V/G$, so τ is the thermal relaxation time of the system. Eq.(79) says the same thing: if the power source W varies rapidly, on time scales short compared to τ (frequencies high compared to $1/\tau$), the amplitude of the temperature change goes down due to the finite relaxation time.

Now we argue that $\left|\widehat{W_{\mathrm{n}}}(\omega, a)\right|^2$ is independent of frequency, that is, that the power associated with the thermodynamic fluctuations has a white noise spectrum. Now integrate $\left|\widehat{\delta T_{\mathrm{n}}}(\omega, a)\right|^2/2a$ over all frequencies, and set the result, $\langle \delta T_{\mathrm{n}}^2 \rangle$, equal to kT^2/c_V:

$$\frac{\left|\widehat{W_{\mathrm{n}}}(\omega, a)\right|^2}{2a} = \frac{kT^2 G}{\pi}. \tag{82}$$

This result is usually quoted in terms of a spectral density of the noise power, W_f, defined such that

$$\langle W_{\mathrm{n}}^2 \rangle \equiv \int_0^{\infty} \mathrm{d}f \; W_f \tag{83}$$

with f the linear frequency ($\omega = 2\pi f$); with that definition

$$W_f = 4kT^2G. \tag{84}$$

For the spectral density of the temperature fluctuations we get

$$\frac{\left|\widehat{\delta T_{\rm n}}(\omega, a)\right|^2}{2a} = \frac{kT^2}{\pi G(1+\omega^2\tau^2)}. \tag{85}$$

Now we need to derive the spectral density of the temperature signal due to the absorption of a photon. Assume that the photon is absorbed at $t = 0$, by an ideal detector. The temperature pulse shape will look like a sharp spike of amplitude $\delta T_{\rm X,0} = E_\gamma / c_V$, with E_γ the photon energy, at $t = 0$, followed by exponential decay as the heat leaks away to the reservoir:

$$\delta T_{\rm X}(t) = \delta T_{\rm X,0}\; e^{-t/\tau}. \tag{86}$$

Its Fourier transform, in the sense of Eq.(74) is

$$\begin{aligned}\widehat{\delta T_{\rm X}}(\omega, a) &= \frac{\delta T_{\rm X,0}}{\sqrt{2\pi}} \frac{1 - e^{-(i\omega + 1/\tau)a}}{i\omega + 1/\tau} \cong \\ &\cong \frac{\delta T_{\rm X,0}}{\sqrt{2\pi}} \frac{1}{i\omega + 1/\tau}\end{aligned} \tag{87}$$

provided $a \gg \tau$. We have

$$\left|\widehat{\delta T_{\rm X}}(\omega, a)\right|^2 = \frac{\delta T^2_{\rm X,0}}{2\pi} \frac{\tau^2}{1+\omega^2\tau^2}. \tag{88}$$

When we take the ratio of Eq.(88) to the spectral density of the noise temperature fluctuations, we get the spectral density of the square of the signal-to-noise ratio, $\rho^2(\omega)$ (dimension: Hz^{-1}):

$$\rho^2(\omega) = \frac{\delta T^2_{\rm X,0} G\tau^2}{2kT^2}. \tag{89}$$

And this is the surprise: because the shape of the spectral densities of the noise and signal temperature variations are identical (actually, not surprising: signal and noise satisfy the same energy conservation equation), the spectral density of the signal to noise ratio is constant. When you integrate $\rho^2(\omega)$ over frequency, you get the total (square of the) signal to noise ratio, in our case equal to the square of the resolving power $\mathcal{R}$, and the integral diverges, which implies that the resolving power is infinite!

You can understand this remarkable result in another way, in the time domain. Superimposed on the randomly fluctuating temperature, you see a sharp spike right when a photon is absorbed. You make a naive measurement of the photon energy by measuring the amplitude of the temperature rise right at the start of the pulse. This number has a statistical fluctuation in it due to the presence of the noise. But as the pulse decays, you can of course make another measurement, and another, and another, correcting the measurements for the known decay of the pulse with time constant τ. The noise averages out, and in fact goes to zero in the limit that you can take infinitely many samples infinitely closely spaced in time. And this is allowed, because we have said that the detector was ideal: it reacts instantaneously to changes in temperature of the absorber (or: it has an infinite frequency bandwidth), so noise samples taken over infinitesimal time intervals are all statistically independent.

Had you designed the circuitry such that you integrated $\rho^2(\omega)$ only up to the cutoff frequency $1/\tau$, you would have found

$$\mathcal{R}^2 = \frac{\delta T^2_{\mathrm{X},0} G\tau^2}{2kT^2} \int_{-1/\tau}^{1/\tau} \mathrm{d}\omega = \\ = \frac{E_\gamma^2}{kT^2 c_V}, \tag{90}$$

and this is precisely the result we derived in the previous section. It is now clear how 'imperfections' in the detector can be incorporated into the calculation. For instance, the presence of another source of noise can be incorporated by adding its spectral density in quadrature to the density of the thermodynamic fluctuations. If this other noise source also has a white power spectrum, the effect will be to just lower the resolving power by a constant number. If there is another time constant in the problem, for instance, associated with a finite time for thermalization of the photon energy, this changes the boundaries of the integral in Eq.(90), again changing the resolving power by a constant factor.

To obtain a *really* complete description of the performance of the microcalorimeter, we need to write down the coupled equations for the thermal and electrical properties of the system, which take the place of the single thermal balance equation Eq.(72). The coupling between these equations of course arises from terms which describe the ohmic heating of the detector, the change in electrical resistance of the thermometer in response to changes in the temperature, etc. The resulting change in the properties of the detector with respect to the simple properties we derived from Eq. (72) is referred to as 'electrothermal feedback'. Moseley, Mather & McCammon (1984) provide such a description in terms of complex impedance theory; Labov et al. (1997) present a derivation based on the differential equations. It is instructive to solve the equations for the simplified case of a constant voltage on the thermometer, assuming that the circuit only contains this thermometer, of resistance R, and the bias power source. Then $W = I^2 R$ in Eq.(72), and $d(IR)/dt \equiv 0$. Taking the time derivative of Eq. (72), and neglecting small terms, you find

$$c_V \frac{\mathrm{d}^2(\delta T)}{\mathrm{d}t^2} + \left(G + \frac{W_0}{T_0}\alpha\right)\frac{\mathrm{d}(\delta T)}{\mathrm{d}t} = 0 \tag{91}$$

that is, the same equation as before, but with an effective time constant

$$\frac{1}{\tau_{\text{eff}}} = \frac{1}{\tau_{\text{th}}}\left(1 + \frac{W_0}{GT_0}\alpha\right) \tag{92}$$

with $\tau_{\text{th}} = c_V/G$ the 'thermal time constant' which characterizes the 'bare' thermal problem. Here, $\alpha \equiv \mathrm{d}(\ln R)/\mathrm{d}(\ln T)$, which measures the steepness of the thermometer response to changes in temperature. This suggests integrating up to a frequency $1/\tau_{\text{eff}}$, in which case the spectral resolving power will exceed the 'thermodynamic limit' by a factor $\left(1 + (W_0/GT_0)\alpha\right)^{1/2}$: a more sensitive thermometer (bigger α) is better.

In general, therefore, the uncertainty in the determination of the photon energy will be approximately

$$\left\langle \delta E^2 \right\rangle = \xi k T^2 c_V \tag{93}$$

with ξ a pure number that depends on the strength of additional noise sources, other relaxation time constants (inclusing electrical), and, implicitly, on the thermometer response and operating conditions of the entire detector system.

Finally, traditionally in detector physics, one does not calculate with the spectral density of the temperature, but of course with that of the (square of the) physical power delivered by the various components of the detector system. You can redo the above calculation in these terms, remembering that the absorption of a photon in an ideal detector delivers a power spike in the form of a delta function (which has a flat power spectrum, just like the noise), normalized such that the time integral over the spike is equal to the photon energy. In this connection, one often uses the concept of the 'noise equivalent power' (NEP), defined such that the NEP is the amount of power delivered to the detector at frequency f equal to the noise power at that frequency in a 1 Hz bandwidth. For the thermodynamic, or phonon, noise, we had the spectral density of the noise power $W_f = 4kT^2G$, so for this noise source,

$$\text{NEP}_{\text{phonons}} \equiv W_f^{\frac{1}{2}} = (4kT^2G)^{\frac{1}{2}} \tag{94}$$

with dimension $\text{W}/\sqrt{\text{Hz}}$. The energy resolution in these terms is

$$\left\langle \delta E^2 \right\rangle = \tau^2 \int_0^{f_{\max}} \mathrm{d}f \; \left|\text{NEP}_{\text{phonons}}(f)\right|^2 . \tag{95}$$

7.4 The microcalorimeter on *Astro-E*

In this section, we will look at some of the features of the operation of a real first-generation X-ray microcalorimeter, the X-ray Spectrometer (XRS) on *Astro-E*. Information on *Astro-E*, a joint Japan-US project, can be found on `http://heasarc.gsfc.nasa.gov/docs/frames/astroe_about.html`; a recent conference paper can be found on `http://wwwvms.mppmu.mpg.de/ltd7/`

`contribute/by_field/C.htm` (Stahle et al. 1997). It carries five telescopes, four with a CCD focal-plane detector, one with the microcalorimeter; there is also a hard X-ray instrument. The microcalorimeter has been developed in collaboration by NASA/Goddard Space Flight Center and the University of Wisconsin (PI: Richard Kelley, GSFC). Launch is in early 2000.

The microcalorimeter consists of 6×6 separate pixels, covering $3' \times 3'$ on the sky (the telescope angular response is of the order of 2 arcmin HEW). The X-ray absorbing element is HgTe, attached to a Si substrate with an implanted thermistor as thermometer. The thermal time constant for the conduction of heat between the absorber and the heat bath is approximately 80μs, so that the detector can count of order 10 counts/sec per pixel without (difficult to discriminate) overlapping pulses. For bright sources, neutral density filters can be inserted in the X-ray beam to reduce count rates.

The microcalorimeter is inside a dewar. The instrument itself is cooled to 65 mK by an Adiabatic Demagnetization Refrigerator, located inside a liquid He dewar, which in turn is located inside a solid Ne dewar. The lifetime is determined by the cryogenics, and is expected to be of order two years. During this initial phase of the mission, priority will be given to microcalorimeter observations.

The energy resolution is approximately 12 eV, while 7 eV has been seen in individual array elements. The microcalorimeter itself has good quantum efficiency (basically X-ray absorption optical depth) up to 10 keV. The low end of the bandpass is determined by the X-ray transmission of a set of filters that prevent radiation from space from heating the instrument, which limits the band to roughly $E \gtrsim 400$ eV.

The performance of the XRS in terms of spectral resolving power, and the spectral diagnostics accessible with it, was plotted in Fig. 8. The XRS will have most of its impact at high energies, specifically in the Fe K band. Novel (for astrophysics) diagnostics that will become available at Fe K are the unambiguous determination of the (distribution of) ionization stages of fluorescing Fe, as long as it has been ionized up into the L-shell, from a direct measurement of the energies of the Fe K fluorescent lines; direct detection of the Compton recoil spectrum of Fe K photons scattered by cold electrons, and possibly even detection of Raman scattered Fe K photons. The powerful hot plasma diagnostics derived from spectroscopy of the dielectronic satellites may just be accessible, provided the spectrum is bright enough.

8 The 21st Century

Astrophysical considerations suggest that after *AXAF*, *XMM*, and *Astro-E*, extending the spectroscopic capability to cosmologically interesting redshifts is the most important thing to do, rather than, for instance, increasing spectral resolving power for its own sake.

X-ray line emission arises predominantly in optically thin plasmas (although significant optical depths may occur in resonance lines), and the emission line power is relatively easy to calculate as a function of the local variables. In addition, the X-ray band is so wide that it simultaneously covers the K transitions of all ionization stages of all the elements from C up to Fe. This makes, for instance, absolute abundance determinations easy, because they don't have to rely on model-dependent corrections for the abundances of unseen ionization stages (as is the case in the optical and UV bands). The importance of X-ray spectroscopy to nucleosynthesis, stellar evolution, and formation and evolution of galaxies is obvious. As another example, detailed information on the physical conditions in clusters at significant redshifts (distribution of temperature, density, and abundances) has direct implications for the evolution of large-scale structure and the cosmological parameters. Galactic astrophysics will of course also benefit. To name just one example, detailed photospheric spectroscopy of hot neutron stars will finally constrain the mass-radius relation for these objects, and hopefully will uniquely constrain the equation of state at supranuclear densities.

Obviously, what is required is a spectrometer, preferably imaging, with resolving power no less than *AXAF*, *XMM*, and *Astro-E* (so at least $\mathcal{R} \gtrsim$ few hundred), and a huge effective area. For the study of redshifted objects, it is crucial that these characteristics extend to low photon energies (the oxygen Lyα line, to name an example, shifts to $\sim$ 300 eV at $z \sim 1$). These requirements naturally suggest lightweight, medium angular resolution optics, with as large a combined area as you can afford, with an imaging spectrometer at the focal plane. The natural (or only) candidate for the latter is, of course, a microcalorimeter array, with significantly improved energy resolution.

As you can see from Fig. 8, the $\sim$ 12 eV characteristic of the *Astro-E* spectrometer is not good enough for detailed spectroscopy at low energies. Hence, work is in progress to construct $\Delta E \sim 2$ eV microcalorimeters. The most important innovation is the use of an extremely sensitive thermometer, based on a superconducting element that is held at a temperature in the middle of the transition between the superconducting and normal states. Hence, its resistance is extremely sensitive to changes in temperature. Thermometers of this type are referred to as Transition Edge Sensors (TES). Recently, a resolution of 4 eV was already demonstrated with such a device by a group at the National Institute of Standards and Technology (NIST, Boulder, Colo.; see *Physics Today*, July 1998, p. 19).

NASA is currently funding work on the *Constellation X* project, a fleet of six identical observatories, with a combined effective area of order 15,000 cm^2 at 1 keV and 6,000 cm^2 at 6 keV. The proposed instrument package includes $\Delta E = 2$ eV microcalorimeter arrays, reflection grating spectrometers, and hard X-ray telescopes that will be sensitive up to at least 40 keV (1500 cm^2 at 40 keV).

The grating spectrometers complement the microcalorimeters at low energies, where their sensitivity declines due to declining resolving power and strong X-ray absorption in the necessary long-wavelength radiation filters. The design of the spectrometers is similar to those on *XMM*, but the gratings could be of a novel design: lightweight thin Si films, etched along crystal planes to achieve an extremely smooth and accurate groove shape. In order to increase the graze angles and allow a higher resolving power (Eq. 31) without loss of reflectivity, multi-layer coatings may be applied to the gratings.

A prime motivation for including sensitive hard X-ray telescopes is the study of AGN spectra. The study of these spectra with spectrometers that cut off at 10 keV has shown that it is important to have the ability to characterize the hard continuum spectrum, to allow for an accurate extrapolation to lower energies, where the continuum is strongly modulated by absorption and the presence of emission line complexes. In addition, one hopes to study heavily absorbed type II objects (Seyfert 2 galaxies and, if they are ever found, the analogous 'type 2 QSO's') with much higher sensitivity. And there are, of course, numerous other novel applications.

The European *XEUS* project (*X-ray Early Universe Spectrometer*) pursues similar goals. The design currently revolves around a truly giant (10 m^2 aperture) single X-ray telescope, which would be assembled in sections over the years at the space station. Operations can begin as soon as the core of this assembly has been placed in orbit. A telescope with such a large area needs to have a very long focal length, in order to keep graze angles small and reflectivity high up to high energies; the current figure is $F \sim 50$ m! With such an extreme focal length, building a 'monolithic' telescope becomes more expensive than simply putting focusing optics and focal plane instrument on two separate satellites that track each other, and this is the current plan. The focal-plane satellite will be designed such that its instrumentation can be exchanged. *XEUS* might start out with CCD detectors, and then move on to microcalorimeter arrays.

Acknowledgements

I would like to express my gratitude to the following people, for conversations, reading of the draft manuscript, or permission to use their data: Duane Liedahl (LLNL), Steve Kahn, Andy Rasmussen, Masao Sako, Joshua Spodek, and Jean Cottam (Columbia University), Bert Brinkman, Theo Gunsing, Jan van Rooijen, Jelle Kaastra, Rolf Mewe, and Piet de Korte (SRON), John Houck, Dan Dewey, John Davis, and the entire HETGS group (MIT), and Steve Holt (GSFC). Finally, I would like to thank the organizers of the School, Johan Bleeker and Jan van Paradijs, for their invitation to teach the lectures, and Jan van Paradijs for his most generous hospitality during the School.

References

Beiersdorfer, P., Schneider, M.B., Bitter, M. & von Goeler, S. (1992): Rev. Sci. Instrum. 63(10), 5029

Bennet, W.R. (1956): Proc. IRE 44, 609; reprinted in M. S. Gupta: *Electrical Noise: Fundamentals and Sources*, (IEEE Press: New York) (1977)

Beuermann, K.P., Bräuniger, H. & Trümper, J. (1978): Appl. Optics 17, 2304

Bixler, J.V. et al. (1991): Proc. SPIE 1549, 420

Born, M. & Wolf, E. (1959): *Principles of Optics*, (Pergamon: London).

Bowyer, S. & Malina, R.F. (1996): *Astrophysics in the Extreme Ultraviolet*, Proc. IAU Coll. 152, Berkeley, March 1995 (Kluwer:Dordrecht)

Brinkman, A.C. et al. (1980): Appl. Optics 19, 1601

Brinkman, A.C. et al. (1987): Astrophys. Letters 26, 73

Brinkman, A.C. et al. (1997): Proc. SPIE, 3113, 181

Canizares, C.R. et al. (1987): Astrophys. Letters, 26, 87

Canizares, C.R. (1990): in *Imaging X-ray Astronomy*, M. Elvis (Ed.), (Cambridge UP: Cambridge), p. 123

Church, E.L., Jenkinson, H.A. & Zavada, J.M. (1977): Opt. Engineering 16, 360

Compton, A.H. & Allison, S.K. (1935): *X-rays in Theory and Experiment*, Second Edition (van Nostrand: New York).

Davis, J.E. (1997): on `http://space.mit.edu/HETG/LRF/scatter.html`

Decaux, V., Beiersdorfer, P., Osterheld, A., Chen, M. & Kahn, S.M. (1995): ApJ 443, 464

Debye, P. (1914): Ann. der Physik 43, 49

Decaux, V., Beiersdorfer, P., Kahn, S.M. & Jacobs, V.L. (1997): ApJ 482, 1076

Deiker, S. et al. (1997): in *Proceedings of the VIIth International Conference on Low Temperature Detectors*, July 1997, Munich, Germany.

Dewey, D. et al. (1997): Proc. SPIE 3113, 144

Done, C., Madejski, G.M. & Smith, D.A. (1996): ApJ 463, L63

Evans, A. (1986): MNRAS 223, 219

Fabian, A. et al. (1994): in *New Horizons in X-ray Astronomy*, F. Makino & T. Ohashi, Eds (Tokyo: Universal Academic Press), p. 573

Fraser, G.W. (1989): *X-ray Detectors in Astronomy*, (Cambridge UP: Cambridge)

Fukazawa, Y. et al. (1994): PASJ 46, L141

Gabriel, A.H. & Jordan, C. (1969): MNRAS 145, 241

Hamann, F. et al. (1998): ApJ 496, 761

Henke, B.L., Gullikson, E.M. & Davis, J.C. (1993): *Atomic Data and Nuclear Data Tables*, Vol. 54

Hettrick, M.C. & Bowyer, S. (1983): Appl. Optics 22, 3921

Houghton, J.T. & Smith, S.D. (1966): *Infra-red Physics*, (Oxford University Press: Oxford)

Illarionov, A., Kallman, T., McCray, R. & Ross, R. (1979): ApJ 228, 279

Iwasawa, K., Yaqoob, T., Awaki, H. & Ogasaka, Y. (1994): PASJ 46, L167

Iwasawa, K. et al. (1997): MNRAS 285, 683

Jark, W. & Nevière, M. (1987): Appl. Optics 26, 943

Kaastra, J.S., Roos, N. & Mewe, R. (1995): A&A 300, 25

Kahn, S.M. & Hettrick, M.C. (1985): in *A cosmic X-ray spectroscopy mission : proceedings of a workshop held in Lyngby, Denmark on 24-26 June 1985*, ESA SP 239 (ESA: Paris), p. 237

Kahn, S.M. et al. (1996): Proc. SPIE 2808, 450
Kitamoto, S. et al. (1994): PASJ 46, L105
Kittel, C. (1958): *Elementary Statistical Physics*, (Wiley:New York).
Kawashima, K. & Kitamoto, S. (1996): PASJ, 48, L113
Klos, R.A. (1985): Proc. SPIE 597, 135
Kunieda, H. et al. (1994): in *New Horizon in X-ray Astronomy*, F. Makino & T. Ohashi (Eds) (Universal Academy Press: Tokyo), p. 317
Labov, S. et al. (1997): in *Proceedings 7th International Workshop on Low Temperature Detectors LTD-7*, July 27-August 2, 1997, Munich (MPI für Physik: Munich) (`http://wwwvms.mppmu.mpg.de/ltd7/welcome.html`).
Lee, P.A., Citrin, P.H., Eisenberger, P. & Kincaid, B.M. (1981): Rev. Mod. Phys. 53, 769
Liedahl, D.A., Kahn, S.M., Osterheld, A.L. & Goldstein, W.H. (1990): ApJ 350, L37
Liedahl, D.A., Kahn, S.M., Osterheld, A.L. & Goldstein, W.H. (1992): ApJ 391, 306
Liedahl, D.A. & Paerels, F. (1996): ApJ 468, L33
Longair, M.S. (1984): *Theoretical Concepts in Physics*, (Cambrdige University Press: Cambridge).
Markert, T.H. et al. (1994): Proc. SPIE 2280, 168
Marshall, H.L. et al. (1997): Proc. SPIE 3113, 160
Martin, P.G. (1970): MNRAS 149, 221
Mather, J.C. (1982): Appl. Optics 21, 1125
Mathur, S., Elvis, M. & Wilkes, B. (1995): ApJ 452, 230
Matt, G. (1996): `astro-ph/9612002`.
McCray, R. (1984): Physica Scripta T7, 73
Mewe, R., Gronenschild, E.H.B.M. & van den Oord, G.H.J. (1985): A&AS 62, 197
Mewe, R. (1991): A&AR 3, 127
Michette, A.G. (1986): *Optical Systems for Soft X-rays* (Plenum Press: New York).
Moseley, S.H., Mather, J.C. & McCammon, D. (1984): J. Appl. Phys. 56, 1257
Nelson, C.S. et al. (1994): Proc. SPIE 2280, 191
Nussbaumer, H., Schmid, H.M. & Vogel, M. (1989): A&A 211, L27
Owens, A. (1997): ApJ 476, 924
Paerels, F. et al. (1994): Proc. SPIE 2283, 107
Paerels, F. (1997): at `http://astro1.nevis.columbia.edu/xmm/documents/index.html`.
Paerels, F., Kahn, S.M. & Wolkovitch, D.N. (1998): ApJ 496, 473
Pais, A. (1982): *Subtle is the Lord...*, (Oxford University Press: Oxford).
Peierls, R.E. (1955): *Quantum Theory of Solids*, (Oxford University Press: Oxford).
Petit, R. (1980): *Electromagnetic Theory of Gratings* (Ed.) (Springer: Berlin)
Predehl, P. et al. (1997): Proc. SPIE 3113, 172
Rando, N. et al. (1996): Nucl. Instr. Meth., A 370, 85
Rasmussen, A.R. et al. (1998): Proc. SPIE 3444.
Ross, R., Weaver, R. & McCray, R. (1978): ApJ 219, 294
Rybicki, G.B. & Lightman, A.P. (1979): *Radiative Processes in Astrophysics* (Wiley: New York)

Sanders, W.T., Edgar, R.J. & Liedahl, D.A. (1996): in *Röntgenstrahlung from the Universe*, H.U. Zimmermann, J. Trümper, & H. Yorke (Eds), MPE Report nr. 263, p. 339
Schattenburg, M.L. et al. (1994): Proc. SPIE 2280, 181
Schnopper, H.W. et al. (1977): Appl.Optics 16, 1088
Sevier, K.D. (1979): *Atomic Data and Nuclear Data Tables* 24, 323.
Seward, F.D. et al. (1982): Appl. Optics 21, 2012
Snow, T.P. & Witt, A.N. (1996): ApJ 468, L65
Spodek, J., Rasmussen, A., Cottam, J., Kahn, S.M. & Paerels, F. (1998): *in preparation*
Stahle, C.K. et al. (1997): *Proceedings 7th International Workshop on Low Temperature Detectors LTD-7*, July 27-August 2, 1997, Munich (MPI für Physik: Munich) (`http://wwwvms.mppmu.mpg.de/ltd7/welcome.html`).
Stroke, G.W. (1967): in *Handbuch der Physik*, Vol. XXIX, S. Flügge (Ed.) (Springer: Berlin), pp. 426-754
Sunyaev, R.A. & Churazov, E.M. (1996): Astronomy Letters 22, 723
Tanaka, Y., Inoue, H. & Holt, S.S. (1994): PASJ 46, L37
Turner, M.J.L. et al. (1996): in *The Next Generation of X-ray Observatories: Workshop Proceedings*, M.J.L. Turner & M.G. Watson (Eds), Leicester X-ray Astronomy Group Special Report, XRA97/02, p. 165 (`http://ledas-www.star.le.ac.uk/ngxo`)
Ueno, S. et al. (1994): PASJ 46, L71
van den Heuvel, E.P.J. (1994): in *New Horizon of X-ray Astronomy – First Results from ASCA*, F. Makino & T. Ohashi (Eds), (Universal Academy Press: Tokyo).
van Kerkwijk, M.H. et al. (1992): Nat 355, 703
Weisskopf, M.C. et al. (1987): Astrophys. Letters 26, nrs. 1 & 2
Woo, J.W. (1995): ApJ 447, L129
Zarnecki, J.C. & Culhane, J.L. (1978): MNRAS 178, 57P

New Developments in X-ray Optics

Richard Willingale

Department of Physics and Astronomy, University of Leicester, University Road, Leicester LE1 7RH, UK

Abstract. To appreciate the new developments in X-ray optics a good understanding of the established foundations of the subject is required. This discourse starts with a substantial introduction to the basic physics of the interaction of X rays with matter, X-ray dispersion theory, reflection, transmission and scattering of mirrors, multi-layers, crystals and gratings. In the light of this the geometries used for X-ray imaging and spectroscopy are described. Finally the latest developments of such optics for use in X-ray astronomy are reviewed.

1 Introduction

1.1 What is or are X-ray optics?

In all experimentation with X rays there is some source, which in the case of astronomy is a distant star, galaxy or whatever in the sky, and some detector. Elsewhere in this series of lectures you will learn much about the sources and detectors but here we are concerned with instrumentation that comes between the two; that is X-ray optics. In general, X rays propagate through such optics, some may be absorbed by accident or design but hopefully most make it to the detector. The optics reflect, focus, image, scatter, disperse or absorb the X rays. We shall be investigating the design and properties of X-ray mirrors, collimators, diffraction gratings and diffracting crystals.

1.2 The fundamental interaction utilised in X-ray optics

Traditionally, X rays occupy the energy band 0.1 to 100 keV corresponding to a frequency range of 2.4×10^{16} to 2.4×10^{19} Hz or a wavelength range of 124 to 0.124 Å. Amazingly, there is just one fundamental interaction responsible for all X-ray optics. That is the coherent scattering and the associated photoelectric absorption of X-ray photons by electrons. X-ray energies are too low to promote pair production and Compton (incoherent) scattering is unimportant below 50 keV. X-ray energies are too high for photon-phonon interactions to dominate although thermal effects do influence the finer details of the response of some X-ray optics.

X rays are particles and waves. We detect them as quanta and it is natural to think of them as photons that are created (emitted), bounce around between collisions with other particles or elements of an optics system and finally get absorbed in some detector. We must use quantum mechanics to describe these processes and indeed we must use relativistic quantum theory to describe the behaviour accurately. However, the propagation of X rays through an optics system is better modelled by a wave and, in fact more often than not, must be described by a wave. Therefore, the physics of X-ray optics is a strange mixture of quantum mechanics, classical electrodynamics and classical optics. Despite this there is, in general, a good agreement between theory and experiment.

1.3 The challenges of X-ray optics in astronomy

Although the fundamental interactions of X-ray optics are known, the challenges of X-ray optics in astronomy are considerable. Because the wavelengths of X rays are of atomic dimensions the engineering tolerances imposed are very severe and present a major technological manufacturing problem. On the other hand, the flux density of cosmic X rays at Earth is rather small. The brightest cosmic X-ray source, Sco X-1, gives about 150 photons cm^{-2} s^{-1} while typical sources are at least 10,000 times fainter. Therefore the collecting optics must have very large areas. So the design and construction of X-ray telescopes must combine engineering skills from the realms of both the very large and the very small.

The electrons responsible for the scattering and absorption of X rays in all optics are bound up in atoms and every orbital of every element gives a unique response as a function of energy. On top of this, the atoms are not isolated but reside in a solid matrix and the electron orbitals overlap, complicating the response still further. Therefore, the prediction and measurement of the detailed response of the electrons and hence the optics, to X rays, is a big problem. The instrumentation required to make these fundamental measurements and to calibrate X-ray optics is hard to construct, difficult to use and expensive.

2 X-ray Dispersion Theory

2.1 The classical electromagnetic theory

We will start with a very brief description of the electromagnetic theory that leads to the definition of a complex refractive index. A comprehensive explanation can be found in Born & Wolf (1970) or any good textbook on EM theory.

In a classical treatment of electromagnetic fields in media we use the following definitions

$$\underline{D} = \varepsilon \underline{E} \qquad \varepsilon = \varepsilon_r \varepsilon_o \tag{1}$$

$$\underline{B} = \mu \underline{H} \qquad \mu = \mu_r \mu_o \tag{2}$$

$$\underline{J} = \sigma \underline{E} \qquad c = 1/\sqrt{\varepsilon_o \mu_o} \tag{3}$$

where $\underline{D}$ is the electric flux density, $\underline{E}$ is the electric field strength, $\underline{B}$ is the magnetic flux density, $\underline{H}$ is the magnetic field strength and $\underline{J}$ is the current density. ε_o and μ_o are the permittivity and permeability of a vacuum and c is the speed of light in a vacuum. Media are characterised by the permittivity ε, the permeability μ and the conductivity σ. If a medium is inhomogeneous, as is often the case for the X-ray regime, then ε and μ will be a function of position and if it is anisotropic they will be tensors. The electric and magnetic fields obey Maxwell's equations

$$\nabla \cdot \underline{D} = \rho_{\mathrm{f}} \tag{4}$$
$$\nabla \cdot \underline{B} = 0 \tag{5}$$
$$\nabla \wedge \underline{E} = -\partial \underline{B}/\partial t \tag{6}$$
$$\nabla \wedge \underline{H} = \underline{J} + \partial \underline{D}/\partial t \tag{7}$$

We can take the curl of Eq. (6), which is Faraday's law in differential form, to give

$$\nabla \wedge (\nabla \wedge \underline{E}) = -\nabla \wedge \partial \underline{B}/\partial t \tag{8}$$

Substituting for $\underline{B}$ from Eq. (2) gives us

$$\nabla(\nabla \cdot \underline{E}) - \nabla^2 \underline{E} = -\mu(\nabla \wedge \partial \underline{H}/\partial t) \tag{9}$$

We can then eliminate $\underline{H}$ using Eq. (7) which is Ampere's law in differential form

$$\nabla(\nabla \cdot \underline{E}) - \nabla^2 \underline{E} = \mu \partial \underline{J}/\partial t - \mu \partial^2 \underline{D}/\partial t^2 \tag{10}$$

If there is no free charge, $\rho_{\mathrm{f}} = 0$, then the divergence of $\underline{D}$ is zero. Substituting for $\underline{D}$ from Eq. (1) gives the wave equation

$$\nabla^2 \underline{E} - \varepsilon\mu\, \partial^2 \underline{E}/\partial t^2 - \mu\sigma\, \partial \underline{E}/\partial t = 0 \tag{11}$$

Plane-wave solutions travelling in the z direction have the form

$$\underline{E} = \underline{E_o} \exp[i(\omega t - kz)] \tag{12}$$

where k is the wavenumber and ω is the angular frequency. Substituting such a solution into Eq. (11) gives us

$$(-k^2 + \varepsilon\mu\omega^2 - i\omega\sigma\mu)\underline{E} = 0 \tag{13}$$

Consequently the wave number is complex:

$$k^2 = \omega^2 \mu(\varepsilon - i\sigma/\omega) \tag{14}$$

The conductivity σ, or more strictly I suppose, the resistivity, introduces absorption and hence we get an imaginary component.

We can define a complex refractive index as

$$n_{\mathrm{c}} = n - iK = ck/\omega \tag{15}$$

Substituting for the wave number gives us

$$(n - iK)^2 = c^2\mu(\varepsilon - i\sigma/\omega) \tag{16}$$

and equating the real an imaginary components we get the following relationships

$$n^2 - K^2 = \varepsilon_r\mu_r \tag{17}$$

$$2nK = \sigma\mu_r/(\varepsilon_o\omega) \tag{18}$$

We can also define a complex dielectric constant of the form

$$\varepsilon_c = \varepsilon_1 - i\varepsilon_2 = 1 - \alpha - i\gamma \tag{19}$$

where $\varepsilon_1 = \varepsilon_r$ and $\varepsilon_2 = \sigma/(\omega\varepsilon_o)$. If $\mu_r = 1$ the material is non-magnetic and we have $\varepsilon_1 = n^2 - K^2$ and $\varepsilon_2 = 2nK$ or

$$\varepsilon_c = (n - iK)^2 \tag{20}$$

If there is no absorption then $K = 0$, $\varepsilon_2 = 0$ and $\varepsilon_1 = n^2$. We can substitute for the complex refractive index in the general plane-wave solution and rearrange giving

$$\underline{E} = \underline{E}_o \exp(-\omega Kz/c)\exp[i(\omega t - nz/c)] \tag{21}$$

The absorption coefficient is therefore given by

$$\alpha = 2\omega K/c = 4\pi K/\lambda_o \tag{22}$$

I have presented this standard analysis because it is usually done with radio waves or possibly visible light in mind, but not X rays. I think it is surprising that such a classical EM wave analysis should apply to X rays at all, but it does. X rays are an EM wave and really do behave like one. It turns out to be sensible to think of a refractive index of materials for X rays although, as we shall see in the subsequent sections, the real part n is a little less than unity and the imaginary part K is often rather important.

2.2 The origin of dispersion - optical constants for X rays

The classical EM treatment above embodies the phenomenon of dispersion in the permittivity (and permeability if magnetic) of the material. You should recall that

$$(\varepsilon - \varepsilon_o)\underline{E} = \underline{P} \tag{23}$$

$$(\mu - \mu_o)\underline{H} = \underline{M} \tag{24}$$

The dielectric constant is directly related to the polarization $\underline{P}$ induced in a material by an applied electric field and the permeability is similarly related to the magnetization $\underline{M}$. The electric and magnetic properties of a material are functions of the atoms. The dielectric constant $\varepsilon_c = \varepsilon_1 - i\varepsilon_2$ must be related to the polarization of the individual atoms, and the magnetization is related to the magnetic dipole moment of the individual atoms. In a static electric field a constant dipole moment is induced and we need only concern ourselves with the ε_1 component. If the field varies with time then the ε_2 part comes into play.

What happens at the atomic level when we apply a varying electric field must be investigated using quantum mechanics. However, a semi-classical treatment can yield equivalent results and provides a simple model of what is happening. This is described in greater detail in Born & Wolf (1970). We can imagine that the electrons are bound to the nuclei by springs. These *springs* arise from the electrostatic potential well of the positive nuclear charge. When an external $\underline{E}$ field is applied these springs are stretched or compressed and the negative charge is displaced with respect to the positive charge, giving a dipole. Using Newton's 2nd law we can set up an equation of motion for the electrons

$$q_e E_o \cos\omega t - m_e\omega_o^2 x - \tau m_e \mathrm{d}x/\mathrm{d}t = m_e \mathrm{d}^2 x/\mathrm{d}t^2 \tag{25}$$

In order the terms are the applied electric force, the restoring force from a *spring*, the damping of a *spring*, and the mass times the acceleration. ω_o is the *natural frequency* and τ is the damping constant of the spring. The solution has the form $x(t) = x_o \exp i(\omega t + \delta)$ which can be substituted into Eq. (25) and solved for x_o and δ.

$$x_o = \frac{q_e E_o}{m_e} \frac{1}{(\omega^2 - \omega_o^2 - i\tau\omega)} \tag{26}$$

The incident EM field sets up a dipole which scatters the radiation. The amplitude of the radiated electric field from an oscillating dipole, moment $x_o q_e$, on the equatorial plane at a distance r is given by $x_o q_e \omega^2/(4\pi\varepsilon_o c^2 r)$ so in this case the scattered amplitude is

$$E_s = \frac{q_e^2 E_o}{4\pi\varepsilon_o m_e c^2 r} \frac{\omega^2}{(\omega^2 - \omega_o^2 - i\tau\omega)} \tag{27}$$

If the electron were free then $\omega_o = 0$ and $\tau = 0$ so the ratio of the amplitude from the bound electron to that of a free electron is given by

$$f = \frac{\omega^2}{\omega^2 - \omega_o^2 - i\tau\omega} \tag{28}$$

This is called the scattering factor of the bound electron.

The polarization of the medium is given by $P = q_e x N$ where there are N electrons per unit volume, so we have

$$P = \frac{q_e^2 N E}{m_e \omega^2} f \tag{29}$$

Hence using the relationship between E and P in Eq. (23) we have

$$\varepsilon = \varepsilon_o + P/E = \varepsilon_o(1 + \frac{q_e^2 N f}{m_e \varepsilon_o \omega^2}) \tag{30}$$

Actually this is not quite correct. Individual atoms see the incident photon field plus a contribution from the induced dipoles in the surrounding atoms. However, in the case of X-ray dispersion the photon scattering from atoms is very small and to a very good approximation the atoms are not coupled in this way. See Born & Wolf (1970) for details of the correct way to proceed if this approximation is not valid. Thus the complex dielectric constant and refractive index are given by

$$\varepsilon_c = 1 + \frac{q_e^2 N f}{m_e \varepsilon_o \omega^2} = n_c^2(\omega) \tag{31}$$

In practice there is more than one *spring* present and there is a series of restoring and damping terms which must be summed to calculate the scattering factors

$$f = \sum_j \frac{\omega^2 g_j}{\omega^2 - \omega_{oj}^2 - i\gamma_j \omega} \tag{32}$$

where g_j are weighting factors or oscillator strengths for each *spring* normalised such that $\sum_j g_j = 1$.

An equivalent formulation can be derived using quantum mechanics. The result is essentially the same but the model is different. We no longer have *springs* but discrete electron energy states, electron transitions between the energy states and transition lifetimes. Instead of the EM field promoting oscillation of the electrons there are X-ray photons that cause electronic transitions. Yet the mathematical form of the result is exactly the same.

So we have shifted the problem of describing dispersion or the refractive index for X rays to that of finding:

- ω_{oj} the characteristic frequencies
- τ_j the damping constants
- g_j the oscillator strengths

We replace electrons on springs by an electron *cloud* represented by a wavefunction ψ_e and we use the Schrödinger equation with a Hamiltonian component that contains an $\underline{E}$ term. We can solve the resulting wave equation for sinusoidally varying $\underline{E}$ using perturbation theory. Now the ω_{oj}'s are the characteristic frequencies for absorption or emission associated with transitions between discrete energy levels.

$$\hbar\omega_j = \epsilon_{j2} - \epsilon_{j1} \tag{33}$$

The g_j's represent the transition probabilites for the $\epsilon_{j2} - \epsilon_{j1}$ transition. These are related to the wavefunctions of the initial and final state by what are called *matrix elements*. The τ_j's are given by the lifetimes of the energy states.

Many transitions are bound-free transitions and the ϵ_{j2}'s form a continuum so the summation in Eq. (32) becomes an integration in which the oscillator strength is replaced by an oscillator strength density.

$$f_j(\omega) = \int_{\omega_j}^{\infty} \frac{\omega^2 (\mathrm{d}g_j/\mathrm{d}\omega_o)\mathrm{d}\omega_o}{\omega^2 - \omega_o^2 - i\tau_j\omega} \tag{34}$$

where ω_j is some threshold energy for the j'th electron. Near absorption edges there is a significant contribution from discrete bound-bound transitions. These are usually included by adding the appropriate average value to the oscillator strength density over a small energy range near the edge.

Summing the contributions from all the electrons in an atom will yield the so-called atomic scattering factor. The cross section of the atom will then be the Thomson cross section multiplied by the square of the atomic scattering factor. It is assumed in this summation of the individual electron contributions that all the electrons in the atom scatter in phase. Since the electrons are in fact spread out in a cloud around the nucleus this is not the case. If θ_s is the scattering angle the finite size of the electron cloud introduces a $\sin\theta_s/\lambda$ dependence into the atomic scattering factors that can be calculated using the electronic wavefunctions. This is called the form factor.

In a solid the outer electrons are often delocalised and take part in bonding or are free to move in the conduction band. At low X-ray energies (<50 eV) these electrons become significant and the concept of an isolated atomic scattering factor becomes compromised. Thus the calculation of the refractive index for EUV photons is notoriously tricky.

The application of dispersion theory to X rays was originally formulated in the 1930's spurred on by the development of Bragg crystallography and is described in the excellent books by Compton & Allison (1935) and later by James (1965). Further details about calculating the refractive index using atomic scattering factors can be found in these texts. It is customary to express atomic scattering factors as a series of terms

$$f = f_1 - \Delta f_{\mathrm{r}} - \Delta f_o + i f_2 \tag{35}$$

The $\Delta f_{\mathrm{r}} = (5/3)E_{\mathrm{tot}}/(m_e c^2)$ is a relativistic correction term derived by Cromer & Liberman (1970a) where E_{tot} is the total energy of the atom and Δf_o is a form factor correction to allow for the finite size of the atom. Tabulations of these scattering-factor components for all atomic types form the bedrock of all physical calculations in X-ray optics.

Of course, many materials contain more than one atomic type and the refractive index must then be calculated from an average, weighting the contribution from each atom by number density. Providing the molecular size is less than the wavelength, the resulting average refractive index is meaningful. If the inhomogeneities become larger than the wavelength then we must be careful as you shall see in subsequent sections.

2.3 The Kramers-Kronig relations - measuring and calculating the refractive index for X rays

The real and imaginary parts of the dielectric constant (or refractive index) are related because they are both calculated from the same transition probabilities, oscillator strengths and transition lifetimes. This is embodied in the Kramers-Kronig relations

$$\varepsilon_1(\omega) = \varepsilon(\infty) + \frac{2}{\pi}\int_0^\infty \frac{\omega'\varepsilon_2(\omega')}{\omega'^2 - \omega^2}\mathrm{d}\omega' \tag{36}$$

$$\varepsilon_2(\omega) = \frac{2\omega}{\pi}\int_0^\infty \frac{\varepsilon_1(\omega')}{\omega^2 - \omega'^2}\mathrm{d}\omega' \tag{37}$$

So if we know either ε_1 or ε_2 as a function of ω we can calculate an approximation to the other by integration.

The complex refractive index for X rays is usually written as

$$n_\mathrm{c} = 1 - \delta - i\beta \tag{38}$$

The decrements δ and β are both very small and so using Eq. (20) we have the following equations relating the decrements of the refractive index and dielectric constant

$$\alpha = 2\delta - \delta^2 + \beta^2 \approx 2\delta \tag{39}$$

$$\gamma = 2\beta(1 - \delta) \approx 2\beta \tag{40}$$

The imaginary decrements γ or β are directly related to the linear absorption coefficient of the medium, Eq. (22), or equivalently the imaginary part of the atomic scattering factor is directly related to the photoelectric absorption. So if the linear absorption (or photoelectric absorption) is measured as a function of frequency (energy) we can calculate $\gamma(\omega)$ or $f_2(\omega)$. Using the integral in Eq. (36) we can approximate the real part of the dielectric constant, providing the frequency (energy) coverage is large enough. If the photoelectric absorption cannot be measured directly it can be calculated theoretically using electronic wavefunctions. There are a number of tabulations of X-ray optical constants which have been compiled by a combination of experimental measurements and theoretical calculation in this way. For example, a tabulation of X-ray scattering factors calculated for use in crystalography can be found in Cromer & Mann (1968). A full tabulation of the anomalous terms for all elements calculated using relativistic wavefunctions was published by Cromer & Liberman (1970b). A comprehensive tabulation was compiled by Henke et al. (1982). The latest best estimates of the optical constants for X rays can be found on the World Wide Web, see for example the references to CXRO (1997) or the X-ray WWW Server Uppsala University (1997).

Such tabulations are being improved all the time by photoelectric absorption and emission measurements at synchrotron sources and quantum

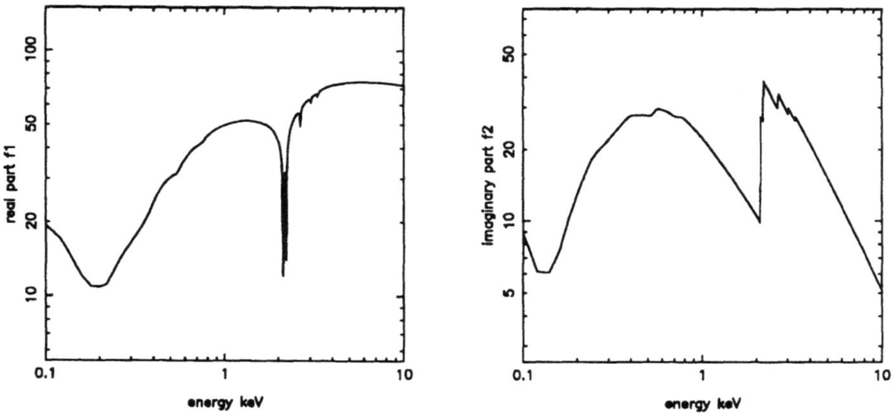

Fig. 1. Atomic scattering factors for Platinum, $Z = 78$.

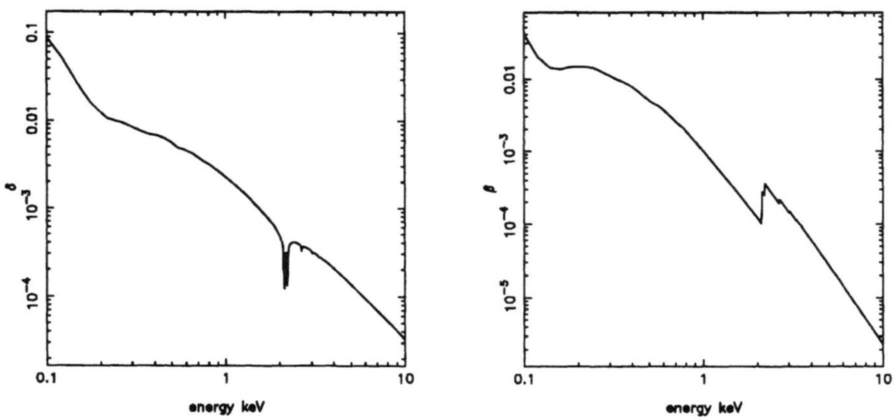

Fig. 2. Refractive index decrements for Platinum, $Z = 78$.

mechanical calculations. In general there is a good agreement between calculation and measurement. Figure 1 shows f_1 and f_2 taken from the tabulation of Henke et al. (1982). Figure 2 shows the refractive-index decrements δ and β calculated from these scattering factors. The refractive-index decrements of all media for X rays are always very small so that refraction is very weak. However, β is often comparable to or greater than δ so that absorption dominates. Therefore conventional refracting lenses can never be constructed for X rays. The difference in curvature required between the entrance and exit surfaces always introduces a material thickness that will substantially absorb the X radiation.

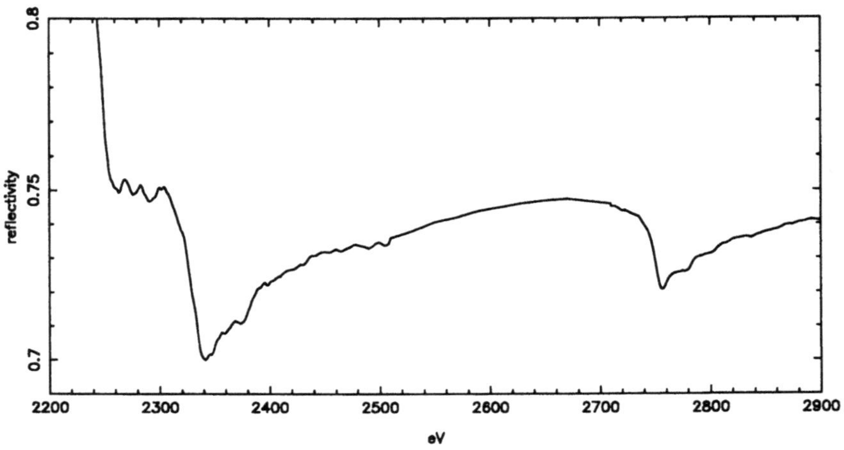

Fig. 3. EXAFS visible in the measured grazing-incidence reflectivity of gold around the M absorption edges.

2.4 EXAFS

It was stated above that the photoelectric absorption and refractive index could be calculated without considering the influence of neighbouring atoms. This is not quite true. X-ray absorption and reflection at energies just above absorption edges exhibits Extended X-ray Absorption Fine Structure, EXAFS. This is due to a second scattering of the out-going primary scattered wave by nearby atoms. The absorption shows a fluctuation as a function of energy with a period $\Delta E \sim 10$ eV, characteristic of the near-neighbour structure of the material. The measurement and calibration of EXAFS can be most important for high resolution X-ray spectroscopy in astronomy, both in the optics and the X-ray detectors. For example, Figure 3 shows structure in the grazing-incidence reflectivity of a gold X-ray mirror introduced by EXAFS measured during the *JET-X* calibration programme (Wells et al. 1988). The sharp steps are three of the M absorption edges of gold, and the undulations just above each edge are due to EXAFS. The theory and interpretation of EXAFS are described in detail by Gurman (1995).

3 The Reflection of X Rays

3.1 Fresnel reflection

The reflection of EM radiation from a plane interface between two media can be calculated using the boundary conditions of the electric and magnetic fields and matching the fields at either side of the boundary. The standard result for the Fresnel amplitude reflection coefficients, see for example Lorrain, Corson & Lorrain (1988), is

$$r_{\parallel} = \frac{Z_2 \cos\theta_2 - Z_1 \cos\theta_1}{Z_2 \cos\theta_2 + Z_1 \cos\theta_1} \tag{41}$$

$$r_{\perp} = \frac{Z_2 \cos\theta_1 - Z_1 \cos\theta_2}{Z_2 \cos\theta_1 + Z_1 \cos\theta_2} \tag{42}$$

where $r_{\parallel}$ is for the polarization with the E-vector parallel to the plane of reflection and $r_{\perp}$ with the E-vector perpendicular to this plane. Z_1 and Z_2 are the complex impedences of the two media which are directly related to the complex refractive index

$$Z = \frac{E}{H} = \frac{c\mu}{n_c} \tag{43}$$

The angles θ_1 and θ_2 are the incidence and refraction angle given by Snell's law, $n_1 \sin\theta_1 = n_2 \sin\theta_2$. Note that since the refractive indices can be complex then so can these angles. The corresponding transmission amplitude coefficients are

$$t_{\parallel} = \frac{2Z_2 \cos\theta_1}{Z_2 \cos\theta_2 + Z_1 \cos\theta_1} \tag{44}$$

$$t_{\perp} = \frac{2Z_2 \cos\theta_1}{Z_2 \cos\theta_1 + Z_1 \cos\theta_2} \tag{45}$$

The coefficients of reflection and transmission concern the flow of energy normal to the surface. Since the angle of incidence equals the angle of reflection this flow is simply given by the square of the modulus of the amplitude coefficents and we have $R = rr^*$ where the asterisk indicates the complex conjugate.

For X rays $\mu = 1$ and we can calculate the complex refractive index n_c as outlined above, and so we can calculate the reflection coefficients. Of course, there is also a refracted or transmitted component that penetrates into the second medium at a refraction angle given by Snell's law. Since the real part of the refractive index is slightly less than 1, above some critical incidence angle given by

$$\sin\theta_1 = \frac{n_2}{n_1} \tag{46}$$

there will be no transmission but Total External Reflection. The critical grazing angle ($\theta_g = \pi/2 - \theta$, when $n_1 = 1$, a vacuum) is $\theta_{gc} = \delta$ radians. Since absorption is present $\beta > 0$ and the reflectivity for grazing angles smaller than θ_{gc} is less than unity. Figure 4 shows calculated reflectivity curves for a pure gold surface which are typical of a dense, high-Z, metallic surface. The left-hand panel shows how the critical grazing angle gets smaller as the photon energy increases. The right-hand panel shows the reduction in reflectivity introduced by absorption edges. In this case there are a series of M-edges for gold starting at 2.1 keV (cf. Figure 3).

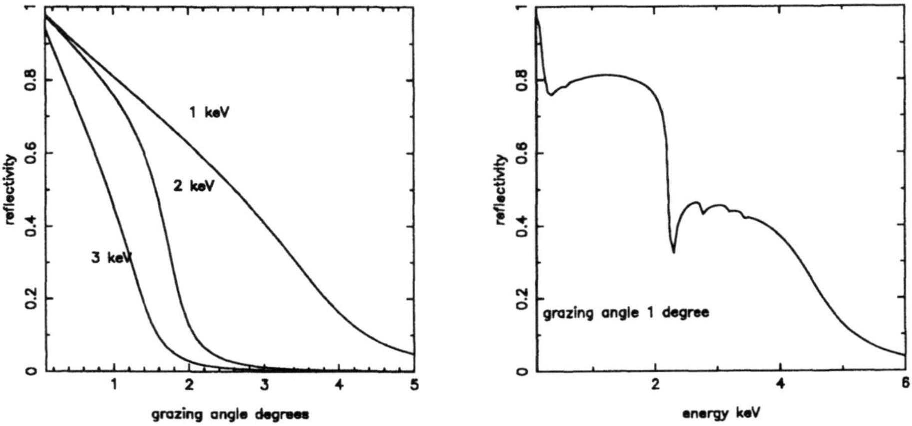

Fig. 4. Calculated grazing-incidence reflectivity curves for gold.

3.2 Reflection from multi-layers

A multi-layer is the modern term for a stratified medium which consists of a series of plane layers of differing refractive index. The simplest periodic multi-layer is constructed using alternate layers of high and low refractive index. Each interface in the multi-layer produces a reflected and a transmitted component as shown in Figure 5.

The reflection from such a structure can be calculated by matching the tangential components of the electric and magnetic fields either side of each interface. See, for example, Crook (1948) or Lee (1981). For the interface between the k'th and $(k+1)$'th layer

$$E_k^+ = t_{k,k+1}^{-1}(E_{k+1}^+ e^{i\psi_{k+1}} + r_{k,k+1} E_{k+1}^- e^{-i\psi_{k+1}}) \tag{47}$$

$$E_k^- = t_{k,k+1}^{-1}(E_{k+1}^+ e^{i\psi_{k+1}} r_{k,k+1} + E_{k+1}^- e^{-i\psi_{k+1}}) \tag{48}$$

where the + and − superscripts indicate waves travelling into and out of the stack, respectively, and $r_{k,k+1}$ and $t_{k,k+1}$ are the Fresnel amplitude reflection and transmission coefficients for the interface between the k'th and $(k+1)$'th layer.

$$\psi_{k+1} = 2\pi d_{k+1} n_{k+1} \cos\theta_{k+1}/\lambda_0 \tag{49}$$

is the complex phase shift through the $(k+1)$'th layer where d_{k+1} is the layer thickness, n_{k+1} is the complex refractive index θ_{k+1} is the complex refraction angle and λ_o is the free-space wavelength. Eqs (47) and (48) are conveniently written as a matrix equation

$$\begin{pmatrix} E_k^+ \\ E_k^- \end{pmatrix} = t_{k,k+1}^{-1} \begin{pmatrix} e^{i\psi_{k+1}} & r_{k,k+1} e^{-i\psi_{k+1}} \\ r_{k,k+1} e^{i\psi_{k+1}} & e^{-i\psi_{k+1}} \end{pmatrix} \begin{pmatrix} E_{k+1}^+ \\ E_{k+1}^- \end{pmatrix} \tag{50}$$

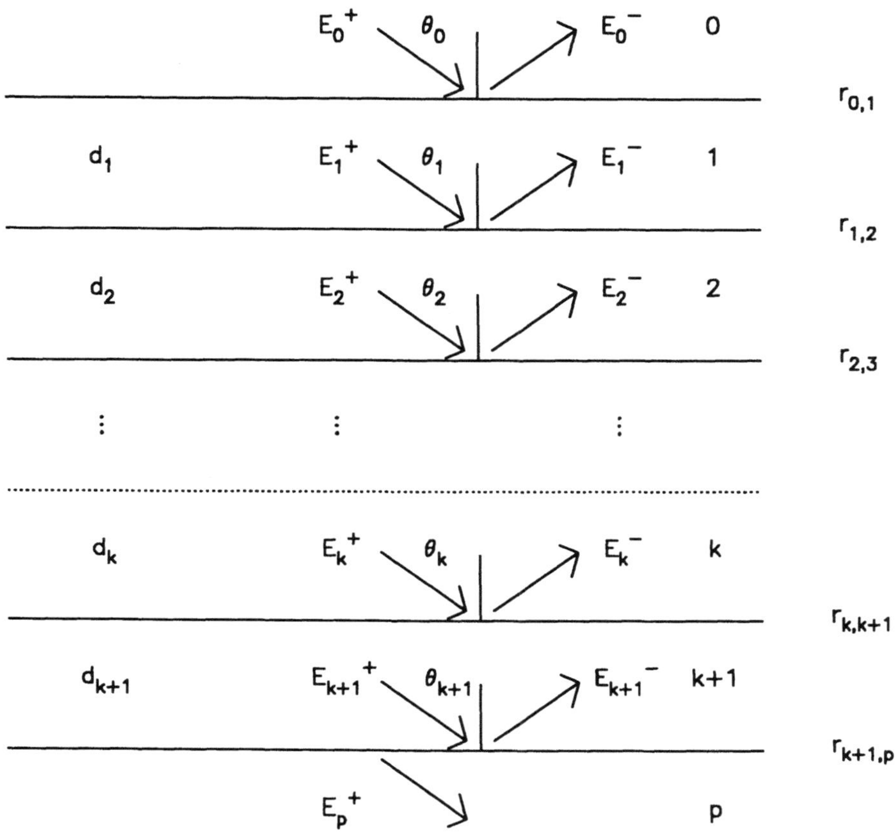

Fig. 5. The wave components in a multi-layer stack

This equation can be applied for each successive layer to relate E_o^{+-} to E_p^{+-} for the p'th layer. If this layer is infinitely (or very) thick there will be no E_p^- component and therefore we can calculate the ratio of E_o^-/E_o^+ which is the amplitude reflection coefficient from the multi-layer. If the absorption is high then the transmission through each successive layer will be low and the fields rapidly diminish. In this case deep layers will have little effect on the reflection coefficient.

If the multi-layer is periodic each period has the same characteristic matrix and the total reflection is found by raising this matrix to the appropriate power. A periodic multi-layer obeys Bragg's law with a peak of reflection at incidence angles given by

$$n\lambda_o = 2d\cos\theta \tag{51}$$

where n is the order and d is the period. Actually, the peak of the reflection is shifted slightly from the position predicted by Eq. (51) because of the effects of absorption and refraction. This is most noticeable at small grazing angles.

Figure 6 shows the reflectivity of a multi-layer calculated by this method. In

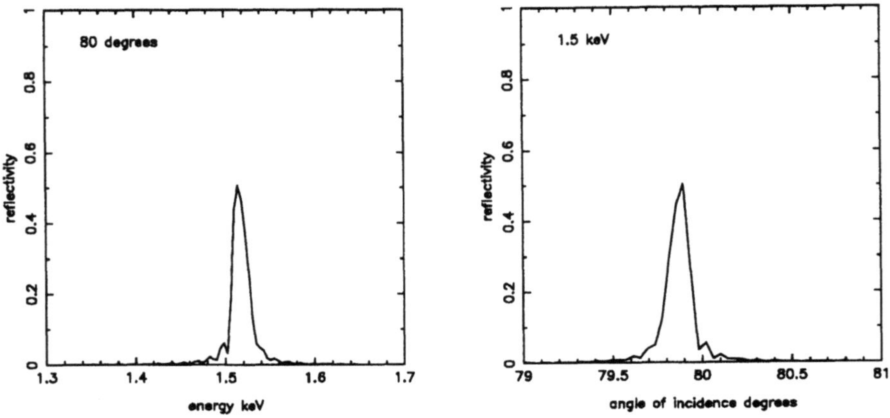

Fig. 6. The reflectivity of a Si-W multi-layer with period 100 Å.

a multi-layer the in-going and out-going waves are uniquely coupled and the solution for the EM waves in the structure is said to be a dynamical solution. The in-going wave suffers primary extinction as power is scattered into the out-going wave at each interface. Both the in-going and out-going wave suffer secondary extinction due to absorption.

3.3 Reflection from crystals

The calculation of the X-ray reflection from crystals is too complicated to be described fully here, but we shall outline the basic principles. The electron density in a crystal is periodic in all 3 axes and can therefore be expressed in terms of a Fourier series.

$$\rho(\underline{r}) = \frac{1}{V}\sum_{H} F_H \exp(-2i\pi \underline{R}_H \cdot \underline{r}) \tag{52}$$

where V is the volume of a unit cell, H are Miller indices, $\underline{r}$ is the position vector, F_H is called the structure factor and $\underline{R}_H$ is a reciprocal lattice vector.

$$F_H = \sum_{m} f_m \exp(2i\pi \underline{R}_H \cdot \underline{r}_m) \tag{53}$$

where the sum is over all the atoms in the unit cell and f_m is the atomic scattering factor for the m'th atom. We can use the electron density to calculate a periodic, complex dielectric constant for the crystal [essentially substituting for Nf in Eq. (31)]

$$\varepsilon(\underline{r}) = 1 + \frac{q_e^2}{\varepsilon_o m_e V \omega^2} \sum_H F_H \exp(-2i\pi \underline{R}_H . \underline{r}) \tag{54}$$

This dielectric constant can be substituted into the wave equation derived from Maxwell's equations

$$\nabla^2 \underline{E} = \varepsilon \varepsilon_o \partial^2 \underline{E} / \partial t^2 \tag{55}$$

A two-wave solution to this equation arises when the wave vectors $\underline{k}_o$ and $\underline{k}_H$ for the two waves obey

$$\underline{k}_H - \underline{k}_o = \underline{R}_H \tag{56}$$

This is the so-called Bragg reflection. The intensity and width of the reflection depend on the structure factor for the reflection, F_H, which, in turn, depends on the scattering factors of the atoms in the unit cell. The full theory is able to predict the reflectivity as a function of angle or energy (frequency) for all values of H. As in the case of multi-layers, the two waves suffer primary and secondary extinction and the solution is a dynamical solution. It is this extinction that gives the reflection curve a finite width in angle or energy. A much more comprehensive description of all this can be found in James (1965) and Azaroff (1974).

3.4 Reflection and transmission gratings

A reflection grating consists of N parallel, identical, grooves formed on an otherwise flat surface. A typical profile of a blazed X-ray grating is shown in Figure 7. The angle of the sawtooth, the so-called blaze angle, can be chosen to give the maximum reflection at a particular angle and hence for a particular X-ray wavelength. A transmission grating consists of N parallel, identical, slits formed in a flat plate. Ideally the bars between the slots completely block the X rays and thus the incident wavefronts are modulated by a square tooth function.

The form of the diffraction pattern from a grating is given by standard Fraunhofer diffraction theory. If a plane wave is incident on the grating at angle θ_o as shown in Figure 7, then the diffracted intensity at angle θ (reflected or transmitted) is

$$I(\theta) = I_o(\theta) \left(\frac{\sin Nx}{\sin x} \right)^2 \tag{57}$$

where, if the period of the grating is d,

$$x = \frac{2\pi}{\lambda} d(\sin\theta - \sin\theta_o) \tag{58}$$

If N is large then there are just principle maxima or diffraction orders at

$$\sin\theta - \sin\theta_o = m\lambda/d \tag{59}$$

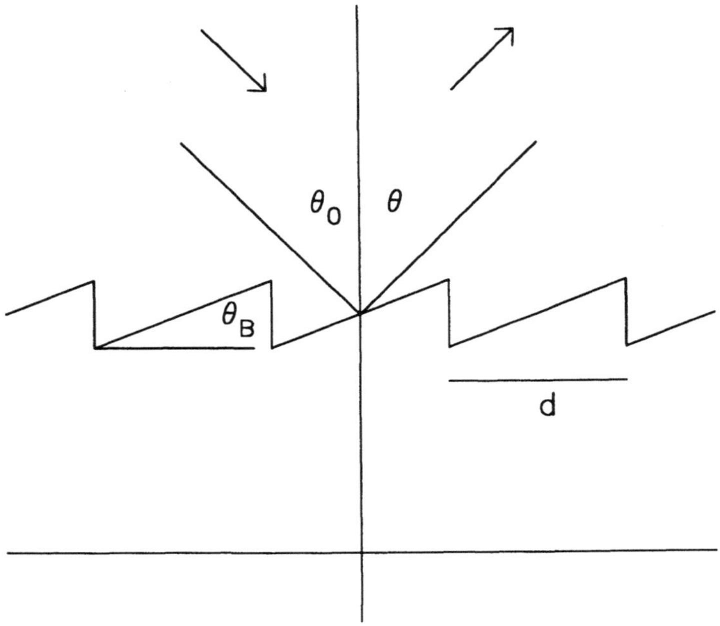

Fig. 7. Profile of a blazed sawtooth reflection grating.

$I_o(\theta)$ is the intensity scattered or transmitted from a single period of the grating. For a reflection grating this depends on the Fresnel reflectivity from the facets or profile and hence depends on the blaze angle. For a transmission grating it depends on the width of the slits and the blocking efficiency (photoelectric absorption) of the bars.

In the X-ray regime reflection gratings must be used at grazing angles of a few degrees or less to achieve a high reflectivity. An alternative geometry is the out-of-plane mount illustrated in Figure 8. Dispersion is now around a semi-circle in a plane perpendicular to the ruling direction. Then the grating equation becomes

$$\sin\gamma\,(\sin\theta - \sin\theta_o) = m\lambda/d \qquad (60)$$

As γ is reduced the effective d spacing is decreased.

3.5 Scattering from surface roughness

In all the above analysis it was assumed that the interfaces, surfaces and grooves etc. were perfect. In practice, any roughness of these elements will introduce small phase changes in the wavefronts. Because the wavelengths of X rays are so small, imperfections of order Å in size are sufficient to perturb the wavefronts and introduce significant scattering. The complete theory of scattering from surface roughness was originally developed to describe the

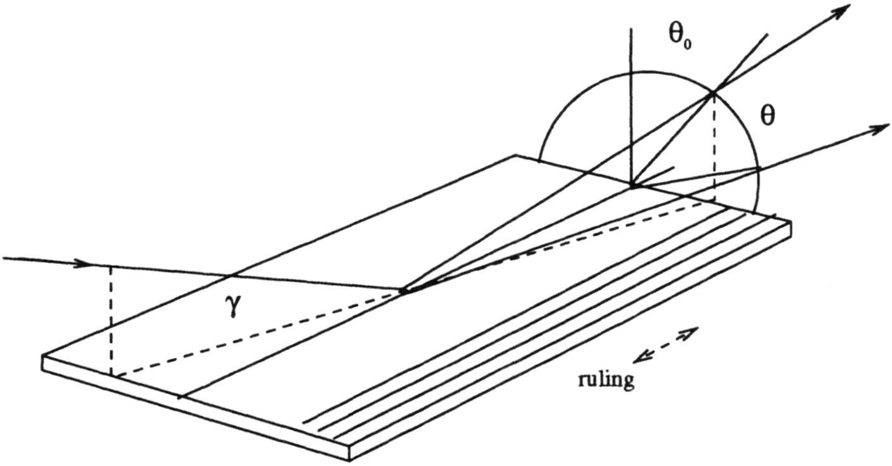

Fig. 8. Reflection grating in out-of-plane configuration.

scattering of radar from land and sea, see Rice (1951) and Beckmann & Spizzichino (1963), but amazingly the results apply equally well to the much shorter wavelength regime of X rays. In essence, the surface roughness is modelled as a 2-D aperiodic diffraction grating. Each Fourier component of the surface roughness produces diffraction orders given by the grating equation (Eq. 59). The sum of all these components gives a scattered distribution. The differential scattering cross section is directly related to the power spectral density $W(p,q)$ of the surface height perturbations. The geometry of the process is shown in Figure 9.

$$\frac{\mathrm{d}I}{\mathrm{d}\Omega} = 4k^2 \cos\theta_\mathrm{i} \cos^2\theta_\mathrm{s} a_\tau W(p,q) \tag{61}$$

where we have $k = 2\pi/\lambda$, $\mathrm{d}\Omega = \sin\theta_\mathrm{s}\,\mathrm{d}\theta_\mathrm{s}\,\mathrm{d}\phi_\mathrm{s}$, $p = k(\sin\theta_\mathrm{s}\cos\phi_\mathrm{s} - \sin\theta_\mathrm{i})$, $q = k(\sin\theta_\mathrm{s}\sin\phi_\mathrm{s})$ and a_τ depends on the polarization state of the incident and scattered components.

If σ is the root mean square of the surface height fluctuations (found by integrating the power spectral density) and $(k\sigma)^2 << 1$ then the Total Integrated Scatter (TIS) fraction is

$$TIS_\tau(\theta_\mathrm{i}) = R_\tau(\theta_\mathrm{i})(2k\sigma\cos\theta_\mathrm{i})^2 \tag{62}$$

where R_τ is the Fresnel reflectivity for polarization τ. This equation can be used to estimate the scattering losses from specular reflecting surfaces, multilayer structures and reflection gratings. Figure 10 shows a typical roughness power spectral density specification of an X-ray mirror taken from the ESA *XMM* mirror programme. The ranges indicate the metrology devices used to measure the surface roughness and figuring errors. Note that if the surface

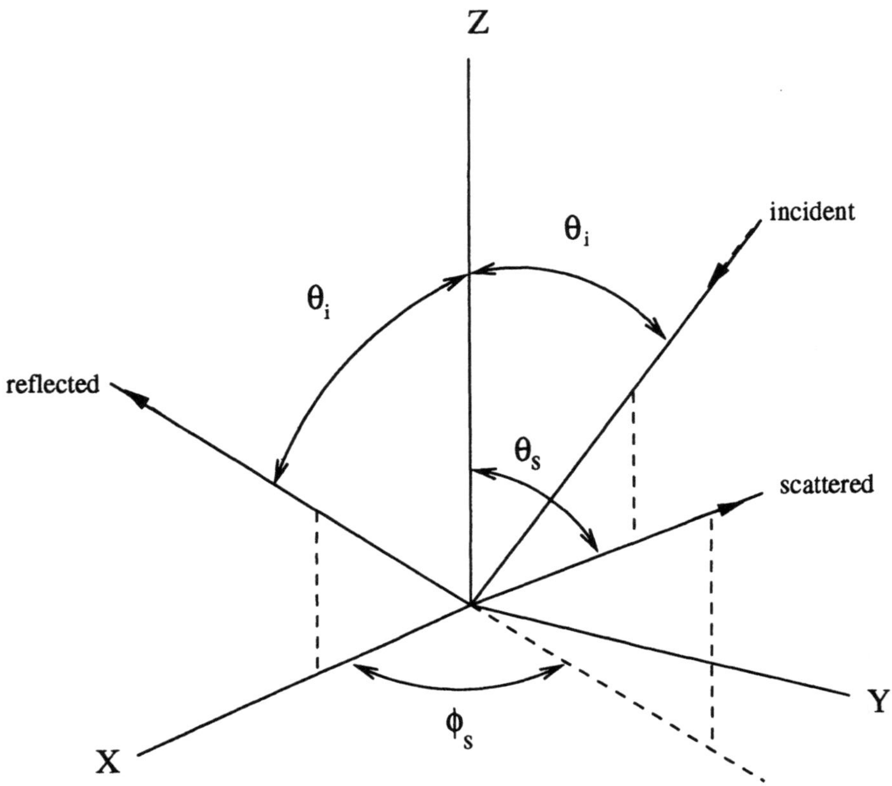

Fig. 9. Scattering geometry from a rough surface.

is periodic then $W(p,q)$ will consist of sharp peaks corresponding to the harmonics of the surface profile. Then the scattering $\mathrm{d}I/\mathrm{d}\Omega$ will also exhibit peaks which are the diffraction grating orders of the surface.

4 Geometries for X-ray Optics

4.1 The geometrical theory of imaging

In order to appreciate the geometries used in X-ray optics you need to know a little about the geometrical theory of imaging. This is described in detail by Born & Wolf (1970). The salient points are as follows:

- An imaging system consists of some object space, some instrument and some image space. A perfect imaging system performs what is known as a *projective transformation* such that lines in the object space are transformed into lines in the image space. The projection is performed by the propagation of light from the object space through the instrument

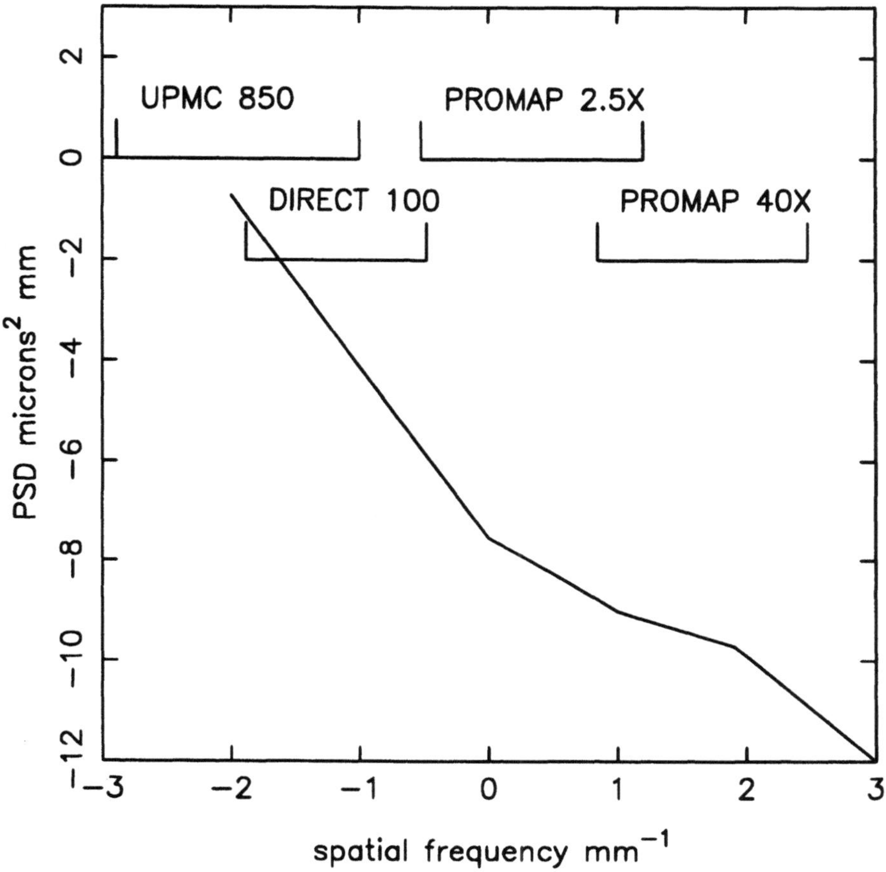

Fig. 10. The surface roughness power spectral density specification for the *XMM* mandrels

to the image space. Such a perfect system cannot be constructed even in theory but it can be approximated.

- Optical systems usually have symmetry about an axis in which case they are called centred systems. The projective transformation in a centred system is defined by just 4 parameters. Such a system is illustrated in Figure 11.

- The system is said to be stigmatic if points in the object space are transformed into points in the image space. In reality a narrow pencil of rays from a region in the object will transform into a near perfect image. In order for this to happen it can be shown that

$$n_1 y_1 \sin \gamma_1 = n_0 y_0 \sin \gamma_0 \tag{63}$$

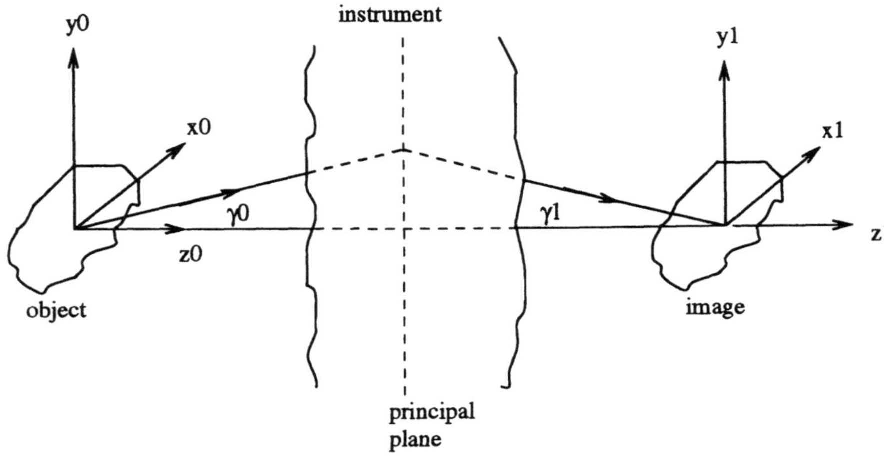

Fig. 11. The centred imaging system.

– If the object lies at (or near) infinity, as is indeed the case in astronomy, then the system looks like Figure 12 and this condition reduces to

$$\frac{h_0}{\sin \gamma_1} = f_1 \tag{64}$$

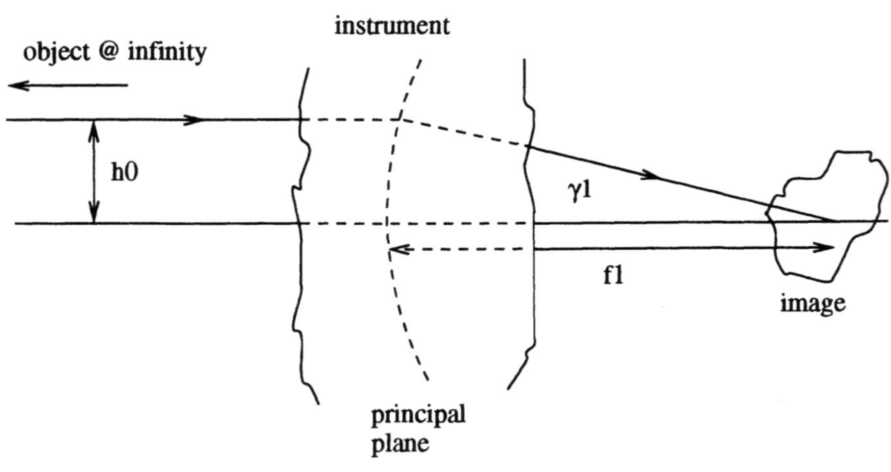

Fig. 12. The centred imaging system, object at infinity.

– This means that stigmatic imaging will be achieved for some pencil of rays over a small region near the axis provided that each incident ray

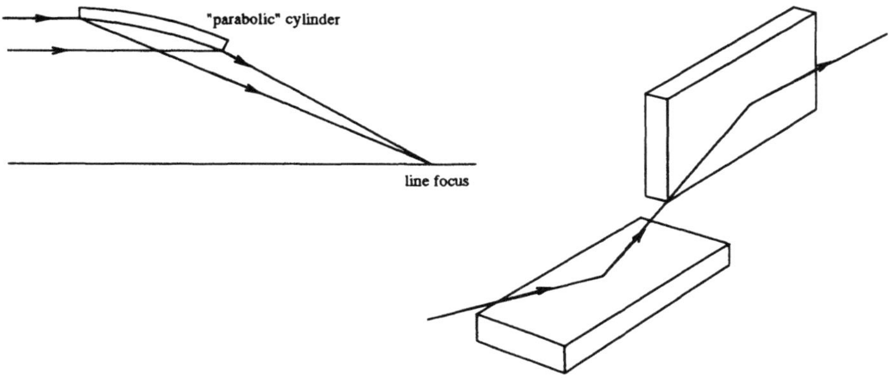

Fig. 13. The Kirkpatrick-Baez geometry.

intersects its conjugate (focused) ray on a sphere of radius f_1 centred at the focus. This sphere is the Principal Surface of the imaging system.
- If the pencil of rays is narrow then $\sin\gamma_1 \approx \gamma_1$ and the sphere becomes a plane and the optics are called Gaussian Optics.
- A near stigmatic telescope system must have a Principal Surface that is a sphere centred on the axial focus. If the F ratio is large enough (small aperture diameter/focal length) then this sphere is well approximated by a plane perpendicular to the optical axis.

4.2 Grazing-incidence telescopes; Wolter type I and II and Kirkpatrick-Baez systems

An X-ray imaging system must rely on Fresnel reflection or reflection from crystals or multi-layers; no other phenomenon is available. Thus for broad-band imaging grazing-incidence reflection is the only viable option.

What is required is a system of grazing-incidence reflections that produce a Principal Surface which is a plane or a sphere as described above. The first system was described by Kirkpatrick & Baez (1948). This consisted of two orthogonally mounted cylindrical mirrors each of which provided a line focus. This is illustrated in Figure 13. The axially symmetric solution to this problem was provided by Wolter (1952). The Wolter type I and II systems, which are of most use in astronomy, are shown in Figure 14. The solid lines are the physical mirrors, the dashed lines indicate the path of on-axis rays and the dotted lines show the principal plane and surface generators. In both Kirkpatrick-Baez and Wolter geometries two grazing incidence reflections are used to place the intersection of the in-coming and out-going rays on a plane perpendicular to the optical axis. Provided there are an even number of reflections this can always be achieved.

The K-B and Wolter I geometries can be nested. In the K-B case a large number of reflecting plates can be stacked together such that they have a

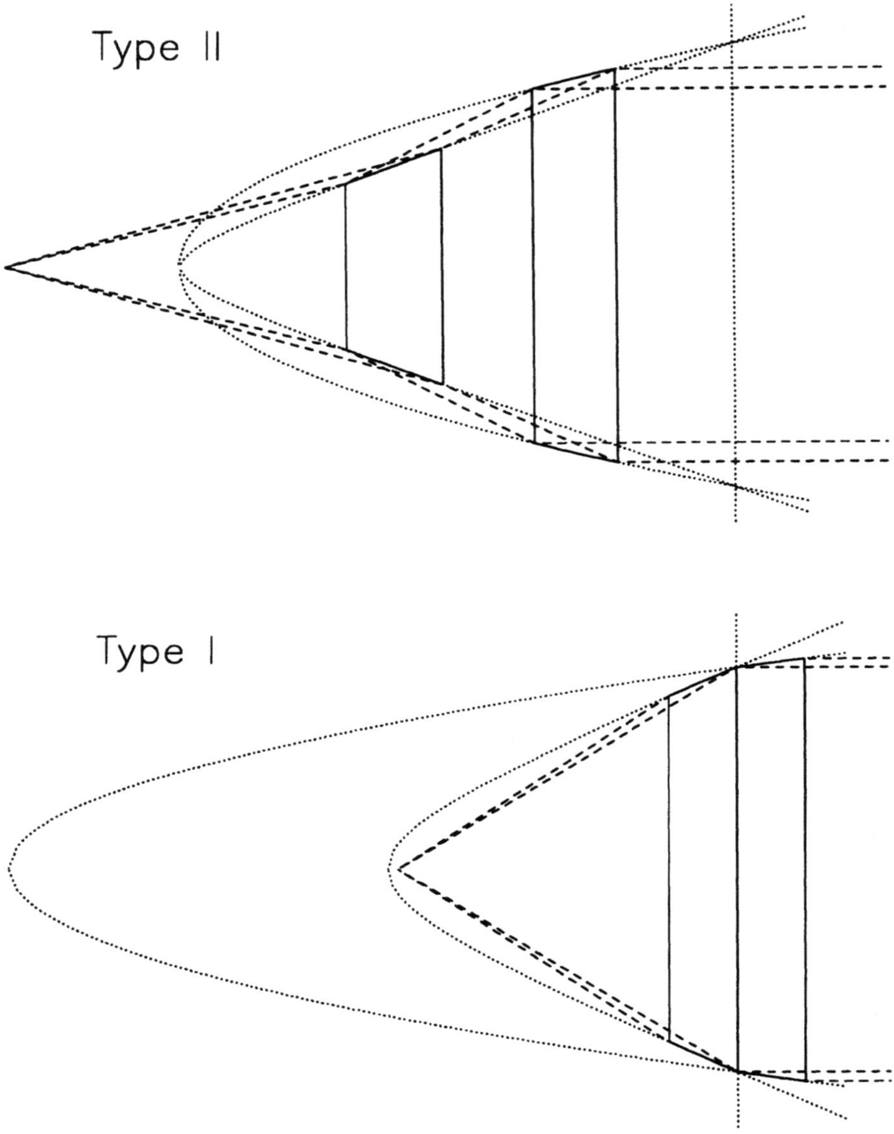

Fig. 14. The Wolter type I and II geometry.

common line focus. For Wolter I systems a series of shells or surfaces of revolution are placed one inside the other with a common axis.

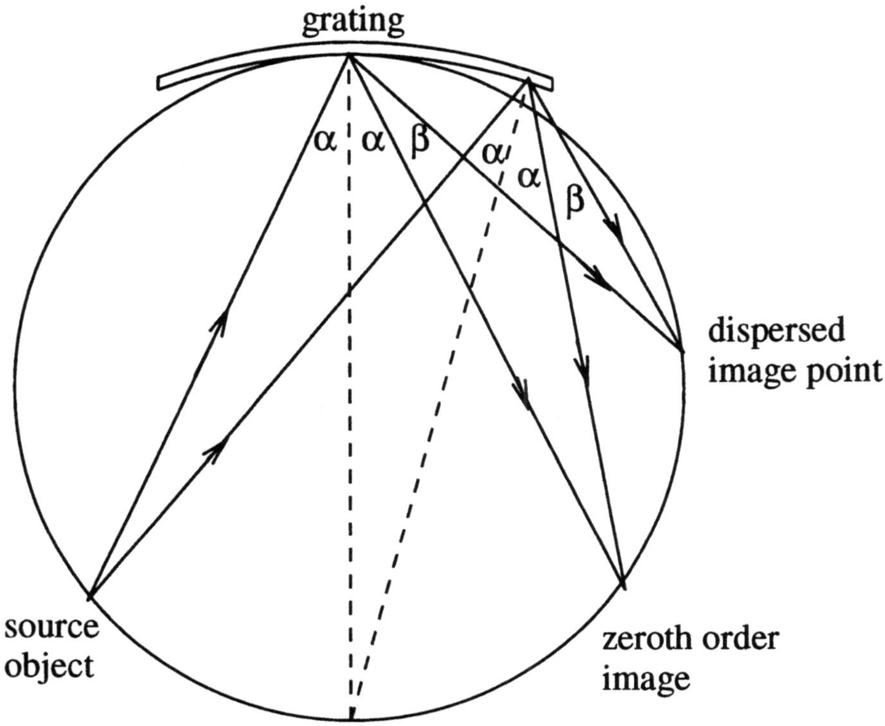

Fig. 15. Focusing with a concave grating and Rowland's circle.

4.3 Grating and crystal spectrometers

In an astronomical X-ray spectrometer a mirror system must be use to collect and focus the light usually to produce an image. Therefore dispersive elements must usually be placed in a converging or diverging beam. A curved grating or crystal can be used to avoid excessive losses from the beam and to form a dispersed image. The geometrical arrangement used to achieve this is called Rowland's circle as shown in Figure 15. Note that the radius of curvature of the grating must be twice that of Rowland's circle to satisfy the necessary reflection condition. If the grating is toroidal then some focusing can also be achieved normal to the dispersion plane.

5 X-ray Telescopes and Spectrometers

5.1 Optimization of the design

The most sensitive and most useful telescope has:

- Large aperture - large collecting area

- High angular resolution over a large field of view
- Large energy bandwidth

Problems with grazing-incidence X-ray telescopes are:

- Low surface utilization - (effective aperture area)/(mirror surface area). For a Wolter I or Kirkpatrick-Baez system with grazing angles θ_g and surface reflectivity $R(\theta_g)$

$$u_s = \frac{1}{2}R^2(\theta_g)\sin\theta_g \tag{65}$$

- The surface roughness must be typically $\sigma \leq 5$ Å r.m.s. The reflecting surfaces must be superpolished.
- The figuring errors must be significantly less than the angular resolution, typically a few arc seconds.
- As a consequence, the axial length of the surfaces must be large and/or the surfaces must be nested. Area is inevitably lost at the front edge of the surfaces.
- The radius of the field of view is limited to θ_g.
- The reflectivity $R(\theta_g)$ must be broad-band. High-Z metallic surfaces impose a grazing-angle cutoff. Multi-layer surfaces can give large reflectivities at large grazing angles, but have a small bandwidth.
- The reflecting surfaces must be kept very clean, free from dust and contamination that would reduce the reflectivity.

The fundamental trade-off hinges on shell thickness. Thick shells are stiff and can be figured and polished giving high angular resolution, low scattering and high reflectivity. Thin shells will give large aperture utilization (collecting area as a fraction of aperture area) and therefore large collecting areas.

5.2 Types of primary X-ray mirror

5.2.1 Foils In such a mirror, Wolter type I surfaces are approximated by thin conical aluminium foil sectors with a foil thickness $\sim$ 130 μm. A very large number of nested sections can give a high-aperture utilization up to $\sim$ 0.6. A high quality surface finish is achieved by laquering or epoxy replication of a smooth mandrel former and high X-ray reflectivity is achieved by coating the surfaces with gold or some other high-Z material. The major problem with foils is that figuring errors limit the resolution to $\sim$ 2 arc minutes. The *ASCA* mirrors which were constructed from foils (Serlemitsos 1981) are shown in schematic cross section in Figure 16. The aperture diameter is 346 mm, the focal length 3.5 m. and the nest contains 119 shells. The mirror module was constructed in quadrants with the paraboloid and hyperboloid surfaces integrated as separate foils. So a total of $4 \times 119 = 476$ conical foil sectors had to be integrated into a single module. The foils are supported along the axial edges and by radial support bars at about every 10 degrees around the aperture.

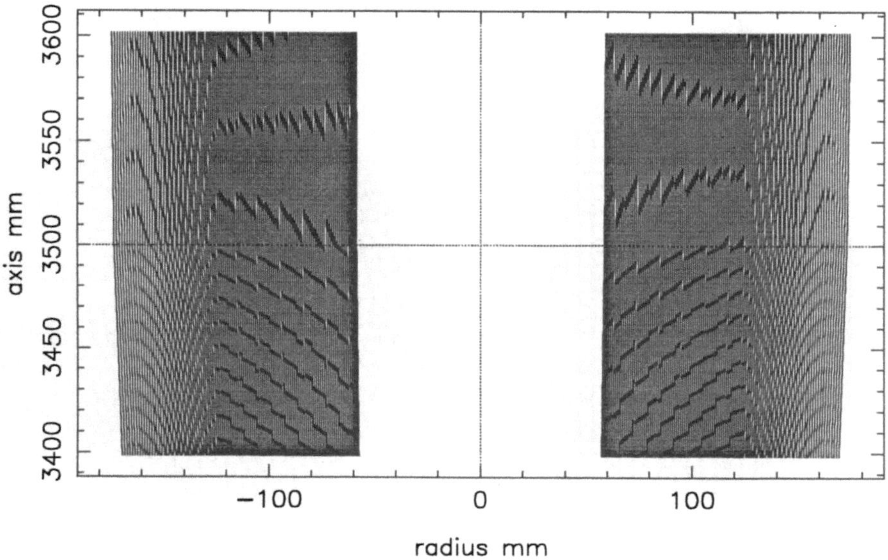

Fig. 16. Schematic cross section of the *ASCA* mirror assembly.

5.2.2 Lobster eyes These optics consist of small square pores or channels packed together to approximate a K-B or Wolter I geometry. The geometrical arrangement of the channels is shown in the upper panel of Figure 17. Cross sections showing the alignment of the channels for the 2 geometries are shown in the lower panel. Present designs use microchannel plate (MCP) technology to produce the pores, see for example Lees et al. (1995). Very small grazing angles arise for the channels near the centre of the plates and therefore a hard X-ray response can be realised and furthermore the geometry can provide a very wide field of view. Problems with such optics include accurate manufacturing of the channels, channel alignment, wall coating to enhance the reflectivity and reducing the surface roughness to lower the scattering.

5.2.3 Replicated shells The process used to produce thin replicated nickel shells is shown in cartoon format in Figure 18. The magic part of the production sequence is the use of gold as a release agent for the electroforming process. The gold is conveniently transfered from the mandrel onto the electroformed nickel shell and therefore after release the shell has a superpolished gold surface ready for use as an X-ray mirror. Shells can be produced with a thickness as little as 0.3 mm and so the aperture utilization can be ~ 0.4. Surfaces of revolution (complete shells) exhibit relatively high stiffness and if handled carefully during integration they can retain the high-quality figure reproduced from the mandrel. Problems with the process include residual stress in the nickel and handling, both of which introduce figure errors. The

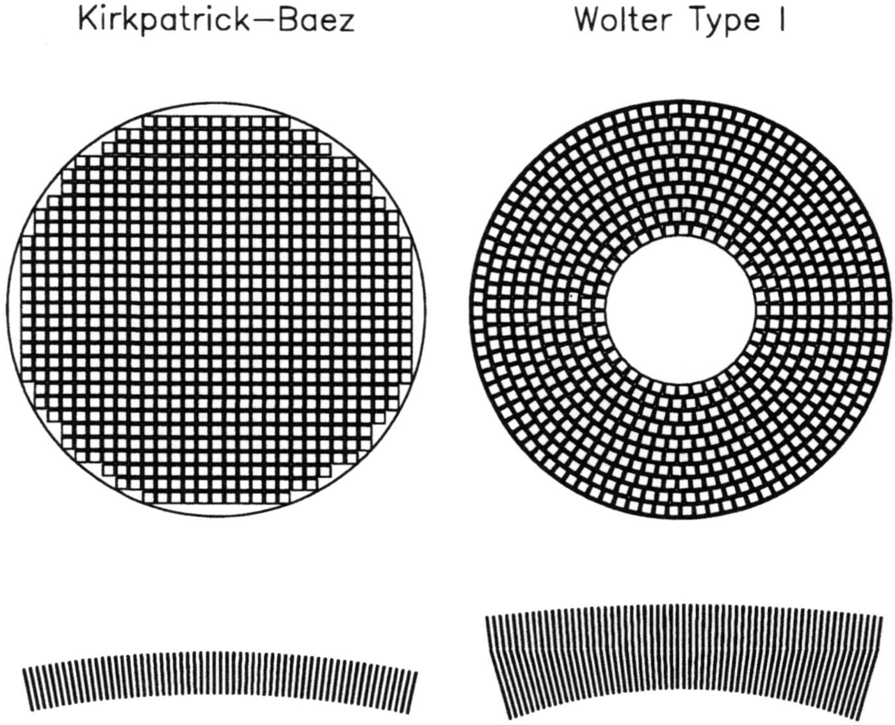

Fig. 17. MCP optics.

shells are difficult to mount in a nest without introducing further distortion and nickel is dense so the final assembly is rather massive. Prime examples of the use of replication technology are the mirrors on *JET-X* and *XMM*.

5.2.4 Monolithic shells This is the traditional way of producing high-quality grazing-incidence X-ray mirrors. Wolter I blanks are drilled out of a block of Zerodur or similar glass ceramic. The blanks are then ground to the correct thickness and figure. Superpolish is achieved by a very fine grinding/polishing process. Providing the shells are thick enough, very high figure tolerances can be met giving as good as 0.5 arc second angular resolution. The major problem with this approach is that the aperture utilization is limited to ~ 0.1 by the shell thickness required and the shells are very massive. The very successful *ROSAT* XRT and the forthcoming *AXAF* telescope both use such monolithic shell technology. Figure 19 shows the main features of the *AXAF* High Resolution Mirror Assembly.

5.2.5 Diffraction limited X-ray mirrors These can probably be realised using high-quality spherical mirrors in a Kirkpatrick-Baez geometry as shown

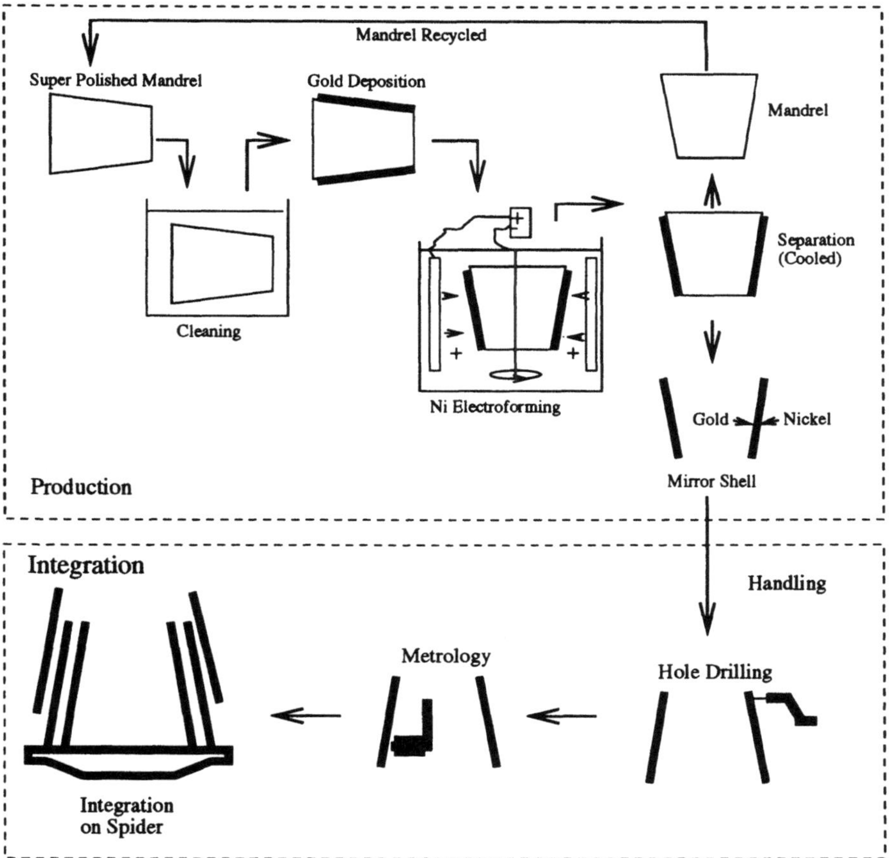

Fig. 18. The manufacture of replicated nickel shells

in Figure 20. In such a system the radii of curvature and placements of the mirror elements are chosen such that the aberrations introduced by the first two reflections are largely cancelled by the second two reflections. The advantage of such an arrangement is that very high quality spherical surfaces can be manufactured relatively easily compared with the aspheric surfaces required for a true K-B system or Wolter optics. Present test systems have achieved ~ 0.1 arc seconds angular resolution at soft X-ray energies. The overriding problem is the small aperture. Nesting is not possible so we can only achieve high collecting areas using many parallel modules. Alignment of the surface elements is also very critical if a diffraction limited performance is to be realised. For further details see Cash (1996) and Gallagher (1996).

5.2.6 Crystal lenses Concentric rings of single crystals can be arranged to bring X rays or γ rays to a common focus using Bragg or Laue reflections, as

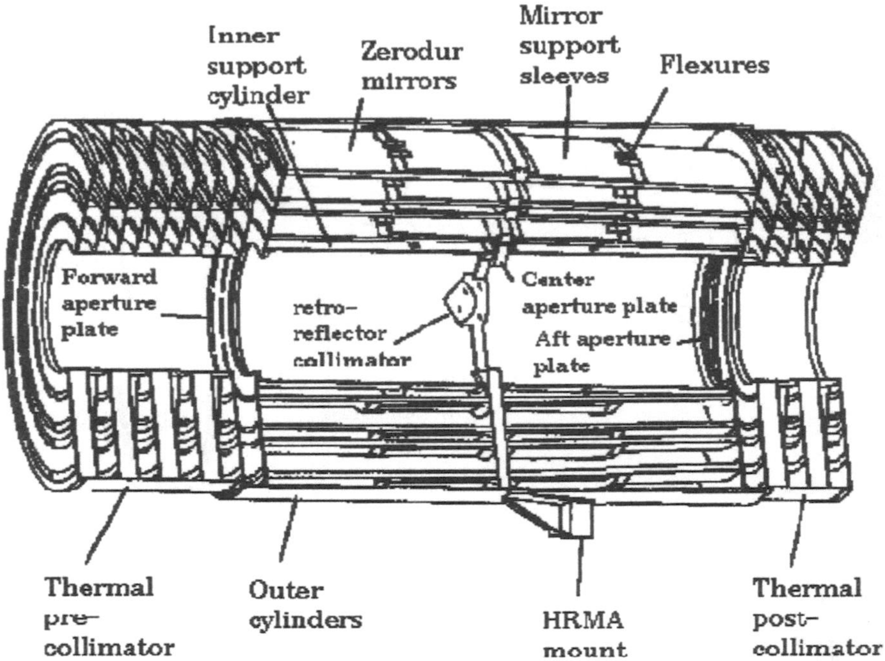

Fig. 19. The *AXAF* High Resolution Mirror Assembly.

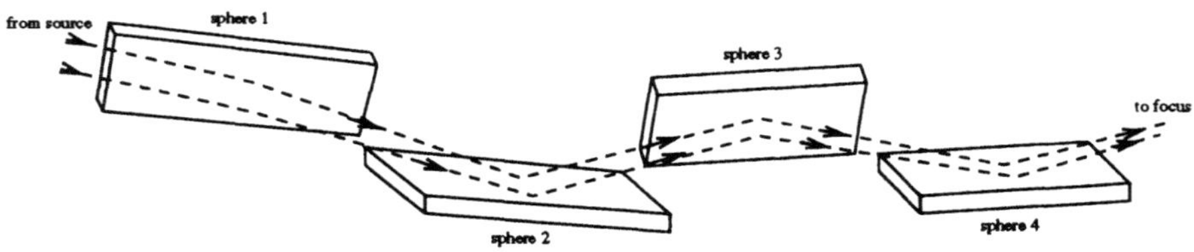

Fig. 20. A four reflection, high resolution, K-B system.

shown schematically in Figure 21. Each ring must consist of a different crystal and/or reflection plane and the ring radius adjusted to obey Bragg's law. The crystals can be used in the Bragg (surface-diffraction) or Laue (volume-diffraction) mode. The size of the focused spot, and therefore the angular resolution, is limited by the size of the individual crystals. Further details can be found in Smither et al. (1995). The problem with crystal lenses is that they operate over a very narrow energy band defined by the width of the diffraction orders and the effective area is very small. However they do operate at high energies > 50 keV which are not accessible to grazing-incidence reflection.

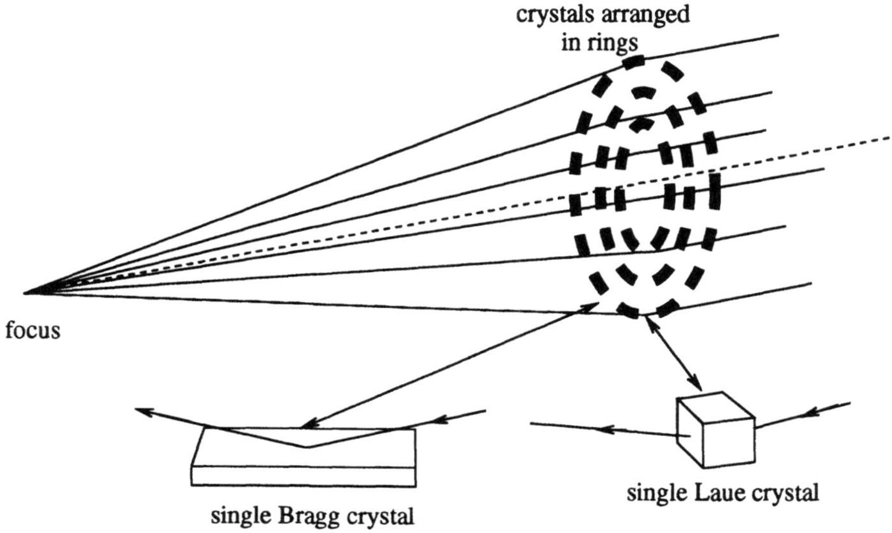

Fig. 21. Crystal lens constructed from rings of single crystals.

5.3 Mirror coatings

The conventional mirror coatings are high-Z metals like gold, nickel, iridium or platinum, since they give high X-ray reflectivities. It is possible to coat a Wolter I or K-B surface with a multi-layer to enhance the reflectivity. At energies below about 15 keV the absorption is high and so only a small number of layers are active and the resulting reflectivity is only enhanced over a small energy band and is likely to be suppressed outside that band. However at higher energies the penetration depth is larger, and graded-period multilayer coatings can be used to enhance the performance of Wolter I system over a wide high-energy band upto $\sim$ 50 keV. See for example the proposed design in Christensen et al. (1991).

5.4 *AXAF* and *XMM*

AXAF and *XMM* are the two large X-ray observatories that will be launched by the turn of the century. They are complementary, both in terms of the X-ray optics employed and in the astrophysics that they afford. The latest information on the status of these two state-of-the-art missions can be found on the WWW sites references XMM (1998) and AXAF (1998).

5.4.1 The *XMM* X-ray optics The optical design of *XMM* is shown in Figure 22. The mirrors are based on the shell replication technology described above and Table 1 summarises the main characteristics of the mirror modules.

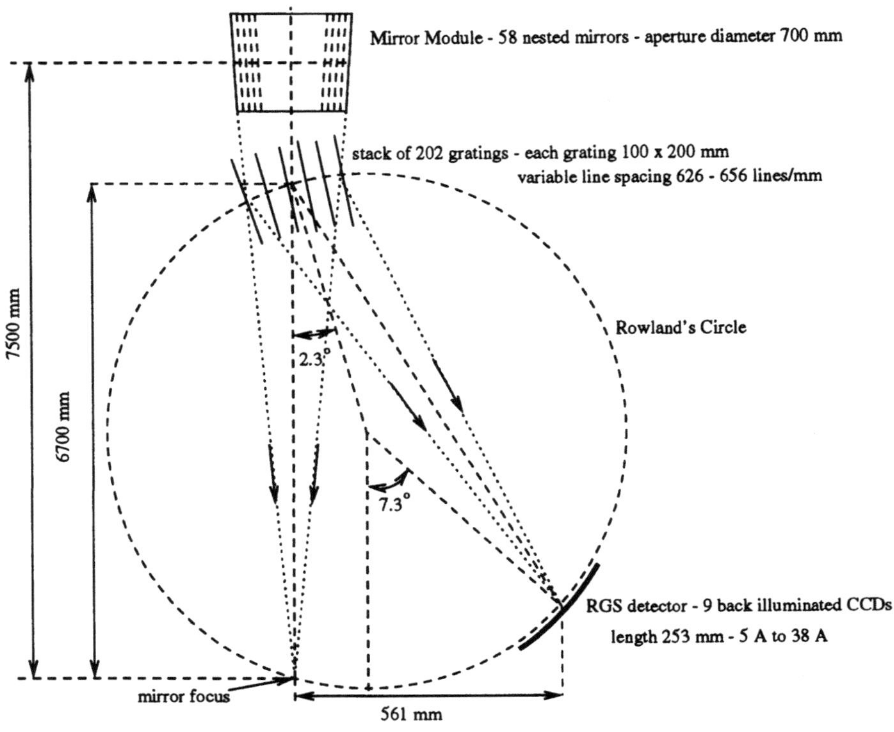

Fig. 22. Optical design of the *XMM* X-ray Telescope including the Reflection Grating Spectrometer.

Note that the ruling spacing of the gratings has to be variable because the grating plates are flat and the top of the plates are outside Rowland's circle while the bottoms are inside. Furthermore the centres of the plates are set parallel to the normal of a toroidal surface so that all the dispersed images of a point source are brought to the same focus. A reflection grating design was chosen because the mirrors have a moderate angular resolution, and high efficiency was a priority. An in-plane reflection grating also shifts all the diffracted flux including the zeroth order away from the primary focus so that flux which does not intersect the gratings is imaged undisturbed. Three mirror modules will be flown, one open and two incorporating a Reflection Grating Array (RGA). For the two incorporating an RGA, 47% of the focused X rays passes through to the primary focus to be detected by an array of MOS CCD's. In the open module all the X rays are brought to the primary focus and detected by a PN CCD detector. The field of view of both the CCD cameras is about 30 $\times$ 30 arc minutes. Further details of the detectors are not included here since they don't constitute part of the X-ray optics of the instrument.

Table 1. *XMM* mirror module characteristics.

Mirror geometry	Wolter Type I
Mirror shell material	nickel
Mirror surface coating	gold
Focal length	7500 mm
Angular resolution HEW	16 arc seconds (0.1 - 10.0 keV)
Angular resolution FWHM	8 arc seconds (0.1 - 10.0 keV)
Effective area 1.5 keV	1475 cm^2
Effective area 8.0 keV	580 cm^2
Outermost mirror diameter	700 mm
Innermost mirror diameter	300 mm
Mirror shell axial length	600 mm (300 mm for each surface)
Mirror shell thickness	0.40 mm to 0.72 mm
Minimum packing distance	1 mm
Number of mirror shells	58
Mirror module mass	425 kg

Table 2 summarises the characteristics of the *XMM* Reflection Grating Spectrometer.

Table 2. *XMM* RGS characteristics (1 module).

Grating plates	100 × 200 mm silicon carbide - replicated
Number of plates	202
Effective groove density	645.6005 lines/mm
Blaze angle/wavelength	0.698871 degrees/ 15 Å
Angle of incidence	1.576191 degrees
Fraction of X rays from mirrors intercepted	0.53
Waveband	5-35 Å (0.35-2.5 keV)
Resolving power 10 Å	287 (1st order) 520 (2nd order)
Resolving power 18 Å	477 (1st order) 817 (2nd order)
Resolving power 35 Å	799 (1st order)
Effective area 10 Å	102 cm^2 (1st order) 38 cm^2 (2nd order)
Effective area 18 Å	64 cm^2 (1st order) 10 cm^2 (2nd order)
Effective area 35 Å	16 cm^2 (1st order)

5.4.2 The *AXAF* X-ray optics The *AXAF* High Resolution Mirror Assembly (HRMA) has been constructed using monolithic shell technology to produce the highest possible angular resolution. Table 3 summarises the characteristics of the *AXAF* HRMA and Figure 19 is a schematic showing the main elements of the assembly.

Table 3. *AXAF* mirror characteristics.

Mirror geometry	Wolter Type I
Mirror shell material	zerodur glass
Mirror surface coating	iridium
Shell diameters (1,3,4,6)	1200, 960, 850, 620 mm
Shell axial length (parabola or hyperbola)	830 mm
Focal length	10069 mm
Angular resolution FWHM	0.5 arc seconds
Effective area 0.25 keV	800 cm^2
Effective area 5.0 keV	400 cm^2
Effective area 8.0 keV	100 cm^2

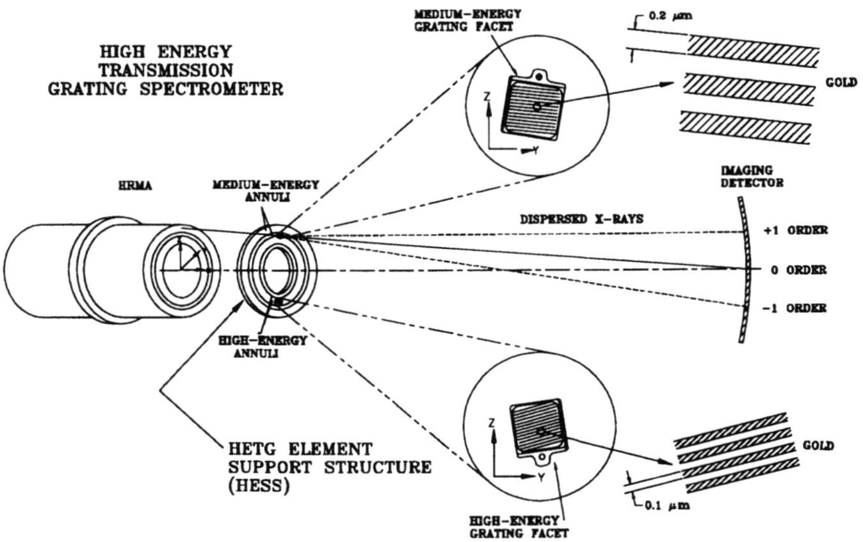

Fig. 23. *AXAF* HETG spectrometer layout.

AXAF has two transmission grating spectrometers, the HETG and the LETG, which can be deployed behind the HRMA when required. The optical layout of the HETG is shown in Figure 23. As you can see, two sets of gratings are used. The geometry is the same for the LETG and is based on the Rowland's circle geometry like the *XMM* RGS. Transmission gratings rather than reflection gratings were chosen to exploit the high angular resolution of the HRMA. The HETG uses the ACIS-C focal plane detector and the LETG uses the HRC-S focal plane detector so that, unlike the *XMM* RGS, separate spectrometer detectors are not required. Tables 4 and 5 list the characteristics of the *AXAF* spectrometers for comparison with the *XMM* RGS details given above.

Table 4. *AXAF* HETG characteristics.

Rowland diameter	8633.69 mm
HEG bar material	Gold
HEG period	2000.81 Å
HEG bar thickness	5100 Å
HEG bar width	1200 Å
HEG support	9800 Å polyimide
MEG bar material	Gold
MEG period	4001.41 Å
MEG bar thickness	3600 Å
MEG bar width	2080 Å
MEG support	5500 Å polyimide
HEG wavelength range	14-1.2 Å (0.9-10.0 keV)
MEG wavelength range	31-2.5 Å (0.4-5.0 keV)
Effective area	5 cm^2 at 0.5 keV
	180 cm^2 at 1.5 keV (first order, with ACIS-S)
	25 cm^2 at 6.5 keV (MEG+HEG, 10″ dia. source)
Resolving power HEG	1070-65 (1000 at 1 keV)
Resolving power MEG	970-80 (520 at 1 keV)

5.5 Assessing the performance of X-ray telescopes

As we have seen above two important factors that impact on the performance of X-ray telescopes are the aperture utilization and the half energy width of the point response (HEW). Figure 24 shows these factors plotted against one another for various Wolter type I modules. The *HIREX* is a proposed improved/larger version of the *XMM* technology. The current devices tend to lie on a diagonal where there is a direct trade off between the two factors. The dotted lines divide the parameter space into quadrants which contain the monolithic shell mirrors (bottom left-hand quadrant) and the replicated shell and foil mirrors (top right-hand quadrant). Improvements like *HIREX* are trying to push the technology up into the top left-hand corner. Figure 25 shows the collecting area of Wolter type I modules at 1 keV plotted as a function of HEW. The *HTXS* is a proposed improved version of the *ASCA* foil mirror optics. The horizontal lines correspond to the area required to collect 100 photons from faint sources of 10^{-14}, 10^{-15} and 10^{-16} ergs cm^{-2} s^{-1} in the energy band $0.5 - 2$ keV in 10^5 seconds. The slanting dotted lines are for a constant *collecting area/HEW*^{2} normalised to give a 5σ detection of the same source fluxes in 10^5 seconds. The vertical lines correspond to the HEW which yields a confusion limit of 1 source in 40 beams for the same fluxes using the $\log N$-$\log S$ derived from deep *ROSAT* PSPC exposures (Hasinger et al. 1993). To be well suited to a given flux limit and not photon limited or confusion limited, telescopes should be near the intersection of the horizontal and vertical lines. We must move into the top left-hand corner of the plot to study very faint sources. Figure 26 shows the source counts required in 10^5

Table 5. *AXAF* LETG characteristics.

Rowland diameter	8633.69 mm
Grating material	Gold
Grating period	9912.5 Å
Bar thickness	5000 Å
Facet frame material	stainless steel
Module material	aluminum
Fine-support structure	Period 25.4μm
	Thickness 2.5μm
	Obscuration <10%
	Material Gold
Coarse-support structure	Triangular height 2mm
	Width 68μm
	Thickness < 30μm
	Obscuration <10%
	Material Gold
Wavelength range	2.0-160 Å (0.08-6.0 keV)
Resolving power	>1000 (60-160 Å) $\approx 16\lambda$ (3-60 Å)
Effective area (with HRC-S)	>8 cm^2 0.1-2.0 keV

for a 5σ detection of faint source plotted against the equivalent source flux. The very faint, distant, sources will not yield sufficient counts to produce a high resolution energy spectrum with good statistics, because the collecting areas of the latest generation of X-ray mirrors are not large enough. Actually the true situation is worse than this, since the effective areas plotted don't include the efficiency of any gratings, crystals or detectors used to perform the spectroscopy. You can see from the listed performance of the *XMM* RGS and *AXAF* LETG and HETG that these grating spectrometers are not very efficient. Therefore, it is even more imperative that the effective areas of the primary mirrors are increased. Figure 27 shows the flux at the confusion limit of 1 source in 40 beams plotted against the flux required to get 100 photons in 10^5 seconds. It is sensible to design telescopes which lie close to the diagonal line shown. For fluxes near the detection limits most of the telescopes are marginally confusion limited. In order to perform high-resolution spectroscopy on a source > 1000 counts are required. This corresponds to shifting all the telescope points on figure 27 by 1 decade to the right. For such a flux all the modules are then photon limited.

Above we have only considered the performance of X-ray telescopes when observing a single point source or a group of distant sources in a small area of sky. For such observations the useful field of view of the mirrors is not very important. Actually the field of view for all the telescopes listed is about the same, of the order 30-60 arc minutes in diameter. This is because they all employ Wolter type I optics in the same energy band and therefore use very similar grazing angles. If we want to survey the sky or try to monitor large

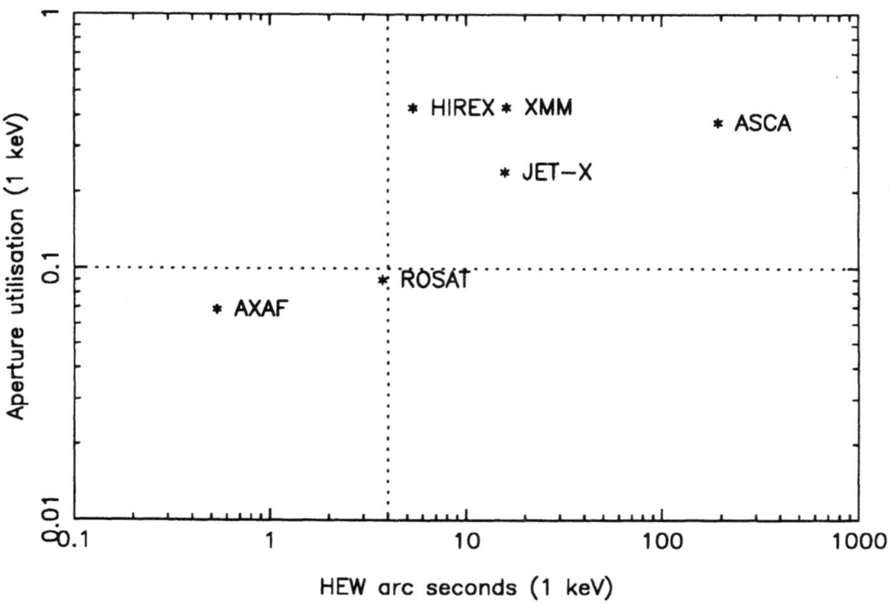

Fig. 24. Aperture utilization vs. HEW of Wolter I modules.

areas of sky then a large field of view is desirable. This can only be achieved with Wolter Type I optics using larger grazing angles and compromising the high-energy response.

5.6 Future X-ray astronomy missions

There are many new and exciting missions under development for the future. *LOBSTER* and *XEUS* are two where new developments in X-ray optics are crucial. The latest information concerning these missions can be found at the WWW sites given in the references LOBSTER (1998) and XEUS (1998).

5.6.1 The *LOBSTER* X-ray optics The K-B geometry utilized in the square pore design shown in Figure 17 is the only X-ray optical focusing system which can offer a very large field of view. D is the width of the square pores and L is the pore length. *LOBSTER* is a proposed instrument that is designed to exploit this, providing all-sky coverage at a reasonable sensitivity in a day. The wide-field characteristics of the geometry are illustrated in Figure 28. The sensitivity in the medium-energy band (0.5-3.5 keV) is expected to be 2×10^{-12} ergs cm^{-2} s^{-1} in a one-day observation with some sensitivity extending into the soft X-ray (carbon) band at around 0.25 keV. This is comparable with the sensitivity of the *ROSAT* all-sky survey which took 6 months to complete and is considerably less sensitive than the typical

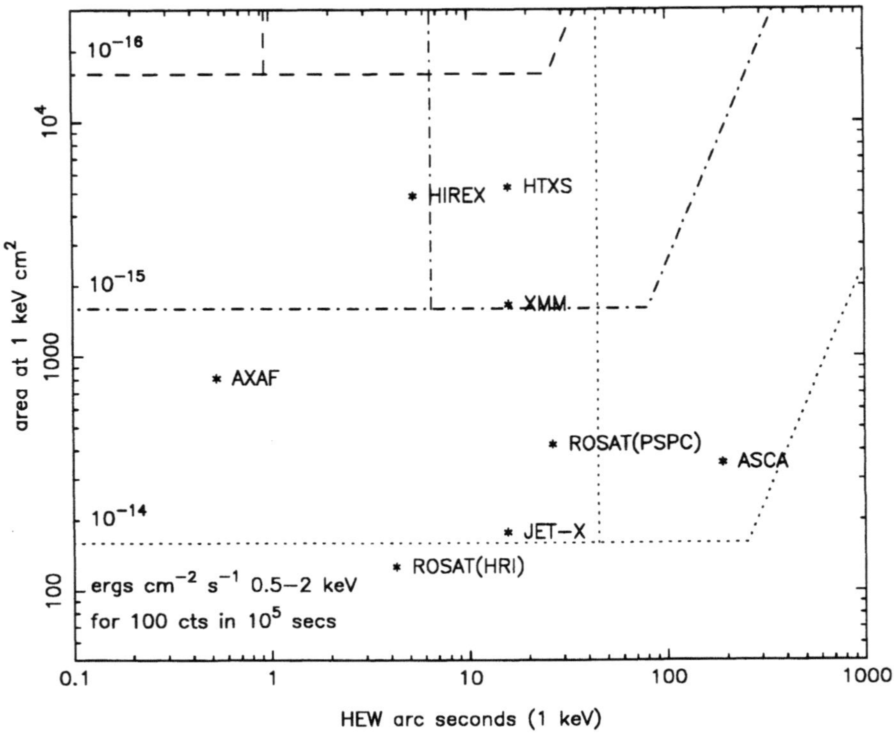

Fig. 25. Collecting area vs. HEW of Wolter I modules.

pointing sensitivities plotted in Figures 26 and 27. The angular resolution is expected to be 3-4 arc minutes limited by the quality of the microchannel plate X-ray optics, which is adequate to avoid source confusion at this sensitivity and will provide positions of ~ 1 arc minute for transient sources which are only seen in a one-day observation. Except for unavoidable dead regions near the Sun all-sky coverage will be possible in every 5 seconds so that bright but very rapidly varying objects like γ-ray bursters can be seen and positioned. Thus *LOBSTER* aims to utilize the recent developments in microchannel plate X-ray focusing optics to provide all-sky coverage similar to *ROSAT* but on a much much shorter time scale.

5.6.2 The *XEUS* X-ray optics The *XEUS* concept being studied by ESA at present aims to push X-ray astronomy into a sensitivity regime the equivalent of which has long been available at lower photon energies. The desired characteristics of the X-ray optics system are summarised as follows:

- Single focal plane
- Effective area 10 m^2 at 1 keV, 1 m^2 at 8 keV, no large dips in between

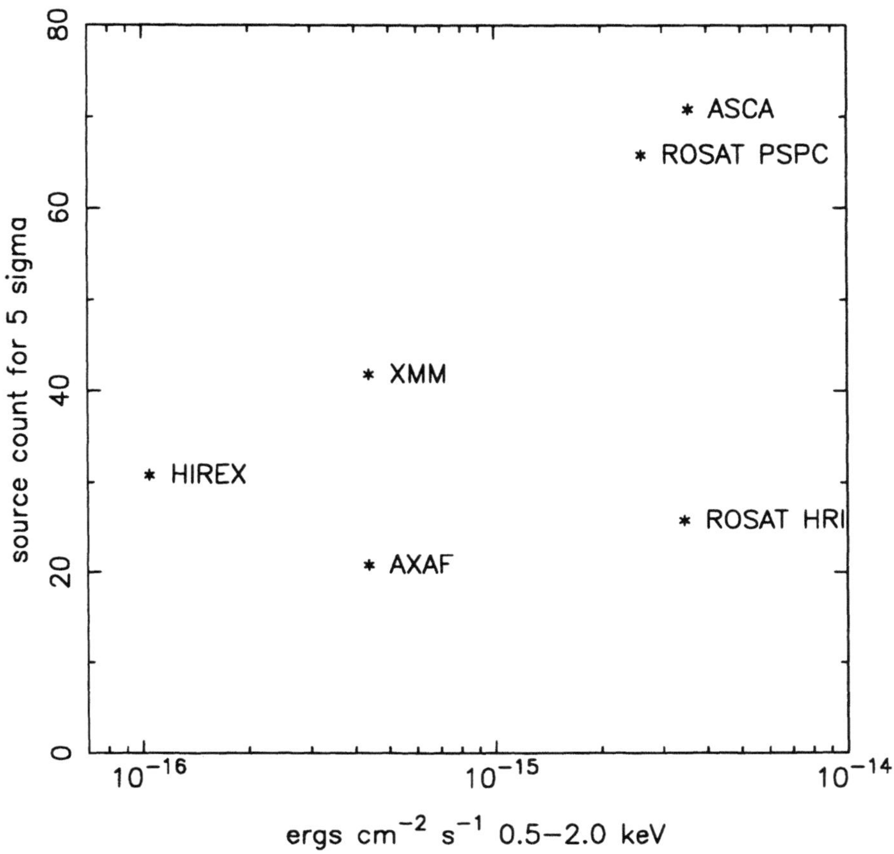

Fig. 26. Limiting source counts and sensitivities of Wolter I modules in 10^5 seconds.

– Angular resolution HEW 2″ at 1 keV, 5″ at 8 keV

The instrument is to be serviced and possibly assembled in space and is designed to optimize mirror surface utilization rather than aperture utilization. Figure 29 illustrates the basic construction of the telescope. A complete highly nested Wolter Type I system will be constructed from petal units. Each petal is about the same size as an *XMM* mirror module, length 1m, mass 500 kg. The crucial technological development required is the manufacture and integration of sectors of large radius Wolter I shells. This may be possible using the *XMM* shell replication technique but without the advantage that closed shells afford. Because of the segmentation of the aperture and the sheer size of the outer radius, the number of individual shells required is truly astronomical so an efficient and very accurate replication/integration scheme is needed.

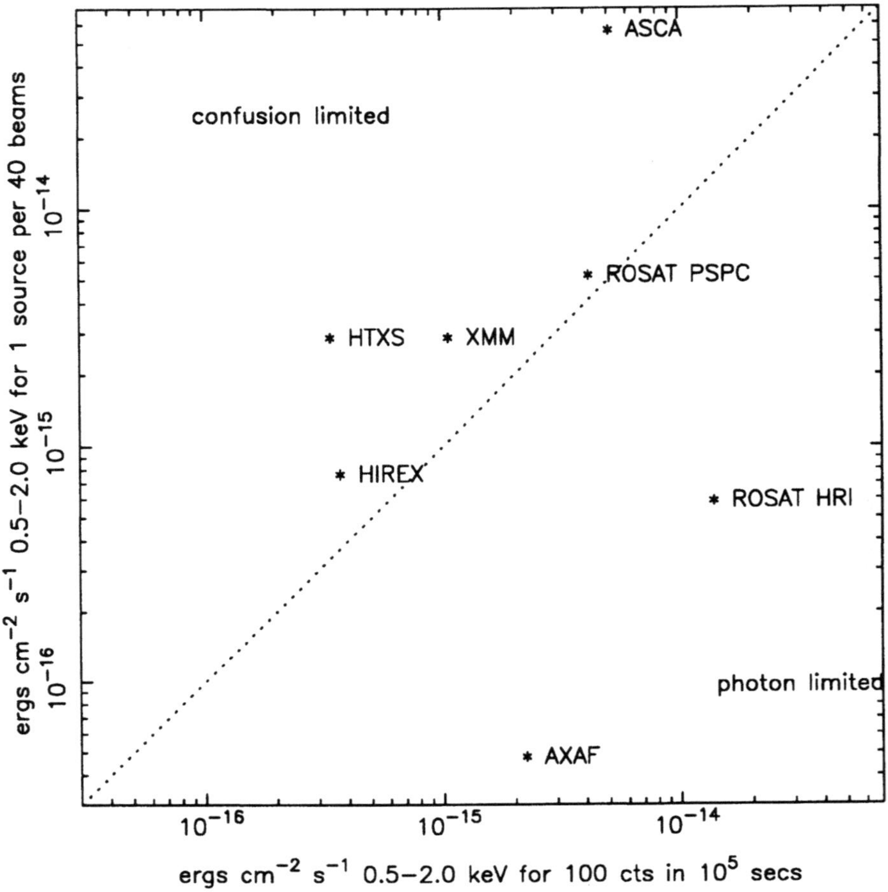

Fig. 27. The flux at the confusion level vs. the flux to gain 100 counts in 10^5 seconds of Wolter I telescope modules

The baseline optics would probably be launched as a single unit with the following characteristics:

- 255 Wolter-I surfaces
- Aperture radius range 0.4 - 2.7 m
- Effective area 10 m^2 at 1 keV, 1.0 m^2 at 8 keV
- Total mass of mirror system $\sim$ 35 tons

Subsequently further petals could be added around the perimeter of the aperture to increase the low-energy effective area:

- 365 Wolter-I surfaces
- Radius range 0.4 - 5.0 m
- Effective area 26 m^2 at 1 keV, 1.0 m^2 at 8 keV

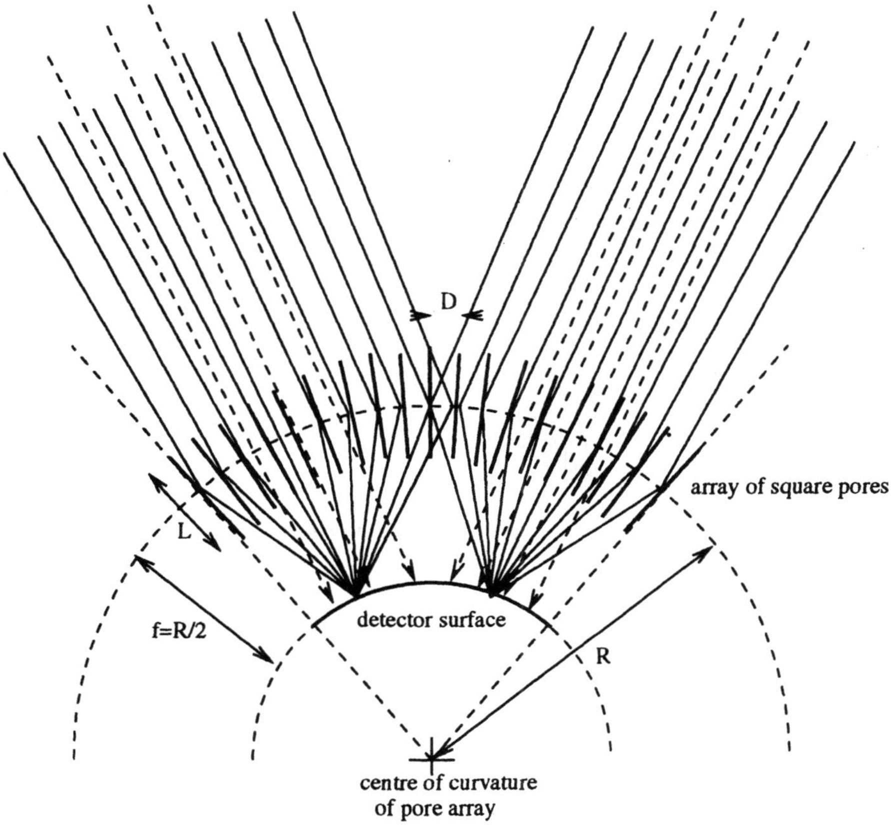

Fig. 28. The lobster-eye geometry allows a large field of view with no preferred axis.

5.6.3 High-resolution X-ray spectroscopy In the current generation of X-ray telescopes, including *XMM* and *AXAF*, high-resolution spectroscopy is provided by X-ray optics in the form of gratings. It is hoped that the new generation of X-ray detectors (bolometers etc.) will be able to provide a few eV energy resolution over the soft X-ray band with a much higher efficiency than the present optics, making the gratings redundant for spectral resolutions < 1000. However very high resolution, ~ 10000, will still only be accessible through the use of diffraction gratings and crystals. Large-area mirrors are required whatever X-ray spectroscopy is attempted, since X-ray astronomy of all but a few very bright sources is always photon starved.

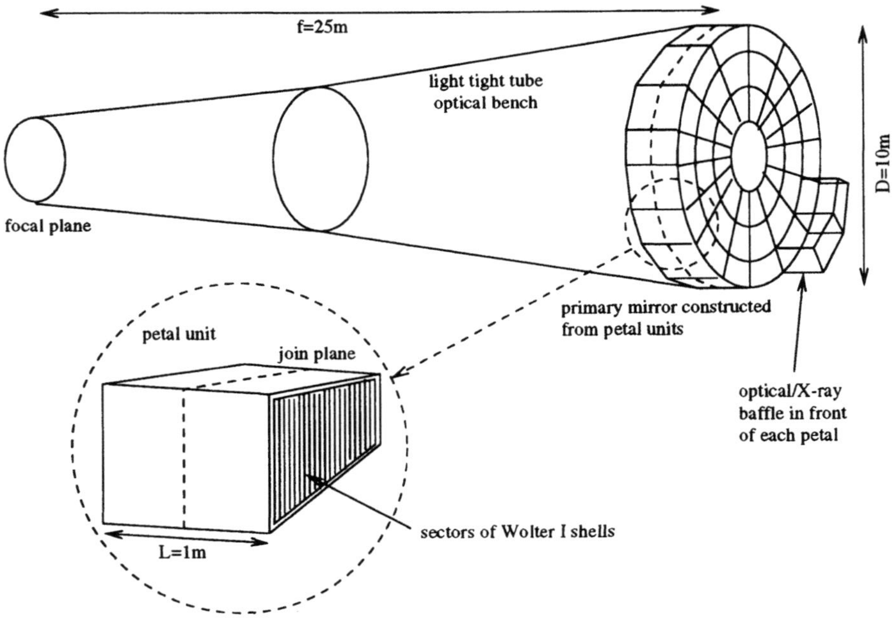

Fig. 29. *XEUS* telescope concept.

References

AXAF Science Center (1997): WWW http://asc.harvard.edu/

Azaroff, L.V. (Ed.) (1974): *X-ray Spectroscopy*, McGraw-Hill Inc.

Beckmann, P. & Spizzichino, A. (1963): *The Scattering of Electromagnetic Waves from Rough Surfaces*, Pergamon, New York

Born, M. & Wolf, E. (1970): *Principles of Optics*, Pergamon Press, Fourth Edition

Cash, W. (1996): private communication

Center for X-ray Optics (1997): WWW, http://www-cxro.lbl.gov/

Christensen, F.E. et al. (1991): SPIE vol. 1546, p. 160

Compton, A.H. & Allison, S.K. (1935): *X-rays in Theory and Experiment*, Van Nostrand, New York

Cromer, D.T. & Liberman, D. (1970a): J. Chem. Phys. 53, 1891

Cromer, D.T. & Liberman, D. (1970b): *Relativistic Calculation of Anomalous Scattering Factors for X-rays*, LASL Report LA-4403

Cromer, D.T. & Mann, J.B. (1968): Acta Cryst. A24, 321

Crook, A.W. (1948): J. Opt. Soc. Am. 38, 954

Gallagher, D., Cash, W, Jelsma, S. & Farmer, J. (1996): Proc Soc Photo-Opt. Instrum. Eng. in press

Gurman, S.J. (1995): J. Synchrotron Rad. 2, 56

Hasinger, G. et al. (1993): A&A 275, 1

Henke, B.L., Lee, P., Tanaka, T.J., Shimabukuro, R.L. & Fujikawa, B.K. (1982): *Low-energy X-ray Interaction Coefficients: Photoabsorption, Scattering, and Reflection*, Atomic Data and Nuclear Data Tables 27, 1

James, R.W. (1965): *The Optical Principles of the Diffraction of X-rays*, 5th Ed. G. Bell & Sons Ltd., London
Kirkpatrick, P. & Baez, A.V. (1948): J. Opt. Soc. Am. 38, 766
Lee, P. (1981): Opt. Comm. 37, 159
Lees, J.E., Fraser, G.W., Brunton, A.N. & Willingale, R. (1995): in *Imaging in High Energy Astronomy*, eds. L. Bassani & G. Di Cocco, Kluwer Academic Publishers, p. 305
Lobster Eye X-ray Telescope (1998): WWW http://www.star.le.ac.uk/mcp/lobster/lobster.html
Lorrain, P., Corson, D.R. & Lorrain, F. (1988): *Electromagetic Fields and Waves*, 3rd Ed., W.H. Freeman & Co., New York
Parratt, L.G. & Hempstead, C.F. (1954): Phys. Rev. 94, 1593
Rice, S.O. (1951): Commun. Pure Appl. Math. 4, 351
Serlemitsos, P.J. (1981): in *X-ray Astronomy in 1980's*, NASA TM 83848, p. 441
Smither, R.K. et al. (1995): Experimental Astronomy 6, 47
Wells, A. et al. (1988): Proceedings of IAU Colloquium 115, Cambridge University Press, 318
Wolter, H. (1952): Ann. Phys. 10, 94 and 286
XEUS Home Page (1998): WWW http://astro.estec.esa.nl/XEUS/
XMM Science Operations Centre (1998): WWW http://astro.estec.esa.nl/XMM/xmm_top.html
X-Ray WWW Server Uppsala University: WWW http://xray.uu.se/

Instrumentation for X-ray Spectroscopy

George W. Fraser

X-ray Astronomy Group, Space Research Centre, Department of Physics and Astronomy, University of Leicester, Leicester LE1 7RH, England.

Abstract. This chapter describes, with an emphasis on physical principles, satellite-borne instrumentation for cosmic X-ray spectroscopy. Both wavelength dispersive spectrometers and energy dispersive detectors are considered; the potential and problems of new forms of cryogenic detector are examined in particular detail.

1 Introduction

As we approach the launches of two powerful X-ray observatories - NASA's Advanced X-ray Astrophysics Facility (*AXAF*) (Weisskopf 1995) and ESA's X-ray Multi Mirror (*XMM*) (de Chambure et al. 1997) - the observational emphasis of cosmic X-ray astronomy is on high-throughput spectroscopy. This emphasis will be even stronger in the major successor missions to *AXAF* and *XMM*, now being planned. This chapter describes the development - past, present and anticipated - of the instrumentation which makes X-ray spectroscopy such a powerful scientific tool.

In Section 2, source plasma diagnostics are briefly reviewed in order to estimate the required spectrometer resolving power R in each of three energy bands:

$$R = \frac{\lambda}{\Delta\lambda} = \frac{E}{\Delta E} \tag{1}$$

where ΔE ($\Delta\lambda$) is the full-width-at-half-maximum (FWHM) of the instrumental response in signal space to a monoenergetic (monochromatic) stimulus. In the literature, medium-resolution spectroscopy commonly refers to R values in the range 50 - 500, while high-resolution spectroscopy signifies resolving power in excess of one thousand.

Section 3 examines a number of important secondary requirements (signal-to-noise ratio, imaging capability, count rate capacity..) in an attempt to completely specify the ideal spectrometer. In Section 4, we introduce instruments which are *wavelength dispersive* as opposed to *energy dispersive*. In the latter class, photon energies are derived from the signal pulse heights at some detector output. In the former, X-ray wavelength is derived from the photon arrival position in the spectrum dispersed by a grating or Bragg crystal. In Sections 5 and 6, we examine, from the instrumental viewpoint, the immediate past of energy-dispersive astrophysical X-ray spectroscopy (represented by gas proportional-counter technology), its present state (semiconductor

CCDs), and its challenging future (where cryogenic microcalorimeters are generally agreed to be the detectors of choice for energies below 10 keV and pixellated high-Z semiconductor arrays for higher-energy X rays). We will concentrate throughout on physical principles, rather than on the engineering detail of specific experiments. We will use X-ray wavelength λ and energy E interchangeably. The following relationship is therefore worth noting:

$$E(\mathrm{keV}) \times \lambda(\text{Å}) = 12.4 \tag{2}$$

2 Astrophysical X-ray Spectra as Measurable Objects

Here we reduce the scientific interest in astrophysical X-ray spectroscopy to its basics. More detail is to be found elsewhere in this volume. We describe the expected spectra as "measureable objects" - benchmarks against which the performance of spectroscopic instrumentation may be judged - in order to answer the simple question: how good must the resolving power be ?

2.1 The primary energy band: 0.1 -10 keV

This band is of principal interest to X-ray astronomers at the present time and is the main subject matter of this review.

The spectroscopy of line emission from an optically thin plasma [represented in the X-ray sky by stellar coronae, clusters of galaxies and the interaction of supernova remnants (SNRs) with the interstellar medium] in this energy band - the energy range accessible to the Wolter Type 1 telescopes of the *XMM* or *AXAF* class - can give us information on :

- the plasma temperature, density and ionisation state;
- elemental abundances, in a range of atomic number Z from about Nitrogen (7) to Nickel (28);
- mass motion;
- any departure from thermal equilibrium;
- the redshift of the emitting object.

A common element in astrophysical plasmas is oxygen ($Z = 8$). From observations of laboratory plasmas, the hydrogen-like (i.e., one electron) oxygen ion (denoted O VIII or O^{7+}) is known to have an emission line energy of 654 eV. Kα emission from neutral oxygen is observed at 525 eV. The spectrum of the helium-like (i.e. two electron) ion (O VII or O^{6+}) has three line components :

1. the resonance line (denoted r or w) at 574 eV;
2. the intercombination line (i or y) at 569 eV;
3. the forbidden line (f or z) at 561 eV.

The intensity ratio f/i turns out to be an important plasma density diagnostic, for electron densities $\sim 10^{10}$ -10^{13} cm^{-3} (see the Chapters by Liedahl and Kaastra). Thus, at a first level of complexity, in order to distinguish H-like and He-like emission from oxygen (and hence gain information on temperature) requires an instrumental "resolution" of better than 654 – 574 = 80 eV. To perform a density investigation, however, requires a resolution of better than 569 – 561 = 8 eV, one order of magnitude better.

Such considerations have been addressed more generally by Holt (1990, 1994). Holt's "Rules of Thumb" give the energy separations (in eV) between pairs of emission lines in an ionised element of atomic number Z, as follows:

- Lyman α analogue line and Resonance line $10Z$
- Resonance line and Intercombination line $0.32\,Z^{4/3} \sim Z$
- Resonance line and Forbidden line $0.77\,Z^{4/3} \sim 2Z$
- Resonance line and Neutral $2.3\,Z^{3/2} \sim 10Z$

The reason for concentrating on the example of oxygen is now obvious; since all the relevant separations scale as some (near unity) power of Z, it is the observation of the low-Z elements which would appear to pose the most severe constraints on energy resolution ΔE - the important figure of merit for an energy-dispersive spectrometer (Section 5).

We might then conclude from these rules-of-thumb that the energy resolution required of the ideal spectroscopic detector in 0.3 - 12 keV band is $\Delta E \sim 10$ eV FWHM. This was indeed Holt's conclusion (perhaps unsurprisingly, since the detector technology he was then championing was capable of just about that level of resolution). Such a conclusion, however, would leave still further spectral features - known to be important from solar X-ray studies - unexplored. To resolve dielectronic satellite lines (see the Chapter by R. Mewe) or mass motion requires resolution of a few eV; thermal Doppler broadening and natural line widths require better than 1 eV. The conclusion of the late 1990's, currently driving the designs of future missions such as the NASA *Constellation-X*, formerly known as the High Throughput X-ray Spectrometer (*HTXS*) mission (White et al. 1997), and the European X-ray Early Universe Spectrometer (*XEUS*) (Palumbo et al. 1997), is that a resolution $\Delta E \sim 2$ eV FWHM is really what is required. Figs. 1 and 2 illustrate this conclusion: the former figure shows a solar-flare X-ray spectrum in the vicinity of the resonance line of He-like iron at 1.85 Å and its long-wavelength satellite lines, blurred, from top to bottom, by increasing instrumental resolutions. At 10 eV, the individual satellite lines are hopelessly blended. The latter figure contrasts the same features in the simulated spectrum of the star AR Lac observed for 80,000 seconds with spectrometers having resolutions of 2 eV (the planned microcalorimeter for *HTXS*) and 10 eV [the cryogenic X-ray Spectrometer (XRS) on the Japanese *Astro-E* satellite, due for launch in 2000 (see Section 7)]. The resolving power ultimately required for "definitive" iron line spectroscopy is therefore ~6700/2 or 3350.

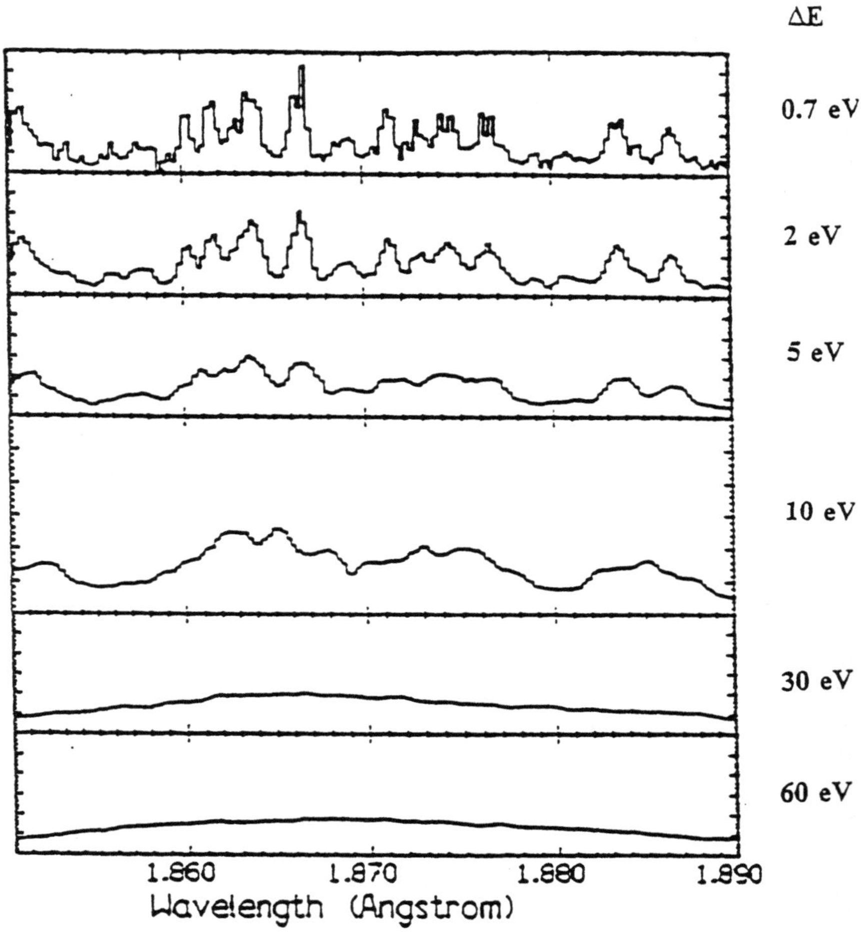

Fig. 1. Solar-flare iron line spectrum blurred by increasing instrumental energy resolutions ΔE. The top panel (0.7 eV resolution) is the original observation by the Solar Maximum Mission (*SMM*) Bragg crystal spectrometer (Schmitt 1990).

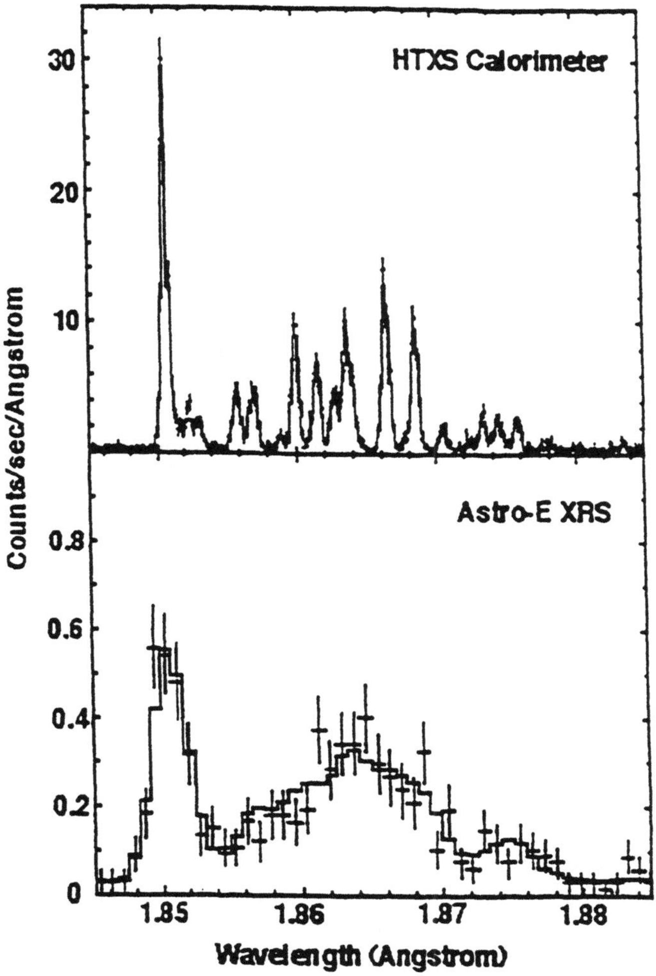

Fig. 2. Simulated He-like iron spectra from the RS CVn variable AR Lac (derived from a two-temperature fit to the spectrum observed with the *ASCA* SIS instrument) seen by spectrometers with 2 eV (top) and 10 eV (bottom) energy resolution (White et al. 1997).

2.2 The EUV band

Spectroscopy in the extreme ultraviolet (EUV) band, usually defined in terms of wavelength as the band 100 - 912 Å (13 - 124 eV), by contrast with the primary band of the previous section, concentrates on (Linsky & Luttermoser 1990):

- the emission line series of neutral (He I) and singly-ionised (He II) helium;
- lines from "metals" such as Fe (ionisation states II - XXIV), O (II), Si (VII - VIII), Ne (VIII);

- absorption lines due to the interstellar medium (ISM) in the line-of- sight to the source.

That self-same ISM absorption largely limits EUV spectroscopy to nearby (~100 pc) galactic objects such as white-dwarf stars. What resolving power is required in the EUV band? Existing spectrometers such as that of the Extreme Ultraviolet Explorer (*EUVE*, Bowyer & Malina, 1990) have wavelength-dependent R values of a few hundred.

One current problem in the study of the hottest Type DA white dwarfs is to confirm the presence of helium in the stellar atmosphere. Modelling of coarse *EUVE* spectra suggests the helium is there, but direct spectroscopic detection is lacking. To resolve this question, the "object" to be measured is therefore the He II Lyman series (304, 256,2 43...228Å). The problem is that this spectral range is confused by the lines of other species (e.g. Fe, Ni). To categorically identify the He series requires an EUV spectrometer with $R = 7000$ in the 225 - 245Å wavelength band, a resolving power roughly twice that identified above in the "primary" X-ray band. That spectrometer is J-PEX (Joint Astrophysical Plasmadynamic Experiment), a Naval Research Laboratory (NRL) /Mullard Space Science Laboratory (MSSL) /University of Leicester sounding rocket experiment scheduled to observe the DA white dwarf G191 - B2B in January 1999. The novel grating technology for J-PEX is described in Section 4.4.

2.3 The hard X-ray band: 10-100 keV

Spectroscopy in the third and last of our X-ray bands is principally concerned, unlike the first two, with the measurement of continuum (rather than line) spectra and the testing of emission models by the observation of "spectral breaks" - changes in spectral slope, especially in the X-ray spectra of Active Galactic Nuclei (AGN). Line features arise in this band not from atomic transitions but from nuclear emission, where lines are expected in young supernova remnants, e.g., ^{44}Ti lines at 68 and 78 keV (Woosley 1987), and from cyclotron emission (lines ~30 keV in neutron stars). Probably an energy resolution $\Delta E \sim 200$ - 500 eV ($R \sim 100$ at 50 keV) is adequate for most spectroscopy in the high-energy band. The key instrument characteristic in this band, where photons are scarce, is effective area (see next section).

Now, having established the resolution required for spectrometers in the various energy bands, we investigate the other essential instrumental properties, without which that resolution is worthless.

3 The Ideal Spectrometer

Suppose that we build a generalised X-ray spectrometer with:
- energy resolution ΔE (keV)
- energy band pass $\delta E = E_{\rm max} - E_{\rm min}$ (keV)
- quantum efficiency Q counts/photon
- effective area for source photons A (cm^2)

and use it, to observe a cosmic source flux F (photons/cm^2 s keV) for an observing time t. The number of detected photons is then:

$$N = F \times A \times Q \times \delta E \times t \tag{3}$$

and the number of counts per energy bin is :

$$n = \frac{N \times \Delta E}{\delta E} = F \times A \times Q \times \Delta E \times t$$

The precision of a narrow-line measurement (the signal-to-noise ratio) is then determined by the uncertainty in n:

$$\frac{\Delta n}{n} = n^{-1/2} = (F \times A \times Q \times \Delta E \times t)^{-1/2} \tag{4}$$

So, for a given precision on a given source in a fixed observing time, the product A times Q times ΔE must be maintained. In other words, what might appear to be the overwhelmingly important development goal of improving energy resolution must be matched in practice by improvements of equal magnitude in effective area or throughput (the product of A and Q).

Thus, better energy resolution requires bigger X-ray mirrors (see the Chapter by R. Willingale). For example, in the NASA programme we now await the launch of *AXAF* early in 1999. The *AXAF* CCD focal plane spectrometer (ACIS) has ~100 eV FWHM resolution and collects photons from the *AXAF* high-resolution mirror assembly (HRMA), with a collecting area of ~1000 cm^2. The successor mission to *AXAF*, *Constellation-X*, has an intended energy resolution of 2 eV FWHM (to be provided by a microcalorimeter array) coupled to 15,000 cm^2 effective area, synthesised by six identical mirror systems on separate, co-pointed satellite buses. Within the ESA programme, the comparable missions are *XMM* (100 eV resolution, 4000 cm^2) and *XEUS* (~2 eV resolution, up to 23,000 cm^2 in its fully-deployed configuration).

This simple analysis yields a second important point. We see from Eq. (3) that if the throughput AQ is indeed increased in proportion to the decrease in ΔE, the detected count rate N/t increases pro rata. Thus, to obtain spectra with higher and higher energy resolution requires detectors of ever increasing count rate capacity.

It is obvious from the definition of wavelength dispersive spectroscopy given in Section 1 that the X-ray detector in such a spectrometer must be able to locate X-ray arrival positions and must therefore have non-zero extent. Less obviously, the same imaging properties are also highly desirable for an energy dispersive detector in the focal plane of an X-ray telescope. Why? There are at least five reasons, the last of which, one might argue, is the study of extended objects (most X-ray sources, with the exceptions of SNRs and clusters of galaxies, are pointlike) why imaging is important. The other reasons are :

(a) to accommodate the point spread function (PSF) of the X-ray telescope;
(b) to accommodate the pointing accuracy of the satellite;
(c) to assist in background rejection;
(d) to alleviate detector aging by allowing "dithering" of the focal image in the detector plane, thus speading accumulated signal over more than one location.

Fig. 3 shows the PSF (encircled energy versus radius) of the *AXAF* HRMA (focal length $f = 10\,\mathrm{m}$, angular resolution, quantified by the half-power diameter, $\Delta\theta \sim 1$ arcsecond) measured with the High Resolution Camera (HRC), a large-area microchannel plate (MCP) camera. In order to avoid gain degradation proportional to accumulated output charge, the HRC has been moved relative to the focused spot in a serpentine path. Incomplete removal of this "serpentine dither" in software has resulted in the higher of the two point spread function curves. In any case, the minimum extent required of a detector for a telescope like *AXAF* is $L \sim 5f\Delta\theta$, or about 250 microns. For longer focal lengths and/or poorer telescope resolution L increases accordingly. This minimum size requirement is a very great concern in the development of modern cryogenic detectors (see Section 6.2, and Twerenbold 1987).

To summarise: the ideal X-ray spectrometer must have, in addition to the resolving power specified in Section 2:

- high quantum detection efficiency Q;
- imaging capability over a minimum field determined by the telescope optics;
- a high count rate capacity,

together with the generic properties of any good instrument: low internal background, linear energy response, stable response on mission time scales (~5-10 years), radiation (proton) tolerance, low cost, low power, small volume, no need for mechanisms and realisable cooling requirements. This last requirement is also vitally important in the development of cryogenic detectors.

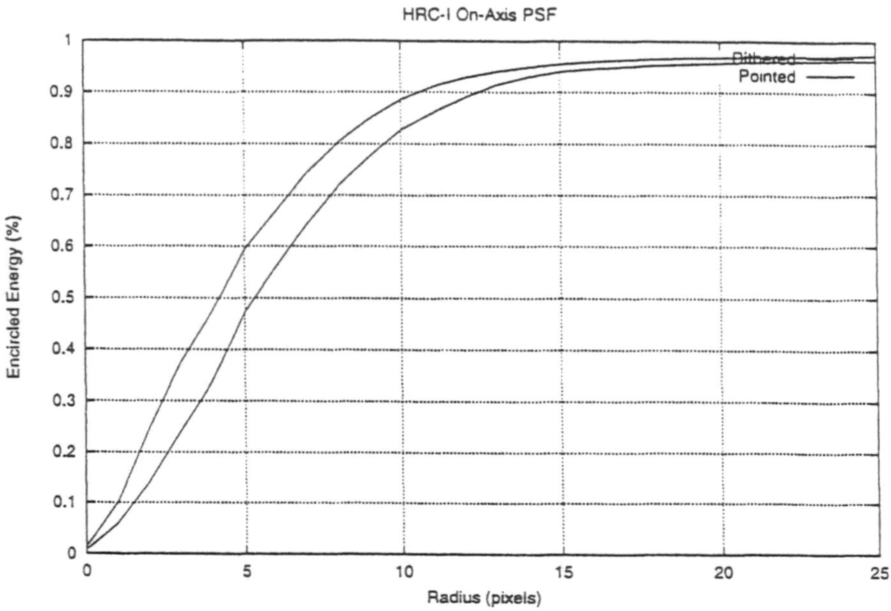

Fig. 3. Encircled energy function for the *AXAF* HRMA/HRC-I (Murray et al. 1997). Pixel size - 5 microns. The diameter enclosing half the power is 1 arcsecond or 49 microns.

4 Wavelength Dispersive Spectrometers

4.1 Operating principles

There are three categories of wavelength dispersive optical element employed in X-ray astronomy :

- transmission gratings (as used on the *Einstein*, *EXOSAT* and *AXAF* missions);
- reflection gratings (*XMM*, *Constellation-X*);
- Bragg crystal spectrometers (*Einstein*, *Spectrum X-Gamma*).

All of these convert wavelength into dispersion angle and, hence, into focal plane position in an imaging X-ray detector. The physical basis of a grating spectrometer is the familiar grating equation:

$$m \times \lambda = p \sin\theta \tag{5}$$

where m is the order of diffraction (0, ±1, ±2), λ is the X-ray wavelength, p is the grating period and θ is the dispersion angle. Eq. (5) assumes normal incidence of the input beam.

Bragg spectrometers depend on the equally familiar Bragg's law :

$$n \times \lambda = 2d \times \sin\theta \tag{6}$$

where n is the order of reflection (1,2..), λ is the X-ray wavelength, d is the crystal lattice spacing and θ is the angle of reflection. The use of Bragg crystals as polarimeters (for $\theta = 45°$, only s-polarised X-rays are reflected) is beyond the scope of this paper.

All these spectrometers are "constant $\Delta\lambda$" devices, a concept most easily illustrated with reference to the transmission grating geometry (Brinkman et al. 1997). If X rays of wavelength λ are incident normally to such a grating, placed immediately behind an X-ray telescope of focal length f and are deviated on exit by an angle θ, the arrival position X relative to the optical axis in the focal plane is :

$$X = f\tan\theta = f\ sin\theta$$

for small values of θ (i.e., for a long focal length). Thus, from the grating equation (5):

$$X = \frac{fm\lambda}{p}$$

Then, assuming that telescope resolution $\Delta\theta$ dominates over detector resolution, grating aberrations and fabrication errors :

$$\Delta X = f\Delta\theta$$

and:

$$\frac{\Delta X}{X} = \frac{\Delta\lambda}{\lambda} = \frac{1}{R}\frac{f\Delta\theta}{fm\lambda/p} = \frac{p\Delta\theta}{m\lambda}$$

implying:

$$\Delta\lambda = \frac{p\Delta\theta}{m} = constant \tag{7}$$

Resolving power improves for longer wavelengths (and for higher orders) for a given grating period. The grating efficiency, however, is highest in first (or minus first) order and operation in these modes is usually preferred. A corollary of this analysis is that gratings are useful spectrometers essentially only for point sources.

4.2 Transmission grating spectrometers : examples from *AXAF*

The value of the constant of Eq. (7) for the *AXAF* Low Energy Transmission Grating Spectrometer (LETGS) is ~ 0.03 Å in first order ($p = 0.992\mu$m, $\Delta\theta = 1$ arcsecond); for its High Energy sister spectrometer (HETGS), the constant is 0.02 Å.

Fig. 4 shows the *AXAF* HETGS (Markert et al. 1994) and its location in the converging X-ray beam from the HRMA. Note the deployment mechanism required to insert and remove the grating ring (holding the 336 individual grating elements, split between 2500 lines/mm and 5000 lines/mm) from its observing position.The resolving power is $R = 800$ for O VII X-rays. The grating efficiency of the "phased" HETGS gratings, in which the bar height is optimised to suppress the zeroth order and maximise higher-order transmission, rises to ~40% at about 5 keV, compared to the energy-independent 20% if the grating bars were infinitely deep. Below 1 keV, the HETGS efficiency falls because of the opacity of the polyimide support film. To reach energies down to 0.1 keV, the *AXAF* LETGS grating elements are free-standing (Predehl et al. 1997), albeit with larger period than the HETGS.

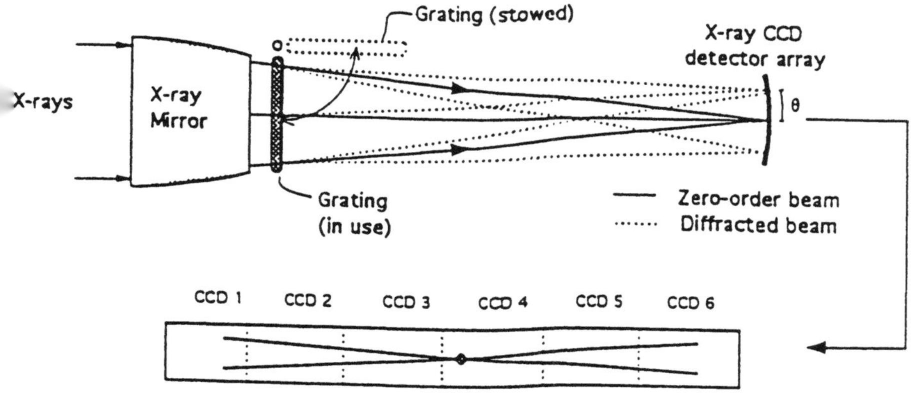

Fig. 4. *AXAF* HETGS (Markert et al. 1994).

4.3 Reflection gratings

Fig. 5 shows the possible illumination geometries for a reflection grating, as used on *XMM* and proposed for *Constellation-X*. In the classical or in-plane geometry, incident and reflected rays are in the plane orthogonal to the grating rulings and the appropriate form of Eq. (5) is :

$$m \times \lambda = p\,(\cos\beta - \cos\alpha) \tag{8}$$

The grazing-incidence geometry reduces the effective grating spacing to $\sim p \sin\alpha$, so that [see Eq. (7)] a given value of resolving power in a given order can be attained with rather lower quality mirrors - the ~10 arcseconds of *XMM* rather than the 1 arcsecond of *AXAF*. Conversely, if the reflection geometry is used with arcsecond quality optics, the resolving power will be higher than for a transmission grating of the same period. Reflection gratings, furthermore, offer added flexibility of design in that the blaze angle (δ) may be chosen to select the wavelength of maximum efficiency λ_B where :

$$\lambda_B = 2p \sin\gamma \sin\delta \tag{9}$$

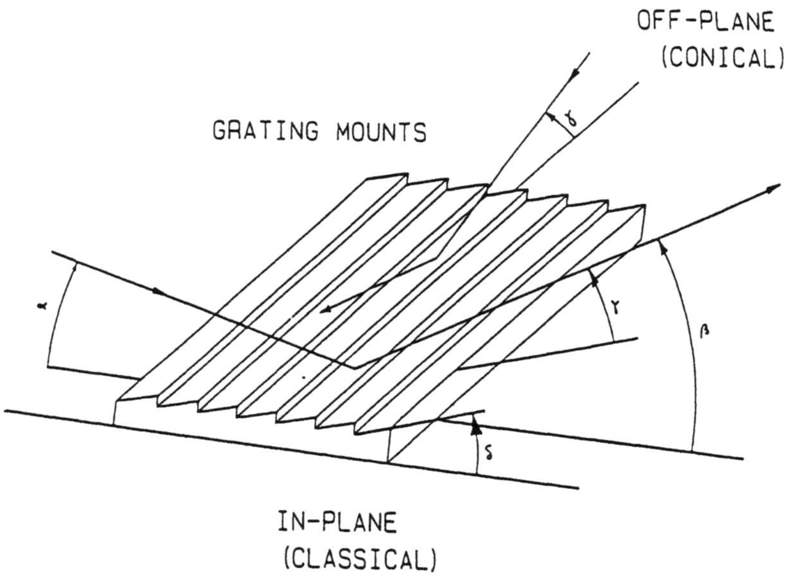

Fig. 5. Reflection grating geometry (Briel et al. 1987).

The *XMM* Reflection Grating Spectrometer (RGS) (Rasmussen et al. 1998), shown schematically in Fig. 6, consists of 200 plane gratings, each 10 cm × 20 cm, with pitch $p = 1.5\mu$m, ruled on a 1 mm thick SiC substrate. The reflection efficiency is ~20% and $\Delta\lambda$ equal to 0.06 Å in first order, corresponding to a resolving power $R = 400$ at 0.5 keV. The overall RGS efficiency is reduced by the non-unity open area fraction of the massive instrument resulting from the non-zero substrate thickness presented to the telescope.

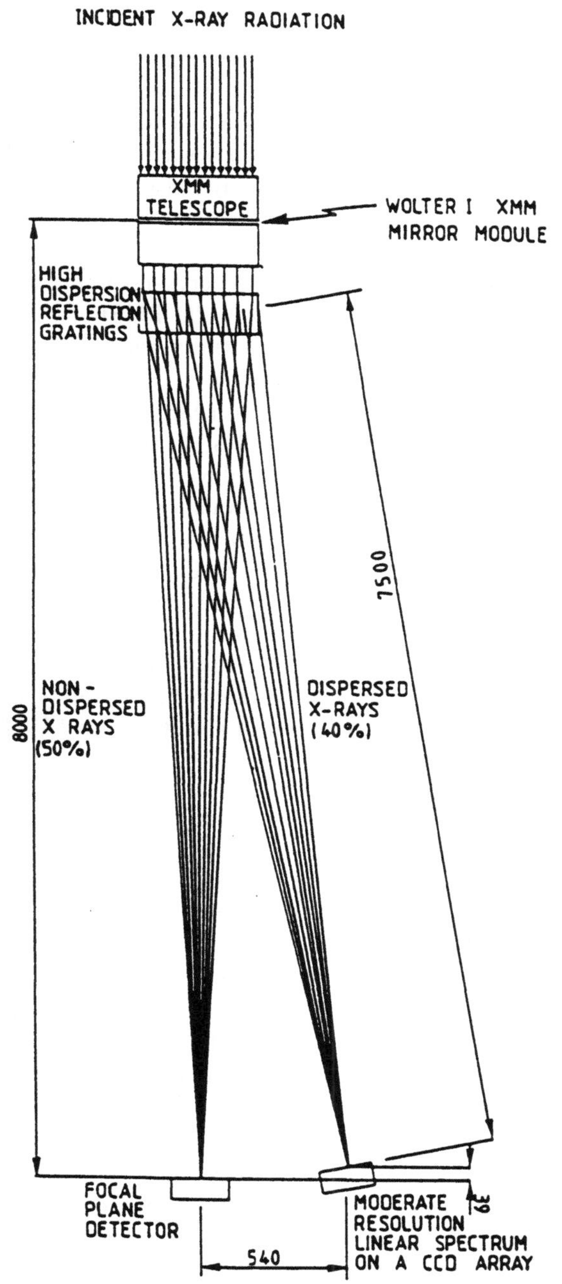

Fig. 6. Schematic view of the *XMM* RGS (Briel et al. 1987).

4.4 Disadvantages of gratings: novel developments

Grating spectrometers are excellent instruments for medium-resolution spectroscopy (resolving powers of a few hundred) but provide inadequate resolving power shortward of $\sim$ 6 Å (at energies above about 2 keV). Where mechanisms are required for deployment there is always the possibility of mechanical failure (the *EXOSAT* TGS stuck midway between its stowed and deployed positions). Eq. (5) show that each position in the detector focal plane will contain a superpositon of wavelengths in different orders; order confusion must then be removed by the use of an energy-resolving imaging detector (a CCD array for the *AXAF* HETGS and *XMM* RGS) or otherwise. Finally, all gratings have difficulty in providing spatially-resolved spectroscopy and are of limited efficiency (Q ~20-30%), especially above about 2 keV.

Gratings, by virtue of Eq. (7), remain largely unchallenged as the spectrometers of choice in the EUV band of Section 2.2. One novel development in this band is the production of multilayer-coated ion etched gratings on normal-incidence spherical mirrors (Kowalski et al. 1996) for the J-PEX experiment described in Section 2.2. The use of the normal-incidence geometry provides access to higher effective areas. The J-PEX mirrors, with a focal length of 2.2 m and a geometric area of 512 cm^2 have an effective area of 9 cm^2 at 235 Å, a factor of ten higher than the conventional gratings on *EUVE*. The grating period is ~0.4 μm, which will yield the required resolving power of Section 2.2, provided the imaging microchannel plate detector recording the spectrum has a spatial resolution of about 20 μm FWHM.

4.5 Bragg crystal spectrometers

Bragg spectrometers may be used in conjunction with a grazing-incidence telescope in one of two modes - following the optic, as in the *Einstein* Focal Plane Crystal Spectrometer (FPCS, Giacconi et al. 1979), or preceding it, as in the Objective Crystal Spectrometer (OXS) instrument for *Spectrum X-Gamma* (Abdali etal. 1997). These geometries are illustrated in Fig. 7.

The *Einstein* FPCS consisted of curved crystals of pentaerythritol (PET), ammonium dihydrogen phosphate (ADP), thallium acid phthalate (TAP), rubidium acid phthalate (RAP), lead stearate and lead laurate on an interchange mechanism. The multiplicity of crystals is required, since a given material 2-D spacing at a fixed angle of incidence essentially selects (see Eq. 6) a single wavelength in the source spectrum; to cover any sort of bandpass requires many materials and a scanning mechanism to vary θ. A typical resolving power for the *Einstein* FPCS was $R \sim 330$ at 1 keV and a typical effective area ($A \times Q$) only ~0.5 cm^2 (Giacconi et al. 1979). Apart from telescope blur and detector resolution, the intrinsic Darwin width and mosaicity of the crystals themselves contribute to the resolving power. Note that the position sensitive proportional counter readout for the FPCS had to move to

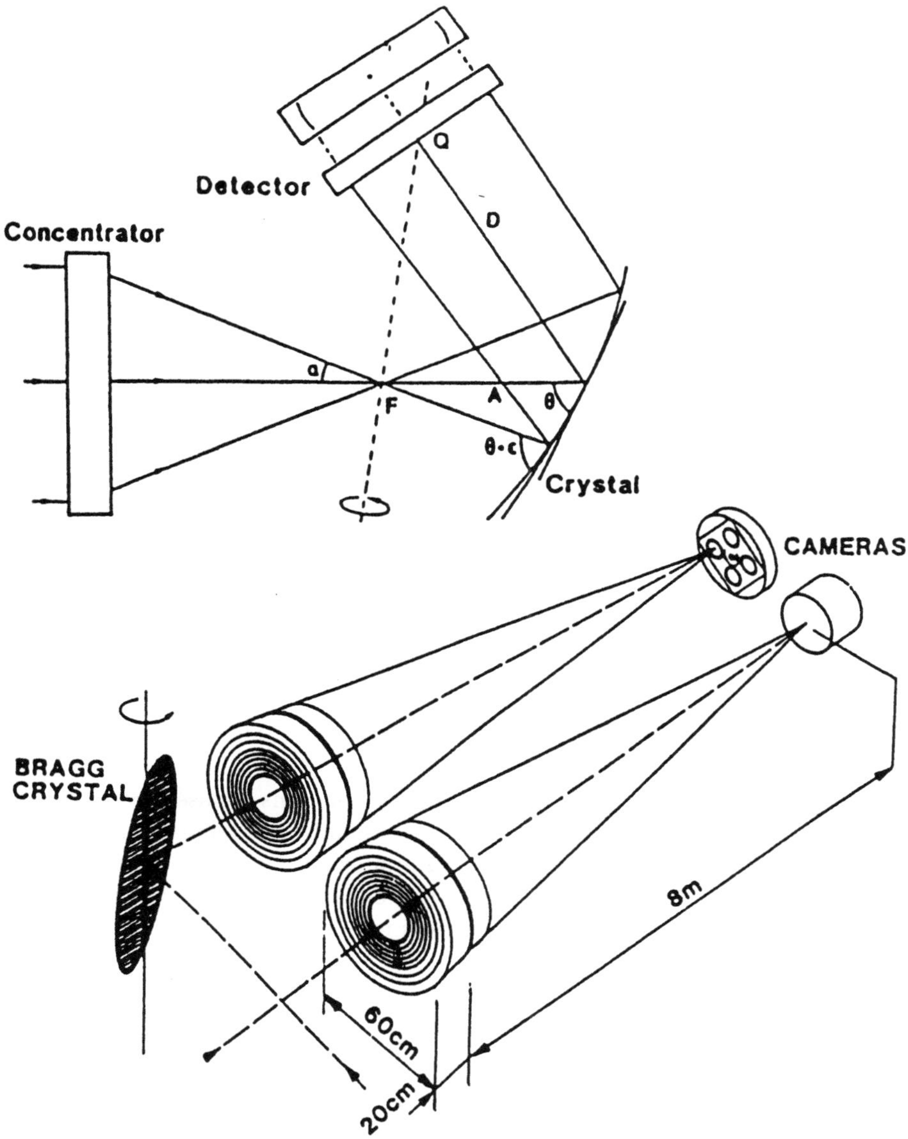

Fig. 7. Bragg crystal spectrometer geometries. Top - crystal following optic. Bottom - crystal preceding optic.

a different position for each crystal, doubling the mechanical complexity of the instrument.

The Objective Crystal Spectrometer (OXS) for *Spectrum X-Gamma* embodies a completely different approach. OXS consists of a large (610 mm × 680 mm), flat crystal panel bearing on one side 212 LiF(220) crystals and on the other 136 RAP(001) and 38 Si(111) elements, the last crystal type overcoated with a Co/C multilayer to produce "bandpass doubling". Bragg reflection from the underlying silicon is at a wavelength of 5.1 Å, tuned to the emission of He- like sulphur, while reflection from the multilayer at 63 Å selects Mg X and Fe XVI line emission. Despite the large size of the OXS crystal panels, the final effective areas are still small; Abdali et al. (1997) quote values for $QA\delta E$ in the range 0.017-0.58 cm^2 keV.

The disadvantages of Bragg crystal spectrometers are very simply summarised. Despite resolving powers of up to 2700 (TAP in the Oxygen band in a design proposed for *XMM*) they require complex mechanisms for scanning and crystal interchange and a multiplicity of crystal types for complete wavelength coverage. They provide spatially-resolved spectroscopy only with difficulty and - the fundamental problem - have an instantaneous bandpass typically only a few eV wide, leading inevitably to low throughput.

The emphasis of modern spectrometer development is now on efficient, energy dispersive systems, in particular cryogenic detectors, which we begin to consider in the next section.

5 Energy Dispersive Spectroscopy: Basic Principles

In this class of spectrometer, we absorb single photons, focused by a large X-ray telescope, in a detector mass, create a population of "signal carriers" and then "count" the signal carriers to obtain information on the photon arrival coordinates, arrival time, linear polarisation and, of principal importance here, on the energy E. The signal carriers may be :

- electron-ion pairs in a gas;
- electron-hole pairs in a semiconductor;
- quasi-particles in a superconductor;
- scintillation photons in a gas, liquid or solid;
- phonons in a calorimeter.

Fig. 8 shows the "family tree" of photon counting X-ray detectors, with branches according to detection medium - solid, liquid or gas. Detectors based on the external photoeffect exhibit very limited X-ray energy resolution, Geiger counters and ionisation chambers none at all. All the others can form the basis of an energy dispersive spectrometer to be placed at the focus of an X-ray telescope. The detailed physics of photon counting X-ray detectors is described by Fraser (1989).

With almost-perfect generality (excepting the calorimeter), we can work out the energy resolution in terms of two parameters - W and F - in a "one-step approximation" to the actual processes of energy partition which follow X-ray absorption in real materials. Let the mean energy required to create a signal carrier in the detector mass be W (assumed to be a material constant). Then the number of signal carriers, our measure of energy, is :

$$N = E/W \tag{10}$$

The variance on N, σ^2, determines the uncertainty in the energy. We write :

$$\sigma^2 = FN$$

where F is the Fano Factor ($0 < F < 1$), i.e., the factor by which the observed output pulse height distribution (PHD) is narrower than that expected from random Poissonian statistics.

In a Gaussian approximation to the PHD, the FWHM intrinsic energy resolution is then :

$$\Delta E = (8 \ln 2)^{1/2} W \left[\frac{F.E}{W}\right]^{1/2} = 2.36 W \left[\frac{F.E}{W}\right]^{1/2} \tag{11}$$

Thus, for most energy dispersive spectrometers, the resolving power, R, increases as the square root of the energy. For calorimeters, the scaling law turns out to be linear with energy (i.e., calorimeters are "constant ΔE" spectrometers - see next section). For a wavelength dispersive spectrometer, conversely, we have already seen that resolving power increases linearly with wavelength λ. It follows that there must always be a crossover energy where a given energy dispersive system has the same resolving power as a given wavelength dispersive system. For the proposed *Constellation-X* mission (White et al. 1997) with a model payload consisting of a $\Delta E = 2$ eV microcalorimeter and a $\Delta\lambda = 0.05$ Å reflection grating, the crossover energy is ~750 eV for the grating in first order.

In the case of the conventional (i.e. avalanche) gas proportional counters with which cosmic X-ray astronomy began, the $N = E/W$ primary photoelectrons each give rise to an avalanche of mean gain G electrons/electron. The output pulse magnitude is $P = NG$. In addition to the Fano statistics of the primary charge cloud creation, there will be a spread in G. Avalanche statistics are described by a relative variance $f = (\sigma_G/G)^2$, whence the uncertainty in P is determined by a sum of variances and :

$$R^{-1} = \frac{\Delta E}{E} = 2.36 \frac{\Delta P}{P} = 2.36 \left[\left[\frac{\sigma}{N}\right]^2 + \left(\frac{1}{N}\right) \times \left[\frac{\sigma_G}{G}\right]^2\right]^{1/2}$$

or:

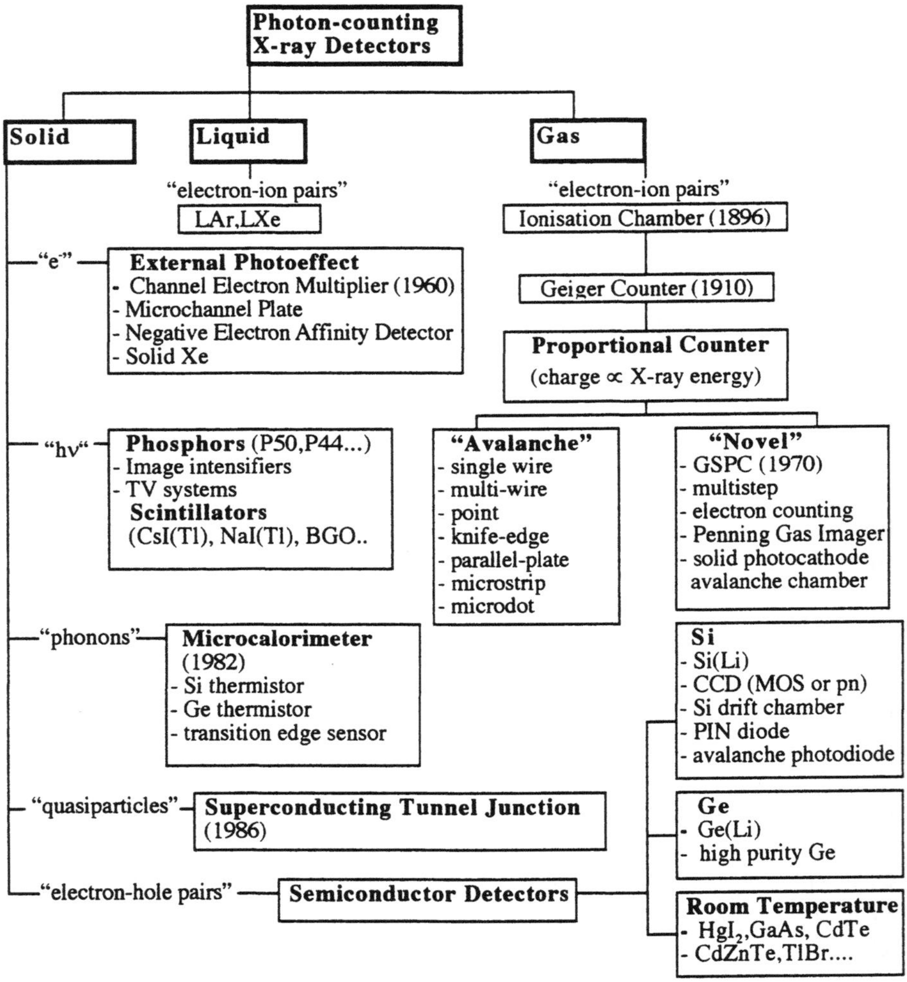

Fig. 8. Family tree of photon counting X-ray detectors. Dates where given refer to the year of first report. Liquid-based detectors are used at very high photon energies ($E > 1$ MeV), beyond the scope of this discussion. The notation "e^-" denotes an output signal mediated by electrons, "$h\nu$" by photons and so on.

$$R^{-1} = \frac{\Delta E}{E} = 2.36 \left[W \frac{F+f}{E} \right]^{1/2} = A\, E^{-1/2} \tag{12}$$

For Xe-based counting gas mixtures: $f \sim 0.67$, $W \sim 25$ eV, $F \sim 0.2$, giving $A \sim 0.35$ (with E expressed in keV). The spectral characteristics of proportional counters at low (99 - 277 eV) X-ray energies is described in detail by Jahoda & McCammon (1988). In the Gas Scintillation Proportional Counter (GSPC, Policarpo et al. 1972), as used on *Spacelab 1*, *EXOSAT* and the *ASCA* GIS instrument, the avalanche variance is essentially eliminated by measuring the UV light produced by the primary electrons and the constant of proportionality A becomes equal to 0.14, implying a factor of two improvement in resolving power.

The present widespread use of Si Charge Coupled Devices (CCDs) in X-ray astronomy (Holland 1997) in preference to gas counters of any kind follows from the much smaller W value (3.68 eV) and Fano factor ($\sim$0.14, Fraser et al. 1994) in silicon (leading to $A = 0.045$). Holland (1997) examines the ongoing developments in CCDs, in particular the requirement for higher readout rates ($\sim$100 kpixels/s to $\sim$1 Mpixels/s) to cope with the counting rates of future X-ray observatories (Section 3). Similarly, the present excitement in the development in cryogenic detectors such as Superconducting Tunnel Junctions (STJs) arises from the fact that the relevant pair creation energy W in such systems is of order meV, leading to resolving powers at last approaching the requirements of Section 2.1. Fig. 9 illustrates the evolution of energy dispersive X-ray detectors in terms of Eq. (11) and other practical factors, such as the attainable image area and ease of manufacture. Figure 9 is strictly relevant only for the primary energy range of Section 2.3. For hard X-ray astronomy, the ongoing transition is one from scintillation counters or high-pressure Xe proportional counters (Ramsey 1995) to the pixellated high-Z semiconductors (especially CdZnTe) described by Gehrels (1995). Pixellation is required to take advantage of the development of novel focusing optics based on grazing-incidence graded d-space multi-layers (Joensen et al. 1995), microchannel plates (Lees et al. 1995) and crystals. 48×48 CdZnTe (Doty et al. 1994) and Ge (Barber et al. 1994) pixel arrays are now available, with inter-element pitches of $125\,\mu$m. Microstrip arrays of 1 cm^2 area have also been fabricated in mercury iodide (Dusi et al. 1994).

Before moving to a detailed discussion of cryogenic detectors, we revisit Eq. (10). The one step "approximation" implicitly assumes that the X-ray detector is perfectly linear. Recently, a number of authors have examined, using Monte Carlo methods, the actual dependence of the factor W on energy E - in gases (Dias et al. 1991), in CsI (Akkerman et al. 1992), in Si (Fraser et al. 1994) and in superconductors (Zehnder 1995). Suppose that W is not now independent of E. Let the number of source photons per unit energy be $A(E)$. The number of counts per output channel x is :

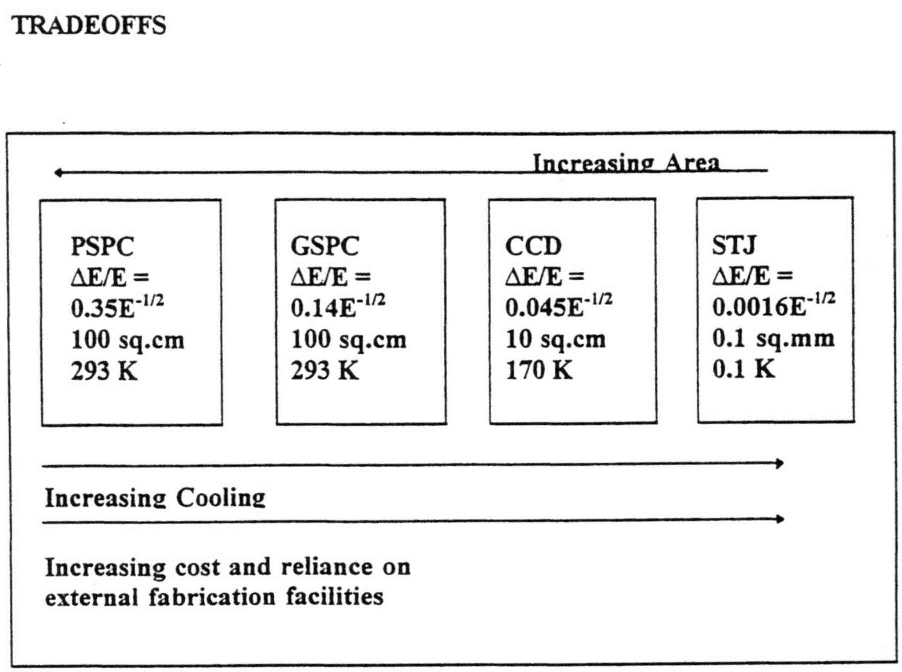

Fig. 9. Tradeoffs in the evolution of energy resolving detectors.

$$M(x) = \left[\frac{A(E)Q(E)}{K}\right] \times \left[W(E)\left[1 - \frac{E}{W(E)}\frac{dW}{dE}\right]^{-1}\right] = \left[A(E).\frac{Q(E)}{K}\right] \times D(E) \tag{13}$$

where Q is the detector quantum efficiency and K is the system gain. The function $D(E)$ should ideally be independent of X-ray energy to avoid distortion of continuum spectra and/or the introduction of spurious emission or absorption features near atomic absorption edges in the detector medium, where $W(E)$ is expected to change discontinuously.

Fig. 10(a) shows the form of $W(E)$ computed for Xe gas by Dias et al. (1991). Note that the sawtooth form of W (left-hand scale) follows the shape of the photoionisation cross-section (right-hand scale). Fig. 10(b) then shows the resulting form of $D(E)$. The small peak in D in the vicinity of the L-shell absorption edges implies the presence of a spurious line feature at $\sim$ 5 keV in the output spectrum of any Xe-based counter illuminated by a featureless input spectrum. Such a spurious feature was first observed in an observation of the Crab Nebula (supposedly a featureless synchrotron spectrum) by Lamb et al. (1985) during the flight in November 1983 on *Spacelab 1* of the ESTEC Space Science Department non-imaging Xe gas scintillation proportional counter (GSPC). The linearity of Xe-based GSPCs at higher energies, in the vicinity of the Xe K edge at 34.6 keV, is described by Tsunemi et al. (1993) and by dos Santos et al. (1994). The synchrotron calibration of all classes of energy resolving detector is now routine, in order to elucidate any deviation from Eq. (10). In solid-state detectors, not only the pair-creation energy W but also the quantum efficiency Q may vary rapidly with energy near atomic absorption edges, due to the effects of XAFS (X-ray Absorption Fine Structure). A discussion of such effects in silicon CCDs is given by Owens et al. (1996).

6 Cryogenic Detectors

Cryogenic X-ray detectors with operating temperatures T in the 0.1-1 K range may be divided into the following main categories :

(a) Superconducting Tunnel Junctions (STJs) - either SIS type (superconducting–insulator–superconductor) or SIN (superconductor–insulator–normal) geometries.
(b) Microcalorimeters - with either Si or Ge thermistors or Superconducting Phase Transition (SPT) thermometry (Ferger et al. 1996) or embodying a Transition Edge Sensor (TES). Magnetic calorimeters, in which X-ray absorption changes the magnetisation of a paramagnetic salt such as Au:Er also appear in the literature (Bandler et al. 1997). The best resolution for a magnetic calorimeter to date is $\Delta E = 529$ eV at 5.9 keV, far poorer than for the other calorimeter variants described below.

Other categories of cryogenic detector have not been seriously developed for astronomy, despite their attraction for imaging. An example is the superheated superconducting granules (SSGs), in which an array of small superconductors are optimally "biased" in the (B, T) plane so that energy absorption promotes a superconducting-to-normal transition for which the disappearance of the Meissner effect is picked up by SQUID magnetometers (Seidel et al. 1987). The energy threshold for an SSG appears to be $\sim$1 keV .

General reviews of cryogenic detectors have been given by Twerenbold (1996), Chardin (1996), Cabrera (1996) and de Korte (1997).

Cooling in the space environment to temperatures of $\sim$10 - 100 mK is in itself a difficult enterprise. The favoured method is the Adiabatic Demagnetisation Refrigerator (ADR), based on the cyclic reorientation of the dipoles of a paramagnetic salt (ferric chromium alum, ferric ammonium sulphate, chromium potassium alum...) in a strong B field ($\sim$ 4 - 6 T) and coupled to a liquid helium bath via a heat switch. The ADR is "gravity independent" (unlike the ^{3}He/^{4}He dilution refrigerator conventionally used to test detectors in the laboratory). The ADR operating cycle is :

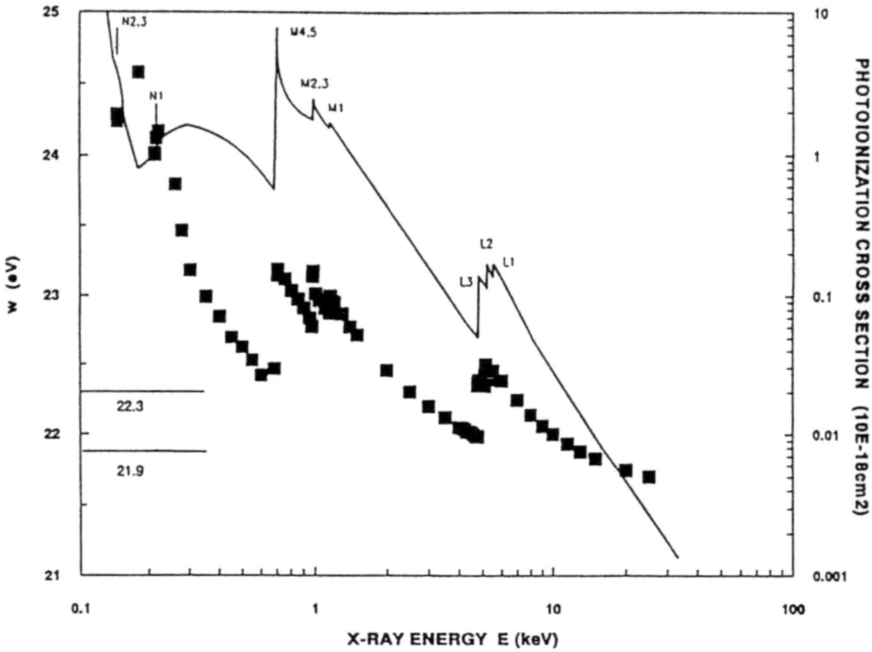

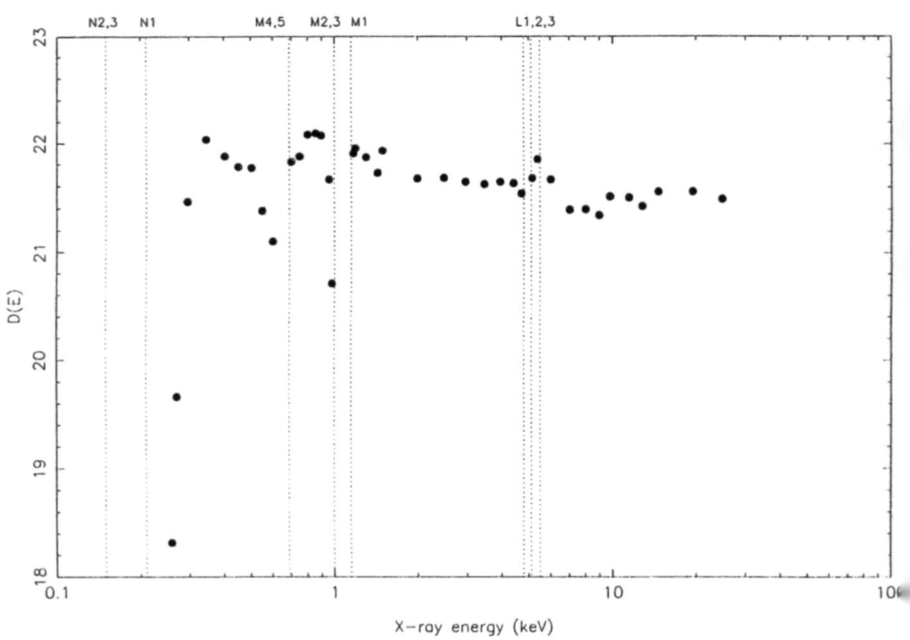

Fig. 10. (a) $W(E)$ for Xe. (b) $D(E)$ for Xe. The broken vertical lines indicate the positions of the named atomic absorption edges.

1. Dipoles randomly aligned, salt pill thermally coupled to bath via heat switch;
2. B-field applied, dipoles align;
3. Heat switch opened, B-field off, dipoles randomise, drawing energy from surroundings - including detector;
4. Repeat.

The achievable cooling power is of order microwatts. The development of ADRs has been described by Bromiley et al. (1997). The related technical problem of producing UV/optical filters working at cryogenic temperatures is addressed by Keski-Kuha (1989).

6.1 Superconducting tunnel junctions (STJs)

A SIS superconducting tunnel junction consists of two thin layers of superconducting metal cooled well below its transition temperature ($T \sim T_c/10$) and separated by a thin (~10 Å) oxide barrier. Absorption of ionising radiation breaks Cooper pairs in the superconductor and gives rise to an excess tunnel current of quasi-particles across the insulating barrier. Because W (Eq. 10) is equal to 1.75Δ in such a system (Kurakado 1982), where 2Δ (~meV, see Table 1) is the superconducting energy gap, about a thousand times more charge carriers are generated per unit X-ray energy than in silicon. High-temperature superconductors, despite the very obvious advantage of less severe device cooling, are not an attractive basis for X-ray detectors. High-T_c materials, such as YBaCuO have much higher Δ values ($\sim 20\,\text{meV}$, Bluzer & Forrester 1994) than elemental superconductors. With a Fano factor for superconducting metals estimated to be ~0.2, STJs have been extensively investigated as high-resolution, energy dispersive spectrometers with the potential for $\Delta E = 4\,\text{eV}$ at $E = 6\,\text{keV}$. Fig. 11 (Kraus 1997) illustrates the various tunneling processes in an STJ with dissimilar metal electrodes (energy gaps Δ_l and Δ_r, bias voltage V_B). In "Process 1", the excited quasiparticle tunnels left to right: electron current also is in the same direction. In "Process 4", the quasiparticle tunnels right to left; the electron current is left to right. This process is the basis for repeated tunneling, signal gain and an excess blur of energy resolution beyond the formula given in Eq. (11) (Goldie et al. 1994). The actual processes governing the time evolution of quasiparticle and phonon numbers in real devices are extremely complicated. Fig. 12 shows a schematic single-pixel STJ. Note the presence of a B field in the plane of the device (~ 100 Gauss) in order to suppress the so-called Josephson "supercurrent" across the barrier - the tunneling of Cooper pairs.

The first reports of the response of STJs to 5.9 keV X rays were by Kraus et al. (1986) and by Twerenbold (1986). Both these groups used tin thin-film junctions at 0.3 K. Twerenbold (1986) reported somewhat the better resolution – $\Delta E = 90$ eV FWHM. Tin turns out not to be a suitable basis for practical detectors (tin junctions cannot be repeatedly cycled from room

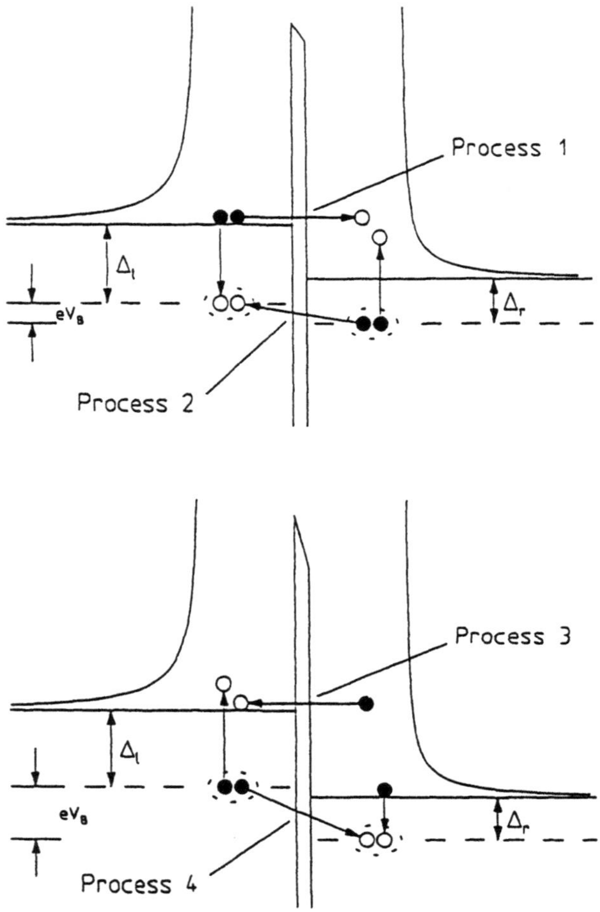

Fig. 11. Tunneling processes in STJ with dissimilar electrodes (Kraus 1997) .

temperature to operating temperatures). Table 1 lists the significant material properties of Sn and other candidate superconductors. The rightmost column shows the metal thickness required to absorb 90% of 10 keV X rays - a necessary (but not sufficient) condition for an efficient detector.

From the viewpoint of in-orbit cooling, the higher the operating temperature the better. The temperature-ordered rank (best material first) is therefore Nb : Ta : Sn : Al

The energy gap determines W via: $W = 1.75\Delta$, so that the energy resolution-ordered rank is therefore Al : Sn : Ta : Nb.

The tunneling probability, and hence signal size for a given energy deposit, is inversely related to junction volume, which is in turn dictated by stopping

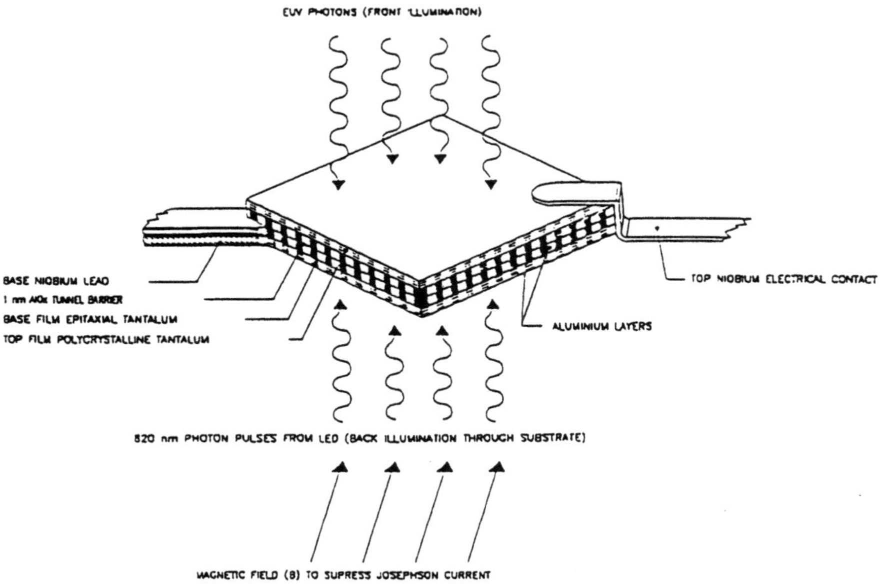

Fig. 12. Schematic Ta single-pixel STJ X-ray detector (Peacock et al. 1997). Both front- and backside illumination (for EUV and optical wavelengths, respectively) are illustrated.

power and quantum efficiency. The rank according to thickness is therefore Ta : Sn : Nb : Al.

Clearly, the development of STJs has to involve a set of non-trivial trade-offs. Much research over the past decade has failed to reach the few eV resolution promised at energies of 6 keV even for single pixels in the size ranges $20 \times 20 \mu m^2$ to $40 \times 40 \mu m^2$. recent activity has concentrated on lower energies - down into the EUV (Peacock et al. 1997) and optical. It appears that hafnium STJs could provide $R < 200$ over the entire EUV band, a level of resolving power marginally competitive with current (*EUVE*) grating instruments.

The current "state of the art" for SIS STJs in the primary band of Section 2.1 is represented in Fig. 13 (Peacock & Voorhoeve 1998). Energy resolutions $\Delta E \sim 5$ eV at sub-keV X-ray energies have also been reported by Labov et al. (1997) using NbAl STJs. The best energy resolution at $E = 6$ keV remains ~ 29 eV (Frank et al. 1996), dominated by spatial non-uniformity within a single pixel. Selective illumination of a 10 μm spot in the centre of a $100 \times 100 \mu m^2$ Ta STJ has produced $\Delta E = 15.7\,\mathrm{eV}$, within a factor ~ 2 of the Fano limited resolution (Verhoeve et al. 1998).

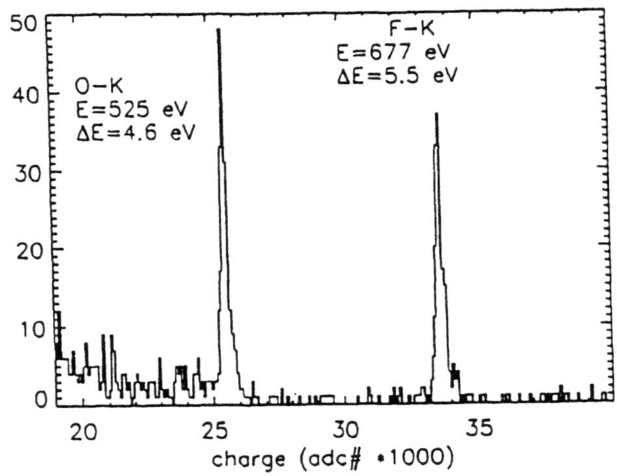

Fig. 13. Response of base film of Ta STJ to oxygen $K\alpha$ and fluor $K\alpha$ X rays (Peacock & Verhoeve 1998).

Radiation damage in Nb STJs due to protons is described by Rando et al. (1997). No changes in the current-voltage characteristic or spectroscopic performance were observed for fluences of 1.4×10^{11} protons cm^{-2}, a dose exceeding that expected for a typical satellite mission. The energy linearity of Nb based STJs is described by Lumb et al. (1995) and Verhoeve et al. 1996). No significant variation of W is observed, either globally or across the L and M edges of niobium. The first 3×3 STJ pixel arrays are described by Peacock et al. (1997) and by Rando et al. (1997b). Very recently, a 6×6 $25 \mu m$ pixel Ta array has been produced for optical work (Peacock et al. 1998).

Table 1. Material parameters of candidate STJ superconductors

Material	T_c (K)	2Δ (meV)	90%/10keV thickness (μm)
Sn	3.72	1.15	23
Al	1.18	0.34	300
Nb	9.25	3.05	34
Ta	4.47	1.4	6

6.2 Microcalorimeters

The principle of operation of any of the various forms of microcalorimeter is simply to convert the energy of the X ray photon into heat in a small absorber mass and measure the resultant temperature rise (see Fig. 14). First proposed in 1984 (McCammon et al. 1984), the energy resolution achieved with calorimeters has improved since then from 270 eV FWHM to better than 7 eV (at 5.9 keV) today.

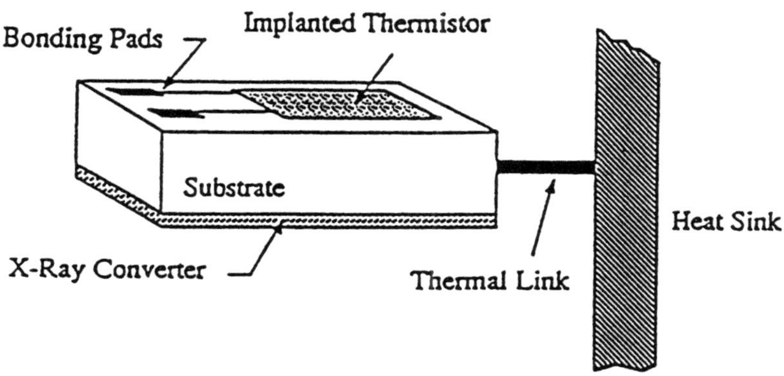

Fig. 14. Schematic view of calorimeter (Holt 1990).

We can roughly estimate the attainable energy resolution for a microcalorimeter as follows. Let:

- C be the heat capacity of the absorber;
- T be the equilibrium temperature of the absorber;
- ΔT be the temperature rise following X-ray absorption;
- E be the X-ray energy.

We assume that E is completely converted into heat (i.e. phonons) and that $\Delta T \sim T$ (a poor approximation in real devices !). Then:

$$E \sim E_{\text{phonons}} = CT \tag{14}$$

The energy of a single phonon is kT, where k is Boltzmann's constant, so the mean number of phonons to be "counted" is:

$$N = \frac{E_{\text{phonons}}}{kT} = \frac{E}{kT} = \frac{CT}{kT} = \frac{C}{k} \tag{15}$$

The standard deviation on N is:

$$\sigma_N = \sqrt{N} = \sqrt{C/k} \tag{16}$$

so that the FWHM uncertainty in the energy estimate is :

$$\Delta E = 2.36\, kT\, \sqrt{C/k} = 2.36\sqrt{kT^2C} \tag{17}$$

In actual fact:

$$\Delta E = 2.36\xi\sqrt{kT^2C} \tag{18}$$

where ξ is a numerical factor of order ~ 2 (McCammon et al. 1984; de Korte 1997). ΔE is independent of the X-ray energy, so that microcalorimeters are "constant ΔE spectrometers" (cf. Section 4). Eq. (14) also shows that the attainable energy resolution is dependent on the absorber size. For a crystalline absorber, the Debye lattice specific heat dominates:

$$C = 1944 \left[\frac{T}{T_D}\right]^3 \text{J/mole/K} \tag{19}$$

so that the energy resolution scales as the bath temperature to the power five-halves. The absorber mass returns to its base temperature with an exponential time constant τ, given by:

$$\tau = C/G \tag{20}$$

where G is the conductance of the link between the absorber and the heat bath. Practical values of τ lie currently in the $\sim 1\,$ms regime, implying maximum count rates of only a few hundred - a problem for future observatories (see Section 3).

We can immediately make rough calculations of the likely size constraints (Section 3) on microcalorimeters. Suppose, for simplicity, we assume a Si absorber whose specific heat is dominated by lattice contribution, operated at 0.1 K. The Debye temperature of silicon $T_{\rm D}$ is 640 K, which leads to:

$$C = 7.42 \times 10^{-9}\text{J/mole/K} \tag{21}$$

Let the mass of the Si absorber be m grams. One mole of silicon is 28 g. The density of Si is 2.33 g cm^{-3}. Supposing further that the induced temperature rise for a 6 keV photon is $\Delta T = 0.1$ K, we can compute m from $E = C\Delta T$:

$$m = \frac{28 \times 9.6 \times 10^{-15}}{7.42 \times 10^{-9}} = 36\,\mu\text{g} \tag{22}$$

equating to a characteristic size of ~250 microns. Note that 10 microns is approximately the 1/e absorption depth for 6 keV X-rays in silicon. Using the narrow-gap semiconductors ("Semi-metals") such as HgTe which are favoured practical absorbers, the absorber dimensions turn out to be $0.7 \times 500 \times 2000$

microns - still uncomfortably small compared to the desired minimum detector size derived in Section 3.

We can also simply estimate how closely the temperature of the heat bath must be controlled (Labov et al. 1990). Let R_0 be the resistance of the thermistor monitoring the absorber temperature. If S is the voltage step corresponding to absorption of a single X-ray photon, energy E:

$$S = I.\Delta R_0 = I \left[\frac{dR_0}{dT}\right] \Delta T = I \left[\frac{dR_0}{dT}\right] \left[\frac{E}{C}\right] \tag{23}$$

Assuming the Debye law dependence of C on temperature T and the hopping conduction law appropriate to Si or Ge thermistors [resistance varying as $\exp(\sqrt{T_0/T})$] then the log temperature coefficient of resistance α is given by:

$$\alpha = -\left[\frac{9}{2} + \sqrt{\frac{T_0}{2}}\sqrt{T}\right] \tag{24}$$

while the resolving power R is determined by:

$$\frac{1}{R} = \frac{\delta S}{S} = \alpha \left[\frac{\delta T}{T}\right] \tag{25}$$

So for a resolving power of 1000, $T = 0.1$ K, $T_0 = 25$ K, we require:

$$\delta T \sim 8\mu K \tag{26}$$

Active control of temperature of the bath is therefore essential.

The observational "state of the art" in energy-dispersive X-ray spectroscopy is currently represented by the first succesful sounding rocket flight of the Goddard Space Flight Center - University of Wisconsin X-ray Quantum Calorimeter to observe the soft X-ray spectrum of the Interstellar Medium (ISM) (Deiker et al. 1997). This instrument, shown schematically in Fig. 15, consisted of a 36 element microcalorimeter array, each element consisting of a 0.5 mm × 2.0 mm ion implanted Si thermistor coupled to a 1 mm^2 HgTe absorber. The composite spectral resolution of the array was 14 eV FWHM at an X-ray energy of 277 eV, 21 eV at 677 eV. The base temperature of the ADR during flight was 60 mK. Fig. 16 shows the spectrum resulting from 240 seconds observation time of a 1sr field-of-view centred on galactic coordinates $l = 90°$, $b = 60°$. The observed count rate was ~ 7 per second. A strong line from He-like O VIII at 0.57 keV is clearly visible, but the constituent f, i, r lines are not resolved (Section 2.1). No significant O VIII flux was observed at 0.654 keV.

The Goddard-Wisconsin sounding rocket experiment is a precursor for the XRS instrument on the Japanese *Astro-E* satellite, due for launch in February 2000. The XRS calorimeter array will have either a 6 × 6 or 2 × 18 array of 0.39 mm^2 HgTe absorber elements (Stahle et al. 1997). The goal is to have ~ 10 eV FWHM resolution over an energy band 0.3-12 keV.

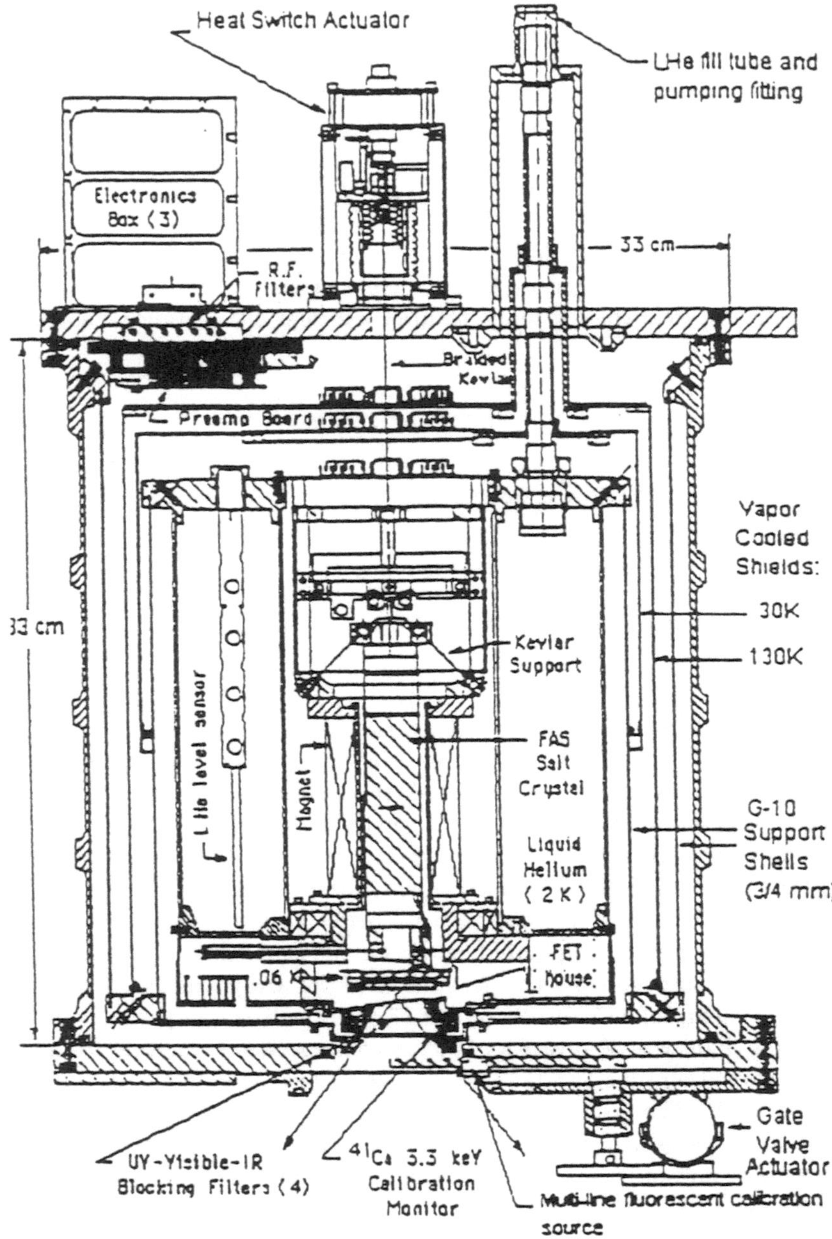

Fig. 15. Cross section of X-ray Quantum Calorimeter (Deiker et al. 1997). The cooling system consists of a ferric ammonium alum ADR inside a 4 T, 8 A superconducting magnet inside a 4 liter He dewar.

For future observatories such as *XEUS* and *Constellation-X*, improvements in energy resolution of $\sim 2-5$ times and, most importantly, improvements in count rate capacity of order $\sim 10^2$ are required.

Using Nuclear Transmutation Doped (NTD) Ge thermistors, Silver et al. (1997) have demonstrated $\Delta E = 7.8\,\mathrm{eV}$ at $E = 1.74\,\mathrm{keV}$. Other groups are developing the "hot electron bolometer", in which a normal metal acts as the absorber and a SIN junction acts as a thermometer. If the absorber temperature is below 1 K, the "hot electrons" created by X-ray absorption are decoupled from the phonon population in the metal. Time constants τ of order $20\,\mu\mathrm{s}$ are estimated for $\sim 0.25\,\mathrm{mm}^2$ detector areas, coupled to energy resolution $\Delta E = 30\,\mathrm{eV}$ at $E = 5.9\,\mathrm{keV}$ (Martinis 1996).

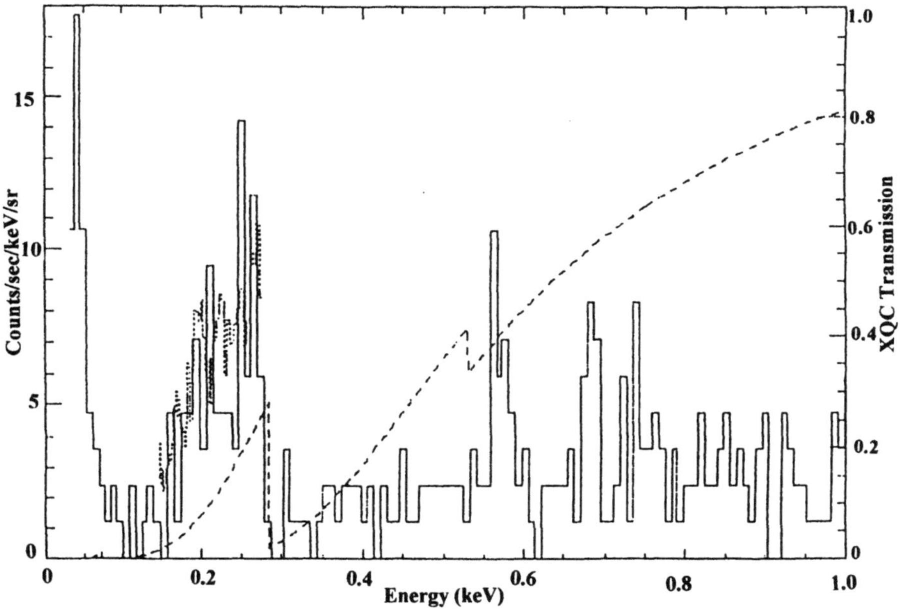

Fig. 16. Sounding rocket microcalorimeter array X-ray spectrum of the ISM (Deiker et al. 1997). The transmission of the aluminised parylene optical/UV/IR filters is shown by the dashed line.

Perhaps the most exciting calorimeter development is the Transient Edge Sensor (TES) with extreme electrothermal feedback (ETF). Here the temperature sensing element is a superconducting film voltage-biased in the center of it superconducting-to-normal transition. Joule heating balances TES cooling to the bath which is at a temperature well below the superconducting transition temperature. When an X ray is absorbed, the TES resistance rises sharply, causing the Joule heating to fall. The decreasein current through the TES is sensed by a SQUID ammeter. Analysis of the TES system (Irwin 1995; Irwin et al. 1998) shows that by virtue of the high attainable value of the parameter $\alpha(\sim 10^3)$, energy resolution below the "thermodynamic limit" (see Eq. (18) and temporal response faster than indicated by Eq. (20) can both be achieved. The best performance so far reported is $\Delta E = 7.2\,\mathrm{eV}$ FWHM at 5.9 keV, coupled to a response time of $\sim 200\,\mu\mathrm{s}$ (Wollman et al. 1997).

Acknowledgments. I acknowledge the assistence of Nigel Bannister and Adrian Martin in the production of this manuscript.

References

Abdali, S. et al (1997): Proc. SPIE 3114, 358

Akkerman, A., Gibrekhterman, Breskin, A., Chechik, R. (1992): J. Appl. Phys. 72, 5429

Bandler, S. et al. (1997): Proc. 7th Intnl. Workshop on Low Temperature Detectors, Ed. S. Cooper, Max Planck Institute, p. 145

Barber, H.B. (1994): Nucl. Instr. Meth. A353, 361

Bluzer, N. & Forrester, M.G. (1994): Opt. Eng. 33, 697

Bowyer, S., Malina, R.F. (1990): Adv. Space. Res. 11, 205

Briel, U. et al. (1987): *The high throughput X-ray spectroscopy mission: report of the instrument working group*, ESA SP-1092

Brinkman, A.C. et al. (1997): Proc. SPIIE 3113, 187

Bromiley, P.A. et al. (1997): Proc. 7th Intnl. Workshop on Low Temperature Detectors, ed. S. Cooper, Max Planck Institute, p. 201

Cabrera, B. (1996): Nucl. Instr. Meth. A 370, 150

Chardin, G. (1996): Nucl. Instr. Meth. A 370, 279

de Chambure, D., Laine, R., van Katwijk, K. (1997): Proc. SPIE 3114, 113

de Korte, P.A.J. (1997): in *The next generation of X-ray observatories*, M.J.L. Turner & M.G. Watson (Eds), University of Leicester Special Report 97/02, p. 67

Dias, T.H.V.T., Santos, F.P., Stauffer, A.D., Conde, C.A.N. (1991): Nucl. Instr. Meth. A 307, 341

Deiker, S. et al. (1997): Proc. 7th Intnl. Workshop on Low Temperature Detectors, Ed. S. Cooper, Max Planck Institute, p. 108

dos Santos, J.M.F., Morgado, R.E., Tavora, L.M.N., Conde, C.A.N. (1994): Nucl. Instr. meth. A 350, 216

Doty, F.P. (1994): Nucl. Instr. Meth. A353, 356

Dusy, F. (1994): Nucl. Instr. Meth. A348, 531

Ferger, P. et al. (1996): Nucl. Instr. Meth. A370, 157

Frank, M. et al. (1996): Nucl. Instr. Meth. A370, 41

Fraser, G.W. (1989): *X-ray detectors in astronomy*, Cambridge University Press.

Fraser, G.W. et al. (1994): Nucl. Instr. Meth. A 350, 368.

Gehrels, N. (1995): in *Imaging in high energy astronomy*, Eds. L Bassani and G. di Cocco, Kluwer Academic Publishers, p. 129.

Giacconi, R. et al. (1970): ApJ 230, 540

Goldie, D.J., Brink, P.L., Patel, C., Booth, N.E., Salmon, G.L. (1994): Appl. Phys. Lett. 64, 3169.

Holland, A.D. (1997): Proc. SPIE 3114, 586.

Holt, S.S. (1990): in *High-resolution spectroscopy of cosmic X-ray Plasmas*, P. Gorenstein & M.V. Zombeck (Eds), Proc. 115th IAU Symposium, Cambridge University Press, p. 350

Holt, S.S. (1994): in Proc. International School of Space Science Course in X-ray Astronomy, Aquila, Italy

Irwin, K.D. (1995): Appl. Phys. Lett. 66, 1998

Irwin, K.D. et al. (1998): J. Appl. Phys. 83, 3978

Jahoda, K., McCammon, D. (1988): Nucl. Instr. Meth. A 272, 800.

Joensen, K.D. et al. (1995): Proc. SPIE 2515, 146

Keski-Kuha, R.A.M. (1982): Appl. Opt. 28, 2965

Kowalski, M.P., Cruddace, R.G., Seely, J.F., Rife, J.C., Hunter, W.R., Barbee, T.W. (1996): J. Electron Spectroscopy 80, 473.

Kraus, H., Peterreins, Th., Probst, F., von Feilitzch, F., Mossbauer, R.L., Umlauf, E. (1986): Europhys. Lett 1, 161.

Kraus, H. (1997): Proc. 7th Intnl. Workshop on Low Temperature Detectors, Ed. S. Cooper, Max Planck Institute, p. 1

Kurakado, M. (1982): Nucl. Instr. Meth. 196, 275

Labov, S.E. et al. (1997): Proc. 7th International Workshop on Low-temperature Detectors, S. Cooper (Ed.), Max Planck Institute, p. 82

Labov, S.E. et al. (1990): in *High-resolution Spectroscopy of Cosmic X-ray Plasmas*, P. Gorenstein & M.V. Zombeck (Eds), Proc. 115th IAU Symposium, Cambridge University Press, p. 357

Lamb, P. et al. (1987): Ap&SS 136, 369

Lees, J.E. et al. (1995): in *Imaging in High-energy Astronomy*, L. Bassani & G. di Cocco (Eds), Kluwer Academic Publishers, p. 305

Linsky, J.L., Luttermoser, D.G. (1990): Adv. Space. Res. 11, 5

Lumb, D.H. et al. (1995): Proc. SPIE 2518, 258

Markert, T.H. et al. (1994): Proc. SPIE 2280, 168

Martinis, J.M. (1996): Nucl. Instr. Meth. A370, 171

McCammon, D. et al. (1984): J. Appl. Phys. 56, 1263

Murray, S.S. et al. (1997): Proc. SPIE 3114, 11

Owens, A. et al. (1996): Nucl. Instr. Meth A 382, 503

Palumbo, G.G.C. et al. (1997): in *All sky monitors in the next decade*, Proc. RIKEN workshop, eds. M. Matsuoka & N. Kawai, p. 271

Peacock, A. et al. (1997): Proc. SPIE 3114

Peacock, A., Rando, N. & Verhoeve, P. (1998): Astronews 35, 3

Peacock, A. & Verhoeve, P. (1998): Astronews 35, 6

Policarpo, A.J.P.L, Alves, M.A.F., Dos Santos, M.C.M., Carvalho, M.J.T. (1972): Nucl. Instr. Meth. 102, 337

Predehl, P. et al. (1997): Proc. SPIE 3113, 172

Ramsey, B.D. (1995): in *Imaging in High Energy Astronomy*, Eds. L. Bassani & G. di Cocco, Kluwer Academic Publishers, p. 119

Rando, N. et al. (1997): Nucl. Instr. Meth. A 394, 173.

Rando, N. et al. (1997): Proc. 7th International Workshop on Low Temperature Detectors, S. Cooper (Ed.), Max Planck Institute, p. 101
Rasmussen, A. et al. (1998): Proc. SPIE 3444, in press
Schmitt, J.H.M.M. (1990): in *High resolution spectroscopy of cosmic X-ray plasmas*, Proc. IAU Coll. 115, Eds. P. Gorenstein & M.V. Zombeck, Cambridge University Press, p. 35
Seidel, W., Oberauer, L. & von Feilitzsch, F. (1987): Rev. Sci. Instrum. 58, 1471
Silver, E. et al. (1997): X-ray Spectrometry 26, 265
Stahle, C.K. et al. (1997): Proc. 7th International Workshop on Low Temperature Detectors, S. Cooper (Ed.), Max Planck Institute, p. 101
Tsunemi, H. et al. (1993): Nucl. Instr. Meth. A 336, 301
Twerenbold, D. (1986): Europhys. Lett. 1, 209
Twerenbold, D. (1987): Nucl. Instr. Meth. A 260, 430
Twerenbold, D. (1986): Reports Prog. Phys. 59, 349
Verhoeve, P., Rando, N. & Verveer, J. (1996): Phys. Rev. B53, 809
Verhoeve, P. et al. (1998): Appl. Phys. Lett., in press
Weisskopf, M.C. (1995): Proc. SPIE 2515, 302
White, N., Tananbaum, H., Kahn, S. (1997): in *The next generation of X-ray observatories*, eds. M.J.L. Turner & M.G. Watson, University of Leicester X-ray Astronomy Group special report 97/02, p. 173
Wollman, D.A. et al. (1997): J. Microscopy 188, 196
Woosley, S.E. (1987): ApJ 318, 664
Zehnder, A. (1995): Phys.Rev. B 52, 12858

AUTHOR/NAME INDEX

SUBJECT INDEX

OBJECT INDEX

:cture Notes in Physics

r information about Vols. 1–482
:ase contact your bookseller or Springer-Verlag

483: G. Trottet (Ed.), Coronal Physics from Radio . Space Observations. Proceedings, 1996. XVII, 226 es. 1997.

484: L. Schimansky-Geier, T. Pöschel (Eds.), chastic Dynamics. XVIII, 386 pages. 1997.

485: H. Friedrich, B. Eckhardt (Eds.), Classical, ni-classical and Quantum Dynamics in Atoms. VIII, pages. 1997.

. 486: G. Chavent, P. C. Sabatier (Eds.), Inverse Prob-s of Wave Propagation and Diffraction. Proceed-s, 1996. XV, 379 pages. 1997.

487: E. Meyer-Hofmeister, H. Spruik (Eds.), Accre-n Disks – New Aspects. Proceedings, 1996. XIII, 356 es. 1997.

. 488: B. Apagyi, G. Endrédi, P. Lévay (Eds.), Inverse l Algebraic Quantum Scattering Theory. Proceed-s, 1996. XV, 385 pages. 1997.

. 489: G. M. Simnett, C. E. Alissandrakis, L. Vlahos ls.), Solar and Heliospheric Plasma Physics. Proceed-s, 1996. VIII, 278 pages. 1997.

. 490: P. Kutler, J. Flores, J.-J. Chattot (Eds.), Fif-nth International Conference on Numerical Methods Fluid Dynamics. Proceedings, 1996. XIV, 655 pages. 7.

. 491: O. Boratav, A. Eden, A. Erzan (Eds.), Tur-lence Modeling and Vortex Dynamics. Proceedings, 6. XII, 245 pages. 1997.

. 492: M. Rubí, C. Pérez-Vicente (Eds.), Complex Be-viour of Glassy Systems. Proceedings, 1996. IX, 467 ges. 1997.

l. 493: P. L. Garrido, J. Marro (Eds.), Fourth Granada tures in Computational Physics. XIV, 316 pages. 1997.

l. 494: J. W. Clark, M. L. Ristig (Eds.), Theory of Spin ttices and Lattice Gauge Models. Proceedings, 1996. , 194 pages. 1997.

l. 495: Y. Kosmann-Schwarzbach, B. Grammaticos, M. Tamizhmani (Eds.), Integrability of Nonlinear stems. VII, 380 pages. 1997.

l. 496: F. Lenz, H. Grießhammer, D. Stoll (Eds.), Lec-res on QCD. VII, 483 pages. 1997.

l. 497: J. P. Greve, R. Blomme, H. Hensberge (Eds.), llar Atmospheres: Theory and Observations. VIII, 2 pages. 1997

l. 498: Z. Horváth, L. Palla (Eds.), Conformal Field eories and Integrable Models. Proceedings, 1996. X, pages. 1997.

l. 499: K. Jungmann, J. Kowalski, I. Reinhard, F. äger (Eds.), Atomic Physics Methods in Modern Re-arch, IX, 448 pages. 1997.

l. 500: D. Joubert (Ed.), Density Functionals: Theory d Applications, XVI, 194 pages. 1998.

l. 501: J. Kertész, I. Kondor (Eds.), Advances in Com-ter Simulation. VIII, 166 pages. 1998.

l. 502: H. Aratyn, T. D. Imbo, W.-Y. Keung, U. khatme (Eds.), Supersymmetry and Integrable Mod-s. Proceedings, 1997. XI, 379 pages. 1998.

Vol. 503: J. Parisi, S. C. Müller, W. Zimmermann (Eds.), A Perspective Look at Nonlinear Media. From Physics to Biology and Social Sciences. VIII, 372 pages. 1998.

Vol. 504: A. Bohm, H.-D. Doebner, P. Kielanowski (Eds.), Irreversibility and Causality. Semigroups and Rigged Hilbert Spaces. XIX, 385 pages. 1998.

Vol. 505: D. Benest, C. Froeschlé (Eds.), Impacts on Earth. XVII, 223 pages. 1998.

Vol. 506: D. Breitschwerdt, M. J. Freyberg, J. Trümper (Eds.), The Local Bubble and Beyond. Proceedings, 1997. XXVIII, 603 pages. 1998.

Vol. 507: J. C. Vial, K. Bocchialini, P. Boumier (Eds.), Space Solar Physics. Proceedings, 1997. XIII, 296 pages. 1998.

Vol. 508: H. Meyer-Ortmanns, A. Klümper (Eds.), Field Theoretical Tools for Polymer and Particle Physics. XVI, 258 pages. 1998.

Vol. 509: J. Wess, V. P. Akulov (Eds.), Supersymmetry and Quantum Field Theory. Proceedings, 1997. XV, 405 pages. 1998.

Vol. 510: J. Navarro, A. Polls (Eds.), Microscopic Quantum Many-Body Theories and Their Applications. Proceedings, 1997. XIII, 379 pages. 1998.

Vol. 511: S. Benkadda, G. M. Zaslavsky (Eds.), Chaos, Kinetics and Nonlinear Dynamics in Fluids and Plasmas. Proceedings, 1997. VIII, 438 pages. 1998.

Vol. 512: H. Gausterer, C. Lang (Eds.), Computing Particle Properties. Proceedings, 1997. VII, 335 pages. 1998.

Vol. 513: A. Bernstein, D. Drechsel, T. Walcher (Eds.), Chiral Dynamics: Theory and Experiment. Proceedings, 1997. IX, 394 pages. 1998.

Vol. 514: F. W. Hehl, C. Kiefer, R.J.K. Metzler, Black Holes: Theory and Observation. Proceedings, 1997. XV, 519 pages. 1998.

Vol. 515: C.-H. Bruneau (Ed.), Sixteenth International Conference on Numerical Methods in Fluid Dynamics. Proceedings. XV, 568 pages. 1998.

Vol. 516: J. Cleymans, H. B. Geyer, F. G. Scholtz (Eds.), Hadrons in Dense Matter and Hadrosynthesis. Proceedings, 1998. XII, 253 pages. 1999.

Vol. 517: Ph. Blanchard, A. Jadczyk (Eds.), Quantum Future. Proceedings, 1997. X, 244 pages. 1999.

Vol. 518: P. G. L. Leach, S. E. Bouquet, J.-L. Rouet, E. Fijalkow (Eds.), Dynamical Systems, Plasmas and Gravitation. Proceedings, 1997. XII, 397 pages. 1999.

Vol. 519: A. Pękalski, K. Sznajd-Weron (Eds.), Anomalous Diffusion. From Basics to Applications. Proceedings, 1998. XVIII, 378 pages. 1999.

Vol. 520: J. A. van Paradijs, J. A. M. Bleeker (Eds.), X-Ray Spectroscopy in Astrophysics. EADN School X. Proceedings, 1997. XV, 530 pages. 1999.

Vol. 521: L. Mathelitsch, W. Plessas (Eds.), Broken Symmetries. Proceedings, 1998. VII, 299 pages. 1999.

Monographs

For information about Vols. 1–10 please contact your bookseller or Springer-Verlag

Vol. m 11: A. D. Yaghjian, Relativistic Dynamics of a Charged Sphere. XII, 115 pages. 1992.

Vol. m 12: G. Esposito, Quantum Gravity, Quantum Cosmology and Lorentzian Geometries. Second Corrected and Enlarged Edition. XVIII, 349 pages. 1994.

Vol. m 13: M. Klein, A. Knauf, Classical Planar Scattering by Coulombic Potentials. V, 142 pages. 1992.

Vol. m 14: A. Lerda, Anyons. XI, 138 pages. 1992.

Vol. m 15: N. Peters, B. Rogg (Eds.), Reduced Kinetic Mechanisms for Applications in Combustion Systems. X, 360 pages. 1993.

Vol. m 16: P. Christe, M. Henkel, Introduction to Conformal Invariance and Its Applications to Critical Phenomena. XV, 260 pages. 1993.

Vol. m 17: M. Schoen, Computer Simulation of Condensed Phases in Complex Geometries. X, 136 pages. 1993.

Vol. m 18: H. Carmichael, An Open Systems Approach to Quantum Optics. X, 179 pages. 1993.

Vol. m 19: S. D. Bogan, M. K. Hinders, Interface Effects in Elastic Wave Scattering. XII, 182 pages. 1994.

Vol. m 20: E. Abdalla, M. C. B. Abdalla, D. Dalmazi, A. Zadra, 2D-Gravity in Non-Critical Strings. IX, 319 pages. 1994.

Vol. m 21: G. P. Berman, E. N. Bulgakov, D. D. Holm, Crossover-Time in Quantum Boson and Spin Systems. XI, 268 pages. 1994.

Vol. m 22: M.-O. Hongler, Chaotic and Stochastic Behaviour in Automatic Production Lines. V, 85 pages. 1994.

Vol. m 23: V. S. Viswanath, G. Müller, The Recursion Method. X, 259 pages. 1994.

Vol. m 24: A. Ern, V. Giovangigli, Multicomponent Transport Algorithms. XIV, 427 pages. 1994.

Vol. m 25: A. V. Bogdanov, G. V. Dubrovskiy, M. P. Krutikov, D. V. Kulginov, V. M. Strelchenya, Interaction of Gases with Surfaces. XIV, 132 pages. 1995.

Vol. m 26: M. Dineykhan, G. V. Efimov, G. Ganbold, S. N. Nedelko, Oscillator Representation in Quantum Physics. IX, 279 pages. 1995.

Vol. m 27: J. T. Ottesen, Infinite Dimensional Groups and Algebras in Quantum Physics. IX, 218 pages. 1995.

Vol. m 28: O. Piguet, S. P. Sorella, Algebraic Renormalization. IX, 134 pages. 1995.

Vol. m 29: C. Bendjaballah, Introduction to Photon Communication. VII, 193 pages. 1995.

Vol. m 30: A. J. Greer, W. J. Kossler, Low Magnetic Fields in Anisotropic Superconductors. VII, 161 pages. 1995.

Vol. m 31 (Corr. Second Printing): P. Busch, M. Grabowski, P.J. Lahti, Operational Quantum Physics. XII, 230 pages. 1997.

Vol. m 32: L. de Broglie, Diverses questions de mécanique et de thermodynamique classiques et relativistes. XII, 198 pages. 1995.

Vol. m 33: R. Alkofer, H. Reinhardt, Chiral Quark Dynamics. VIII, 115 pages. 1995.

Vol. m 34: R. Jost, Das Märchen vom Elfenbeinernen Turm. VIII, 286 pages. 1995.

Vol. m 35: E. Elizalde, Ten Physical Applications of Spectral Zeta Functions. XIV, 224 pages. 1995.

Vol. m 36: G. Dunne, Self-Dual Chern-Simons Theories. X, 217 pages. 1995.

Vol. m 37: S. Childress, A.D. Gilbert, Stretch, Twist, Fold: The Fast Dynamo. XI, 406 pages. 1995.

Vol. m 38: J. González, M. A. Martín-Delgado, G. Sierra, A. H. Vozmediano, Quantum Electron Liquids and High-Tc Superconductivity. X, 299 pages. 1995.

Vol. m 39: L. Pittner, Algebraic Foundations of Non-Com-mutative Differential Geometry and Quantum Groups. XII, 469 pages. 1996.

Vol. m 40: H.-J. Borchers, Translation Group and Particle Representations in Quantum Field Theory. VII, 131 pages. 1996.

Vol. m 41: B. K. Chakrabarti, A. Dutta, P. Sen, Quantum Ising Phases and Transitions in Transverse Ising Models. X, 204 pages. 1996.

Vol. m 42: P. Bouwknegt, J. McCarthy, K. Pilch, The W3 Algebra. Modules, Semi-infinite Cohomology and BV Algebras. XI, 204 pages. 1996.

Vol. m 43: M. Schottenloher, A Mathematical Introduction to Conformal Field Theory. VIII, 142 pages. 1997.

Vol. m 44: A. Bach, Indistinguishable Classical Particles. VIII, 157 pages. 1997.

Vol. m 45: M. Ferrari, V. T. Granik, A. Imam, J. C. Nadeau (Eds.), Advances in Doublet Mechanics. XVI, 214 pages. 1997.

Vol. m 46: M. Camenzind, Les noyaux actifs de galaxies. XVIII, 218 pages. 1997.

Vol. m 47: L. M. Zubov, Nonlinear Theory of Dislocations and Disclinations in Elastic Body. VI, 205 pages. 1997.

Vol. m 48: P. Kopietz, Bosonization of Interacting Fermions in Arbitrary Dimensions. XII, 259 pages. 1997.

Vol. m 49: M. Zak, J. B. Zbilut, R. E. Meyers, From Instability to Intelligence. Complexity and Predictability in Nonlinear Dynamics. XIV, 552 pages. 1997.

Vol. m 50: J. Ambjørn, M. Carfora, A. Marzuoli, The Geometry of Dynamical Triangulations. VI, 197 pages. 1997.

Vol. m 51: G. Landi, An Introduction to Noncommutative Spaces and Their Geometries. XI, 200 pages. 1997.

Vol. m 52: M. Hénon, Generating Families in the Restricted Three-Body Problem. XI, 278 pages. 1997.

Vol. m 53: M. Gad-el-Hak, A. Pollard, J.-P. Bonnet (Eds.), Flow Control. Fundamentals and Practices. XII, 527 pages. 1998.

Vol. m 54: Y. Suzuki, K. Varga, Stochastic Variational Approach to Quantum-Mechanical Few-Body Problems. XIV, 324 pages. 1998.

Vol. m 55: F. Busse, S. C. Müller, Evolution of Spontaneous Structures in Dissipative Continuous Systems. X, 559 pages. 1998.

MIX
Papier aus verantwortungsvollen Quellen
Paper from responsible sources
FSC® C105338

If you have any concerns about our products,
you can contact us on
ProductSafety@springernature.com

In case Publisher is established outside the EU,
the EU authorized representative is:
Springer Nature Customer Service Center GmbH
Europaplatz 3, 69115 Heidelberg, Germany

Printed by Libri Plureos GmbH
in Hamburg, Germany